Springer Tracts in Civil Engineering

Springer Tracts in Civil Engineering (STCE) publishes the latest developments in Civil Engineering - quickly, informally and in top quality. The series scope includes monographs, professional books, graduate textbooks and edited volumes, as well as outstanding PhD theses. Its goal is to cover all the main branches of civil engineering, both theoretical and applied, including:

- Construction and Structural Mechanics
- Building Materials
- Concrete, Steel and Timber Structures
- Geotechnical Engineering
- Earthquake Engineering
- Coastal Engineering; Ocean and Offshore Engineering
- Hydraulics, Hydrology and Water Resources Engineering
- Environmental Engineering and Sustainability
- Structural Health and Monitoring
- Surveying and Geographical Information Systems
- Heating, Ventilation and Air Conditioning (HVAC)
- Transportation and Traffic
- Risk Analysis
- Safety and Security

Indexed by Scopus

To submit a proposal or request further information, please contact:
Pierpaolo Riva at Pierpaolo.Riva@springer.com (Europe and Americas)
Wayne Hu at wayne.hu@springer.com (China)

Hashem Jahangir · Harish Chandra Arora ·
José Viriato Araújo Dos Santos · Krishna Kumar ·
Aman Kumar · Nishant Raj Kapoor
Editors

Damage Detection and Structural Health Monitoring of Concrete and Masonry Structures

Novel Techniques and Applications

Springer

Editors
Hashem Jahangir
Department of Civil Engineering
Research Group of Novel Technologies
in Civil Engineering
University of Birjand
Birjand, Iran

José Viriato Araújo Dos Santos
IDMEC
Instituto Superior Tecnico
Universidade de Lisboa
Lisboa, Portugal

Aman Kumar
AcSIR-Academy of Scientific
and Innovative Research
Ghaziabad, India

Harish Chandra Arora
AcSIR-Academy of Scientific
and Innovative Research
Ghaziabad, India

Krishna Kumar
Department of Water Resources
Engineering and the Center for Advanced
Middle Eastern Studies at Lund University
Lund, Sweden

Nishant Raj Kapoor
AcSIR-Academy of Scientific
and Innovative Research
Ghaziabad, India

ISSN 2366-259X ISSN 2366-2603 (electronic)
Springer Tracts in Civil Engineering
ISBN 978-981-97-8974-0 ISBN 978-981-97-8975-7 (eBook)
https://doi.org/10.1007/978-981-97-8975-7

This Springer imprint is published by the registered company Springer Nature Singapore Pte Ltd.
The registered company address is: 152 Beach Road, #21-01/04 Gateway East, Singapore 189721, Singapore

Contents

Part I
Artificial Intelligence-Driven Structural Health Monitoring

Chapter 1
A Comprehensive Synopsis of Artificial Intelligence-Driven Structural Health Monitoring of Concrete Structures: A Novel Approach Towards Sustainability

Navdeep Mor, Pawan Kumar, Daniel Chukwuemeka, and Madhu

Abstract Civil engineering structures including masonry and concrete structures, seen as buildings, bridges, and even in other applications like monuments are prone to deterioration revealed as cracks, spalling, efflorescence, etc. The amount of money invested in building, time taken to build, quality of the original works, and most importantly the safety and life of the people inhabiting and using the structures, warrant such structures to be monitored and constantly assessed to confirm their utility safety to life. In this direction, a new technique called Damage detection enables us identification and recognition of defects in the structure, quantify them, and even propose measures by which these defects may be managed including predictions of the effects on the remaining structural life. Artificial intelligence (AI) uses data to machine and ensure human-like efficient results have been achieved. In the present study, the detailed study uncovers the present applications and uses of AI, and ML in the identification and recognition of damage present in masonry and concrete structures in the field of structural health monitoring and the utility of AI in damage detection from a different point of view. The findings from the study could help civil engineers about the different aspects of how AI and ML might help in the detection

N. Mor (✉)
Department of Civil Engineering, Guru Jambheshwar University of Science and Technology, Hisar, Haryana, India
e-mail: navdeepmor@gjust.org

P. Kumar
Department of Computer Science and Engineering, Guru Jambheshwar University of Science and Technology, Hisar, Haryana, India

D. Chukwuemeka
Pan African University Institute of Basic Sciences Technology and Innovation, Juja, Kenya

Madhu
School of Chemistry, University of Hyderabad, Hyderabad, India

H. Jahangir et al. (eds.), *Damage Detection and Structural Health Monitoring of Concrete and Masonry Structures*, Springer Tracts in Civil Engineering,
https://doi.org/10.1007/978-981-97-8975-7_1

and monitoring of problems by monitoring structural health using simulations of real-life situations. It will be like a detailed theoretical solution to a practical problem. After thorough study, it can be deduced that AI is a powerful tool applied to the detection of damage in concrete structures, extending also to historical civil works and monuments, yielding quick solutions and reliable results with heightened accuracy.

Keywords Damage detection · Structural health monitoring · Artificial intelligence · Machine learning · Deep learning · Neural networks · Concrete structure · Masonry structure

1.1 Introduction

Masonry construction entails building with clay, brick, stone, or concrete blocks, these individual units assembled and bound by mortar make up the masonry. A common example of this is in the walls of conventionally built buildings involving the application of masonry for the partitioning and mostly concrete as the other prevalent composite of such buildings, for the other structural members. Concrete works are built consisting of pebbles, broken stone, and gravel, in a cement or mortar mix [1]. Thus, concrete and masonry structures are designed for load carrying and resistance of external forces even wind and other natural elements, designed to cater for axial compression, shear, bending, and combined states of forces. The lintels, beams, slabs, foundations, tanks, and retaining wall, are made of concrete with steel reinforcements to cater for the tensile stresses against which concrete resistance is deficient, and thus increase the structures' resistance to the load applied on it, the ductility and strength also. Hence, concrete and masonry structures relate the condition of structural elements made of reinforced concrete, while the partitioning is composed of singular units laid and bound in sync by cement mortar, which actualizes a structure of a composite assemblage, as air and moisture leakages are sealed and anchors, ties, and reinforcement find joint for bonding, ensuring the performance of all the elements as a whole unit [2]. The resulting structure adequately resists attacks is strong, durable, sound-attenuated, and energy-efficient. These structures take different colours, shapes, and sizes, and ensure the achievement of whatever functions and appearances through the architectural finishes.

1.1.1 Type of Structures

Masonry may also be used in lintels, floor and roof beams, and retaining walls, which are all flexural applications; but it is mainly applied to primarily resist compression loads and thence used as columns or load-bearing walls. However, masonry structures may be prone to deterioration or damage as a result of changes in temperature and being exposed to moisture alone or alongside other environmental factors. Environmental friendliness, economy, adoption of locally available materials, and non-combustion, are among the merits of using masonry units, while low tensile strength, the requirement of a large foundation caused by the huge weight of the structure, susceptibility to settlement and cracking, degradation due to extreme weather, are among the demerits [3]. Masonry unit could be made of several types of a brick wall: (i) English bond which is made by having in every heading course, a queen closer to immediately follow the first header (ii) Flemish bond is made by staring each alternative course by placing a header at the corner (iii) Stretcher bond which is used in the partitioning of wall and is made by laying stretcher of bricks along the wall face. (iv) Header bond though unsuitable to serve load-bearing structure, is formed by laying head of bricks along the wall's face and providing a three-quarter bat at the corner of every alternative course (v) Zigzag bond is mainly employed in flooring with bricks and laying same in a zigzag fashion.

1.1.2 Categories of Masonry Structures [4]

Masonry structure could be distinguished as.

(i) Non-load bearing and load bearing masonry: In non-load bearing masonry easy removal is assured without potential danger to the building, the weights of the walls are the only loads carried by the walls, the structural dead load is reduced due to the lighter weight of the build, and they mostly serve as a partition wall. Brick walls, hollow brick walls, hollow concrete brick walls, and façade brick walls are the main categories of non-load-bearing masonry. Load-bearing masonry is mostly used for the construction of residential dwellings, loads are carried to the ground through the foundations by the wall elements, and this type of masonry is the most used technique as well as the oldest, they possess high aesthetic attraction, no need of fabrication, and highly resist fire, however, the construction may consume large amount of time, require manpower, and the structure behaves poorly in an earthquake. Stone walls, retaining walls, brick walls, and precast concrete walls are categories of load-bearing masonry.

(ii) Unreinforced and reinforced masonry: Unreinforced masonry consists of a weak mortar doing the bonding, it contains no reinforcing bars and is prone to damage in the occurrence of an earthquake. Reinforced masonry increases not just compressive strength but particularly tensile strength, and is made up by embedding or placing reinforced concrete bands, steel bars, timber, and bamboo in holes and eventually filling with concrete. Reinforced masonry may be classified as (a) reinforced grouted cavity masonry which is made up of masonry units in two leaves that are separated by an opening into which both the horizontal and vertical reinforcement are placed and grouted through the application of concrete grout and infill (b) reinforced hollow unit masonry which is made by inserting vertical reinforcement into block or brick and eventual filling with grout or concrete (c) reinforced packet type wall with are built with filling with grout the pockets consisting vertical bars.

1.1.3 Other Classifications of Masonry Structures

Masonry buildings may further be classified based on the raw material homogeneity, piers continuity, construction period, restoration work, and continuity of the regular floors, all of which are parameters of construction. The consideration of the vertical loads acting on the structure and transferred by the foundations to the ground generates classes of the building based on whether the components are made of the same raw material, characteristics of the structures, load bearing, and transmission capability.

1.1.4 Concrete and Reinforced Concrete Structures

Concrete structure largely amounts to civil infrastructure but could be exposed to a service environment that is complex which causes a reduction in reliability [5]. Concrete can be readily available, environmentally friendly, economical, and suitable for various applications. It is a good resistant to fire, has fair shear strength, good in compression and durability, but poor in tension. Steel at elevated temperature loses strength and thus is poorly resistant to fire, good in shear, possesses good compressive strength though it buckles if a slender member, has fair durability since if left unprotected it would corrode, but has good tensile strength. These complimentary characteristics cause the use of these two materials as composite in structure. Reinforced concrete (RC) structure is made by embedding steel reinforcement in concrete to make up a composite that resists load. Reinforced concrete design is a safety and economy-focused process that deals with specifying the steel area and sizes of the concrete members to achieve good performance when such structure is loaded. RC is applied to structures including industries, high-rise buildings, residential buildings, swimming pools, rigid pavement in highways, and bridges.

1.1.5 Concrete Structural Elements

The elements consisting of concrete structure are basically [6]: (i) foundations and bases which are supported on the ground and spread the loads received from walls or columns to the supporting surface to ensure that the structure is added by the ground with no excessive settlement (ii) columns are vertical structural members which are generally made to resist the axial moment and load but primarily carries vertical members (iii) beams are structural members which are horizontal in existence and carry lateral loads (iv) walls are plate elements vertically positioned and which resists lateral, vertical, or loads that are in-plane. (v) Slabs are horizontal plate members that are subjected to lateral load. These components can be stated to consist of the substructure: fundamentally all forms such as foundation or basement that are situated beneath the surface of the soil; superstructure: the part of the structure which is above-ground, and it includes the roof and main building. Specifically, the structure takes dead loads: the weight of any material that is rested on the structure and the weight of the structure itself; live loads: the weight of items or people in the building consisting of the structure, including traffic or occupancy loads; environmental loads: stress induced on the structure by earthquakes, wind, snow, rain, and extreme temperatures, soil pressure against the substructure.

1.1.6 Concrete's Merits and Demerits [7, 8]

Concrete swells in usage due to the affordability and accessibility of its consisting materials, flexibility to be made into any shape and to take the shape of the mold being used to contain it, works well with reinforcement to cater for tension and bending forces, serves well in withstanding compressive loads due to its high compressive strength, minimized self-weight and section sizes due to newer findings such as prestressed concrete, serve to retain water as used in Dam construction due to concrete's permeability to water, high unit weight, and massive nature. However, concrete structures also show some shortcomings: almost no scrap value, brittleness, at least 28 days before being ready for use, weakness in tension, development of creep in the structure due to prolonged loadings, efflorescence due to the presence of salts in the concrete, poor resistance to seismic attacks due to the high self-weight, need of construction joints to avoid cracks which are caused by moisture expansion and drying shrinkage. Cracking in concrete normally occurs when it undergoes flexural or tensile stress, however, the crack width must be within that which is tolerable for the structure to still be considered safe. More steel reinforcement is needed to add to the concrete's capacity to carry load especially when the concrete's strength is exceeded by shearing or compressive forces. The exposure of steel-reinforced concrete structures to an environment that is aggressive, for a long period, causes more susceptibility to corrosion. In the monitoring of the condition of concrete structures while in service, a monitoring technology suitable for these structures is needed. Cost, sensitivity, durability, and compatibility are among the factors to be considered.

1.2 Artificial Intelligence

Artificial intelligence (AI) could be related to the broad science encompassing any kind of technique employed to render machines capable of solving a task like how humans solve tasks by the use of rules that are pre-defined. AI is seen as programs possessing the ability to learn and reason like humans, mimicking the cognitive functions of humans, simulating reasoning used by people to acquire knowledge from new information, and finally make decisions [9]. AI utilizes data, algorithms, and models to recreate human behaviours and intelligence in the machine to complete tasks. The prediction, automation, and completion of tasks enjoyed by the use of AI are comparable to greater precision and accuracy, reduced cost, bias, and time-saving.

1.2.1 Machine Learning

Machine learning (ML) is an AI technique subset that employs statistical methods to have machines improve as they undergo experience, the computer learns and acquires knowledge using data-driven mathematical models, thus the computer system is enabled to continually learn and improve solely based on experience. ML are algorithm that permits computers to acquire knowledge and learn from instances without an explicit programmed condition. ML is one of the set of algorithms on which AI is constructed. Data is consumed by ML and statistics is employed to improve in learning of the data. Supervised learning, is the acquisition of knowledge from a dataset that is labeled, and gives enablement to base on former data to produce outputs, while unsupervised learning, is the acquisition of knowledge and discovery of unknown data patterns from unlabelled data [10].

Methods used in ML are derived from operations research, statistics, neural networks, and also physics, to discover insights hidden in data without the programming of what the machine is to conclude or where the machine is to look. Structured data which are used in machine learning are quantitative data such as product names, customers' names, phone numbers, and dates. ML is beneficial in automation, accurate forecasting, and recognition of trends and patterns, while it can be applied to make educational tools, streaming platforms, and chatbots. Arthur Samuel came up with the expression Machine Learning (ML) in 1959. The AI and ML are interrelated and the relationship between AI and ML is presented in Fig. 1.1.

1.2.2 How Artificial Intelligence Works [11]

Sequentially AI and ML work thus: first, the AI system is constructed by employing techniques including machine learning, next, the models of machine learning are made by the study of data patterns, afterwards the machine learning models are optimized by data scientists through the use of the data patterns, finally there is a repetition of the process and a refinement until the accuracy of the models is sufficiently high for the tasks needed to be executed. The procedure related to how AI works with Machine Learning is presented in Fig. 1.2.

Together, ML and AI are connected and used to transform products and processes through natural language understanding and speech recognition, recommendation

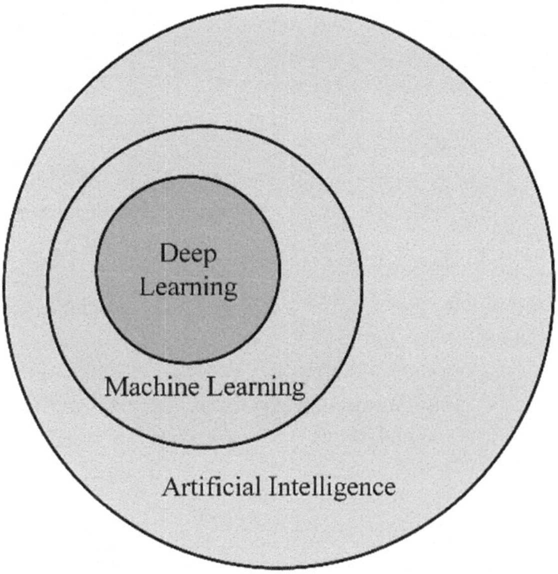

Fig. 1.1 Relationship between artificial intelligence, machine learning, and deep learning

engines, predictive analytics, sentiment analysis, and video and image processing. Collaboratively, ML and AI ensure faster and better decision-making, expanded sources of data input, and increased efficiency of operations. As such they can be applied to healthcare, banking and finance, retail, cybersecurity, transportation, customer service, marketing and sales, and manufacturing, exhaustively areas dealing with products and processes.

1.2.3 Artificial Neural Networks

When machine learning models are brain-inspired, that is, to assume a line of processing information similar to the human brain, they are referred to as artificial neural networks (ANN) [12]. A neural network is simply one of the ways of training a computer to imitate human reasoning through a sequence of algorithms modelled in resemblance to the human brain, as such, it is a system of computing that composes units like neurons, interconnected, and which by response to external inputs processes information and relays information between the units.

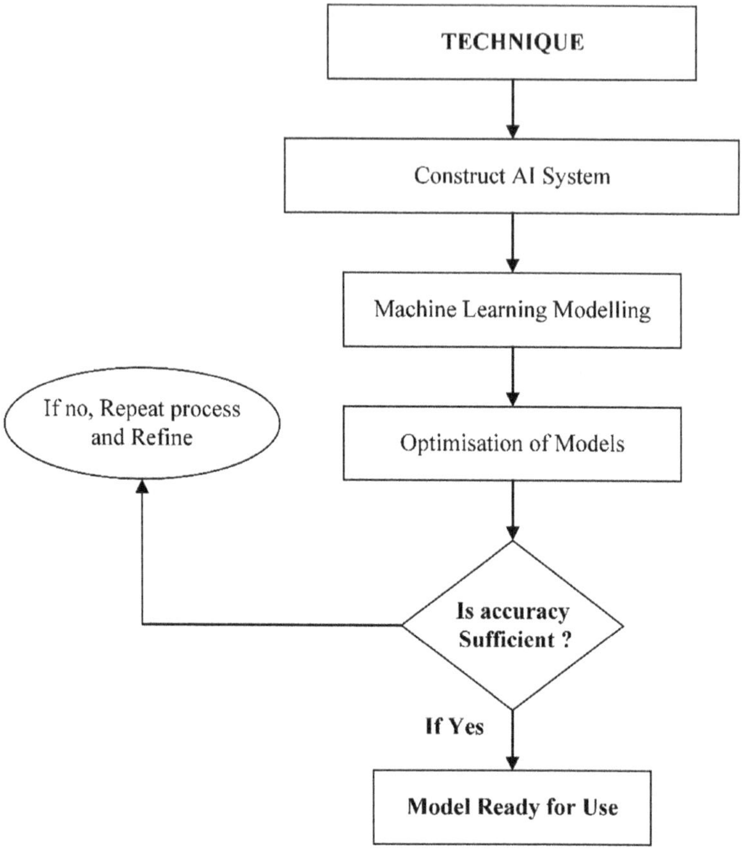

Fig. 1.2 Working mechanism of AI with ML

1.2.4 Deep Learning

The subset of ML that makes use of algorithms that are complex and artificial neural networks that are deep, as models, and builds a power structure of data representations automatically, is termed deep learning (DL). Deep learning is the methodology through which the computer system is enabled to actualize deep learning with the aid neural network. It enables the accomplishment of the multi-layer neural network, especially as the ANN adapts and gains knowledge from an immense quantity of data, the software trains to adequately perform tasks, and the extraction from input data increases the quality of characteristics progressively [13].

DL is also referred to as deep neural learning as it involves creating patterns and making connections by processing data through hidden layers in a neural network which is biological-inspired, and unto which data is inputted. Models are trained and complex tasks completed using data that is unstructured, specifically, qualitative data such as audio, photos, and video files. DL is used for problems that are complicated like defect detection, image processing, and facial recognition, image classification is a common task in which deep learning is used, with immense accuracy. DL is beneficial as it is scalable, utilizes unstructured data efficiently, and involves the use of algorithms that are distributed and also parallel. DL is applied to self-driving vehicles, virtual assistants, and manufacturing.

1.2.4.1 Deep Learnings and Dependents

Computer vision depends on deep learning and pattern recognition to recognize and identify what is in a video or an image. The ability of machines to process, analyse, and make meanings of images results in such machines being able to capture videos or images in real-time coupled with the interpretation of their existing surrounding [14].

Natural Language Processing (NLP) attributes to the capability of computers to not only analyze and understand human language with speech included but also to generate them. Succeeded by NLP is the enablement of communication between humans and computers by the use of normal and mainstream language in performing tasks, this stage is termed natural language interaction.

1.2.5 Synoptic Artificial Intelligence

Early work in AI with neural networks, 1959–1980; ML grows in popularity from 1980–2010, Deep Learning 2010–Present and beyond. A computer scientist, John McCarthy, coined the expression 'Artificial Intelligence', creating the AI terminology in 1955. However, research in AI as existing at present, formally started in 1956, specifically in a United States research university, private Ivy League Dartmouth College, New Hampshire, at a workshop, during a research project that sought to explore topics connected to symbolic methods and problem-solving. Development in AI could be attributed to technology evolution and the immense increase in the availability of large quantities of data [15].

1.3 Damage Detection

Defining damage as system variation that reduces its ability to execute its intended function, whether at present or later in the future. Degradation which induces damage in a structure is caused by earthquakes, overloading, or aging, and makes itself obvious as rebar corrosion, rebar exposure, segregation, spalling, and efflorescence. The increasing attention to damage detection may be related to the increasing amount of natural disasters and deteriorating structures, hence structural safety becomes peculiar. The nature of structural damages may pose constraints preventing it from being inspected by humans, and the location of the area of defect may be dangerous if it is accessed through normal methods.

Damage existence identification through correlation analysis, damage location identification through sensitivity analysis and error localization, damage magnitude estimation through model updating, and further estimation of the structure residual lifetime. The success of damage detection depends on test data quality and amount, the application of damage detection may be found in rotation machinery and robots, marine and land structures, space structures, missile and aircraft structures, and civil infrastructure including but not limited to power plants, bridges, highway systems, and buildings [16].

Damage detection routines can be combined with test data acquired at pinpointed time intervals while the structure exists or continuously through the structure's lifetime. Operational modal analysis, which is a way of analysing structural knowledge though no knowledge of the input force is possessed, has enabled the novel test post-processing methods to be available and caused the damage detection techniques to increase in acceptance.

1.3.1 Categorizing Damage Detection and SHM

Structural Health Monitoring (SHM), a process of applying a strategy of detecting damage in a structure, comprises monitoring structural components and whole structures in real-time, and also through the life cycle, even during the manufacturing process of such structure, with no hampering of the structural integrity as it is to meet the initial design specification of the structure. SHM identifies geometric changes and materials behaviour-induced structural vibrations. Methods of numerical simulation, sensor technology, and technology for damage diagnosis. It may be more profitable to monitor and repair structures than it is to reconstruct new structures. SHM involves implementing a strategy that would aid in the detection of damage by including an embedded measurement system in sensing technologies [17].

The monitoring of structural health, which can yield reliable quantitative information, utilizes both destructive and non-destructive methods, of which the non-destructive methods extend to local and global methods, while the global methods further extend to static-based methods and dynamic/vibration-based methods, lastly the vibration/dynamic based methods branches to signal-based and model-based methods. Local methods of detecting structural damage include leak tests, optical methods; magnetic particle testing, magnetic flux leakage; laser testing methods, infrared thermography; digital image correlation, ultrasonic testing; ground penetrating radar, radiographic testing; liquid penetrant testing, acoustic emissions; and visual testing [18]. The detailed methods used in structural health monitoring are presented in Fig. 1.3.

Global or vibration-based methods include nonparametric methods: Machine learning-based methods, and other methods, while parametric methods consist

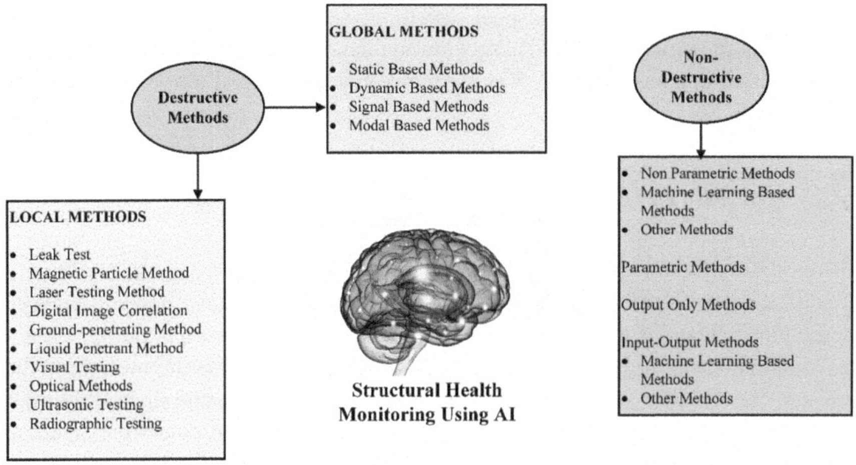

Fig. 1.3 Categorizing damage detection methods and SHM

of output-only-methods and input–output methods, under which are Machine learning-based methods and other methods.

The classification of damage identification methods of structures based on vibration characteristics is suggested by Rytter to be: firstly, determining the presence of damage in the structure, detection; secondly, determining the structural damage geometric location; third level, quantifying the structural damage severity, quantification; finally, predicting the structure's service life that is left, prediction.

Vibration-based detection of damage, advantageous for identifying internal structural problems in civil engineering structures, is sequenced from the monitored structure to the hardware component which is made up of the sensors, data acquisition system, processing unit, and the eventual damage detection algorithm which outputs damage localization and detection. The use of on-site measurements to detect changes in the structure's vibration response can be connected to the condition of the building. Shakers and impact hammers may serve to generate controlled vibrations which enables the accomplishment of vibration testing, and also the capturing of the system's responses over time. A transform is utilized to make the time history responses into the frequency domain, with the application of techniques for processing signals, used to estimate damping ratio, mode shapes, and natural frequencies which are vibration parameters. The characteristics of frequency response aid in the application of SHM to modern structures [19].

Implementing structural health monitoring in buildings connects sensors which are installed strategically in the building, to a data acquisition and analysis system, and afterwards, the damage detection is actualized by drift computation, base shear computation, and shear computation. The computations are made possible by the inputted data and the eventual output could be an alarm that is fed to the internet and can be accessed even on computers and smartphones as alerts. Smartphones, cameras, satellites, unmanned aerial vehicles, and ultrasonic devices, are also among the number of sensors used to inspect structures, specifically, these do not come in contact with the structure [20]. These store the data directly on themselves, or may be transmitted to a base station for subsequent analysis.

1.3.2 Introduction to Damage Detection in Concrete

Masonry construction works may experience either structural degradation due to movement of supports, applied loads, or environmental impacts-induced material degradations. There is the possibility of acquiring knowledge of a building's structural health continuously or just once [21]. When key parameters are monitored the structure conservation state or performance level can be known, and such result from this finding may be used to retrofit and upgrade, evaluate, and assess damage and vulnerability, while the structure's evolution through the time of life and service may be assessed by monitoring the structural health continuously. Data collection is what is initially done to conclude rightly if a structure requires intervention and what the type of the required intervention is [21]. As such prognostics, SHM yields warnings on the need for corrective action to be taken for a structure in a critical health state, and divulges predictions on the functionality duration of the structure, while diagnostic methods cover flaws identification, site pinpointing, and severity measurement.

1.3.2.1 Types of Monitoring

The comprehensive process entails the implementation of a strategy to detect damage by measuring under operational or environmental conditions the structural response attributable to certain key parameters. Identification of the right health metrics, and development of serviceability, safety, and durability criteria to measure the non-exceeding limit states of the structure's life cycle. Sensors, and devices, alongside other supporting tools and systems including, communication, processing of data, measurement, acquisition, decision-making, and modeling or identification system.

Dynamic-based methods used in the monitoring of the health of structures are the harnessing of vibration characteristics possessed by a structure to assess its state of health, while static methods are founded on the measurement of variation in the response through static. The proper identification of the dynamic behaviour of a structure depends on the structure's dynamic properties may be grouped into three parameters, namely, mode shapes, damping ratios, and natural frequencies [22].

1.3.2.2 Reasons of Monitoring

Masonry buildings may experience progressive damages caused by ambient vibrations that are induced by vehicles, winds, or construction equipment that are heavy

[23]. That the building is safe for inhabitation, increasing the safety of the occupants, and fitting modern standards are some of the reasons for which is structure should be monitored, and to ensure its operational, reliability, and durability utility. The lack of regular maintenance of a structure to ensure the structural integrity is intact may result in sudden loss even of integrity and such without warning, which could yield partial collapse, increasing reconstruction cost, and danger to lives.

The acquisition of knowledge on the state of the structure, information on the response, deterioration, and identification during both the design and construction, with the understanding of the consequence to the structure reliability and bearing capacity caused by damage; decisions on operations and maintenance of the structure together with warnings on the potential damage to be caused by overload and damage.

The life span of a structure can be extended by engaging in continuous monitoring as decisions on strengthening and repair can more adequately be taken [24]. Paying attention to the safety and reliability conditions of structures is also important, especially during the service life [25].

1.3.3 Methodology of SHM

The algorithms used in the monitoring of structural health should be developed to detect, localize, and quantify structural damage through the appreciation of data received from sensing systems. Data-driven and model-based structural monitoring are two types, while validated mathematical models are required for model-based techniques, developed using the finite element method, solving problems leads to damage identification. Statistical models instead of mathematical models aid in establishing data-driven methods, and experimental data, leading to techniques for recognizing patterns by a knowledge of not only the damaged structure but also the structure still intact, this is termed supervised learning, while the case where the knowledge possessed is that alone of the intact structure, is unsupervised learning, which may be restricted to initial point contained in the identification process for detecting damage. In the management of the damage, the process may include: detecting, locating, characterizing, and assessing severity, reporting of damage, and structure's useful residual life assessment.

The set of elements that adequately represents the substance of implementing the SHM systems are those by which they are developed [26]. The inspection of a structure using the sensory network, with sensors installed concerning the structure's geometry and size, for continuous and automated observations, that are linked to the structural component or the structure, leads to the generation of a large amount of data which may be more timely analysed by data-processing facilities that are well optimized, to be finally made worked on by appropriate algorithms. Thus, the process involves the initial acquisition of data, and data processing to pinpoint the severity and location of damage that exists, indicate maintenance, and state possible limitations.

1.3.4 Artificial Intelligence and Structural Monitoring

Due to the availability of abundant information and increasing processor powers, the AI field has been progressing and doing so significantly. Farrar in the year 1999 seems to have first begun the discussion of taking up AI algorithms in SHM, by the attempt to consider unsupervised learning and supervised learning for detection of damage. ML and AI use in damage detection is increasing in popularity because of its efficacy, especially, in how SHM systems perform [27]. A neural network is fitting for handling issues with big data and extracting the existing features which are multi-dimensional and the non-linearity existing among input data, specifically for generating regional deterioration models [28]. However, challenges may occur because of the inadequacy of relevant information, and difficulties linked with measuring, modelling, and processing signals. Thus, predictions of structural deterioration, abutment conditions, simulation of structural conditions, and deterioration through bridge inspection feature extraction, are accomplishable using neural network models.

Neural networks and other machine learning techniques enable the early detection of potential damages, through the judicious use of huge quantities of sensor data retrieved from structures that have been monitored, as depicted in Fig. 1.4. Helps us to save time for multiple decisions and activities, money, and human labour effectively and efficiently.

Defining structural damage to be changes introduced into the structural system that results in a reduction of either present, future, or both present and future performance, would appreciate damage detection to be noted as detection of the changes occurring in the structure due to damage that has been introduced therein. Reflecting these changes in damage detection that are based on vibration responses is understood by indexes which are arrived at from either the use of statistics through the modelling of

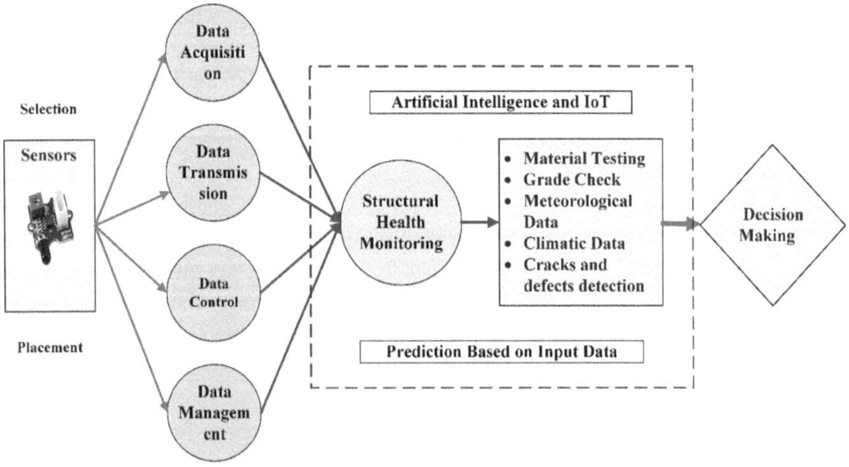

Fig. 1.4 Structural depiction of sensor-based SHM with AI

structural response to vibration when in ambient excitation, or physical laws which are constituted in the modal characteristics derived from the structure's responses to vibration when set to ambient excitation. The change in the modal characteristics that have been monitored indicates the damage.

The limit in conventional algorithms' computational efficiency is needed for real-time decision-making as monitoring data is received continuously, and deficiency in the use of supervised methods results in applying AI to cater to the gaps [25]. Preparation of data, statistical modelling by ANN, and decision-making also by ANN.

Monitoring infrastructure of civil works and assessing them based on a vision for inspection, is assisted by uniquely modelled artificial intelligence methods, which detect even subtle patterns and differences in data, ensuring damage detection by data points correlations, and measurements of structural response. The structural response data make up the input data [26].

The aim of building systems which would diagnose self and signal damage or fault by giving an alarm. Damage detection models aim to distinguish a damaged structure from a healthy one. They are performing damage detection by the discovery of spatial and temporal correlation existing in the data containing the independent variables, with the consistency of the input data being the key.

1.3.5 Steps in an SHM System that Incorporates ML

According to Zinno, eight different stages are needed in the implementation of an SHM system that is data-driven: Excitation, acquisition of data, normalization data, data cleaning, compression of data, extraction of features, data fusion, and recognition of pattern and evaluation. This is summarized in Fig. 1.5. Excitation deals with the arousal of the existing structure to collect deterioration signs, while considering the state of the structure, and comparing actual measurements with numerical models. The method of data acquisition is vital as different requirements may be sought for the monitoring of static and dynamic characteristics [29]. Data normalization is pertinent in situations when the data is composed of a wide scale range, as it is an act of having all the data points to be of a consistent scale and strike out deficiencies resulting from different sensors' data collection. Recognition and removal of duplications and errors to create a valid dataset are composed in data cleaning, which addresses data quality potentially harmed by extraneous influences and sensors that have been loosely positioned. Feature size reduction, or alteration of data to reduce its size, by rearranging, encoding, or any other means, is referred to as data compression, which aids in focusing on aspects that are damage sensitive and with the most significance statistically. Feature extraction is a technique by which the feature space is minimized through subset selection, it is also the transformation of data to allow for easy identification and analysis by the ML algorithm. The gathering together of information collected using various sources to increase usefulness, accuracy, and consistency, is data fusion. Pattern recognition ML techniques enable the accomplishment of automatic identification of regularities and patterns, determining structural health.

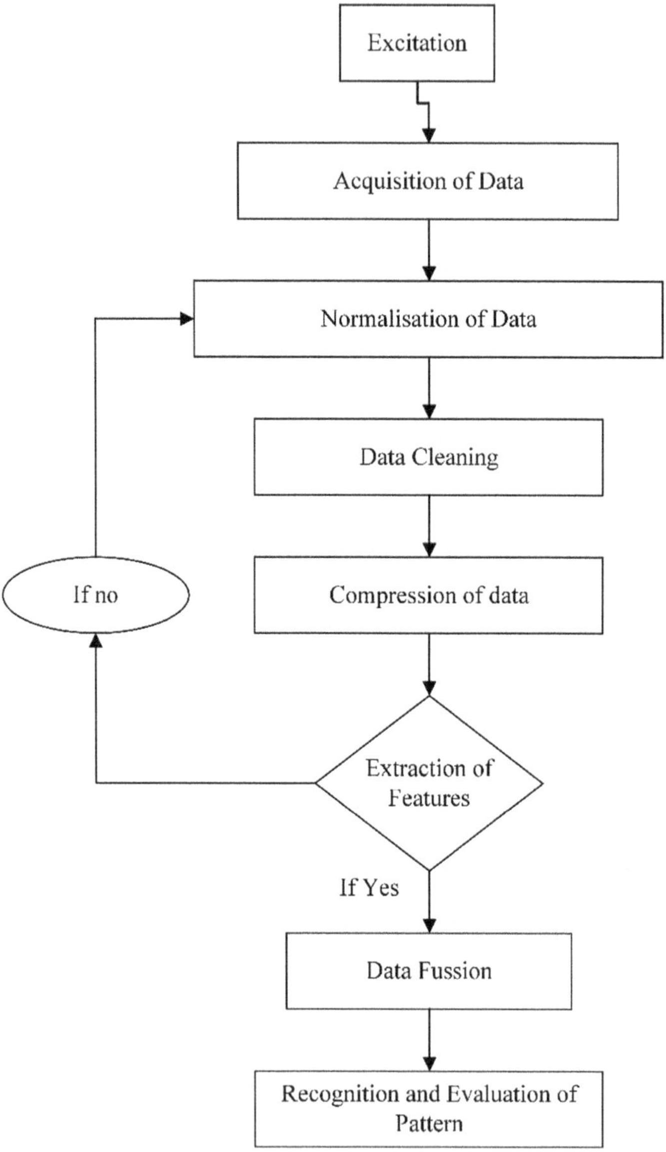

Fig. 1.5 Stages in implementing the SHM system

1.3.6 Artificial Intelligence and Damage Detection

It is preferred that neural networks be trained using standardized data to, cater to magnitude variations. Figure 1.6 illustrates the steps embedded in using AI for detecting damage. Data is to be cleaned efficiently and over-fitting monitored. The data used is suitable to be split into a training set, validation set, and testing set. Tuning of the modelling parameters is accomplished by the use of training data, the condition at which the training process is to be terminated is made known by the employment of the validation data which aids in preventing over-fitting as the data are not precisely used for parameter tuning. The data used for training and validation of the model is to be retrieved from a structure in its healthy state. However, a damaged structure or even a healthy structure is to divulge the data to be used to test the model's performance and estimate the likelihood of retraining the model. A healthy structure with known recorded history gives the data to be used in the training of a prediction model, as such, an input outside the inputted data range would indicate a damaged structure.

Though past works in the development of SHM algorithms have focused on physical models, unsupervised machine learning methods and approaches, data-driven, and not requiring physical models which are based on physics, is the current direction of research [30]. AI to assist in developing software to control the route and direction of drones in either fully autonomous or semi-autonomous systems, in the conversion of obtained images and data and transfer of the data to sensors, in the analysis of the enormous amount of datasets generated from health monitoring. AI is useful in processing data retrieved from the site, and the imagery outcomes obtained through drones or cameras. The algorithm is used to automatically detect the extent and position of such cracks posing a threat to the building's structural health [31], yielding less error and faster detection of cracks.

1.3.7 Pattern Recognition in SHM and AI

Detection of failures using algorithms that are established on pattern recognition which considers the state of an investigated structure and available range of patterns to examine the similarities and verdict on the presence and type of damage. Statistical or historical data is used to find patterns with large sample numbers considered. Pattern extraction and mining aid not only in effectively and accurately grasping service conditions and longer-term characteristics but also give prompt warning information and decision-making on strengthening, repair, and inspection. Challenges include user settings during the collection of data, and environmental variability when pattern recognition systems are to operate in the real world [28]. Data employed in recognizing patterns with the aid of AI includes video and image analysis, signal processing, and statistical data analysis.

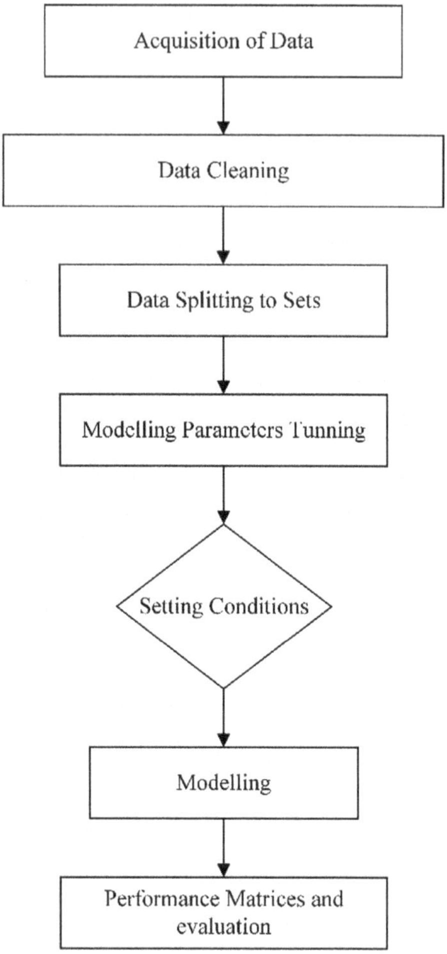

Fig. 1.6 Steps taken in modelling for damage detection

1.3.7.1 Difference Between SHM Done Using Traditional and AI Techniques

There are generally two ways to do structural health monitoring i.e. construction using traditional and artificial intelligence techniques. The major differences between both the techniques on different parameters have been compiled in Table 1.1.

1.3.8 Machine Learning and Damage Detection

Machine learning models, do not exhaustively support vector machines, genetic algorithms, or Gaussian process regression, but may be used in the detection and

Table 1.1 Difference between SHM using traditional and AI techniques

Sr. no	Factor(s)	Construction using traditional technique	Construction using AI technique
1	Price	Low price due to manual observations	High price due to precise calculations at initial stage
2	Time spent	Consumption of time is high	Consumption of time is low
3	Detailed design	Not available	Available
4	Risk and cost examination	Moderate analysis for risk and cost is present	Detailed cost and risk analysis
5	Productivity range	Moderate-high	High-improved
6	Requirement of supervision	Moderate	High
7	Requirement of type of team	Moderate	Highly skilled
8	Building analysis	Detailed analysis of traditional methods is used	Exact analysis using advanced 3D modelling

recognition of damage by the use of motion data to discover as modelling outputs if damage has occurred or not, expression of frame structure damage. Experimental data was employed to investigate damage detection, and restricted by the difficulty of specimen preparation to also capture a larger representation of damage geometries. Considering such limitations and generating training data from the finite element method with damage patterns featured in the numerical simulations, employed a framework of supervised machine learning, the Random Forest approach, to recognize not just the location and type of damage that is obtainable from the features, but also the damage extent. Supervised machine learning is the mode through which the machine learns from the input and output data contained in the training dataset to predict the correct output for input cases with unknown output. Hence, the developed algorithm receives vibration features as inputs and gives damage location, kind, and extent as outputs.

Identification of failure mode through rapid assessments based on models created from experimental databases, failure mode classification, derivation of structural fragility extent, infrastructure seismic vulnerability, and assessing damage in infrastructure systems, are also afforded by machine learning models. Evaluation of tall frame damage and revealing structural conditions using the status of nodes actualized by a trained neural network adopted more machine learning models with leading outcomes on interpretability, accuracy, and efficiency, gradient boosting methods including categorical boosting, light gradient boosting machine, extreme gradient boosting, and adaptive boosting, to strongly predict the mode of failure in shear walls. The recent significant studies carried out using AI and ML are presented in Table 1.2.

Also, recognition, classification, and assessment of seismic damage response by civil structures with the aid of photographic records, prediction of seismic damage

Table 1.2 Significant research on SHM using AI and machine learning

Name of author and year	Technique used	Dataset used	Remarks/conclusion
Melvin et al. [32]	AI and ML	Single building	AI-ML algorithms helped in predicting accurate cracks
Padmapoorani et al. [33]	Decision tree, neural network	Critically reviewed	AI-ML techniques give highly accurate and real results
Altabey et al. [34]	AI, linear regression	Multiple datasets	AI algorithm showed accurate results for SHM
Mondal and Chen [35]	AI	Single building	AI technique enhanced the analysis carried out for SHM
Daniel et al. [36]	AI and ML techniques	Multiple buildings	AI-ML can be trusted for all analysis related to SHM
Yuan et al. [37]	ML algorithms	Critical review	ML algorithms could be used for all analyses in the field of SHM
Diego et al. [38]	AI- piezo sensor	Critical review	Helped in the delamination and cracking of the structure
Kay et al. [39]	Artificial neural networks	Critical review	Improved reliability and accuracy of structural health monitoring systems
Kay et al. [40]	AI and computational intelligence	Multiple buildings	Robust, redundant, environmentally aware system was developed with highly accurate results
Keith and Manson [41]	ML algorithms	Single structure	Utility of the data-driven approach was presented

mechanism of historical monuments, use of damage records from earthquake occurrence for the classification of building damage, automatic detection of spalled regions on the surface of columns from assessment of images collected after an earthquake impact on structure, collaborative inclusion of digital maps and satellite imagery in detecting and classifying building damage, identification of structural damage as a result of natural hazards, is broadly achievable by AI and ML, including ANN.

The images collected from the use of photogrammetric methods even in models and products of 3D can be employed in detecting and mapping building damage. The methods developed are founded on classification methods incorporating both supervised and unsupervised methods of classifying images. Machine learning based on object identification may also be used for quick detection of damage.

ML is used in managing damage detection in bridges. ANN collecting vibration frequencies and temperature measurements, generates a model capable of detecting bridge damages when subjected to temperature variations. Detection of unusual behaviour in bridge through machine learning model by identification of elements which are damaged, comparison with normal joints.

1.3.9 Deep Learning Damage Detection

Deep learning (DL) algorithms richly collect features from data through the traverse of layers of concepts. DL algorithms enjoy usage in SHM because they possess statistical data, videos, and image analysis abilities. Inspection process of brick masonry surfaces using deep learning, machines are trained to detect cracks by going through photos of structures. This is accomplished by dividing the image into patches and assigning a non-crack or crack label, drawing a rectangle around any crack that has been detected, and labelling the pixel as either non-crack or crack. DL developments embed simpler and greater approaches to data processing employed DL for distinguishing the component of the building affected and cracks present, by assessing processed images, with the use of labels to recognize pixels that indicate the damaged area or the object. Gopalakrishnan in the year 2018 put forward the use of deep learning models that have been prior trained and coupled with transfer learning, to automatically detect crack damage on assessing civil infrastructure images from SHM executed using Unmanned Aerial Vehicles (UAV), which affords navigation flexibility and accessibility even when complex structure is being considered for quality data collection. Challenges may include generating image datasets of good quality for the model training.

1.3.10 Convolutional Neural Network Damage Detection

Deep learning (DL) models in their variety can through digital images automatically achieve feature representation of structural damage in concrete. The convolutional neural network (CNN) pattern of connectivity between neurons is motivated by animal visual cortex organization. Due to capability in increased accuracy, adaptability regarding image classification, and robustness, CNN is maximizing performance improvement. A feed-forward ANN is an advanced technique for analysing images applicable to various domains with strong performance when addressing complex problems through recognition of images, as it possesses a convolutional filter with only slight influence by variations in the environment and accurate image classification and identification results. The cortex's neurons are organized to respond to overlapping regions that fit the field of vision exactly together. Several hidden layers and input–output layers are consistent in CNN. The existence of convolutional layers that purposefully detect patterns while considering the input data, is the main characteristic of a CNN. The various forms of damage include rebar surface exposure, leakage, delamination, and not just crack damage. Shin et al. [51] applied CNNs to identify damage of various types by analysing the images containing them, thus providing a multi-damage identification network to automatically classify damage types. Faster region CNN was utilized by Ali for not just automatic detection but also localization of damages occurring in masonry structures, involving images of heightened complexity, SVM and NN were employed for pattern recognition, and CNN for extracting features automatically. Extraction of feature types and map, detection of the morphology of cracks, general detection and localization of cracks, analysis of crack width and length, segmentation of cracks, detection of damage even of various forms including efflorescence and spalling, and also focusing on historic masonry structures, are achievable with CNN, with outcome proving high accuracy. These applications were also demonstrated in the work of automated structural inspections. The major significant research done in the field of SHM using Neural Networks is presented in Table 1.3.

1.3.11 Other Applications and Crack Detection

To recognize crack damage close to bolts on structures of steel from extracting features of images. Identify features of linear crack from pavement images, extract a minimized map of spal damage, detect potholes in images of asphalt pavement, transverse and longitudinal cracks, analysis of pavement cracks on images [51]. Outliers detection is used in damage identification. Generally the identification and detection of specific damages that occur in structures, these structures include bridges, asphalt pavement, and buildings, with the geometric patterns, damage type, and other variations being important considerations in model building.

1.3.11.1 Adaptive-Neuro-Fuzzy Inference System Damage Detection

The mode shape of a civil structure can reveal the damaged condition of the structure by a comparison of the state of the shape before and after damage has occurred. Since AI can make human-like decisions and can adapt well to situations, artificial intelligence systems for damage identification in civil engineering include adaptive-neuro-fuzzy inference systems (ANFIS), fuzzy logic, and artificial neural networks [52]. Huge datasets, and computer time, are needed by ANN which is a reference to its outputs and inputs in a black box. Numerical data and linguistic knowledge, non-requirement of cycles of data training, lack of ability to learn to adjust to a new environment, and ease of understanding characterize fuzzy logic systems [10]. Whereas ANFIS is the amalgamation of fuzzy systems and ANNs, which aids in

Table 1.3 Significant research on SHM using neural network

Name of author and year	Technique used	Dataset used	Remarks/conclusion
Siyu and Ziang [42]	ANN algorithms	Various dataset	Concluded that the proposed neural network model can identify structural damage with high accuracy
Tran et al. [43]	ANN, CNN	Models were developed	Highly accurate results were found in SHM for crack detection
Tran-Ngoc et al. [44]	ANN, GD	5 buildings	The proposed model showed highly accurate results
Xue and wen [45]	CNN	Single building	Accurate results in the detection of various damage parameters
Kareem and Liang [46]	Unsupervised learning	Various buildings	Developed framework showed precise results in SHM
Farias et al. [47]	ANN and MPC Algorithm	Various buildings	Damage in the plate-like structures was detected for the first time
Gen [48]	Reviewed ANN techniques	Review	All neural network techniques were found statistically significant
Luca et al. [49]	Convolutional networks, ANN	Statistically significant models were used	Automatically identified the damage-sensitive features of concrete structures from the data
Wu et al. [50]	ANN	Various buildings	The behaviour of trained and untrained structures was analysed

determining parameters of fuzzy logic, the advantages of both systems are utilized in one framework [53].

Feeding dynamic and spring parameters into ANFIS to identify cracks in a beam, and comparison of the identified with the measured natural frequencies so as to determine the damage, and predict both the size and location of damage [54]. Hakim et al. [10] employed the ANFIS model in the identification of damage location through mode shape parameters considering a bridge girder.

1.3.11.2 Acoustic Emission, SHM, and AI

The natural phenomenon related to the generation of stress wave when there is an occurrence of rapid elastic energy release that takes place when damages or cracks are formed in materials is termed acoustic emission [55], the corresponding sensors detect and collect signals while placed on the object surface, monitoring aids to record and process signals which are diagnosed to discover the structural health status. AI serves in this monitoring process as automatic and intelligent damage localization. AI techniques, deep and machine learning algorithm, which does data processing intelligently though supervised, learn the data extracted features, understand the pattern recognized in the data, and finally make a decision. DL learns features existing in complex datasets, automatically, and with no extractions of features.

1.3.11.3 Drones, High Rise Buildings, SHM, AI, Crack and Spall

Since high-rise structures in civil engineering works possess areas difficult to inspect with increasing inaccessibility, traditional methods of detecting cracks by visual examination techniques prove insufficient. This can be overcome by the use of machine learning, drone usage to increase digital image processing prowess, and also the employment of deep learning strategies that are advanced [56]. Thus, while drones are used for spall and crack detections, not only machine learning but also deep learning techniques are being used for the identification and detection of damage when digital images of concrete surfaces are processed. Algorithms are used, classification of the received images of damages, feature extraction, and real-time output is produced, as needed damage-sensitive information is that which the system transmits. The challenges may only dwell in variations in surface texture and lighting of the exterior.

1.3.11.4 Fuzzy Genetic System, and Bridge Damage Detection

The idea of pinpointing damage location and intensity by the examination of structural dynamic response after an occurrence of damage, to be compared with the response before the damage took place, reveals the variation in the structure's physical characteristics. Neuro-fuzzy systems of computing methods consist of neural networks used for recognizing patterns and adjusting to changes in the environment. Whereas the systems of fuzzy inference are employed with the neural systems for the description of human knowledge and arriving at proper decisions from the deductions made. Neuro-fuzzy is thus a combination of these two methods without differentiation but with optimization techniques. The systems can be used to detect failure from structural dynamic response, with accuracy of failure location and intensity determination [57].

Fuzzy logic collects inputs in numeric and gives out verbal output, but is incapable of learning from the data inputted into it. The amalgamation of a genetic algorithm with a fuzzy logic method yields a genetic fuzzy system (GFS). GFS when applied to the monitoring of bridge structures, yields precise damage detection. It is also cost-effective, needing no re-evaluation after having provided outputs, features high precision, and generated responses and analyses that are accurate [58].

Damage detection of bridges, to determine not just the location of the bridge damage but also the extent of damage, utilizes large quantities of data from sensors, and depends on improved sensing techniques to collect the bridge's morphology, external loads, and structural dynamic response [59]. The collected sensing data supports the damage identification of bridges using artificial intelligence and reported by Zhang and Yuen in the year 2022 to be recently combined with ground-penetrating radar and infrared thermography, to overcome some drawbacks by using optical and vibration response sensing technologies, ensuring more accurate surface damage identification, classification, and segmentation.

1.4 Conclusions

The use of artificial intelligence in identifying, recognizing, detecting damage that occurs or has occurred in not just concrete structures but also masonry structures is important as this developing technology critically performs otherwise tedious and repetitive tasks with high accuracy and thus keeping both human specialists, the occupants or users of the concrete structure, in safety. Concrete and masonry structures may suffer damage due to external factors on the structure and the environment, even due to natural disasters, including earthquakes. The present study explored the use of Artificial intelligence to give human-like outputs, make predictions, and analyse data with incredulous speed, while the reliability remains intact and maximum. Thus, its use in damage detection has so far included, comprehensive analysis of images with damages, using convolutional neural networks, and deep learning, through which images with damage are identified and the location of the damage pinpointed; analysis of data collected through sensors and other methods, analysis using machine learning techniques corroborated with data from healthy structure to aid in identifying discrepancies and thus detect damage; use of data received from modelling simulations to accurately determine damage effects on buildings when such simulated events occur; and many more applications of AI algorithms in not only analysing the current state of the structure, but also predict for the future. In a nutshell, it can be revealed very easily how artificial intelligence is extremely helpful for ensuring the exact information of damage detection in structures and output in terms of results, achieving high accuracy at low cost helps various stakeholders associated with it. The study strongly recommends that the incorporation of AI and ML techniques with the traditional methods would definitely increase the productivity, objectivity, and life-long scope of the concrete structures.

References

1. Kumar A, Mor N (2021) An approach-driven: use of artificial intelligence and its applications in civil engineering. Artif Intell IoT Smart Converg Eco-Friendly Topogr 201–221
2. Mor N, Sood H, Goyal T (2020) A statistical model for prediction of road accidents in the State of Punjab. J Interdiscip Math 23(1):229–236
3. Abedin M, Mokhtari S, Mehrabi AB (2021) Bridge damage detection using machine learning algorithms. In: Health monitoring of structural and biological systems XV. SPIE, vol 11593, pp 532–539
4. Chun PJ, Yamane T, Izumi S, Kuramoto N (2020) Development of a machine learning-based damage identification method using multi-point simultaneous acceleration measurement results. Sensors 20(10):2780
5. Mangalathu S, Jang H, Hwang SH, Jeon JS (2020) Data-driven machine-learning-based seismic failure mode identification of reinforced concrete shear walls. Eng Struct 208:110331
6. Motaghed S, Shahid Zadeh MS, Khooshecharkh A, Askari M (2022) Implementation of AI for the prediction of failures of reinforced concrete frames. Int J Reliab Risk Saf Theory Appl 5(2):1–7
7. Stepinac M, Gašparović M (2020) A review of emerging technologies for an assessment of safety and seismic vulnerability and damage detection of existing masonry structures. Appl Sci 10(15):5060
8. Mor N, Sood H, Goyal T (2020) Application of machine learning technique for prediction of road accidents in Haryana-A novel approach. J Intell Fuzzy Syst 38(5):6627–6636
9. Kumar A, Mor N (2020) Prediction of accuracy of high-strength concrete using data mining technique: a review. In: Proceedings of international conference on IoT inclusive life (ICIIL 2019), NITTTR Chandigarh, India. Springer Singapore, pp 259–267
10. Hakim SJS, Ibrahim MHW, Mohammadhassani M, Yeoh D, Jaini ZM, Chik TNT (2022) Implementing fuzzy-based artificial intelligence approach for location of damage in structures. Civil Eng Archit 10(4):1564–1573
11. Banerjee A, Pohit G, Panigrahi B (2017) Vibration analysis and prediction natural frequencies of cracked Timoshenko beam by two optimization techniques-cascade ANN and ANFIS. Mater Today Proceed 4(9):9909–9913
12. Mor N, Sood H, Goyal T (2020) A critical review on use of data mining technique for prediction of road accidents. Comput Methods Data Eng Proceed ICMDE 2020 2:141–149
13. Nandal M, Mor N, Sood H (2021) An overview of use of artificial neural network in sustainable transport system. Comput Methods Data Eng Proceed ICMDE 2020 1:83–91
14. Zinno R, Haghshenas SS, Guido G, Rashvand K, Vitale A, Sarhadi A (2022) The state of the art of artificial intelligence approaches and new technologies in structural health monitoring of bridges. Appl Sci 13(1):97
15. Rytter A (1993) Vibration based inspection of civil engineering structures. PhD thesis, University of Aalborg, Aalborg, Denmark
16. Mor N, Sood H, Goyal T (2020) A statistical model to prioritize selected Northern-Indian States/UT of India based on accident data. J Discret Math Sci Cryptogr 23(1):305–312
17. Grosse CU, Ohtsu M (2008) Acoustic emission testing. Springer Science & Business Media
18. Pallarés FJ, Betti M, Bartoli G, Pallarés L (2021) Structural health monitoring (SHM) and nondestructive testing (NDT) of slender masonry structures: a practical review. Constr Build Mater 297:123768
19. Saidin SS, Jamadin A, Abdul Kudus S, Mohd Amin N, Anuar MA (2022) An overview: the application of vibration-based techniques in bridge structural health monitoring. Int J Concr Struct Mater 16(1):1–17
20. Kumar P, Batchu S, Kota SR (2021) Real-time concrete damage detection using deep learning for high rise structures. IEEE Access 9:112312–112331
21. Kim S, Kim JY, Hwang IH (2015) The construction of initial analytical models structural health monitoring of a masonry structure. Int J High-Rise Build 4(3):191–198. Causes of damages to masonry buildings

22. Doa'ei Y, Jahan AM (2018) Application of artificial intelligence and meta-heuristic algorithms in civil health monitoring systems. Civil Eng J 4(7):1653–1666
23. Dais D, Bal IE, Smyrou E, Sarhosis V (2021) Novel invisible markers for monitoring cracks on masonry structures. Constr Build Mater 300:124013
24. Zhang Y, Yuen KV (2022) Review of artificial intelligence-based bridge damage detection. Adv Mech Eng 14(9):16878132221122770
25. Shi S, Du D, Mercan O, Kalkan E, Wang S (2022) A novel unsupervised real-time damage detection method for structural health monitoring using machine learning. Struct Control Health Monit 29(10):e3042
26. Ali L, Khan W, Chaiyasarn K (2019) Damage detection and localization in masonry structure using faster region convolutional networks. GEOMATE J 17(59):98–105
27. Farrar CR, Duffey TA, Doebling SW, Nix DA (1999) A statistical pattern recognition paradigm for vibration-based structural health monitoring. In: 2nd International workshop on structural health monitoring. Standford, CA, USA
28. Xia Y, Lei X, Wang P, Sun L (2021) Artificial intelligence based structural assessment for regional short-and medium-span concrete beam bridges with inspection information. Remote Sens 13(18):3687
29. Ferreira PM, Machado MA, Carvalho MS, Vidal C (2022) Embedded sensors for structural health monitoring: methodologies and applications review. Sensors 22(21):8320
30. Ubertini F, D'Alessandro A, Downey A, García-Macías E, Laflamme S, Castro-Triguero R (2017) Recent advances on SHM of reinforced concrete and masonry structures enabled by self-sensing structural materials. In: Proceedings, vol 2, no 3. MDPI, p 119
31. Musella C, Serra M, Menna C, Asprone D (2021) Building information modeling and artificial intelligence: advanced technologies for the digitalisation of seismic damage in existing buildings. Struct Concr 22(5):2761–2774
32. Shibu M et al (2023) Structural health monitoring using AI and ML based multimodal sensors data. Meas Sens 27:100762
33. Altabey WA, Noori M (2022) Artificial-intelligence-based methods for structural health monitoring. Appl Sci 12(24):12726
34. Mondal TG, Chen G (2022) Artificial intelligence in civil infrastructure health monitoring—historical perspectives, current trends, and future visions. Front Built Environ 8:1007886
35. Luckey D et al (2022) Explainable artificial intelligence to advance structural health monitoring. Struct Health Monit Based Data Sci Tech 331–346
36. Smarsly K, Lehner K, Hartmann D (2007) Structural health monitoring based on artificial intelligence techniques. Comput Civil Eng 2007:111–118
37. Padmapoorani P, Senthilkumar S, Mohanraj R (2023) Machine learning techniques for structural health monitoring of concrete structures: a systematic review. Iran J Sci Technol Trans Civil Eng 1–13
38. Worden K, Manson G (2007) The application of machine learning to structural health monitoring. Philos Trans Royal Soc A Math Phys Eng Sci 365(1851):515–537
39. Yuan F-G et al (2020) Machine learning for structural health monitoring: challenges and opportunities. Sens Smart Struct Technol Civil Mech Aerosp Syst 11379:1137903
40. Tibaduiza D et al (2018) A damage classification approach for structural health monitoring using machine learning. Complexity (2018):1–14
41. Smarsly K, Dragos K, Wiggenbrock J (2016) Machine learning techniques for structural health monitoring. In: Proceedings of the 8th European workshop on structural health monitoring (EWSHM 2016), Bilbao, Spain
42. Zhu S, Xiang T (2023) Structural damage identification considering uncertainties in nonuniform measurement conditions based on convolution neural networks. Struct Control Health Monit (2023)
43. Tran V-L, Vo T, Nguyen T (2023) One-dimensional convolutional neural network for damage detection of structures using time series data. Asian J Civil Eng 1–34
44. Tran-Ngoc H et al (2022) Damage assessment in structures using artificial neural network working and a hybrid stochastic optimization. Sci Rep 12(1):4958

45. Xue Z, Xu C, Wen D (2022) Structural damage detection based on one-dimensional convolutional neural network. Appl Sci 13.1:140
46. Eltouny KA, Liang X (2023) Large-scale structural health monitoring using composite recurrent neural networks and grid environments. Comput Aided Civil Infrastruct Eng 38(3):271–287
47. Farias SV et al (2021) A damage detection method using neural network optimized by multiple particle collision algorithm. J Sens 1–14
48. Kondo G (2021) Review of development on convolution neural network based structural health monitoring on bridges. J Phys Conf Ser 1. IOP Publishing
49. Rosafalco L et al (2020) Fully convolutional networks for structural health monitoring through multivariate time series classification. Adv Model Simul Eng Sci 7.1:1–31
50. Wu X, Ghaboussi J, Garrett JH Jr (1992) Use of neural networks in the detection of structural damage. Comput Struct 42(4):649–659
51. Shin HK, Ahn YH, Lee SH, Kim HY (2020) Automatic concrete damage recognition using multi-level attention convolutional neural network. Materials 13(23):5549
52. Tanveer M, Kim B, Hong J, Sim SH, Cho S (2022) Comparative study of lightweight deep semantic segmentation models for concrete damage detection. Appl Sci 12(24):12786
53. Wu CS, Peng YX, Zhuo DB, Zhang JQ, Ren W, Feng ZY (2022) Energy ratio variation-based structural damage detection using convolutional neural network. Appl Sci 12(20):10220
54. Wang N, Zhao Q, Li S, Zhao X, Zhao P (2018) Damage classification for masonry historic structures using convolutional neural networks based on still images. Comput Aided Civil Infrastruct Eng 33(12):1073–1089
55. Gopalakrishnan K, Gholami H, Vidyadharan A, Choudhary A, Agrawal A (2018) Crack damage detection in unmanned aerial vehicle images of civil infrastructure using pre-trained deep learning model. Int J Traffic Transp Eng 8(1):1–14
56. Ye XW, Jin T, Yun CB (2019) A review on deep learning-based structural health monitoring of civil infrastructures. Smart Struct Syst 24(5):567–585
57. Punia A, Mor N (2021) For a better tomorrow: smart and sustainable cities using artificial intelligence. Handbook of green engineering technologies for sustainable smart cities. CRC Press, pp 173–189
58. Mor N, Sood H, Goyal T (2019) An overview of challenges and opportunities in development and use of accident prediction models
59. Hassan EM, Mor N, Sood H (2019) Study of level of service (LOS) criteria for measuring traffic congestion-a critical review

Chapter 2
Artificial Intelligence in Damage Detection of Concrete Structures: Techniques, Integration and Future Directions

Salim Barbhuiya and Bibhuti Bhusan Das

Abstract The chapter thoroughly explores the pivotal role played by Artificial Intelligence (AI) in the identification of damages in concrete structures. It delves into conventional methods, their limitations, and how AI can effectively complement these approaches. The basics of AI, encompassing machine learning and deep learning, are elucidated within the specific context of damage detection. Additionally, the chapter examines data acquisition and pre-processing techniques tailored for AI models. It sheds light on AI-driven damage detection methodologies, such as the utilization of convolutional neural networks for image analysis, vibration analysis, and AI-enhanced non-destructive testing methods, highlighting their precision in identifying structural issues. Moreover, the chapter investigates the integration of AI into structural health monitoring systems, providing in-depth discussions on data fusion and real-time monitoring. Emphasis is placed on the significance of performance assessment and model validation to ensure the reliability of AI algorithms. The chapter also addresses future trends, including the integration of AI with the Internet of Things (IoT), and delves into ethical considerations in the sphere of infrastructure development. In summary, the chapter underscores AI's transformative potential in revolutionizing damage detection and structural health assessment, contributing to the creation of more resilient and sustainable concrete structures.

Keywords Artificial intelligence · Damage detection · Concrete structures · Structural health monitoring · Machine learning

---

S. Barbhuiya (✉)
Department of Engineering and Construction, University of East London, London, UK
e-mail: s.barbhuiya@uel.ac.uk

B. B. Das
Department of Civil Engineering, NIT Karnataka, Mangalore, Surathkal, India
e-mail: bdas@nitk.edu.in

H. Jahangir et al. (eds.), *Damage Detection and Structural Health Monitoring of Concrete and Masonry Structures*, Springer Tracts in Civil Engineering,
https://doi.org/10.1007/978-981-97-8975-7_2

2.1 Introduction

Concrete structures serve as the backbone of modern infrastructure, constituting a diverse range of essential elements such as towering skyscrapers, expansive bridges spanning vast rivers, intricate road networks connecting cities, and massive dams holding back immense bodies of water. The inherent attributes of concrete, including its robustness, load-bearing capacity, and adaptability, have solidified its position as a favoured construction material in civil engineering. Despite their stalwart nature, these concrete structures are not impervious to the relentless forces of nature and the wear and tear of everyday use. Over time, they are subject to a multitude of external and internal factors that can gradually erode their integrity and structural soundness. Environmental elements, such as temperature fluctuations, freeze–thaw cycles, moisture ingress, and chemical exposure, can initiate a slow but steady process of degradation [1–4]. Simultaneously, operational factors like heavy loads, dynamic forces, and vibrations induced by vehicular traffic or human activities can further contribute to wear and tear.

These cumulative effects of exposure can manifest as various forms of damage, such as cracks, spalling, corrosion of reinforcing steel, and material deterioration. Left undetected and unaddressed, these damages can escalate, jeopardizing the structural stability of the concrete elements and posing significant safety risks to the occupants and the surrounding environment. To preserve the longevity and safety of concrete structures, it is imperative to implement a rigorous and proactive approach to damage detection and monitoring [5, 6]. Early detection of damages is the key to preventing their propagation and mitigating potential risks. By identifying issues at their inception, engineers and asset managers can implement timely maintenance and repair measures, averting further deterioration and avoiding costly remedial actions.

Traditional methods of damage detection have long been employed for assessing the health of concrete structures. Visual inspections, performed by skilled engineers, involve a meticulous examination of the visible surfaces for signs of distress or deterioration. Additionally, non-destructive testing (NDT) techniques, such as ultrasonic testing, radiography, and ground-penetrating radar, are utilized to assess the internal condition of concrete elements without causing harm [7–10]. While these conventional methods have proven their worth over the years, they do have inherent limitations. Visual inspections may not identify subtle internal damages, and NDT techniques can be time-consuming, expensive, and require specialized equipment and expertise.

Artificial Intelligence (AI) has emerged as a transformative solution for detecting damage in concrete structures, as supported by research [11–13]. AI encompasses a diverse set of advanced algorithms, notably machine learning and deep learning, tailored to analyse extensive datasets, discern intricate patterns, and make well-informed decisions based on available information. AI-powered damage detection leverages data from various sources, including sensor networks, cameras, acoustic signals, and other non-destructive testing methods. Through the processing of this data, AI models can detect complex damage patterns and anomalies within concrete structures, often surpassing human capabilities in terms of accuracy and sensitivity [14–16].

The significance of AI in damage detection lies in its ability to detect even minor damages that may go unnoticed by human inspectors or traditional methods [17, 18]. Its high precision and sensitivity enable early detection, empowering engineers and asset managers to proactively address structural issues before they escalate into more severe problems. This data-driven approach to damage detection facilitates predictive maintenance, wherein historical data is analysed to forecast potential future damages and maintenance needs. By optimizing resource allocation and scheduling maintenance activities based on real-time conditions, AI contributes to more efficient and cost-effective infrastructure management. Moreover, AI-based damage detection reduces the dependence on human intervention for routine monitoring tasks. Engineers and inspectors can concentrate on complex analyses, critical decision-making, and strategic planning, while AI continuously monitors the health of structures, providing comprehensive insights and alerts when anomalies are detected.

The motivation behind writing this chapter is driven by the growing importance of infrastructure safety and sustainability. Concrete structures are vital components of modern society, and their integrity is critical to ensuring the safety of people and the smooth functioning of economies. Artificial Intelligence has emerged as a ground-breaking technology with the potential to revolutionize various industries, including civil engineering. By harnessing the power of AI, we can leverage its ability to process massive amounts of data, identify intricate patterns, and provide real-time insights for proactive decision-making.

The aim of this chapter is to explore the significant role that AI plays in detecting damages in concrete structures. By presenting an overview of AI-based damage detection techniques, their integration with existing structural health monitoring systems, and their impact on infrastructure maintenance, this chapter seeks to empower engineers, researchers, and stakeholders with valuable knowledge to enhance the safety and longevity of concrete structures. Through a comprehensive exploration of AI algorithms, data acquisition techniques, and case studies of real-world applications, this chapter aims to foster an understanding of AI's transformative potential in the field of structural health assessment. By shedding light on the benefits and challenges associated with AI-based damage detection, we hope to inspire further research, development, and adoption of these innovative technologies.

The limitations of the chapter include its inability to cover all specific AI algorithms comprehensively due to the vastness of the topic. It may not delve deeply into the technical intricacies of AI implementation or address all emerging challenges in the rapidly evolving field of AI. Moreover, the chapter may not provide an exhaustive examination of the ethical and societal implications of AI in infrastructure development. Despite these limitations, the chapter offers valuable insights into the significance of AI in damage detection, serving as a foundation for further exploration and research in this transformative domain.

2.2 Traditional Methods of Damage Detection in Concrete Structures

Traditional methods of damage detection in concrete structures involve visual inspections and non-destructive testing (NDT) techniques. Visual inspections are conducted by skilled engineers who visually examine the concrete surfaces to identify visible signs of distress, such as cracks, spalling, and discoloration. NDT techniques, like ultrasonic testing, radiography, and ground-penetrating radar, are used to assess the internal condition of concrete without causing damage. These methods have proven valuable in assessing structural health, but they have limitations. Visual inspections may not detect internal or hidden damages, while NDT techniques can be time-consuming, expensive, and require specialized equipment and expertise.

2.2.1 *Visual Inspection*

Visual inspection is a fundamental method of damage detection in concrete structures that involves a comprehensive examination of the visible surfaces of the structure by trained engineers or inspectors. This method is employed to assess the overall condition of the concrete elements, identify visible signs of distress, and detect any apparent damages or defects. During a visual inspection, the inspectors carefully inspect the exposed surfaces of the concrete, including walls, columns, beams, slabs, and other components. They look for various indicators of potential damage, such as cracks, spalling, surface discoloration, efflorescence, and signs of corrosion on reinforcement bars (rebars). Additionally, they observe for evidence of uneven settlement, joint separation, and other visible indications of structural issues.

The process of visual inspection begins with a preliminary visual survey of the entire structure to identify any apparent damages that may require closer examination. The inspectors then conduct a more detailed and systematic examination of specific areas where damages or anomalies were observed during the initial survey or where they are likely to occur based on experience and knowledge of the structure. Inspectors may use handheld tools, such as magnifying lenses, flashlights, or borescopes, to get a better view of hard-to-reach areas and narrow cracks. They may also take photographs or video recordings to document the findings for further analysis and reference.

Advantages of Visual Inspection:

- Accessibility: Visual inspection is readily accessible and does not require specialized equipment or advanced training, making it a cost-effective method.
- Quick Assessment: Engineers can perform visual inspections relatively quickly, allowing for rapid initial evaluations of the structure's health.
- Early Detection: Visual inspection can identify visible damages at an early stage, enabling timely maintenance and repair interventions.

Limitations of Visual Inspection:

- Limited to Surface-Level Defects: Visual inspection may not detect internal or hidden damages that could be critical to the structural integrity.
- Subjectivity: The effectiveness of visual inspection depends on the expertise of the inspector, leading to potential subjectivity in detecting damages.
- Weather Conditions: Visual inspections can be hindered by adverse weather conditions, reducing visibility and accuracy.

2.2.2 *Non-Destructive Testing (NDT)*

Non-Destructive Testing (NDT) methods are invaluable techniques for detecting internal defects and damages in concrete structures without causing any harm to the structure itself. These methods provide crucial insights into the condition of the concrete elements, allowing engineers and inspectors to assess the extent of damages, identify potential issues, and make informed decisions regarding maintenance and repair.

2.2.2.1 Ultrasonic Testing (UT)

Ultrasonic testing utilizes high-frequency sound waves to assess the internal condition of concrete. A transducer emits ultrasonic pulses into the concrete, and the waves travel through the material until they encounter any defects, such as cracks, voids, or delamination. The reflected waves are recorded, and the time taken for them to return to the transducer is measured. Based on the travel time, engineers can determine the depth and size of internal defects. Figure 2.1 shows the healing and simultaneous ultrasonic monitoring of cracks in concrete.

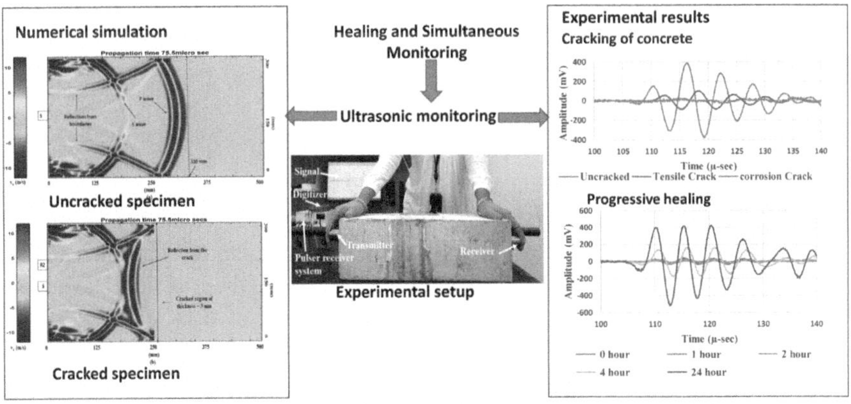

Fig. 2.1 Healing and simultaneous ultrasonic monitoring of cracks in concrete [19]

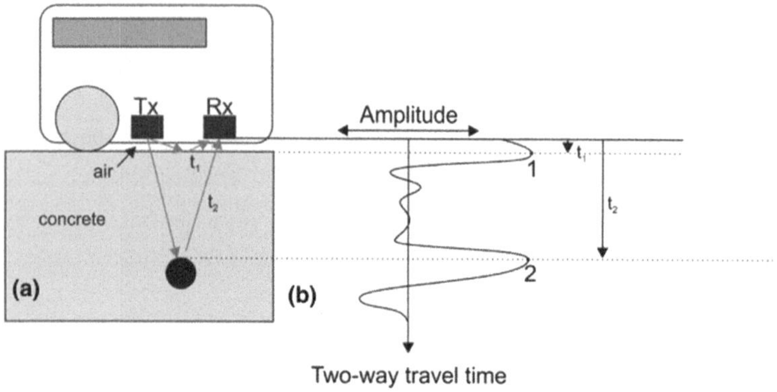

Fig. 2.2 The principle of ground penetrating radar [20]

2.2.2.2 Ground-Penetrating Radar (GPR)

Ground-Penetrating Radar is a non-intrusive technique that uses electromagnetic waves to image the subsurface of the concrete. A GPR system emits radar pulses into the concrete, and the waves bounce back when they encounter different materials or anomalies within the concrete (Fig. 2.2). This method can detect variations in material properties, identify voids, locate rebar and post-tension cables, and assess the presence of moisture or delamination.

2.2.2.3 Rebound Hammer Test

The rebound hammer test is a convenient and portable non-destructive testing (NDT) technique employed to evaluate the surface hardness of concrete. It involves striking the concrete surface with a spring-loaded mass, and then measuring the rebound distance, as illustrated in Fig. 2.3. The rebound value is directly related to the compressive strength of the concrete, thereby offering insights into potential damage or structural weaknesses.

2.2.2.4 Half-Cell Potential Mapping

This technique is utilized to identify areas susceptible to corrosion in reinforced concrete structures. A half-cell probe is placed on the concrete surface, and an electrical potential is measured between the reinforcing steel and the concrete. Deviations from the expected potential indicate areas of potential corrosion, which may lead to concrete spalling and structural degradation. The schematic of the half-cell potential measurement setup is shown in Fig. 2.4.

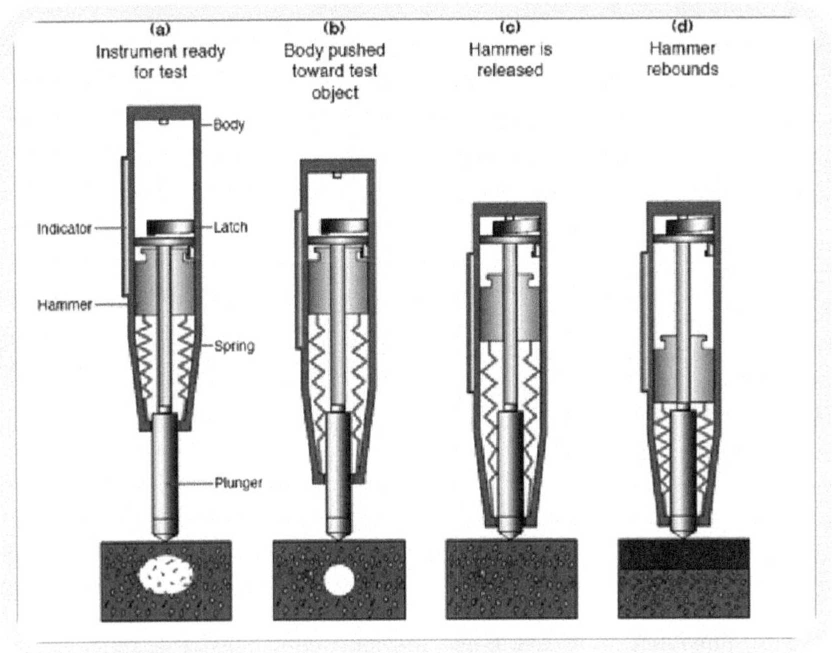

Fig. 2.3 Rebound hammer [21]

Fig. 2.4 Schematic of the half-cell potential measurement setup [22]

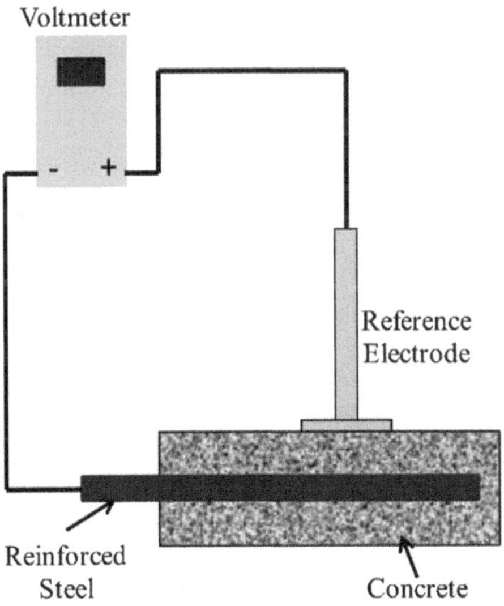

2.2.2.5 Impact-Echo Testing

Impact-echo testing uses stress waves generated by impacting the concrete surface with a hammer or mechanical device. The resulting stress waves travel through the concrete and reflect back when they encounter defects, such as voids, cracks, or delamination. The reflected waves are recorded and analysed to detect internal flaws in the concrete. The schematic diagram of the shock-echo method is shown in Fig. 2.5.

Advantages of NDT Methods:

- Internal Assessment: NDT methods provide insights into the internal condition of concrete structures, detecting defects that may not be visible externally.
- Non-Destructive: NDT techniques do not cause damage to the structure, preserving its integrity during assessment.
- Greater Accuracy: NDT methods can provide quantitative data and detailed imaging, offering more precise information about the extent of damages.

Limitations of NDT Methods:

- Equipment and Expertise: NDT methods require specialized equipment and trained personnel, making them costlier and more time-consuming than visual inspections.
- Depth Limitations: Some NDT methods have limitations in detecting defects beyond certain depths within the concrete.
- Interpretation Complexity: Interpreting NDT results may require expertise, and false positives or false negatives can occur, leading to potential inaccuracies.

Non-Destructive Testing methods play a critical role in assessing the health of concrete structures. By providing valuable information about internal defects and

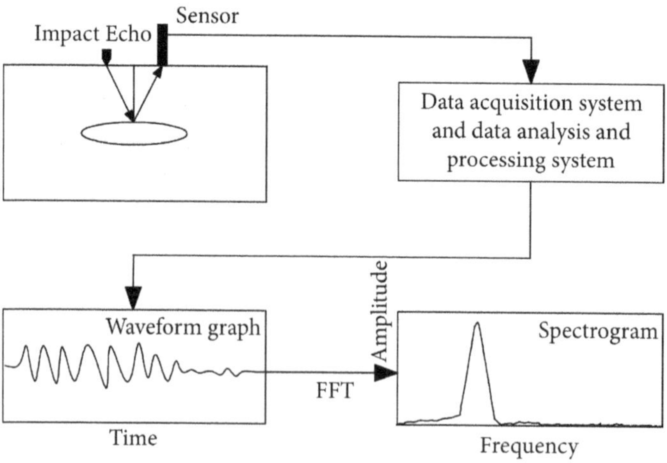

Fig. 2.5 Schematic diagram of the shock-echo method [23]

damages, NDT techniques enable engineers and asset managers to take proactive measures to maintain and enhance the longevity and safety of concrete structures. When combined with other inspection methods, NDT offers a comprehensive approach to structural health assessment, ensuring the continued functionality and resilience of infrastructure for the benefit of society.

2.2.3 Limitations of Traditional Methods

Traditional methods of damage detection, including visual inspection and non-destructive testing (NDT) techniques, have been the cornerstone of assessing the health and integrity of concrete structures for decades. While these methods have proven valuable, they do come with inherent limitations that may hinder their ability to provide a comprehensive assessment of the structural condition.

Visual inspection, being a manual process, is subject to the expertise and experience of the inspectors. Different inspectors may interpret damages differently, leading to potential discrepancies in identifying defects and assessing their severity. Moreover, visual inspection is primarily limited to identifying surface-level defects and visible signs of distress. It may not detect internal damages, such as cracks, voids, and corrosion, which are critical to the structural integrity but not visible externally. Additionally, visual inspections can be impacted by adverse weather conditions, reducing visibility and accuracy, especially in outdoor or hard-to-reach areas.

Non-destructive testing (NDT) methods, such as ultrasonic testing (UT), ground-penetrating radar (GPR), and rebound hammer test, provide valuable insights into the internal condition of concrete structures without causing damage. However, these methods have their limitations. Some NDT techniques may have depth limitations and may not be effective in detecting damages beyond certain depths within the concrete [24–26]. This limitation can be particularly critical in large and deep concrete structures, where internal defects may be located at significant depths. Furthermore, NDT methods require specialized equipment and trained personnel, making them more expensive and time-consuming than visual inspections. The accessibility and operation of the necessary equipment can be challenging, especially in remote or difficult-to-reach locations. Additionally, interpreting NDT results can be complex and may involve subjective judgment, leading to the possibility of false positives (indicating damage where none exists) or false negatives (failing to detect actual damages). These inaccuracies can impact decision-making regarding maintenance and repair strategies, potentially leading to inadequate or unnecessary actions. Another limitation of traditional methods is their inability to detect damages at their early stages. Early detection is crucial for implementing timely maintenance and repair interventions, preventing the escalation of issues and avoiding more extensive and costly rehabilitation work.

To overcome these limitations, a complementary approach is essential, combining traditional methods with advanced technologies. For instance, artificial intelligence (AI) can augment the capabilities of visual inspection and traditional NDT methods by providing more accurate and objective analysis of data. AI algorithms can process vast amounts of data, recognize intricate patterns, and identify anomalies associated with damages in concrete structures, often with a level of precision that surpasses human capabilities. Moreover, modern NDT techniques, such as GPR and impact-echo testing, offer deeper insights into internal damages, allowing for a more comprehensive assessment of concrete structures' health. Integrating these advanced technologies with traditional methods can enhance the effectiveness and efficiency of damage detection, leading to better-informed decision-making, more proactive maintenance strategies, and ultimately, safer and more durable concrete structures.

2.2.4 *Role of Artificial Intelligence in Augmenting Traditional Approaches*

The role of Artificial Intelligence (AI) in augmenting traditional approaches to damage detection in concrete structures is a transformative one. AI brings a new dimension to the field by leveraging its capabilities to process vast amounts of data, identify intricate patterns, and make accurate predictions. When integrated with traditional methods, AI enhances their effectiveness, efficiency, and reliability, leading to more comprehensive and proactive structural health assessments.

2.2.4.1 Data Analysis and Pattern Recognition

AI algorithms excel at analysing large datasets, including sensor data from non-destructive testing, structural monitoring systems, and historical maintenance records [27]. By processing this data, AI can identify subtle patterns and correlations that may be challenging for human inspectors to recognize. AI can flag potential damage indicators, such as crack patterns or corrosion trends, enabling early detection and precise assessment of damages.

2.2.4.2 Real-Time Monitoring

AI-driven systems enable real-time monitoring of concrete structures, providing continuous feedback on their health and performance [28–31]. Traditional methods may involve periodic inspections, leaving gaps in data between assessments. With AI, sensors and data streams can be integrated into a Structural Health Monitoring (SHM) system, enabling continuous data acquisition, analysis, and immediate alerts in case of anomalies or damages.

2.2.4.3 Predictive Maintenance

AI's ability to analyse data and detect patterns allows for predictive maintenance strategies. By understanding the degradation trends of concrete structures, AI can predict the future condition of the infrastructure and recommend optimal maintenance schedules [32, 33]. This proactive approach helps prevent costly and unexpected damages, leading to more efficient use of resources and longer-lasting structures.

2.2.4.4 Automation and Efficiency

AI-powered systems automate various tasks, reducing human involvement and increasing inspection efficiency. For instance, AI can process images from visual inspections, extract relevant information, and prioritize areas that require further inspection. This streamlines the overall inspection process, allowing engineers to focus on critical areas and complex assessments.

2.2.4.5 Data Fusion and Integration

AI facilitates the integration of data from multiple sources, including NDT results, visual inspections, weather conditions, and structural performance data [34, 35]. By fusing these diverse datasets, AI can provide a holistic view of the structure's health, helping engineers make informed decisions based on a comprehensive analysis of the available information.

2.2.4.6 Enhanced Decision Support

AI-powered systems act as decision support tools, providing engineers and asset managers with evidence-based insights and recommendations. By combining AI's analytical capabilities with human expertise, informed decisions can be made regarding maintenance priorities, repair strategies, and resource allocation.

2.2.4.7 Adaptability and Learning

AI models can be continuously trained and refined using new data, making them adaptable to evolving conditions and structural behaviours. As more data is collected and analysed, AI models can improve their accuracy and performance, ensuring the continuous enhancement of the damage detection process.

While AI offers numerous advantages in damage detection, it also faces challenges. Ensuring data privacy, avoiding biases in model training, and addressing interpretability and explainability concerns are crucial considerations in implementing

AI-powered damage detection systems. Ethical handling of sensitive data, eliminating algorithmic biases, and making AI models interpretable and explainable are essential to maintain trust, fairness, and accountability in the use of AI for infrastructure assessment.

2.3 Introduction to Artificial Intelligence (AI)

Artificial Intelligence (AI) represents a field within computer science that strives to create intelligent machines capable of emulating human-like cognitive functions. These machines possess the ability to perceive their surroundings, engage in reasoning, learn from data, and make decisions with the purpose of accomplishing specific objectives. AI encompasses various subdomains, including machine learning, natural language processing, computer vision, and robotics. The applications of AI are remarkably diverse, spanning from virtual assistants such as Siri and Alexa to self-driving vehicles, medical diagnostic systems, and advanced data analysis tools. The ongoing advancements in AI algorithms and technology have resulted in its extensive integration across various industries, establishing it as a transformative force that is shaping the future of technology and human interactions.

2.3.1 Basics of AI and Machine Learning

The fundamental concept underpinning AI revolves around the development of machines and systems that can mimic and reproduce human intelligence and cognitive functions. This encompasses a wide range of human-like capabilities, such as problem-solving, logical reasoning, experiential learning, sensory perception, and data-driven decision-making. AI systems are typically categorized into two primary types: Narrow AI and General AI. Narrow AI, also referred to as Weak AI, is engineered for specific tasks and is tailored to excel within a defined domain. Examples of narrow AI include voice assistants like Siri or Alexa, recommendation systems, and image recognition applications. While these systems excel in their designated tasks, they lack the capacity to perform functions beyond their specialized area.

Figure 2.6 illustrates the comprehensive AI framework. The AI development process consists of three fundamental components: Perceptual intelligence, Cognitive intelligence, and Decision-making intelligence. Perceptual intelligence empowers machines to perceive their surroundings using human-like senses such as vision, hearing, and touch. Cognitive intelligence represents a higher-order skill, enabling machines to engage in reasoning, learning, and knowledge acquisition, drawing inspiration from cognitive science and brain-like intelligence. Once machines possess both perception and cognition, they are poised to make optimal decisions, much like humans, benefiting various facets of life, industry, and society. Decision intelligence relies on a combination of data science, decision theory, and managerial

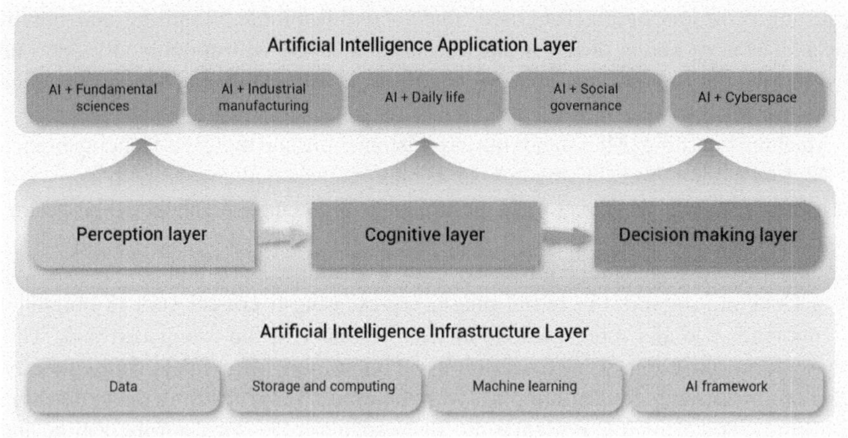

Fig. 2.6 The general framework of AI [36]

science to facilitate informed machine decision-making. To attain these objectives, AI necessitates a robust infrastructure layer supported by data, storage, computational capacity, machine learning algorithms, and AI frameworks. Training models allow machines to comprehend underlying data patterns, culminating in the realization of AI applications. These applications are increasingly integrated into a multitude of fields, encompassing fundamental sciences, industrial manufacturing, human existence, social governance, and cyberspace, exerting a significant impact on both our work and lifestyle.

Conversely, General AI, also known as Strong AI or Artificial General Intelligence (AGI), represents machines endowed with human-like intelligence, capable of comprehending, learning, and executing a wide array of intellectual tasks spanning various domains. General AI would possess the capacity to adapt to novel situations, learn from experiences, and apply knowledge to resolve problems in diverse contexts.

However, at present, General AI remains a theoretical concept and has not been fully actualized. The majority of real-world AI applications belong to the category of Narrow AI, showcasing exceptional proficiency in specific tasks but lacking the comprehensive cognitive capabilities of human intelligence. Nevertheless, even Narrow AI demonstrates transformative potential in numerous industries, including healthcare, finance, transportation, entertainment, manufacturing, and more. AI-driven technologies have brought about revolutionary advancements in medical diagnosis, financial analysis, autonomous vehicles, content recommendations, and various other domains [37–39].

AI's potential is vast and continues to advance rapidly with ongoing research and development. As technology evolves, AI is expected to make significant contributions in addressing complex challenges and providing innovative solutions across various sectors. However, achieving General AI remains an ambitious and challenging goal, requiring profound breakthroughs in understanding human cognition and intelligence. The continuous progress and exploration of AI will undoubtedly open up new frontiers of possibilities, shaping the future of technology and human–machine interaction.

Machine Learning (ML) stands out as an astonishing and rapidly advancing branch of AI, fundamentally reshaping how machines perceive and engage with the world. At its essence, ML empowers machines to learn from data and enhance their performance without the need for explicit programming. The ML process comprises a series of pivotal stages that collectively bestow machines with intelligence [40, 41]. Data Collection marks the initial phase, representing a critical step in obtaining a comprehensive and representative dataset pertinent to the designated task. This dataset forms the cornerstone for training and fine-tuning ML models. Subsequently, Data pre-processing comes into play, with the objective of ensuring data precision, uniformity, and error-free presentation. Tasks such as cleansing, normalization, and feature engineering contribute to elevating data quality, thereby facilitating effective learning.

Training the Model constitutes the heart of ML, where models employ algorithms and mathematical techniques to analyse data, identify patterns, and create internal representations to make predictions or decisions based on new inputs. Evaluation is vital, ensuring the model's competence during assessment on a separate dataset, displaying reliable and accurate predictions for unseen data. Model Deployment concludes the process, enabling the model to operate autonomously, making predictions, and automating decisions in real-world scenarios. ML encompasses diverse learning algorithms, tailored to data characteristics and use cases. Supervised Learning involves learning from labelled data, where input samples are associated with output labels, enabling predictions on new data. Unsupervised Learning involves training on unlabelled data, discovering hidden patterns or structures, such as clustering or data dimension reduction. Reinforcement Learning focuses on agents learning from interactions with the environment, receiving rewards or penalties to maximize cumulative rewards.

ML finds application across diverse industries [42–45]. In natural language processing, it enables sentiment analysis, speech recognition, and language translation. In computer vision, it facilitates image recognition, object detection, and facial recognition. ML powers recommendation systems, suggesting personalized content or products to users based on preferences and behaviours. One of ML's most compelling attributes is adaptability, as models continuously learn and refine predictions with new data, becoming more accurate and efficient in solving complex problems. This adaptability positions ML as a driving force behind innovation in AI, contributing to ground-breaking advancements and shaping the future of technology across industries. As research and development progress, the potential for ML continues to expand, heralding a new era of intelligent machines and transformative applications in various domains.

As illustrated in Fig. 2.7, the early exploration of concrete has adhered to three distinct scientific paradigms: empiricism, theory, and computation. Achieving the desired concrete properties for specific design requirements involves the adjustment

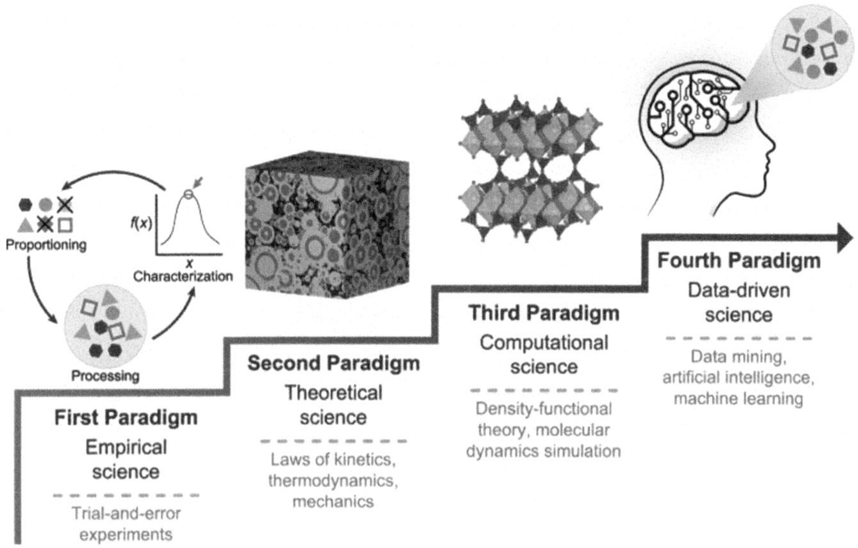

Fig. 2.7 Four paradigms of concrete science: empirical, theoretical, computational and data-driven [46]

of mixture components such as cement, water, aggregate, and admixtures. Traditional concrete mixture design typically relies on trial-and-error techniques and iterative processes until the intended properties are attained (empirical science, Paradigm 1). Starting in the 1980s, the advent of microstructural models for cement hydration ushered in a profound comprehension of the relationships between microstructure and material properties (theoretical science, Paradigm 2), harnessing principles from kinetics, thermodynamics, and mechanics. In the 2000s, the utilization of density-functional theory (DFT) and molecular dynamics (MD) simulations marked the emergence of the computational science paradigm (Paradigm 3). These approaches integrated first-principle models to elucidate the properties and hydration of cementitious materials. In recent years, a fourth paradigm (data-driven science, Paradigm 4) has gained prominence, with the integration of artificial intelligence and machine learning (ML) techniques. This innovative paradigm provides valuable insights and practical solutions to expedite advancements in the design and development of cementitious materials.

2.3.2 *Deep Learning and Neural Networks*

Deep Learning, a subset of Machine Learning, has emerged as a revolutionary force in the realm of Artificial Intelligence, ushering in ground-breaking advancements and reshaping numerous industries. Its success can be attributed to the remarkable

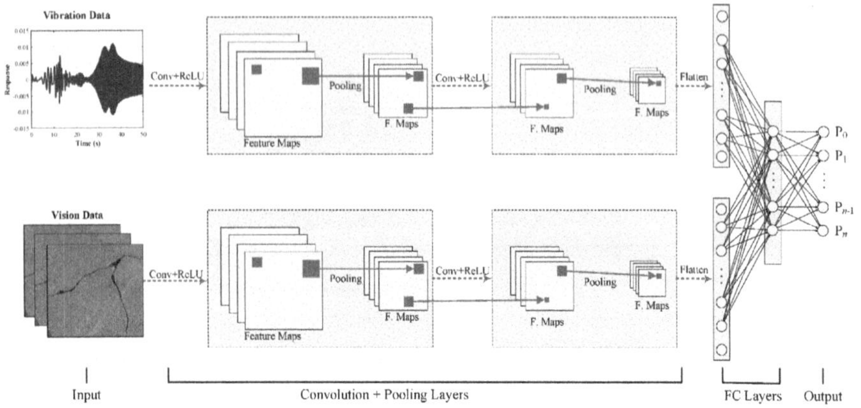

Fig. 2.8 A multi-headed deep neural network for different input data [47]

capacity of artificial neural networks to emulate the interconnectedness and adaptive nature of the human brain, enabling them to grasp intricate patterns and features from vast datasets.

The term "deep" in Deep Learning alludes to the network's depth, signifying the presence of multiple hidden layers that hierarchically process data. Each layer progressively extracts abstract representations of the input data, empowering the network to encapsulate intricate and high-level abstractions. These deep architectures play a pivotal role in the network's ability to generalize and make accurate predictions on unseen data. Figure 2.8 illustrates a multi-headed deep neural network designed for diverse input data.

Training deep neural networks presents a complex challenge that demands substantial computational resources. The procedure entails supplying the network with labelled data and iteratively updating its internal parameters through a mechanism known as backpropagation. This optimization process refines the network's weights and biases, effectively adapting its connections to minimize the disparity between its predictions and the true labels. Deep Learning has demonstrated remarkable achievements across diverse domains. In the realm of computer vision, deep neural networks have outperformed human-level capabilities in tasks such as image recognition, object detection, and segmentation. In the field of natural language processing, deep learning models have driven significant advancements in machine translation, sentiment analysis, chatbots, and text generation.

Speech recognition systems powered by Deep Learning have become integral components of virtual assistants and voice-controlled devices, providing natural and intuitive interactions with machines [48, 49]. Recommendation systems, used by platforms like Netflix and Amazon, utilize deep learning techniques to suggest personalized content or products based on users' past behaviours and preferences. Despite its tremendous successes, Deep Learning also faces challenges. Large-scale training of deep neural networks requires considerable computational resources, and the demand for data can be a limiting factor in certain applications. Moreover, the inherent complexity of deep neural networks often results in a lack of interpretability, making it difficult to understand how a particular decision is reached by the model.

Researchers and practitioners in the field are actively working on addressing these challenges [50, 51]. Hardware advancements, like the use of GPUs and specialized accelerators, have expedited training times and made deep learning more accessible. Transfer learning techniques enable leveraging pre-trained models on large datasets to tackle tasks with limited labelled data. Additionally, researchers are exploring methods for explaining the decisions made by deep learning models, striving to make them more transparent and interpretable. The ongoing progress and innovations in Deep Learning continue to reshape the landscape of Artificial Intelligence, with tremendous potential for further advancements and applications in diverse domains. As the field evolves, it holds the promise of solving even more complex problems, unlocking new frontiers, and ultimately empowering humanity with more intelligent and sophisticated AI systems.

2.3.3 Supervised, Unsupervised, and Reinforcement Learning

Supervised, Unsupervised, and Reinforcement Learning are three fundamental categories of Machine Learning algorithms, each serving distinct purposes and suited for different types of data and tasks (Fig. 2.9).

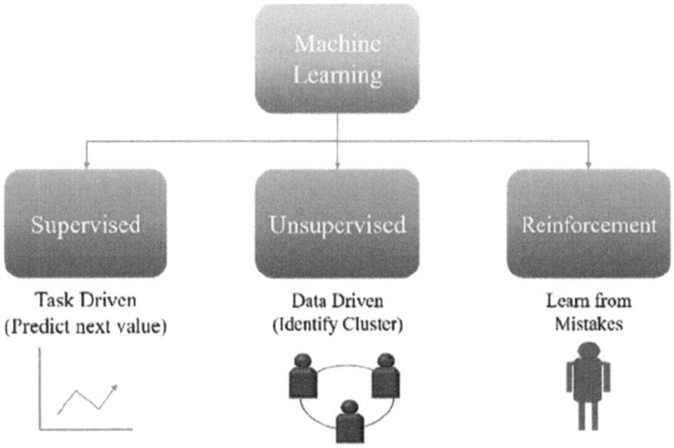

Fig. 2.9 Type of machine learning [44]

2.3.3.1 Supervised Learning

Supervised Learning stands as the most prevalent and widely applied form of Machine Learning [52–54]. In this methodology, the model is furnished with a labelled dataset, wherein each data point is paired with the corresponding correct output label. The primary objective of the model is to grasp the inherent relationship between the input features and the output labels, enabling it to furnish precise predictions for new, previously unseen data. The training process in supervised learning entails minimizing the disparity between the model's predictions and the actual labels, typically gauged using a loss function. Common instances of supervised learning tasks encompass image classification, where the model assigns labels to images based on their content, and regression, where the model forecasts continuous values like house prices based on various features. Supervised Learning is ideally suited for situations where labelled data is accessible, and the task involves mapping inputs to specific outputs. However, the effectiveness of this approach is heavily contingent on the quality and magnitude of the training dataset.

2.3.3.2 Unsupervised Learning

Unsupervised Learning involves the training of a model on an unlabelled dataset, indicating the absence of predefined output labels for the data points [55–57]. The primary objective of unsupervised learning is to unveil latent patterns, structures, or representations within the data. One prominent unsupervised learning task is clustering, where the model organizes similar data points into clusters based on their inherent attributes. Another task involves dimensionality reduction, which trims the number of features while retaining critical information, rendering data more accessible for visualization and analysis. Unsupervised learning proves especially valuable when dealing with unstructured data, revealing concealed relationships, or investigating the inherent properties of the dataset. Its applications extend to areas like customer segmentation, anomaly detection, and feature learning.

2.3.3.3 Reinforcement Learning

Reinforcement Learning sets itself apart from supervised and unsupervised learning by involving an agent in direct interaction with an environment, where it learns from feedback in the form of rewards or penalties [58–60]. The primary objective for the agent is to maximize cumulative rewards over time, achieved by selecting actions that result in favourable outcomes while avoiding actions that lead to negative consequences. The learning process in reinforcement learning relies on a trial-and-error approach, where the agent explores the environment, receives feedback, and adapts its policy or strategy to make improved decisions in subsequent interactions. Reinforcement Learning has proven to be highly effective in domains that necessitate sequential decision-making, such as robotics, game playing, and autonomous vehicles. In these domains, the agent learns to navigate complex environments, formulate optimal strategies, and adjust to dynamic changes in its surroundings.

Each type of learning–supervised, unsupervised, and reinforcement learning–has its strengths and limitations. Choosing the appropriate learning paradigm depends on the specific problem, the availability of labelled data, and the nature of the task at hand. By leveraging the strengths of each approach, Machine Learning continues to push the boundaries of AI, driving advancements across various fields and transforming industries worldwide.

2.3.4 Overview of AI Algorithms Used in Damage Detection

In damage detection of concrete structures, a variety of AI algorithms are employed to analyse data and identify signs of deterioration, cracks, or structural anomalies. These algorithms play a crucial role in enhancing the efficiency and accuracy of damage detection processes. Some of the key AI algorithms used in this domain include.

2.3.4.1 Convolutional Neural Networks (CNNs)

Convolutional Neural Networks (CNNs) represent a remarkable advancement in the field of image-based damage detection, especially in concrete structures [61–64]. These deep learning algorithms have shown exceptional proficiency in recognizing patterns and features within images, making them a powerful tool for identifying various types of structural damage. CNNs operate by employing a series of convolutional layers that extract different features at increasing levels of complexity. This hierarchical approach allows them to identify not only simple patterns but also more intricate details present in images of concrete surfaces.

When applied to concrete structures, CNNs are capable of detecting an array of damage types, including cracks, spalling, corrosion, and even subtle anomalies in the surface. Their ability to learn from a large dataset of images equips them with the skills to generalize and accurately identify damage in various conditions, lighting, and angles. This adaptability ensures that they can effectively operate in real-world scenarios, enhancing the overall efficiency and accuracy of damage detection processes. The convolutional neural network for predicting crack pattern and stress-crack width curve of air-void structure in 3D printed concrete is depicted in Fig. 2.10. CNNs have the potential to revolutionize the field of structural health monitoring by enabling automated and rapid assessment of concrete surfaces. Their utilization not only reduces the dependence on manual inspections but also facilitates early detection of damage, minimizing the risk of structural failures and ensuring the longevity of concrete structures. As technology continues to evolve, further advancements in

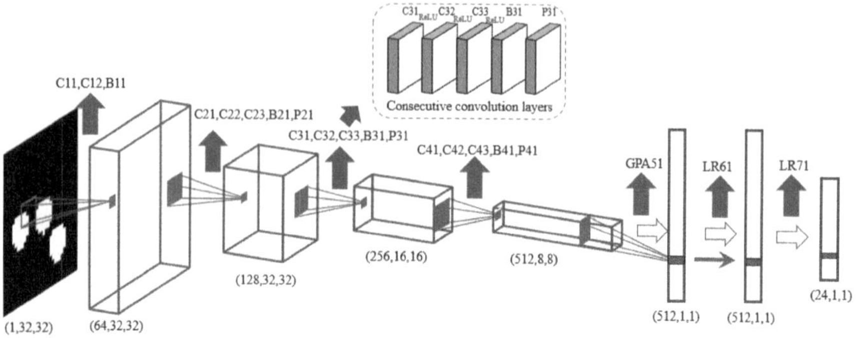

Fig. 2.10 CNN architecture (C#: Convolutional layer; B#: Batch normalization layer; P# Max pooling layer; GPA: Global average activation function; LR# activation layer) [64]

CNN-based damage detection techniques are expected, contributing to safer and more durable infrastructure development.

2.3.4.2 Recurrent Neural Networks (RNNs)

Recurrent Neural Networks (RNNs) excel in analysing sequential data, rendering them invaluable for identifying damage trends over time in structural health monitoring [65, 66]. These networks are particularly adept at processing data from sensors, enabling them to identify abnormal vibrations or acoustic emissions that may signify underlying damage (Fig. 2.11). RNNs' capacity to consider the temporal aspect of data equips them to uncover subtle patterns and changes in structural behaviour, facilitating timely and precise damage detection. This capability makes RNNs a vital tool in enhancing the reliability and safety of structures through continuous monitoring and early warning systems.

2.3.4.3 Support Vector Machines (SVMs)

Support Vector Machines (SVMs) play a pivotal role in damage detection due to their versatility in handling both classification and regression tasks [68–70]. These algorithms excel in supervised learning scenarios where labelled data is available. Their capability to effectively separate data points into different classes makes them particularly well-suited for detecting various types of structural damage or categorizing the severity of defects. In classification tasks, SVMs can accurately classify data points into predefined classes, aiding in identifying specific types of damage, such as cracks, corrosion, or spalling, based on distinctive features.

Additionally, SVMs can be employed in regression tasks to predict numerical values, which proves valuable in assessing the extent or severity of structural defects. The success of SVMs hinges on their ability to establish a clear decision boundary

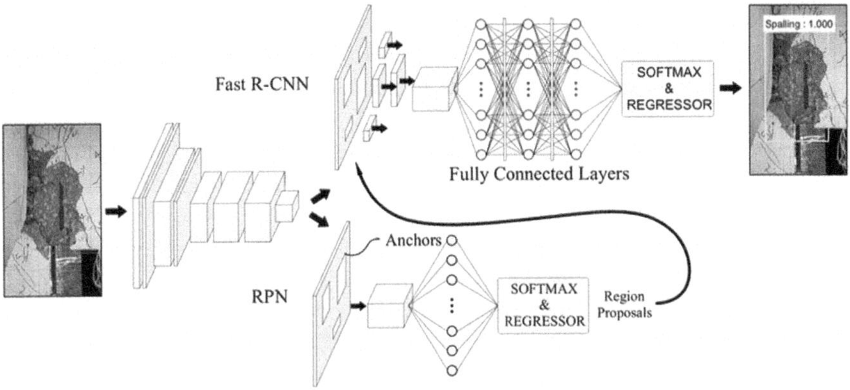

Fig. 2.11 Faster R-CNN-based damage detection [67]

between different classes or predict accurate values within a regression context. This makes them a robust tool for damage detection across different industries, from civil engineering to manufacturing. SVMs' reliance on labelled data and their proficiency in handling complex datasets enable them to contribute significantly to the enhancement of structural health monitoring and maintenance strategies, ultimately ensuring the safety and longevity of various infrastructure systems. An example of damage detection with an OC-SVM is shown in Fig. 2.12.

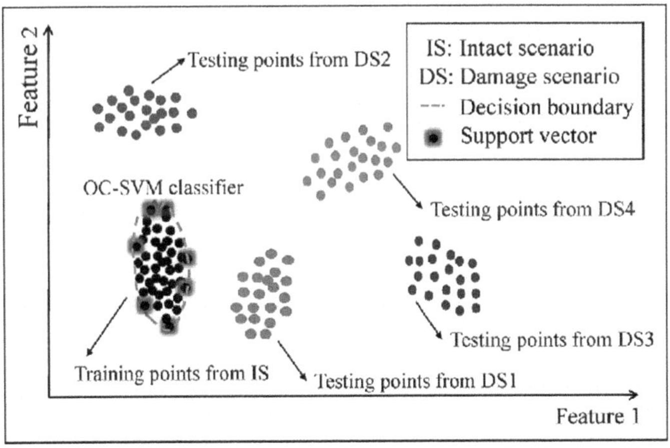

Fig. 2.12 Example of damage detection with an OC-SVM [71]

2.3.4.4 Random Forests

Random Forests stand as a sophisticated ensemble learning technique that harnesses the power of multiple decision trees to deliver enhanced predictive capabilities. This approach has proven particularly effective for feature selection and classification tasks, making it a valuable asset in the realm of identifying various types of damage in concrete structures [72–74]. The inherent strength of Random Forests lies in their ability to aggregate the insights and predictions from a multitude of individual decision trees. Each tree is trained on a different subset of the data and makes its own predictions. By amalgamating the outputs of these trees, the ensemble generates a more accurate and robust prediction than any single decision tree could achieve. The conceptual schematic of the Random Forest method is shown in Fig. 2.13.

In the context of damage detection in concrete structures, Random Forests excel in recognizing complex patterns and relationships that may signify structural issues. They can identify subtle variations and anomalies present in the data, aiding in categorizing different types of damage, whether it be cracks, corrosion, or other structural defects. One of the remarkable features of Random Forests is their capability to handle high-dimensional datasets with numerous features. This not only makes them efficient for concrete damage assessment but also grants them the ability to extract essential features that contribute most to the classification process. Incorporating Random Forests into the realm of structural health monitoring bolsters the accuracy

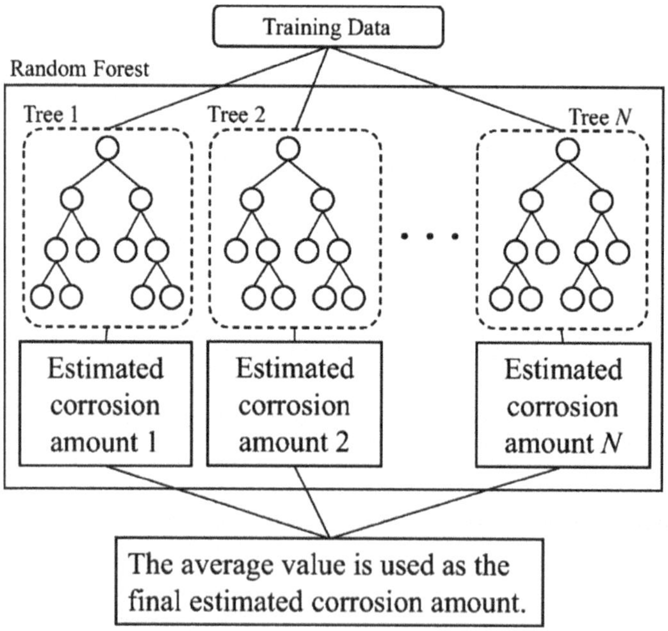

Fig. 2.13 Conceptual schematic of the random forest method [73]

of damage detection and ensures that even nuanced or hidden patterns are identified. By leveraging the collective intelligence of multiple decision trees, this ensemble approach offers an effective solution for enhancing the reliability and precision of concrete structure analysis, contributing to the longevity and safety of infrastructure.

2.3.4.5 Autoencoders

Autoencoders represent a potent class of unsupervised deep learning algorithms that hold significant potential in the domain of structural health monitoring, particularly in concrete structures [75–78]. These algorithms are adept at dimensionality reduction and feature learning, thereby aiding in the detection of subtle changes and anomalies within the data. At the core of autoencoders lies the concept of learning compact representations of the input data. This is achieved through a two-stage process: encoding and decoding. During encoding, the algorithm compresses the input data into a lower-dimensional representation, capturing its essential features. Subsequently, the decoding phase reconstructs the original data from this compressed representation.

The allure of autoencoders in concrete structure analysis lies in their ability to uncover latent patterns and variations that might indicate damage or changes over time. By learning these compact representations, autoencoders can capture even nuanced deviations from the norm, such as slight shifts in material properties, surface conditions, or structural integrity. This capacity for detecting subtle changes is particularly valuable in the context of concrete structures, where even minor shifts can indicate potential concerns. Autoencoders' unsupervised nature allows them to learn from unlabelled data, which is often abundant in real-world scenarios, minimizing the need for manual annotation. An example of a deep autoencoder model is shown in Fig. 2.14.

Furthermore, autoencoders can contribute to efficient data pre-processing by reducing the dimensionality of high-dimensional datasets, making subsequent analysis more manageable. This, in turn, enhances the speed and accuracy of damage detection processes.

2.3.4.6 Long Short-Term Memory (LSTM) Networks

Long Short-Term Memory (LSTM) networks stand as a specialized variant of Recurrent Neural Networks (RNNs), distinguished by their proficiency in capturing long-range dependencies within sequential data. This unique capability holds immense promise in the realm of structural health monitoring, especially when it comes to the intricate task of damage detection [80–82]. The Hybrid CNN-LSTM network for damage detection and classification is shown in Fig. 2.15. LSTMs are engineered to overcome a key limitation of traditional RNNs: the vanishing gradient problem. This issue hindered RNNs from effectively considering distant relationships within sequential data, a critical requirement for comprehending the behaviour

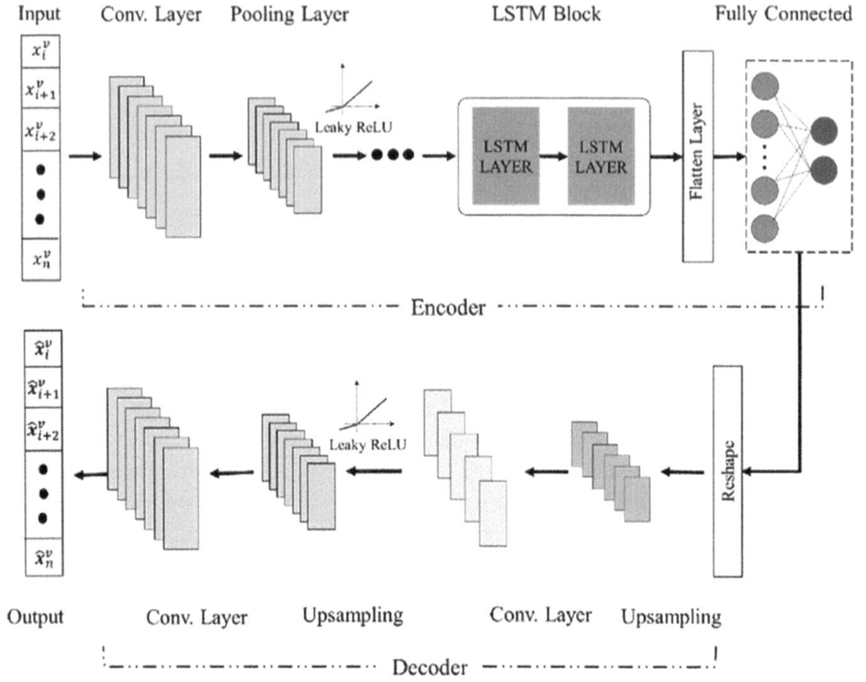

Fig. 2.14 Example of a deep autoencoder model [79]

of complex systems over time. LSTMs address this shortcoming by introducing a memory cell that stores and controls the flow of information, ensuring that relevant past information can influence predictions even after multiple time steps.

In the domain of structural health monitoring, particularly with concrete structures, time-series data are often the foundation for damage detection. Vibrations, strain measurements, temperature changes, and other metrics form sequential data that exhibit temporal patterns. LSTMs thrive in capturing these intricate temporal dependencies, enabling them to discern subtle variations that could signify damage or shifts in structural integrity. The capacity to model long-range dependencies equips LSTMs to analyse time-series data holistically, taking into account historical trends and patterns that could indicate anomalies. For instance, a gradual change in vibration characteristics might indicate underlying structural damage, and LSTMs can pick up on these trends, allowing for early detection and timely intervention. Incorporating LSTMs into structural health monitoring systems facilitates the automatic identification of irregularities, offering an efficient way to enhance safety and maintenance strategies. By learning from historical data and accounting for long-term trends, LSTMs contribute to a comprehensive understanding of concrete structure behaviour, making them a pivotal tool for ensuring infrastructure longevity and resilience.

These AI algorithms are often combined with data acquisition techniques like sensors, cameras, and acoustic emission devices to capture relevant information

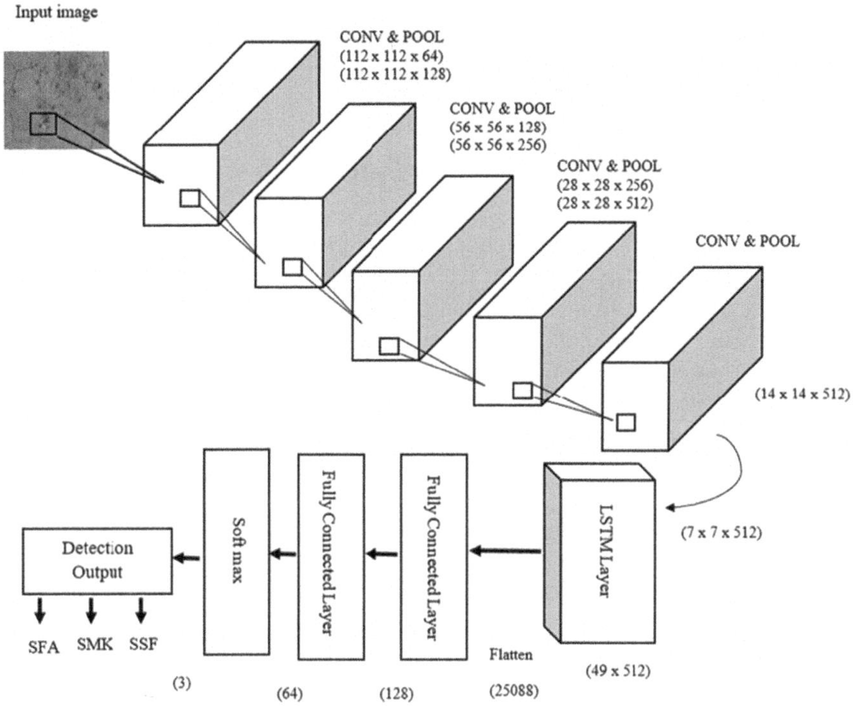

Fig. 2.15 Hybrid CNN-LSTM network for damage detection and classification [83]

about the concrete structure's condition. By analysing the data collected from various sources, AI algorithms can effectively identify and predict potential damage, enabling timely maintenance and repair, thus ensuring the safety and longevity of concrete structures. As AI technology continues to advance, the effectiveness and applicability of these algorithms in damage detection are expected to improve, offering new opportunities for enhancing infrastructure assessment and management.

2.4 Data Acquisition and Pre-processing

Data acquisition and pre-processing are critical steps in the process of damage detection of concrete structures using AI algorithms. Data acquisition involves the collection of relevant data from various sources, such as sensors, cameras, and other monitoring devices, capturing important information about the structural condition. Pre-processing is essential for cleaning, transforming, and preparing the data to ensure its quality and suitability for AI analysis. This step includes data cleaning to remove noise and outliers, normalization to scale the data within a specific range, and feature engineering to extract meaningful features for the AI models. Proper data acquisition and pre-processing are crucial for the accuracy and effectiveness of AI-based damage detection systems.

2.4.1 Data Collection for Damage Detection

Data collection for damage detection in concrete structures is a crucial process that forms the foundation for successful AI-based analysis and prediction. It involves gathering relevant data from various sources to provide insights into the structural health and condition of the concrete elements. The data collected should encompass a wide range of parameters that influence the structural integrity, such as physical properties, environmental factors, and any potential damage indicators. The conceptual diagram of concrete damage inspection process is depicted in Fig. 2.16.

2.4.1.1 Sensors and Instrumentation

Various sensors and instrumentation are deployed to monitor different aspects of the concrete structure. These may include strain gauges, accelerometers, temperature sensors, displacement sensors, corrosion sensors, and acoustic emission devices. Strain gauges measure deformation, accelerometers capture vibrations, temperature sensors track temperature changes, and acoustic emission devices detect active damage processes like crack propagation.

2.4.1.2 Visual Inspection

Visual inspection is an essential component of data collection, providing valuable information about surface cracks, spalling, efflorescence, and other visible signs of deterioration. Experienced inspectors use visual assessments to identify potential areas of concern that can be further analysed with AI algorithms.

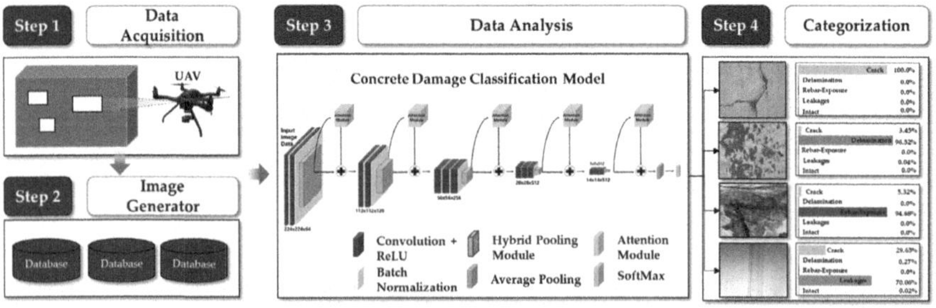

Fig. 2.16 Conceptual diagram of concrete damage inspection process [84]

2.4.1.3 NDT

NDT techniques are employed to assess the internal integrity of concrete structures without causing damage. Common NDT methods include ultrasonic testing, ground-penetrating radar (GPR), magnetic particle inspection, and infrared thermography. These techniques help in identifying subsurface defects and delaminations.

2.4.1.4 Structural Health Monitoring (SHM) Systems

SHM systems continuously monitor the structural behaviour of concrete structures using a network of sensors, capturing real-time data on structural responses to loads, environmental conditions, and other dynamic forces. SHM systems are valuable in understanding the structural performance and detecting changes over time.

2.4.1.5 Historical Maintenance and Repair Data

Historical data on maintenance and repair work done on the concrete structure can provide valuable insights into past damage events, contributing to the understanding of potential patterns or recurring issues.

2.4.1.6 Environmental Data

Environmental factors such as temperature, humidity, rainfall, and exposure to chemical agents can influence the degradation of concrete structures. Incorporating environmental data into the analysis can enhance the accuracy of damage detection.

Data collection for damage detection should be comprehensive, covering both temporal and spatial aspects of the concrete structure. It is essential to have a well-designed data acquisition plan that ensures data is collected at appropriate intervals and locations. Additionally, data quality control measures must be implemented to minimize errors and inaccuracies during the data collection process. Once the data is collected, it undergoes pre-processing to clean, normalize, and prepare it for AI-based analysis. By combining data from diverse sources and applying AI algorithms, damage detection systems can efficiently identify and predict structural issues, enabling proactive maintenance and ensuring the safety and longevity of concrete structures.

2.4.2 Types of Data in Concrete Structure Assessments

Concrete structure assessments involve the collection and analysis of a wide range of data to comprehensively evaluate the condition and integrity of the structure. Each type of data provides unique insights into different aspects of the structure's health, performance, and potential damage indicators. By combining and analysing these diverse data types, engineers and inspectors can gain a holistic understanding of the structure's behaviour and make informed decisions about its maintenance and repair needs.

Physical properties data, such as the concrete's mix design, compressive strength, and modulus of elasticity, are fundamental parameters that dictate the material's load-carrying capacity and overall performance. These properties provide essential information about the concrete's quality and suitability for specific structural applications. Compressive strength, for example, indicates the concrete's ability to withstand compression forces and is critical for assessing the structural safety and capacity. Modulus of elasticity represents the concrete's stiffness and is crucial for understanding its deformation under load. By analysing physical properties data, engineers can ensure that the concrete used in a structure meets the required specifications and design standards, thus ensuring its durability and integrity throughout its service life.

Structural geometry data provides information about the dimensions and layout of the concrete elements. Understanding the structure's geometry is crucial for evaluating its load-carrying capacity and determining potential weak points or areas vulnerable to stress concentrations. For instance, irregularities in the geometry, such as uneven slab thickness or variations in column heights, can lead to localized stress concentrations and structural vulnerabilities. By analysing structural geometry data, engineers can identify design flaws and potential areas of concern that may require additional reinforcement or modifications to ensure the structure's stability and longevity.

Load and stress data capture the forces and loads that the structure is subjected to during its service life. Dead loads, which include the weight of the structure itself and fixed equipment, provide a constant load on the structure. Live loads, on the other hand, are variable and can change based on occupancy or usage. Environmental loads, such as wind, snow, seismic activity, or temperature variations, can exert dynamic forces on the structure. Analysing load and stress data are essential for assessing the structure's safety margins and determining whether it can withstand the various loads it will encounter during its lifetime.

Environmental data includes factors like temperature, humidity, rainfall, and exposure to chemical agents. These environmental conditions can significantly impact the structural durability and integrity of concrete. For example, exposure to aggressive agents like chloride and sulphate can lead to concrete deterioration through processes like corrosion of reinforcing steel. Extreme temperature variations can cause thermal stress and affect the structure's long-term performance. Monitoring and analysing environmental data helps engineers understand the specific environmental challenges a structure faces and make informed decisions about material selection, protective coatings, and maintenance strategies to enhance its durability and resilience.

Non-Destructive Testing (NDT) data provides valuable insights into the internal condition of the concrete without causing damage. Techniques like ultrasonic testing, ground-penetrating radar (GPR), and infrared thermography allow engineers to assess the presence of defects, voids, and delaminations that may not be visible on the surface. NDT data aids in the detection of hidden flaws and potential sources of damage that visual inspection alone may not reveal. By analysing NDT data, engineers can assess the health of the concrete structure, identify areas requiring repair or reinforcement, and monitor changes over time to ensure timely maintenance and avoid potential structural failures.

Visual inspection data is gathered through direct observations by inspectors and engineers. It includes the identification of surface cracks, spalling, efflorescence, and other visible signs of damage. Visual inspection is an essential first step in identifying areas of concern and determining the need for further assessment using other data types. By recording and documenting visual inspection data, engineers can track the evolution of visible damage, identify potential causes, and prioritize areas that require further investigation or repair. Visual inspection is particularly valuable for regular maintenance checks and for identifying changes that may indicate structural health issues before they escalate into more severe problems.

Vibration and dynamic data are obtained through structural health monitoring, capturing the structural response to external forces or dynamic loading. Analysing these data helps identify potential weaknesses, changes in behaviour over time, and the structure's ability to withstand dynamic events such as earthquakes or vibrations from nearby construction. Vibration and dynamic data are critical for understanding the structure's dynamic behaviour and can provide early warning signs of possible damage or structural degradation. Engineers can use this data to assess the structure's overall health and develop appropriate strategies to mitigate dynamic stresses and ensure its long-term stability and safety.

Strain and deformation data are typically obtained through strain gauges and displacement sensors, providing insights into how the structure responds to loads and indicating potential overstressing or deformation. By measuring the strains and deformations in different components of the structure, engineers can assess whether the structure is within the acceptable range of deformations and stress limits. Excessive strains can indicate potential structural weaknesses or deformations that may lead to cracks or failure. Strain and deformation data are essential for understanding the structural behaviour under varying loads and ensuring that the structure remains within safe operating conditions.

For reinforced concrete structures, corrosion monitoring data becomes crucial as corrosion can significantly affect the structural integrity. Data on the corrosion condition of steel reinforcement helps identify potential areas of concern and guide corrosion prevention and repair strategies. Corrosion monitoring techniques, such as half-cell potential measurements and corrosion rate monitoring, provide real-time data on the corrosion activity within the concrete. By analysing corrosion monitoring data, engineers can assess the extent of corrosion-induced damage, evaluate the effectiveness of corrosion protection measures, and implement targeted repair and maintenance interventions to mitigate the impact of corrosion on the structure's longevity and performance.

Finally, historical maintenance and repair data provide valuable insights into past interventions, their effectiveness, and any recurring issues. Understanding the history of maintenance and repairs helps engineers plan for future maintenance and improve the structure's long-term performance. By analysing historical data, engineers can identify patterns of deterioration, assess the long-term effectiveness of past repair strategies, and tailor maintenance plans to address specific vulnerabilities. Historical maintenance and repair data also aid in the development of predictive maintenance models, allowing for more proactive and cost-effective strategies for managing the structure's health and mitigating potential damage.

By integrating and analysing these diverse data types, AI-powered systems can assist engineers and inspectors in making data-driven decisions about maintenance, repair, and reinforcement strategies. This comprehensive approach ensures the safety, durability, and long-term performance of concrete structures, leading to more efficient and cost-effective infrastructure management. As AI technology continues to advance, the integration and analysis of these data types will become even more powerful, enabling more accurate and proactive structural assessments.

2.4.3 Data Pre-processing Techniques for AI Models

Data pre-processing is a crucial step in preparing the collected data for analysis using AI models. It involves cleaning, transforming, and normalizing the data to make it suitable for training and evaluating AI algorithms. The quality of the data pre-processing directly impacts the accuracy and effectiveness of AI models in damage detection of concrete structures. An example flowchart of a data-driven structural damage detection system is shown in Fig. 2.17. Some key data pre-processing techniques used in this domain include.

2.4.3.1 Data Cleaning

Data cleaning refers to the procedure of recognizing and addressing missing values, outliers, and inaccuracies within the gathered data. In the specific context of identifying damage in concrete structures, instances like sensor data or inspection records might exhibit missing values due to sensor malfunctions or incomplete inspections [86–88]. Outliers can arise from measurement errors or infrequent events, while errors may stem from inaccuracies in data entry. Data cleaning techniques encompass the process of imputing missing values through statistical methods or machine learning algorithms, identifying and managing outliers using resilient statistical approaches, or eliminating them if necessary, and rectifying errors through validation checks and data verification.

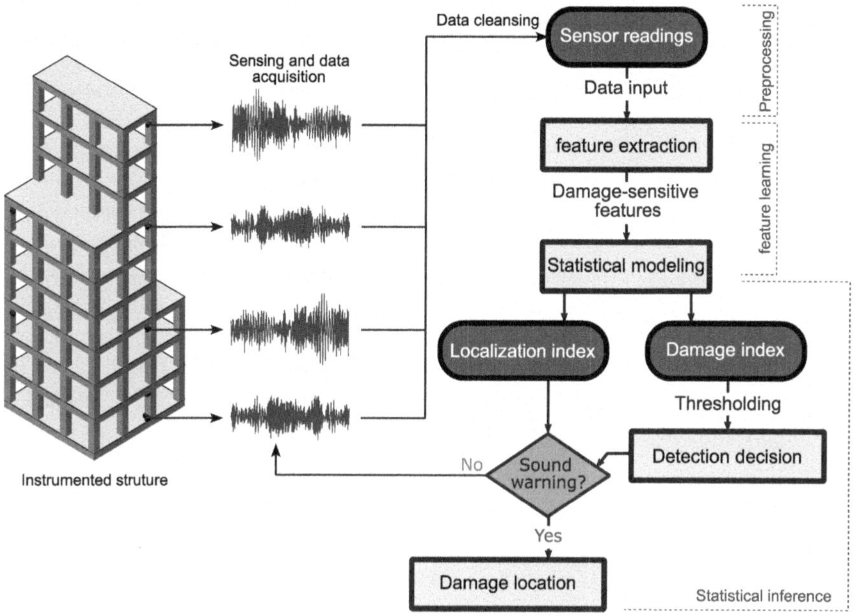

Fig. 2.17 Example flowchart of a data-driven structural damage detection system [85]

2.4.3.2 Data Transformation

Data transformation serves as a crucial step in preparing data to align with the assumptions of AI models. When it comes to assessing concrete structures, specific AI algorithms, like linear regression, presume that the data adheres to a normal distribution [89, 90]. To achieve a distribution that resembles the Gaussian curve, data transformation techniques like logarithmic or square root transformation are employed. Additionally, the Box-Cox transformation offers an alternative method suitable for various data distributions. These transformations play a pivotal role in stabilizing variance, linearizing relationships, and enhancing the model's performance by ensuring that the data conforms to the algorithms' assumptions.

2.4.3.3 Feature Scaling

The purpose of feature scaling is to standardize all features to a consistent scale, thereby preventing any single feature from exerting undue influence on the model's training process due to variations in magnitudes. AI algorithms, such as gradient descent used in training neural networks, exhibit accelerated convergence when applied to scaled data [91]. Common techniques for feature scaling encompass Min–Max scaling, which scales features to a specific range (e.g., [0, 1]), and z-score normalization, which standardizes features to possess a mean of zero and a variance of one. Feature scaling ensures that each feature contributes evenly to the learning process, resulting in enhanced model performance.

2.4.3.4 Feature Engineering

Feature engineering is the process of generating new features or modifying existing ones to enrich the data's representation, thereby enhancing the model's capability to discern underlying patterns and associations. When applied to the realm of detecting damage in concrete structures, feature engineering may entail the computation of derived parameters from sensor data, like strain ratios or damage indices. These derived parameters can offer valuable insights into the structural health [92, 93]. Profound domain expertise is indispensable in guiding the identification and engineering of pertinent features that hold relevance for the precise problem of damage detection.

2.4.3.5 Dimensionality Reduction

High-dimensional data can result in extended computation times and the risk of overfitting, where the model becomes overly focused on memorizing the training data, hindering its ability to generalize effectively to unseen data [94, 95]. Techniques for dimensionality reduction seek to decrease the number of features while retaining the most vital information. One common approach is Principal Component Analysis (PCA), which transforms the original features into a lower-dimensional space while capturing the maximum variance in the data. On the other hand, t-SNE is often employed to visualize high-dimensional data in two or three dimensions. Dimensionality reduction serves to diminish noise, enhance model performance, and streamline the model's representation.

2.4.3.6 Data Splitting

Data splitting involves the segmentation of the amassed data into distinct sets designated for training, validation, and testing purposes [96, 97]. The training set is employed to instruct the AI model, the validation set is utilized to refine hyperparameters and gauge the model's performance throughout the training process, and the testing set serves to appraise the model's capacity to generalize when faced with unfamiliar data. Prudent data partitioning is of paramount importance to avert overfitting and to secure the model's ability to effectively generalize to novel, unobserved data.

2.4.3.7 Handling Class Imbalance

In specific scenarios, data may exhibit an imbalance, where one class (e.g., undamaged structures) substantially outweighs the samples in comparison to the other class (e.g., damaged structures). This class imbalance can potentially lead to the development of models that exhibit a bias toward the majority class. Approaches such as oversampling the minority class, undersampling the majority class, or generating synthetic samples using techniques like SMOTE can effectively rectify the class distribution imbalance. Through addressing this class imbalance, the AI model gains enhanced accuracy and resilience, particularly when dealing with infrequent events, such as instances of structural damage.

2.4.3.8 Data Augmentation

Within the realm of image-based damage detection, data augmentation techniques find widespread use for broadening the diversity of training data without the necessity of collecting more samples. These techniques involve the application of random transformations, such as rotation, flipping, and zooming, to the existing images. By subjecting the model to a wider spectrum of data variations, data augmentation serves to enhance the model's proficiency in recognizing and generalizing to new examples from the same category. This, in turn, leads to improved model performance and mitigates the potential for overfitting.

Data pre-processing is a crucial and iterative process that requires careful consideration of the data characteristics and the specific requirements of the AI model. Properly pre-processing the data enhances the model's ability to learn and extract meaningful patterns, leading to accurate and reliable damage detection results for concrete structures. Effective data pre-processing ensures that the AI model can make robust predictions and contribute to more efficient and informed decision-making in infrastructure maintenance and safety.

2.5 AI-Based Damage Detection Techniques

AI-based damage detection techniques leverage the power of artificial intelligence to detect and assess damages in concrete structures. These techniques utilize various AI algorithms, such as deep learning and neural networks, to analyse data from different sources, including visual images, vibration signals, acoustic emissions, and sensor data. Image-based damage detection techniques employ convolutional neural networks (CNNs) to identify cracks and other visible damages from images. Vibration analysis and structural health monitoring using AI help identify structural anomalies and irregularities. Ultrasonic and acoustic emission testing, enhanced by AI, can pinpoint internal defects and damage. AI-based damage detection techniques offer improved accuracy, efficiency, and automation in assessing structural health and ensuring the safety of concrete structures.

2.5.1 Image-Based Damage Detection Using Convolutional Neural Networks (CNNs)

Image-based damage detection using Convolutional Neural Networks (CNNs) is a powerful technique that leverages the capabilities of deep learning to automatically identify and assess damages in concrete structures from visual images. CNNs are a type of neural network specifically designed for image recognition tasks, and they have proven to be highly effective in detecting patterns and features in images. The process of image-based damage detection using CNNs involves the following steps (Fig. 2.18).

2.5.1.1 Data Collection

Visual images of the concrete structures are collected using cameras or other imaging devices. These images may capture the surface of the structure, including walls, columns, and slabs.

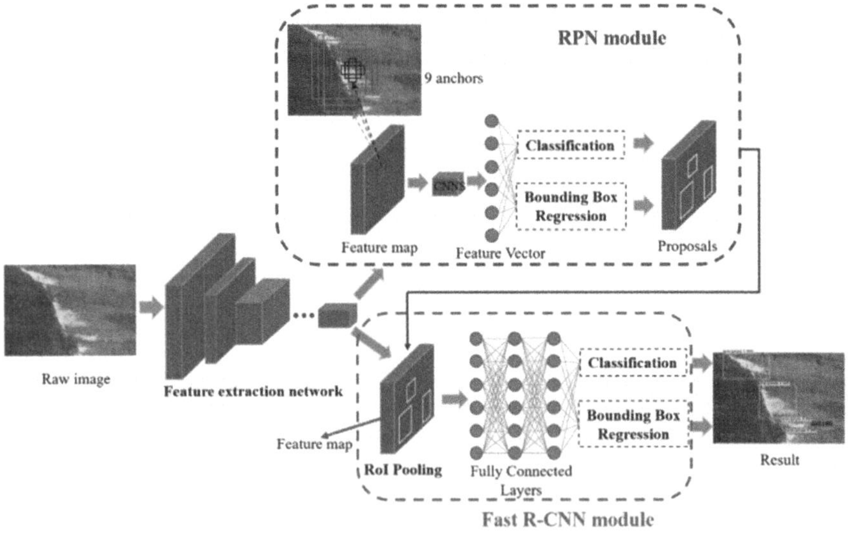

Fig. 2.18 Image-based automatic multiple-damage detection of concrete dams using region-based convolutional neural networks [63]

2.5.1.2 Data Pre-processing

The collected images undergo data pre-processing to ensure uniformity and consistency. This may include resizing the images to a fixed size, normalizing pixel values, and removing any unnecessary information.

2.5.1.3 Model Training

A CNN model is designed and trained using a large dataset of labelled images. The dataset includes both images of undamaged and damaged concrete structures, allowing the CNN to learn to differentiate between the two.

2.5.1.4 Feature Extraction

During training, the CNN learns to extract relevant features from the images that are indicative of damage. The network identifies patterns, edges, textures, and other visual cues that can help distinguish between normal and damaged areas.

2.5.1.5 Classification

Once the CNN is trained, it can be used to classify new images as either undamaged or damaged. The model assigns a probability score to each class, indicating the likelihood of damage in the given image.

2.5.1.6 Post-processing

The output of the CNN can be further post-processed to refine the detection results. This may involve thresholding the probability scores or applying additional filtering techniques to improve accuracy.

The advantages of image-based damage detection using CNNs include:

- Automation: CNNs enable automated and rapid analysis of large numbers of images, reducing the need for manual inspection and saving time and resources.
- Accuracy: CNNs are capable of achieving high accuracy in detecting even subtle damages that may be difficult for human inspectors to identify.
- Versatility: The technique can be applied to various types of concrete structures and different types of damages, making it a versatile tool in structural health monitoring.

However, there are some challenges with image-based damage detection using CNNs. The availability of labelled training data can be a limitation, as creating a large and diverse dataset may require significant effort and expertise. Additionally, the CNN's performance may be affected by variations in lighting, weather conditions, and camera angles, which can impact the quality of the images. Despite these challenges, image-based damage detection using CNNs holds great promise for enhancing the efficiency and accuracy of structural health monitoring in concrete structures, contributing to safer and more sustainable infrastructure management. Ongoing research and advancements in AI and deep learning will likely lead to even more sophisticated and reliable techniques in the future.

2.5.2 Vibration Analysis and Structural Health Monitoring with AI

Vibration analysis and structural health monitoring (SHM) with AI is a cutting-edge technique that combines the principles of structural dynamics, sensor technology, and artificial intelligence to assess the health and integrity of concrete structures. Vibration analysis involves studying the response of structures to external forces or dynamic loading, typically using accelerometers or other sensors to measure vibrations. Figure 2.19 Shows the framework of the early fault detection method based on deep learning and dynamics identification. The process of vibration analysis and SHM with AI includes the following steps.

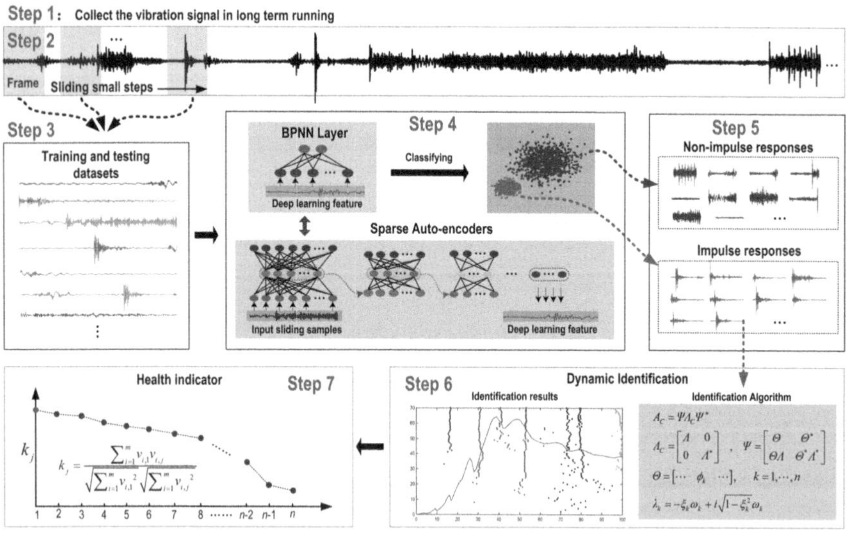

Fig. 2.19 Framework of the early fault detection method based on deep learning and dynamics identification [98]

2.5.2.1 Data Collection

Vibration data is collected from sensors strategically placed on the concrete structure. These sensors measure the accelerations, velocities, or displacements of the structure in response to various loading conditions.

2.5.2.2 Data Pre-processing

The collected vibration data undergoes pre-processing to remove noise, filter out irrelevant frequencies, and ensure data quality. Pre-processing techniques may involve filtering, resampling, and data smoothing.

2.5.2.3 Feature Extraction

AI-based techniques, such as machine learning algorithms, are used to extract meaningful features from the vibration data. These features represent the structural behaviour, resonant frequencies, and dynamic characteristics.

2.5.2.4 Model Training

Machine learning models, such as Support Vector Machines (SVM), Random Forests, or Neural Networks, are trained using a labelled dataset of vibration data. The dataset includes both normal and damaged vibration patterns, allowing the AI model to learn to differentiate between the two.

2.5.2.5 Anomaly Detection

Once the AI model is trained, it can be used for anomaly detection in real-time vibration data. It compares the measured vibration response to the learned patterns and identifies deviations that may indicate potential damage or structural abnormalities.

2.5.2.6 Condition Monitoring

Vibration analysis with AI allows continuous condition monitoring of the concrete structure. The AI model can detect early signs of damage or structural degradation, enabling timely interventions and maintenance to prevent further deterioration.

The advantages of vibration analysis and SHM with AI include:

- Early Detection: AI-powered SHM enables early detection of structural issues, allowing for proactive maintenance and avoiding costly repairs.

- Real-Time Monitoring: The technique provides real-time monitoring, making it suitable for critical structures such as bridges, dams, and high-rise buildings.
- Scalability: AI-based SHM can handle large volumes of data and monitor multiple sensors simultaneously, making it scalable for complex structures and extensive infrastructure networks.
- Cost-Effectiveness: By identifying and addressing potential issues early on, AI-powered SHM can save costs associated with major repairs and downtime.

However, there are challenges in implementing vibration analysis and SHM with AI. One challenge is the complexity of the AI models, which may require significant computational resources for real-time analysis. Additionally, accurate labelling and training data are crucial for the AI model's performance, which may be difficult to obtain for certain types of damages.

In conclusion, vibration analysis and structural health monitoring with AI represent a transformative approach to ensuring the safety and durability of concrete structures. The combination of AI algorithms and sensor technology enables timely and accurate detection of structural issues, supporting efficient infrastructure maintenance and enhancing the overall resilience of civil engineering projects. Continued research and development in AI and SHM will likely lead to further advancements in this field, providing even more sophisticated and reliable solutions for damage detection and structural health assessments.

2.5.3 Ultrasonic and Acoustic Emission Testing Using AI

Ultrasonic and Acoustic Emission Testing (AET) are advanced non-destructive testing (NDT) techniques used to detect and assess internal defects, cracks, and structural anomalies in concrete structures. When combined with artificial intelligence (AI), these testing methods become even more powerful and efficient for damage detection and structural health monitoring (Fig. 2.20).

2.5.3.1 Ultrasonic Testing

Ultrasonic testing involves the propagation of high-frequency sound waves through the concrete structure. When the waves encounter defects or irregularities, they reflect or scatter, creating echoes that are detected by receivers. The time taken for the sound waves to travel through the material and the amplitude of the received signals provide valuable information about the structure's internal condition.

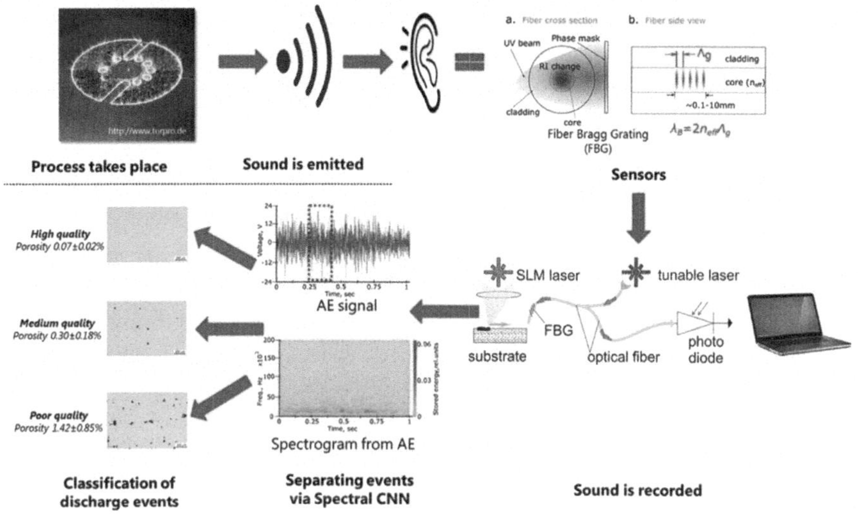

Fig. 2.20 Acoustic emission for in situ quality monitoring in additive manufacturing using spectral convolutional neural networks [99]

2.5.3.2 Acoustic Emission Testing

Acoustic Emission Testing, on the other hand, is based on the detection of transient elastic waves emitted by the structure during stress or deformation. When concrete undergoes changes due to damage or loading, microcracks and other internal defects emit stress waves. These waves are captured by sensors and analysed to identify the location and severity of the damage.

The integration of AI in Ultrasonic and Acoustic Emission Testing brings several advantages:

- Automated Analysis: AI algorithms can automate the analysis of complex ultrasonic and acoustic emission data, reducing the reliance on manual interpretation and increasing the efficiency of damage detection.
- Pattern Recognition: AI models can be trained to recognize specific patterns associated with different types of damage or defects in concrete structures. This enables accurate and consistent identification of damage across various testing scenarios.
- Data Fusion: AI can integrate data from multiple sensors and testing methods, such as ultrasonic and acoustic emission sensors, along with other sensor data like temperature, humidity, and strain, to provide a comprehensive and holistic view of the structure's health.
- Real-time Monitoring: AI-powered AET systems can provide real-time monitoring of concrete structures, enabling immediate action in response to changes or damage, and reducing the risk of catastrophic failures.

- Predictive Maintenance: By continuously monitoring the structure's health and identifying early signs of damage, AI can facilitate predictive maintenance strategies, allowing engineers to address potential issues before they escalate.

Challenges in implementing AI in Ultrasonic and Acoustic Emission Testing include the need for high-quality and calibrated sensors, as well as accurate and comprehensive training datasets for AI models. Additionally, data pre-processing and feature extraction can be complex and require expertise to ensure accurate and reliable results. Despite these challenges, the combination of AI and Ultrasonic/ Acoustic Emission Testing holds great promise for the efficient and accurate assessment of concrete structures' health. It offers a proactive approach to infrastructure maintenance and safety, enabling engineers to make data-driven decisions and ensure the longevity and reliability of concrete structures. As AI technology continues to advance, these testing techniques are expected to become even more robust and widespread in the field of structural health monitoring and damage detection.

2.5.4 AI-Enhanced NDT Methods for Crack Detection

AI-enhanced Non-Destructive Testing (NDT) methods for crack detection are revolutionary approaches that combine the power of artificial intelligence (AI) with advanced NDT techniques to improve the accuracy, efficiency, and reliability of crack detection in concrete structures. Cracks are one of the most common forms of damage in concrete, and early detection is crucial to prevent structural integrity issues and ensure the safety and durability of the infrastructure.

Traditional NDT methods, such as visual inspection and ultrasonic testing, have been effective in detecting cracks but are limited in their ability to analyse large datasets and identify subtle cracks in complex structures. Figure 2.21 shows the autonomous concrete crack semantic segmentation using deep fully convolutional encoder–decoder network in concrete structures inspection. AI-enhanced NDT methods address these limitations and offer several significant advantages.

2.5.4.1 Data-Driven Approach

AI models can process and analyse vast amounts of NDT data, including images, acoustic emission data, and ultrasonic signals, enabling a more comprehensive and data-driven approach to crack detection.

2.5.4.2 Feature Extraction

AI algorithms can automatically extract relevant features from NDT data that are indicative of crack presence, such as crack width, length, orientation, and severity.

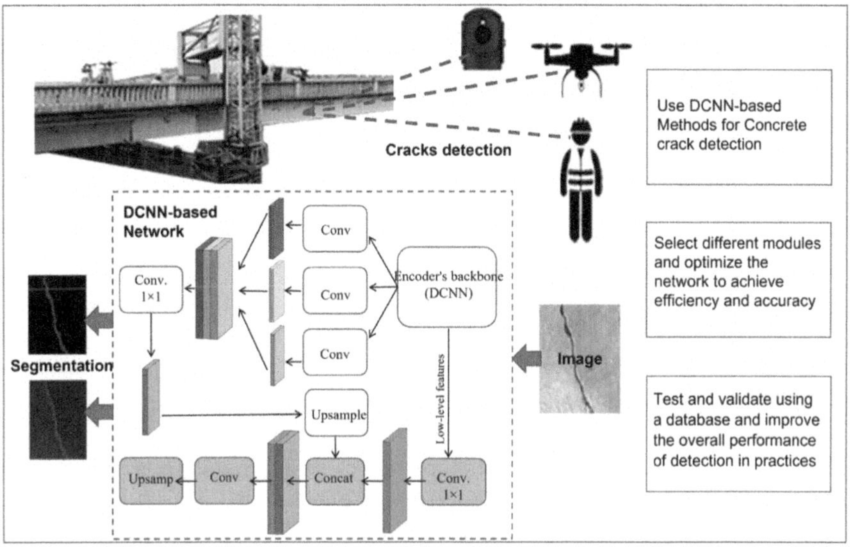

Fig. 2.21 Autonomous concrete crack semantic segmentation using deep fully convolutional encoder–decoder network in concrete structures inspection [100]

This reduces the reliance on manual feature extraction and increases the accuracy of crack identification.

2.5.4.3 Pattern Recognition

AI models can be trained on large datasets of NDT images and signals containing both cracked and intact structures. This allows the AI to recognize patterns and characteristics associated with different types of cracks, enabling more accurate and consistent crack detection.

2.5.4.4 Real-Time Monitoring

AI-enhanced NDT methods can provide real-time monitoring of concrete structures, continuously analysing NDT data and detecting new cracks as they occur. This enables early detection of cracks, allowing for timely interventions and maintenance.

2.5.4.5 Integration with SHM Systems

AI-enhanced NDT methods can be integrated with SHM systems, combining multiple types of sensor data and NDT techniques to provide a comprehensive assessment of the structure's health.

2.5.4.6 Predictive Maintenance

By continuously monitoring and analysing NDT data, AI models can predict the future development of cracks, enabling engineers to implement proactive maintenance strategies and avoid costly repairs.

Challenges in AI-enhanced NDT methods for crack detection include the need for high-quality training datasets that encompass a wide range of crack types, sizes, and orientations. Additionally, the AI models require extensive training and validation to ensure their accuracy and reliability. Overall, AI-enhanced NDT methods for crack detection represent a significant advancement in the field of structural health monitoring. By leveraging the capabilities of AI, these methods offer a more efficient and effective approach to crack detection, improving infrastructure safety and maintenance practices. As AI technology continues to evolve, AI-enhanced NDT methods are expected to become even more sophisticated, contributing to the ongoing efforts in ensuring the integrity and longevity of concrete structures.

2.6 Integration of AI in Structural Health Monitoring Systems

The integration of AI in Structural Health Monitoring (SHM) systems revolutionizes the way we monitor and assess the health of civil infrastructure. AI enables SHM systems to process and analyse vast amounts of sensor data, including data from accelerometers, strain gauges, and other sensors, in real-time. Machine learning algorithms in AI can detect and interpret structural anomalies, damage, or performance degradation, providing early warnings of potential issues. This integration enhances the accuracy and efficiency of SHM, enabling proactive maintenance and timely decision-making to ensure the safety and resilience of concrete structures. AI-driven SHM is a critical step towards smarter and more sustainable infrastructure management.

2.6.1 Designing an AI-Driven SHM System

Designing an AI-driven Structural Health Monitoring (SHM) system involves careful planning, integration of sensors, data processing techniques, and AI algorithms to create an efficient and reliable system for monitoring the health of concrete structures. The following steps outline the key components of designing such a system.

2.6.1.1 Sensor Selection

The first step is to select appropriate sensors to capture relevant data about the structure's behaviour. Common sensors used in SHM include accelerometers, strain gauges, displacement sensors, temperature sensors, and acoustic emission sensors. The choice of sensors depends on the specific parameters to be monitored and the types of potential damages.

2.6.1.2 Data Acquisition

Once the sensors are installed, data acquisition systems are used to collect and record the sensor data. Data acquisition may involve real-time data streaming, depending on the application. The data acquisition system should be robust, accurate, and capable of handling large volumes of data.

2.6.1.3 Data Pre-processing

The raw data collected from sensors often require pre-processing to remove noise, calibrate sensors, and ensure data quality. Pre-processing techniques may include filtering, resampling, and data alignment.

2.6.1.4 Data Storage and Management

The pre-processed data needs to be stored in a suitable database for easy access and retrieval. A well-designed data management system ensures data integrity, security, and efficient querying for analysis.

2.6.1.5 AI Algorithms Selection

Based on the specific monitoring objectives, appropriate AI algorithms are selected. Machine learning algorithms, such as neural networks, support vector machines, or decision trees, can be used for classification, anomaly detection, and predictive modelling.

2.6.1.6 Model Training

The AI models need to be trained using labelled data to learn patterns and relationships in the data. For example, supervised learning may be used for damage classification based on historical data with labelled damage instances.

2.6.1.7 Real-Time Monitoring

The AI-driven SHM system should provide real-time monitoring and analysis of the collected data. This allows for early detection of structural issues and timely responses to potential damages.

2.6.1.8 Visualization and Interpretation

The system should present monitoring results in a user-friendly interface, enabling engineers and stakeholders to interpret and understand the data easily. Visualization tools can help identify trends, anomalies, and critical events.

2.6.1.9 Automated Alerts and Reporting

The system can be configured to generate automated alerts when potential issues or anomalies are detected. This ensures timely notifications to relevant personnel for appropriate action.

2.6.1.10 Continuous Improvement

An AI-driven SHM system should be designed to continuously learn from new data and adapt to changing conditions. Regular updates and retraining of AI models are essential to maintain accuracy and efficiency.

Figure 2.22 illustrates the contrast between conventional and deep learning (DL)-based Structural Health Monitoring (SHM). In conventional data-driven SHM techniques, several distinctions emerge: (1) the need for expert knowledge to design features; (2) a sequential training process; and (3) models that might prove less efficient for larger structures and could lack suitability for vision-based SHM applications. Conversely, DL-based SHM showcases distinct advantages: (1) an end-to-end training approach with automated feature extraction; (2) simultaneous training of hyper-parameters; and (3) models adept at large-scale structures and efficient for

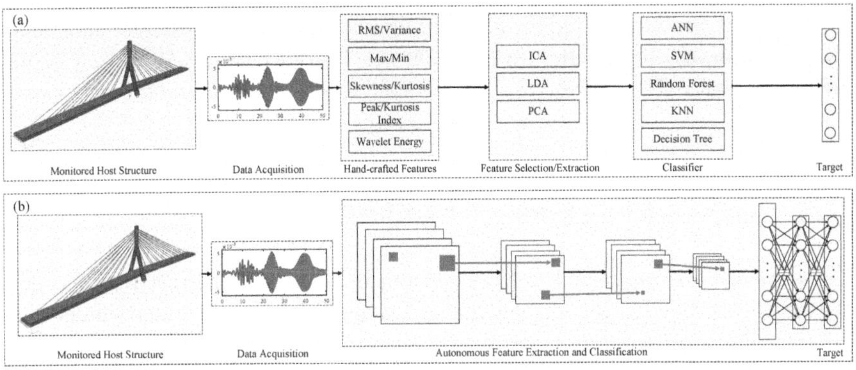

Fig. 2.22 **a** Conventional data-driven SHM versus **b** Deep learning-based SHM [47]

both vision- and vibration-based SHM, even when managing compressed or extensive datasets.

The successful design and implementation of an AI-driven SHM system require multidisciplinary expertise, including civil engineering, data science, and AI. The system's effectiveness lies in its ability to provide accurate and timely information for decision-making, contributing to safer and more reliable concrete structures and infrastructure. As AI technology continues to evolve, the potential for more sophisticated and intelligent SHM systems will grow, further enhancing the capabilities of structural health monitoring in the future.

2.6.2 Data Fusion and Integration with Existing Infrastructure

Data fusion and the integration of existing infrastructure constitute pivotal elements within an AI-powered Structural Health Monitoring (SHM) system. Data fusion amalgamates insights from diverse sensors and sources, amplifying the accuracy of the structure's health assessment. Concurrently, harmonizing the SHM system with the pre-existing infrastructure guarantees smooth operation and compatibility with the structure's monitoring and management mechanisms.

2.6.2.1 Data Fusion

Multi-Sensor Data Integration: Diverse sensors like accelerometers, strain gauges, temperature and acoustic emission sensors are harnessed by the SHM system. Data fusion algorithms unify this heterogeneous data, creating a composite dataset that furnishes a comprehensive understanding of structural behaviour.

Temporal and Spatial Fusion: Fusion encompasses data across disparate time intervals and locations, culminating in a continuous, spatially extended portrait of the structure's health. This facilitates spotting spatial trends and correlations between sensors.

2.6.2.2 Integration with Existing Infrastructure

Communication Protocols: The SHM system is tailored to interact with the pre-existing infrastructure and management systems. This could entail integration with SCADA (Supervisory Control and Data Acquisition) systems or building management frameworks.

- Data Storage and Management: Alignment with the prevailing data storage and management framework ensures seamless data exchange and retrieval. This harmonizes with databases, cloud storage, or on-premises servers.
- Real-time Monitoring: Integration with the standing infrastructure empowers the SHM system to tap into real-time data streams from other monitoring systems, enhancing its prowess in early anomaly detection and potential damage assessment.
- Alarms and Alerts: Integration with prevailing alarm systems enables notifications to pertinent personnel or auto-triggered responses during critical events or damages.

2.6.2.3 Benefits of Data Fusion and Integration

- Holistic Insight: Data fusion broadens the horizon, capturing the structure's health through diverse lenses for a more accurate comprehension.
- Operational Efficiency: Integration with existing systems curbs redundancy, optimizing data handling for an efficient and economical SHM system.
- Proactive Maintenance: Real-time monitoring, coupled with alarm integration, accelerates responses to potential issues, facilitating proactive maintenance strategies.
- Informed Decisions: This integrated SHM system fosters data-informed decision-making, offering pertinent and timely insights to engineers and stakeholders.

2.6.2.4 Challenges

- Data Compatibility: Enforcing compatibility and interoperability amidst varied sensors and systems may necessitate standardized communication formats and protocols.
- Data Quality: Integrating varied sources may entail issues regarding data accuracy, demanding meticulous data pre-processing and validation.

Data fusion and integration epitomize quintessential aspects of a robust AI-powered SHM system. By synergizing data from assorted sensors and melding seamlessly with existing infrastructure, the SHM system furnishes a comprehensive, real-time evaluation of concrete structures' health. This underpins enlightened decision-making and proactive maintenance paradigms, with continuous technological advancements in AI and data integration fortifying SHM systems, thereby ensuring the enduring safety and longevity of civil infrastructure.

2.6.3 Real-Time Monitoring and Predictive Maintenance

Real-time monitoring and predictive maintenance are essential elements in the field of Structural Health Monitoring (SHM), which aims to ensure the safety, integrity, and performance of civil infrastructure, particularly concrete structures. By combining cutting-edge sensor technology, data processing techniques, and artificial intelligence (AI) algorithms, real-time monitoring and predictive maintenance have the potential to revolutionize infrastructure management and maintenance practices.

2.6.3.1 Real-Time Monitoring

Real-time monitoring involves the continuous and automated collection of data from various sensors installed on concrete structures. These sensors can measure parameters such as vibrations, strains, displacements, temperatures, and acoustic emissions. The data collected is transmitted to data processing systems in real-time, where AI algorithms analyse the information promptly. The key benefits of real-time monitoring include early detection of anomalies or damages, allowing engineers to take immediate actions to prevent further deterioration. It also enhances safety by providing timely warnings of potential hazards, especially in critical structures such as bridges, dams, and high-rise buildings. Real-time monitoring can also facilitate a deeper understanding of the structural behaviour under dynamic loading or environmental changes, aiding in the development of improved design and maintenance practices.

2.6.3.2 Predictive Maintenance

Predictive maintenance is a data-driven approach that leverages historical sensor data and AI algorithms to predict the future condition of a structure and anticipate maintenance needs. By analysing the historical behaviour of the structure, the AI models can identify patterns and trends associated with damages or deterioration. This allows engineers to make informed decisions regarding maintenance activities, scheduling repairs or replacements only when necessary, rather than on a fixed time schedule.

Predictive maintenance offers several advantages, including cost savings by optimizing maintenance budgets and reducing unnecessary repairs. It improves asset management by extending the lifespan of structures through proactive interventions. By addressing potential issues before they become critical, predictive maintenance enhances the reliability and performance of concrete structures, ultimately leading to enhanced infrastructure sustainability.

Challenges in Implementation: Implementing real-time monitoring and predictive maintenance systems requires addressing several challenges. Data management becomes critical as large volumes of sensor data need to be collected, stored, and processed in real-time. High-quality and reliable sensors are essential to ensure accurate and consistent data collection. Additionally, AI algorithms need to be carefully developed and validated to provide accurate predictions and insights. Furthermore, successful implementation requires multidisciplinary expertise, bringing together civil engineering, data science, computer science, and domain-specific knowledge to create robust and effective systems.

Real-time monitoring and predictive maintenance represent the future of infrastructure management. By harnessing the power of AI and advanced sensor technology, these approaches enable a proactive and data-driven approach to infrastructure maintenance, ensuring the safety, reliability, and longevity of concrete structures. Continued research and advancements in AI and sensor technology will further enhance the capabilities of real-time monitoring and predictive maintenance, driving the evolution of smarter and more sustainable infrastructure management practices.

2.7 Performance Assessment and Model Validation

Performance assessment and model validation are crucial steps in the development and implementation of AI-based damage detection systems for concrete structures. After training the AI models on historical data, they must be rigorously tested to evaluate their accuracy and reliability. Performance assessment involves using evaluation metrics to measure the models' effectiveness in detecting and classifying damages. Model validation ensures that the AI models generalize well to new and unseen data, verifying their robustness and avoiding overfitting. Properly assessing the performance and validating the AI models are essential to ensure the accuracy and effectiveness of the damage detection system in real-world applications.

2.7.1 Evaluation Metrics for AI Models in Damage Detection

Assessment metrics for AI models in damage detection play a crucial role in gauging the accuracy and efficacy of these models in precisely identifying and categorizing damage within concrete structures. These metrics offer quantitative gauges of the model's performance, enabling engineers and researchers to compare different models, fine-tune parameters, and make informed decisions regarding their application.

- Accuracy: Accuracy stands as the fundamental and widely used evaluation metric, quantifying the proportion of correctly classified instances among all instances in the dataset. While it delivers an overall assessment of the model's performance, it may prove inadequate in scenarios involving imbalanced datasets, where one class (e.g., undamaged structures) significantly outnumbers another (e.g., damaged structures).
- Precision and Recall: Precision and recall are metrics employed in binary classification problems. Precision quantifies the proportion of true positives (accurate damage identification) among all instances predicted as positive (designated as damages). Conversely, recall measures the proportion of true positives among all actual positive instances. These metrics carry particular significance in damage detection, where precise damage identification is of utmost importance.
- F1 Score: The F1 score, as the harmonic mean of precision and recall, strikes a balance between the two metrics. It proves especially valuable in cases where precision and recall exhibit a trade-off. A higher F1 score indicates a superior equilibrium between accurate damage identification and the minimization of false positives.
- Area Under the Receiver Operating Characteristic (ROC) Curve (AUC-ROC): AUC-ROC assesses the model's capacity to differentiate between various classes in a binary classification problem. It charts the true positive rate (recall) against the false positive rate at various thresholds. A higher AUC-ROC score denotes enhanced performance in distinguishing damages from undamaged instances.
- Mean Average Precision (mAP): mAP is commonly utilized in object detection tasks, where the identification of multiple classes is necessary. It calculates the average precision for each class and then computes the mean across all classes. In damage detection, this metric proves beneficial when multiple types of damages need identification.
- Intersection over Union (IoU) or Jaccard Index: IoU assesses the overlap between the predicted bounding box or region and the ground truth bounding box or region. It is frequently employed in object detection tasks and can be adapted for segmenting damages within concrete structures.

The selection of appropriate evaluation metrics hinges on the specific problem and the objectives of the AI model in damage detection. A comprehensive understanding of the strengths and limitations of each metric is essential, enabling the selection of metrics that best align with the application's demands. Rigorous evaluation and validation of AI models using these metrics bolster the models' reliability and effectiveness in real-world scenarios.

2.7.2 Cross-Validation and Model Generalization

Cross-validation stands as a pivotal and widely embraced practice in the realms of machine learning and AI model development, particularly when faced with limited data or the necessity to evaluate model performance on fresh, unseen data. The significance of cross-validation arises from its capacity to supply a more precise and impartial assessment of a model's performance by recurrently training and testing the model on diverse data subsets. In the paradigm of k-fold cross-validation, the dataset is partitioned into k equitably sized folds, often adopting values of 5 or 10 for k. The model is then iteratively trained and validated k times, where each fold serves as the validation set once, while the remaining k-1 folds constitute the training set. This cyclic process guarantees that the model undergoes testing on various data subsets, yielding a more comprehensive appraisal of its performance.

The outcomes gleaned from each fold, encompassing metrics like accuracy, precision, recall, or the F1 score, are amalgamated through averaging, yielding a more robust assessment of the model's comprehensive performance. This method acts as a safeguard against the perils of overfitting, where the model fixates on the noise within the training data instead of discerning the fundamental patterns. By validating the model across multiple diverse data subsets, cross-validation provides a more dependable evaluation of the model's proficiency in generalizing to untapped, unseen data.

The ultimate objective in machine learning and AI is model generalization, signifying that the patterns and relationships the model acquires are pertinent to novel data instances, thus rendering the model dependable in real-world scenarios. Proficient generalization calls for the delicate equilibrium between model complexity and simplicity. Overly intricate models endowed with significant capacity may overfit the training data, resulting in subpar performance when dealing with unseen data. Conversely, excessively simplistic models may underfit, failing to capture the core data patterns. Cross-validation aids in the identification of the optimal balance between model complexity and generalization efficacy. Commonly employed regularization techniques, such as L1 or L2 regularization, are utilized to govern model complexity and avert overfitting. These techniques append penalty terms to the model's optimization function, discouraging excessive weights or intricate feature interactions. Cross-validation plays a significant role in the fine-tuning of regularization parameters, facilitating the pinpointing of optimal model configurations.

By leveraging cross-validation and thoroughly assessing model generalization, engineers and researchers can confidently deploy AI models for the purpose of detecting damage within concrete structures. Such models provide precise and trustworthy appraisals of structural health, empowering proactive maintenance, heightening safety, and fostering cost-effective infrastructure management. Progress in the evolution of cross-validation practices and regularization techniques contributes to the formulation of more potent and resilient AI models, well-suited to grapple with real-world challenges in structural health monitoring and damage detection. Through an enriched grasp of model performance and generalization, AI-fueled damage detection systems stand poised to play a pivotal role in securing the safety, endurance, and sustainability of concrete structures in a diverse array of applications.

2.7.3 Ensuring Robustness and Reliability of AI Algorithms

Ensuring the robustness and reliability of AI algorithms is a continuous and multi-faceted process that requires careful consideration of various factors. Building dependable AI models for damage detection in concrete structures involves the following key aspects:

- Data Diversity and Quality: A diverse and high-quality dataset is fundamental for training AI algorithms. The data should encompass a wide range of concrete structures, environmental conditions, and types of damages. An extensive dataset reduces the risk of bias and ensures that the algorithm learns representative patterns. Data quality, including accuracy, completeness, and consistency, is essential to prevent erroneous conclusions.
- Data Pre-processing: Proper data pre-processing is vital to eliminate noise, handle missing values, and standardize data formats. This step improves the algorithm's robustness by reducing the impact of irrelevant information and ensuring uniformity in the data.
- Regularization and Hyperparameter Tuning: Implementing regularization techniques, such as L1 and L2 regularization, helps prevent overfitting and enhances generalization. Fine-tuning hyperparameters optimizes model performance and ensures that the algorithm adapts well to different data distributions.
- Cross-Validation and Model Evaluation: Conducting thorough cross-validation ensures that the AI model's performance is not dependent on specific subsets of the data. Employing multiple evaluation metrics allows a comprehensive assessment of the algorithm's strengths and weaknesses, aiding in model selection and refinement.
- Bias Mitigation: Identifying and addressing biases in the data and algorithms is critical, especially in damage detection, where biased predictions can have severe consequences. Techniques like data augmentation and fairness-aware learning can help mitigate biases and improve model fairness.
- Interpretability and Explainability: Ensuring transparency in AI algorithms is crucial, particularly for critical applications like infrastructure health monitoring. Employing interpretable models or generating explanations for predictions enhances user trust and allows engineers to validate the model's decisions.
- Regular Updates and Monitoring: AI algorithms should be regularly updated to accommodate new data and evolving conditions. Continuous monitoring of the model's performance helps identify any degradation or drift in accuracy, enabling timely maintenance or retraining.

- Robustness Testing: Subjecting AI models to robustness testing involves evaluating their performance under various challenging conditions, such as noisy inputs, outliers, or adversarial attacks. Identifying vulnerabilities and addressing them improves the algorithm's resilience.
- Ensemble Techniques: Ensembling multiple AI models through techniques like bagging or boosting can enhance robustness and reliability. Combining diverse models reduces the risk of individual model weaknesses affecting overall performance.
- Error Analysis: Analysing model errors helps identify common misclassifications or failure patterns. Understanding these errors guides targeted improvements and ensures continuous refinement.

By carefully implementing these strategies, AI algorithms for damage detection in concrete structures can achieve higher levels of robustness and reliability. A well-validated and trustworthy AI system empowers engineers, construction professionals, and infrastructure managers to make informed decisions, promote proactive maintenance, enhance safety, and optimize resource utilization in civil engineering projects. The ongoing pursuit of robust and reliable AI models in the domain of structural health monitoring contributes to the advancement and sustainability of modern infrastructure.

2.8 Future Perspectives and Emerging Trends

Future perspectives and emerging trends in AI for damage detection in concrete structures are promising and exciting. As AI technologies continue to advance, they will play a vital role in transforming the field of structural health monitoring. Key trends include the integration of AI with IoT devices, enabling real-time data collection and analysis. Additionally, advancements in deep learning and transfer learning will enhance the accuracy and efficiency of damage detection algorithms. The use of AI-driven robotic systems for automated inspections and repairs is another emerging trend. Furthermore, addressing ethical and societal implications will remain a critical focus to ensure responsible and equitable implementation of AI in infrastructure development.

2.8.1 Advancements in AI Technology for Concrete Structure Assessments

Advancements in AI technology for concrete structure assessments have brought about significant improvements in the field of structural health monitoring and damage detection. These advancements have been driven by breakthroughs in machine learning algorithms, computing power, and the availability of large datasets.

Deep learning, particularly Convolutional Neural Networks (CNNs), has emerged as a powerful tool for image-based damage detection in concrete structures. CNNs can automatically learn and extract complex features from images, enabling accurate identification of various types of damages, such as cracks, corrosion, and spalling. This capability has revolutionized the inspection process, allowing for faster and more precise assessments. Transfer learning is another noteworthy advancement that addresses the challenge of limited labelled data for specific damage types. By leveraging pre-trained models on large, general datasets, such as ImageNet, and fine-tuning them on concrete structure data, transfer learning enables effective damage detection with reduced data requirements. This approach boosts the practical applicability of AI in real-world scenarios.

Reinforcement learning is increasingly being explored for optimizing maintenance and repair strategies. By enabling AI systems to learn through interactions with the environment, reinforcement learning algorithms can make decisions to maximize the long-term structural health of concrete structures while minimizing maintenance costs. This approach holds promise for proactive and cost-efficient infrastructure management.

The integration of AI with Internet of Things (IoT) devices and edge computing has also revolutionized data acquisition and analysis. IoT devices, such as sensors, can collect real-time data on structural behaviour, temperature, and moisture, among other parameters. Edge computing processes this data locally, reducing latency and enabling immediate responses to structural changes or damages, critical in ensuring the safety and integrity of concrete structures. Moreover, the development of explainable AI methods is addressing concerns about the interpretability of AI models. Explainable AI techniques provide insights into the decision-making process of the models, allowing engineers and stakeholders to understand how the AI system arrives at its conclusions. This transparency fosters trust in AI models and helps validate their predictions, making them more practical for critical infrastructure applications.

Big data analytics is playing a transformative role in handling vast amounts of data generated by various sources, such as sensors, satellites, and historical records. AI can analyse and process this data, providing a comprehensive understanding of the structural health and behaviour of concrete structures over time. This data-driven approach aids in detecting early signs of damage and assists in making informed decisions for maintenance and repair. Multimodal sensing, which combines data from multiple sensing modalities such as visual, acoustic, and vibrational data, enriches the data used in AI models. This holistic approach ensures a more comprehensive view of concrete structures' health, allowing for more accurate and reliable damage detection.

AI-powered autonomous robots and drones are becoming increasingly prevalent in structural inspections. These robots can navigate complex environments and perform remote inspections, reducing the need for human intervention in hazardous or hard-to-reach areas. Autonomous drones, equipped with AI algorithms, can efficiently inspect large structures, providing cost-effective and rapid assessments. Additionally, the application of AI in prognostics and predictive maintenance enables engineers to forecast the future degradation of concrete structures. By predicting potential failures or deterioration, proactive maintenance can be planned, minimizing downtime and mitigating risks associated with structural failures. The advancements in AI technology have significantly transformed concrete structure assessments, making them more efficient, accurate, and proactive. These innovations are revolutionizing the way engineers and researchers approach structural health monitoring, leading to safer, more resilient, and sustainable infrastructure around the world. As AI continues to evolve, it will undoubtedly play an increasingly crucial role in shaping the future of civil engineering and construction practices.

2.8.2 AI and Internet of Things (IoT) Integration

The integration of Artificial Intelligence (AI) and the Internet of Things (IoT) has paved the way for transformative advancements across various industries, including infrastructure monitoring and damage detection in concrete structures. This powerful combination leverages the strengths of both technologies to create smart and interconnected systems capable of real-time data collection, analysis, and decision-making.

In the context of concrete structure assessments, AI-IoT integration involves deploying a network of IoT sensors throughout the infrastructure to continuously monitor various parameters such as temperature, humidity, vibrations, strain, and corrosion. These sensors collect a vast amount of data, creating a comprehensive digital twin of the structure. AI algorithms play a pivotal role in making sense of this data. Machine learning and deep learning models are employed to analyse the incoming data streams, detect anomalies, and identify patterns indicative of potential damages or structural health issues. The AI algorithms can learn from historical data and adapt to changing conditions, enabling predictive maintenance and early detection of potential problems. The benefits of AI-IoT integration in concrete structure assessments are numerous:

- Real-time Monitoring: AI-IoT systems enable continuous, real-time monitoring of concrete structures, providing immediate insights into their health and performance. This real-time monitoring allows for rapid responses to emerging issues and potential hazards.
- Proactive Maintenance: By detecting early signs of damages, AI-IoT systems facilitate proactive maintenance, addressing issues before they escalate into costly repairs or safety hazards. This approach improves the overall lifespan of the infrastructure and reduces maintenance costs.
- Data-Driven Insights: The abundance of data collected by IoT sensors enables data-driven decision-making. AI algorithms process and analyse this data, extracting valuable insights that guide engineering and maintenance strategies, optimizing resource allocation.
- Remote Inspections: AI-IoT integration enables remote inspections, reducing the need for physical inspections in hazardous or hard-to-reach locations.

Autonomous drones equipped with AI algorithms can conduct aerial inspections, providing detailed and efficient assessments.

- Scalability and Flexibility: AI-IoT systems are scalable and can be easily deployed in various concrete structures, from buildings and bridges to dams and tunnels. The flexibility of the system allows it to adapt to the specific requirements of each structure.
- Predictive Analytics: AI-IoT systems can predict future structural behaviour and potential issues based on historical data. This predictive analytics capability empowers engineers to make data-driven decisions, maximizing the performance and safety of concrete structures.
- Reduced Downtime: By continuously monitoring the structure's health and detecting anomalies early on, AI-IoT integration minimizes unplanned downtime, leading to increased operational efficiency and productivity.
- Improved Safety: Timely detection of damages and structural issues through AI-IoT integration enhances the safety of occupants and passersby, reducing the risk of accidents and ensuring compliance with safety regulations.

However, the integration of AI and IoT also poses challenges, such as managing the massive volume of data generated by IoT sensors, ensuring data security and privacy, and addressing interoperability issues between different IoT devices and AI systems. Nevertheless, the benefits of AI-IoT integration in concrete structure assessments far outweigh these challenges, making it a transformative approach to enhancing the safety, durability, and sustainability of critical infrastructure. As technology continues to advance, the synergy between AI and IoT is expected to revolutionize the field of structural health monitoring and contribute to more resilient and smarter cities.

2.8.3 Ethical and Societal Implications of AI in Infrastructure Development

The integration of Artificial Intelligence (AI) in infrastructure development has the potential to bring significant benefits, such as improved efficiency, enhanced safety, and better resource management. However, it also raises ethical and societal implications that need careful consideration to ensure responsible and equitable deployment.

- Privacy and Data Security: AI applications in infrastructure development often rely on vast amounts of data collected from various sources, including IoT sensors and surveillance cameras. Ensuring the privacy and security of this data is paramount to prevent unauthorized access, misuse, or potential breaches that could compromise individuals' sensitive information.
- Bias and Fairness: AI algorithms can inadvertently perpetuate biases present in the data used for training. In infrastructure development, biased AI systems may

lead to unequal distribution of resources or services, reinforcing existing inequalities. It is essential to address and mitigate biases to ensure fairness and equal opportunities for all communities.
- Transparency and Explainability: As AI systems become more complex, ensuring transparency and explainability of their decisions becomes challenging. In critical infrastructure projects, stakeholders need to understand how AI algorithms arrive at their conclusions. Providing explanations and making AI models interpretable is crucial for building trust and accountability.
- Accountability and Liability: The use of AI in infrastructure development raises questions of accountability and liability in the event of system failures or accidents. Determining responsibility when AI is involved in decision-making becomes complex, and establishing clear guidelines and regulations is essential to allocate liability appropriately.
- Job Displacement and Economic Impact: The automation potential of AI in construction and maintenance tasks could lead to job displacement in certain sectors. Preparing the workforce for AI integration and ensuring a just transition for affected workers is critical to avoid social and economic disruption.
- Cybersecurity Risks: AI-powered infrastructure systems may become vulnerable targets for cyberattacks. The integration of AI must be accompanied by robust cybersecurity measures to protect critical infrastructure from potential threats.
- Equity and Access: Implementing AI in infrastructure development should not exacerbate existing social inequalities or create digital divides. Ensuring equitable access to AI-powered infrastructure and services is essential to promote inclusive growth and development.
- Ethical Decision-Making: AI systems often operate based on algorithms designed to optimize specific objectives. Ensuring that the objectives align with ethical principles and societal values is crucial to avoid unintended negative consequences.
- Dual-Use Concerns: AI applications in infrastructure development may have dual-use implications, meaning the same technology could be used for both beneficial and harmful purposes. It is necessary to establish guidelines and regulations to prevent the misuse of AI technologies.
- Environmental Impact: AI-powered infrastructure may increase energy consumption and electronic waste. Striking a balance between the benefits of AI and its environmental impact is vital for promoting sustainable development.

To address these ethical and societal implications, collaboration among policymakers, engineers, developers, and communities is essential. Implementing transparent and inclusive decision-making processes, adopting ethical AI principles, and conducting regular impact assessments are steps toward ensuring that AI in infrastructure development benefits society responsibly and sustainably.

2.9 Concluding Remarks

The utilization of Artificial Intelligence in the detection of damage in concrete structures marks a significant advancement in the realm of structural health monitoring and infrastructure development. The integration of AI technologies presents unparalleled opportunities to elevate the safety, efficiency, and durability of critical infrastructure, ensuring a more sustainable and resilient future. Through the utilization of AI and machine learning algorithms, concrete structures can undergo continuous monitoring, facilitating real-time assessments and immediate responses to potential damages. This enables the implementation of proactive maintenance strategies, addressing issues at their early stages, thereby preventing costly repairs and prolonging the lifespan of infrastructure assets.

While the potential advantages of AI are substantial, it is equally crucial to address the ethical and societal implications that accompany its widespread adoption. Ensuring data privacy, fairness, and transparency in AI algorithms is paramount for building trust and promoting responsible AI use. Moreover, efforts to mitigate biases and ensure equitable access to AI-powered infrastructure solutions are essential for creating inclusive and fair smart cities. Additionally, collaboration among researchers, engineers, policymakers, and the public is indispensable for navigating the intricate challenges and opportunities posed by AI in infrastructure development. Establishing comprehensive regulations and ethical guidelines is imperative to guide the responsible implementation of AI technologies.

As technology continues to progress, AI algorithms will become more sophisticated and capable, leading to further innovations in the assessment of concrete structures and infrastructure management. Embracing these advancements with a strong emphasis on ethics and social responsibility will shape a future where AI contributes to sustainable development, heightened infrastructure resilience, and an improved quality of life for communities globally.

References

1. Tang SW, Yao Y, Andrade C, Li ZJ (2015) Recent durability studies on concrete structure. Cem Concr Res 78:143–154
2. Hooton RD (2019) Future directions for design, specification, testing, and construction of durable concrete structures. Cem Concr Res 124:105827
3. Ting MZY, Wong KS, Rahman ME, Meheron SJ (2021) Deterioration of marine concrete exposed to wetting-drying action. J Clean Prod 278:123383
4. Zhao H, Hu Y, Tang Z, Wang K, Li Y, Li W (2022) Deterioration of concrete under coupled aggressive actions associated with load, temperature and chemical attacks: a comprehensive review. Constr Build Mater 322:126466
5. Ko JM, Ni YQ (2005) Technology developments in structural health monitoring of large-scale bridges. Eng Struct 27(12):1715–1725
6. Daniyal M, Akhtar S (2020) Corrosion assessment and control techniques for reinforced concrete structures: a review. J Build Pathol Rehabil 5:1–20
7. Büyüköztürk O (1998) Imaging of concrete structures. NDT E Int 31(4):233–243

8. Lee S, Kalos N, Shin DH (2014) Non-destructive testing methods in the US for bridge inspection and maintenance. KSCE J Civ Eng 18:1322–1331
9. Zheng Y, Wang S, Zhang P, Xu T, Zhuo J (2022) Application of nondestructive testing technology in quality evaluation of plain concrete and RC structures in bridge engineering: a review. Buildings 12(6):843
10. Joshaghani A, Shokrabadi M (2022) Ground penetrating radar (GPR) applications in concrete pavements. Int J Pavement Eng 23(13):4504–4531
11. Bakhary N, Hao H, Deeks AJ (2007) Damage detection using artificial neural network with consideration of uncertainties. Eng Struct 29(11):2806–2815
12. Tran-Ngoc H, Khatir S, De Roeck G, Bui-Tien T, Wahab MA (2019) An efficient artificial neural network for damage detection in bridges and beam-like structures by improving training parameters using cuckoo search algorithm. Eng Struct 199:109637
13. Tan ZX, Thambiratnam DP, Chan TH, Gordan M, Abdul Razak H (2020) Damage detection in steel-concrete composite bridge using vibration characteristics and artificial neural network. Struct Infrastruct Eng 16(9):1247–1261
14. Salehi H, Burgueño R (2018) Emerging artificial intelligence methods in structural engineering. Eng Struct 171:170–189
15. Sun L, Shang Z, Xia Y, Bhowmick S, Nagarajaiah S (2020) Review of bridge structural health monitoring aided by big data and artificial intelligence: from condition assessment to damage detection. J Struct Eng 146(5):04020073
16. Gordan M, Ismail Z, Razak HA, Ghaedi K, Ibrahim Z, Tan ZX, Ghayeb HH (2020) Data mining-based damage identification of a slab-on-girder bridge using inverse analysis. Measurement 151:107175
17. Wang N, Zhao X, Zhao P, Zhang Y, Zou Z, Ou J (2019) Automatic damage detection of historic masonry buildings based on mobile deep learning. Autom Constr 103:53–66
18. Ai D, Jiang G, Lam SK, He P, Li C (2023) Computer vision framework for crack detection of civil infrastructure—a review. Eng Appl Artif Intell 117:105478
19. Kaur NP, Shah JK, Majhi S, Mukherjee A (2019) Healing and simultaneous ultrasonic monitoring of cracks in concrete. Mater Today Commun 18:87–99
20. Tosti F, Ferrante C (2020) Using ground penetrating radar methods to investigate reinforced concrete structures. Surv Geophys 41(3):485–530
21. Jedidi M (2020) Evaluation of the quality of concrete structures by the rebound hammer method. Curr Trends Civil Struct Eng
22. Yodsudjai W, Pattarakittam T (2017) Factors influencing half-cell potential measurement and its relationship with corrosion level. Measurement 104:159–168
23. Yang YX, Chai WH, Liu DC, Zhang WD, Lu JC, Yang ZK (2021) An impact-echo experimental approach for detecting concrete structural faults. Adv Civil Eng 8141015:8
24. Colombo M, Felicetti R (2007) New NDT techniques for the assessment of fire-damaged concrete structures. Fire Saf J 42(6–7):461–472
25. Cassidy NJ, Eddies R, Dods S (2011) Void detection beneath reinforced concrete sections: the practical application of ground-penetrating radar and ultrasonic techniques. J Appl Geophys 74(4):263–276
26. Janků M, Cikrle P, Grošek J, Anton O, Stryk J (2019) Comparison of infrared thermography, ground-penetrating radar and ultrasonic pulse echo for detecting delaminations in concrete bridges. Constr Build Mater 225:1098–1111
27. Elwood E, Corotis RB (2015) Application of fuzzy pattern recognition of seismic damage to concrete structures. ASCE-ASME J Risk Uncertain Eng Syst Part A Civil Eng 1(4):04015011
28. de Souza CPG, Kurka PRG, Lins RG, de Araujo Junior JM (2021) Performance comparison of non-adaptive and adaptive optimization algorithms for artificial neural network training applied to damage diagnosis in civil structures. Appl Soft Comput 104:107254
29. Geetha GK, Sim SH (2022) Fast identification of concrete cracks using 1D deep learning and explainable artificial intelligence-based analysis. Autom Constr 143:104572
30. Lutz É, Coradi PC (2022) Applications of new technologies for monitoring and predicting grains quality stored: sensors, internet of things, and artificial intelligence. Measurement 188:110609

31. Zhang Y, Xie X, Li H, Zhou B, Wang Q, Shahrour I (2022) Subway tunnel damage detection based on in-service train dynamic response, variational mode decomposition, convolutional neural networks and long short-term memory. Autom Constr 139:104293
32. Asghshahr MS, Rahai A, Ashrafi H (2016) Prediction of chloride content in concrete using ANN and CART. Mag Concr Res 68(21):1085–1098
33. Pan Y, Zhang L (2021) Roles of artificial intelligence in construction engineering and management: a critical review and future trends. Autom Constr 122:103517
34. Zhang Y, Yuen KV (2022) Review of artificial intelligence-based bridge damage detection. Adv Mech Eng 14(9):16878132221122770
35. Wang H, Barone G, Smith A (2023) Current and future role of data fusion and machine learning in infrastructure health monitoring. Struct Infrastruct Eng 1–30
36. Xu Y, Liu X, Cao X et al (2021) Artificial intelligence: a powerful paradigm for scientific research. Innov 2(4):100179
37. Dwivedi YK, Hughes L, Ismagilova E, Aarts G, Coombs C, Crick T, Galanos V (2021) Artificial Intelligence (AI): multidisciplinary perspectives on emerging challenges, opportunities, and agenda for research, practice and policy. Int J Inf Manage 57:101994
38. Sarker IH (2022) AI-based modeling: techniques, applications and research issues towards automation, intelligent and smart systems. SN Comput Sci 3(2):158
39. Anagnostou M, Karvounidou O, Katritzidaki C, Kechagia C, Melidou K, Mpeza E, Peristeras V (2022) Characteristics and challenges in the industries towards responsible AI: a systematic literature review. Ethics Inf Technol 24(3):37
40. Sacha D, Kraus M, Keim DA, Chen M (2018) Vis4ml: an ontology for visual analytics assisted machine learning. IEEE Trans Visual Comput Gr 25(1):385–395
41. Giannopoulos A, Spantideas S, Kapsalis N, Gkonis P, Sarakis L, Capsalis C, Trakadas P (2022) Supporting intelligence in disaggregated open radio access networks: architectural principles, AI/ML workflow, and use cases. IEEE Access 10:39580–39595
42. Sarker IH (2021) Machine learning: algorithms, real-world applications and research directions. SN Comput Sci 2(3):160
43. Choudhury A (2021) The role of machine learning algorithms in materials science: a state of art review on industry 4.0. Arch Comput Methods Eng 28(5):3361–3381
44. Sharma N, Sharma R, Jindal N (2021) Machine learning and deep learning applications-a vision. Glob Transit Proceed 2(1):24–28
45. Tiwari S, Dhanda N, Dev H (2022) A systematic review of adoption of blockchain and machine learning technology and its application. Soft Comput Secur Appl Proceed ICSCS 2021:73–95
46. Li Z, Yoon J, Zhang R et al (2022) Machine learning in concrete science: applications, challenges, and best practices. npj Comput Mater 8(1):127
47. Azimi M, Eslamlou AD, Pekcan G (2020) Data-driven structural health monitoring and damage detection through deep learning: state-of-the-art review. Sensors 20(10):2778
48. Hinton G, Deng L, Yu D, Dahl GE, Mohamed AR, Jaitly N, Kingsbury B (2012) Deep neural networks for acoustic modeling in speech recognition: the shared views of four research groups. IEEE Signal Process Mag 29(6):82–97
49. Izbassarova A, Duisembay A, James AP (2020) Speech recognition application using deep learning neural network. Deep Learn Classif Memristive Netw Theory Appl 69–79
50. Fourati H, Maaloul R, Chaari L (2021) A survey of 5G network systems: challenges and machine learning approaches. Int J Mach Learn Cybern 12:385–431
51. Monteiro JP, Ramos D, Carneiro D, Duarte F, Fernandes JM, Novais P (2021) Meta-learning and the new challenges of machine learning. Int J Intell Syst 36(11):6240–6272
52. Cunningham P, Cord M, Delany SJ (2008) Supervised learning. In Machine learning techniques for multimedia: case studies on organization and retrieval. Springer Berlin Heidelberg, Berlin, Heidelberg, pp 21–49
53. Verma R, Nagar V, Mahapatra S (2021) Introduction to supervised learning. Data Anal Bioinform Mach Learn Perspect 1–34
54. Tyagi K, Rane C, Manry M (2022) Supervised learning. In Artificial intelligence and machine learning for EDGE computing. Academic Press, pp 3–22

55. Hastie T, Tibshirani R, Friedman J, Hastie T, Tibshirani R, Friedman J (2009) Unsupervised learning. Elem Stat Learn Data Min Inference Predict 485–585
56. Ige AO, Noor MHM (2022) A survey on unsupervised learning for wearable sensor-based activity recognition. Appl Soft Comput 109363
57. James G, Witten D, Hastie T, Tibshirani R, Taylor J (2023) Unsupervised learning. An introduction to statistical learning: with applications in python. Springer International Publishing, Cham, pp 503–556
58. Heuillet A, Couthouis F, Díaz-Rodríguez N (2021) Explainability in deep reinforcement learning. Knowl-Based Syst 214:106685
59. Fayaz SA, Jahangeer Sidiq S, Zaman M, Butt MA (2022) Machine learning: an introduction to reinforcement learning. Mach Learn Data Sci Fundam Appl 1–22
60. Matsuo Y, LeCun Y, Sahani M, Precup D, Silver D, Sugiyama M, Morimoto J (2022) Deep learning, reinforcement learning, and world models. Neural Netw 152:267–275
61. Ren Y, Huang J, Hong Z, Lu W, Yin J, Zou L, Shen X (2020) Image-based concrete crack detection in tunnels using deep fully convolutional networks. Constr Build Mater 234:117367
62. Sony S, Dunphy K, Sadhu A, Capretz M (2021) A systematic review of convolutional neural network-based structural condition assessment techniques. Eng Struct 226:111347
63. Huang B, Zhao S, Kang F (2023) Image-based automatic multiple-damage detection of concrete dams using region-based convolutional neural networks. J Civ Struct Heal Monit 13:413–429
64. Chen L, Chen W, Wang L, Zhai C, Hu X, Sun L, Jiang L (2023) Convolutional neural networks (CNNs)-based multi-category damage detection and recognition of high-speed rail (HSR) reinforced concrete (RC) bridges using test images. Eng Struct 276:115306
65. Barzegar V, Laflamme S, Hu C, Dodson J (2022) Ensemble of recurrent neural networks with long short-term memory cells for high-rate structural health monitoring. Mech Syst Signal Process 164:108201
66. Eltouny KA, Liang X (2023) Large-scale structural health monitoring using composite recurrent neural networks and grid environments. Comput-Aided Civil Infrastruct Eng 38(3):271–287
67. Beckman G, Polyzois D, Cha YJ (2018) Deep learning-based automatic volumetric damage quantification using depth camera. Autom Constr 99:114–124
68. Liu CW, Huang XH, Miao JJ, Ba GZ (2019) Modification of finite element models based on support vector machines for reinforced concrete beam vibrational analyses at elevated temperatures. Struct Control Health Monit 26(6):e2350
69. Kang F, Li J, Dai J (2019) Prediction of long-term temperature effect in structural health monitoring of concrete dams using support vector machines with Jaya optimizer and salp swarm algorithms. Adv Eng Softw 131:60–76
70. Bisheh HB, Amiri GG (2023) Structural damage detection based on variational mode decomposition and kernel PCA-based support vector machine. Eng Struct 278:115565
71. Wang Z, Cha YJ (2020) Unsupervised deep learning approach using a deep auto-encoder with a one-class support vector machine to detect structural damage. Struct Health Monitor
72. Zhou Q, Ning Y, Zhou Q, Luo L, Lei J (2013) Structural damage detection method based on random forests and data fusion. Struct Health Monit 12(1):48–58
73. Chun PJ, Ujike I, Mishima K, Kusumoto M, Okazaki S (2020) Random forest-based evaluation technique for internal damage in reinforced concrete featuring multiple nondestructive testing results. Constr Build Mater 253:119238
74. Pamuncak A, Zivanovic S, Adha A, Liu J, Laory I (2023) Correlation-based damage detection method using convolutional neural network for civil infrastructure. Comput Struct 282:107034
75. Chow JK, Su Z, Wu J, Tan PS, Mao X, Wang YH (2020) Anomaly detection of defects on concrete structures with the convolutional autoencoder. Adv Eng Inform 45:101105
76. Parziale M, Lomazzi L, Giglio M, Cadini F (2022) Vibration-based structural health monitoring exploiting a combination of convolutional neural networks and autoencoders for temperature effects neutralization. Struct Control Health Monit 29(11):e3076

77. Li L, Morgantini M, Betti R (2023) Structural damage assessment through a new generalized autoencoder with features in the quefrency domain. Mech Syst Signal Process 184:109713
78. Pollastro A, Testa G, Bilotta A, Prevete R (2023) Semi-supervised detection of structural damage using variational autoencoder and a one-class support vector machine. IEEE Access
79. Sarwar M, Cantero D (2021) Deep autoencoder architecture for bridge damage assessment using responses from several vehicles. Eng Struct 246:113064
80. Ahmed B, Mangalathu S, Jeon JS (2022) Seismic damage state predictions of reinforced concrete structures using stacked long short-term memory neural networks. J Build Eng 46:103737
81. Ranjbar I, Toufigh V (2022) Deep long short-term memory (LSTM) networks for ultrasonic-based distributed damage assessment in concrete. Cem Concr Res 162:107003
82. Sony S, Gamage S, Sadhu A, Samarabandu J (2022) Vibration-based multiclass damage detection and localization using long short-term memory networks. In structures, vol 35. Elsevier, pp 436–451
83. Andrushia A, Anand N, Neebha M, Naser MZ, Lublóy É (2022) Autonomous detection of concrete damage under fire conditions. Autom Constr 140:104364
84. Shin HK, Ahn YH, Lee SH, Kim HY (2020) Automatic concrete damage recognition using multi-level attention convolutional neural network. Materials 13(23):5549
85. Eltouny K, Gomaa M, Liang X (2023) Unsupervised learning methods for data-driven vibration-based structural health monitoring: a review. Sensors 23(6):3290
86. Watase A, Birgul R, Hiasa S, Matsumoto M, Mitani K, Catbas FN (2015) Practical identification of favorable time windows for infrared thermography for concrete bridge evaluation. Constr Build Mater 101:1016–1030
87. Gordan M, Sabbagh-Yazdi SR, Ismail Z, Ghaedi K, Carroll P, McCrum D, Samali B (2022) State-of-the-art review on advancements of data mining in structural health monitoring. Measurement 193:110939
88. Soleymani A, Jahangir H, Nehdi ML (2023) Damage detection and monitoring in heritage masonry structures: systematic review. Constr Build Mater 397:132402
89. Silva M, Santos A, Figueiredo E, Santos R, Sales C, Costa JC (2016) A novel unsupervised approach based on a genetic algorithm for structural damage detection in bridges. Eng Appl Artif Intell 52:168–180
90. Taffese WZ, Espinosa-Leal L (2022) Prediction of chloride resistance level of concrete using machine learning for durability and service life assessment of building structures. J Build Eng 60:105146
91. Karaaslan E, Bagci U, Catbas FN (2021) Attention-guided analysis of infrastructure damage with semi-supervised deep learning. Autom Constr 125:103634
92. Xu Y, Wei S, Bao Y, Li H (2019) Automatic seismic damage identification of reinforced concrete columns from images by a region-based deep convolutional neural network. Struct Control Health Monit 26(3):e2313
93. Avci O, Abdeljaber O, Kiranyaz S, Hussein M, Gabbouj M, Inman DJ (2021) A review of vibration-based damage detection in civil structures: from traditional methods to machine learning and deep learning applications. Mech Syst Signal Process 147:107077
94. Advani MS, Saxe AM, Sompolinsky H (2020) High-dimensional dynamics of generalization error in neural networks. Neural Netw 132:428–446
95. Tian Y, Zhang Y (2022) A comprehensive survey on regularization strategies in machine learning. Inf Fus 80:146–166
96. Dizaji MS, Mao Z, Haile M (2023) A hybrid-attention-ConvLSTM-based deep learning architecture to extract modal frequencies from limited data using transfer learning. Mech Syst Signal Process 187:109949
97. Dang HV, Tran-Ngoc H, Nguyen TV, Bui-Tien T, De Roeck G, Nguyen HX (2020) Data-driven structural health monitoring using feature fusion and hybrid deep learning. IEEE Trans Autom Sci Eng 18(4):2087–2103
98. Toh G, Park J (2020) Review of vibration-based structural health monitoring using deep learning. Appl Sci 10(5):1680

99. Shevchik SA, Kenel C, Leinenbach C, Wasmer K (2018) Acoustic emission for in situ quality monitoring in additive manufacturing using spectral convolutional neural networks. Addit Manuf 21:598–604
100. Pu R, Ren G, Li H, Jiang W, Zhang J, Qin H (2022) Autonomous concrete crack semantic segmentation using deep fully convolutional encoder–decoder network in concrete structures inspection. Buildings 12(11):2019

Chapter 3
Deep Learning-Enabled Health Assessment for Sustainable Maintenance of Existing Concrete Structures: A Review

Punita Panwar, Khushi Goyal, and Jatin Kumar Shandilya

Abstract With growing emphasis on sustainable infrastructure, the maintenance and durability of existing concrete structures are critical factors. This paper proposes a novel approach leveraging Deep Learning (DL) Techniques for finding and estimating of the health of concrete buildings. The offered approach integrates advanced image processing and machine learning algorithms to analyze visual data obtained through non-destructive testing methods. Our framework focuses on first capturing and then interpreting the refined indicators of structural degradation, such as cracks, spalling, and surface irregularities, through the application of convolutional neural networks (CNNs). By training the model on a varied dataset of pictures of buildings under several environmental conditions, the system learns to identify and classify different levels of damage with high accuracy. The proposed DL-based approach offers several advantages over traditional inspection methods, including real-time analysis, cost-effectiveness, and reduced dependence on human interpretation. Having said that, with the collaboration of Artificial Intelligence (AI), the identification, detection and characterization of degradation and damage in all types of engineering structures consumes less time. Computational techniques (CT) are playing progressively important role in structural health monitoring, allowing for more accurate and efficient analysis of data from sensors and providing valuable insights into the health of structures. Machine Learning (ML) and DL models are taught from data without being specifically programmed. In other words, they take

P. Panwar (✉) · K. Goyal · J. K. Shandilya
Manipal University Jaipur, Jaipur, India
e-mail: punitapanwar7@gmail.com

K. Goyal
e-mail: khushigoyal.cse24@jecrc.ac.in

J. K. Shandilya
e-mail: jatinkumarshandilya.cse24@jecrc.ac.in

P. Panwar
Jaipur Engineering College and Research Centre, Jaipur, India

H. Jahangir et al. (eds.), *Damage Detection and Structural Health Monitoring of Concrete and Masonry Structures*, Springer Tracts in Civil Engineering,
https://doi.org/10.1007/978-981-97-8975-7_3

the help of algorithms to automatically learn patterns and make predictions based on the data is fed into the system. In this chapter, the researchers discuss how AI can be useful for civil engineers and mechanical engineers to monitor the health of buildings and day-by-day implementation of advanced deep learning methods is profitable to civil engineers in terms of cost as well as reduced human-efforts and people in terms of their safety.

Keywords Sustainable infrastructure · Deep learning techniques · Structural health monitoring · Non-destructive testing · Computational techniques

3.1 Introduction

Detecting and repairing cracks, in engineering structures is crucial to ensure their functionality and longevity. With the progress in industry, there is an increase in count of engineering structures. In the past cracks were primarily identified through inspection by examiners, which heavily relied on their judgment. This approach has been instrumental in repairing structures and understanding the nature of crack occurrences over the years. Therefore, there is a need for an approach that's both efficient and accurate. To address the limitations of inspection researchers have explored the usage of Image Processing Methods that is based on ML and DL algorithms, like Random Forest, SVM, RNNs, and CNNs.

By leveraging these networks, it becomes possible to extract essential features, from real-world images and develop the precision of crack detection significantly.

Image processing techniques allows for crack detection along with their characteristics with minimal time and efforts. The algorithm development and its implementation in this field are also straightforward. Identifying damage in buildings holds importance in ensuring overall safety. Structural Health Monitoring (SHM) involves observing a building over a period to develop a healthier understanding of its behavior. Its primary objective is to assess both the qualitative deterioration of a building system during its usage or when subjected to loads with safety and performance considerations in mind. It's crucial to investigate where, when and how this damage occurs.

Recent advancements in materials and sensor technologies have revolutionized the construction industry worldwide offering tools for enhancing building processes. While many buildings remain unchanged for decades it is essential to evaluate their behavior starting from the manufacturing phase and, throughout their lifespan. The key components of SHM includes sensors, machine learning, data acquisition, signal processing, as well as visualization. The general scope of SHM comprises structural assessment, monitoring, and control (SAMCo). It is used to confirm that a structure is safe in its day-to-day use. In other words, it is used to monitor the condition of buildings to detect any signs of structural damage or deterioration. It can also help in understanding how best to maintain a structure and extend its design life. SHM is applicable for various structures like tunnels, roads, bridges, buildings as well as steel structures but with reference to buildings particularly, we use SHM with buildings where safety and security are a significant concern such as historical structures and skyscrapers.

Non-destructive tests (NDTs) are the methods availed to check the qualities of matter, parts, or systems without harming structures. Different NDT approaches, like radiographic and electromagnetic testing, and liquid penetrant testing, have been used to spot physical flaws in various applications. There are countless ways to evaluate materials or components. One significant type is NDTs which offers multiple implementation methods. They involve identifying and characterizing both internal material damage without cutting or harming any part of the material. In times there has been a growing interest in using NDTs to assess the quality of concrete structures. The advantages of NDTs include reduced labor requirements, minimized structural risks during drilling and the use of testing equipment, like core testing devices. Though, these benefits can only be realized if the test results are accurate, reflective and closely resemble the strength of the evaluated section of the system.

NDTs encompasses a range of techniques used to assess the condition of materials or components. NDTs offers an effective way to perform these assessments with its advantage lying in the fact that it enables examination without causing any harm or changes to the composition or form of the object under inspection. In times, there has been growing attention towards intrusive methods, particularly within the fields of civil and mechanical engineering. Coring, for instance, is a destructive testing (NDTs) technique availed to accumulate and estimate models of concrete constituents in order to define the properties of concrete in its original location. Coring alters the appearance. Slightly compromises the integrity of the part. However, when carried out correctly coring ensures that the structure remains functional without any damage, which is why it is referred to non-destructive. This type of testing plays a role in evaluating and estimating factors related to current buildings or structures such, as crack detection defect detection imperfections, resilience, and concrete absorption characteristics.

Despite its advantages, there are limitations associated with these methods. For instance, immediate performance cannot always be accurate. The properties of specimens can differ from those found in actual construction projects. Additionally, concrete strength properties can vary depending on their size and shape. To overcome these limitations several NDTs methods have been developed to address challenges. These methods focus on determining whether the strength and stability of buildings can be attributed to the chemical properties of concrete [1] (Figs. 3.1 and 3.2).

The early focus on aircraft structures eventually covered the path for the improvement of SHM as a field of study and practice, with broader applications in the assessment and management of structures in various industries. The early methods of SHM involved simple visual inspections and non-destructive testing techniques [2] such as X-rays and ultrasounds. In the 1980s and 1990s, progress in computer and sensor technology was a reason of development of more sophisticated methods of SHM (Fig. 3.3).

AI techniques have the power to revolutionize the field of SHM. By leveraging ML-DL, and neural networks, vast volumes of data from instruments like sensors can be analyzed to categorize patterns or anomalies that may indicate potential structural

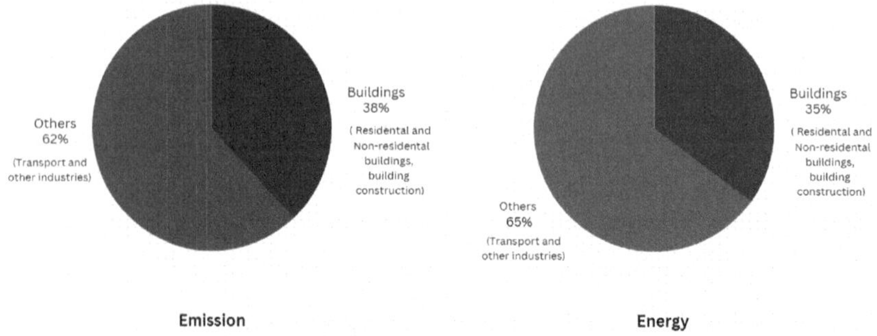

Fig. 3.1 Plot of comparison of emission and energy due to buildings, transport and other industries

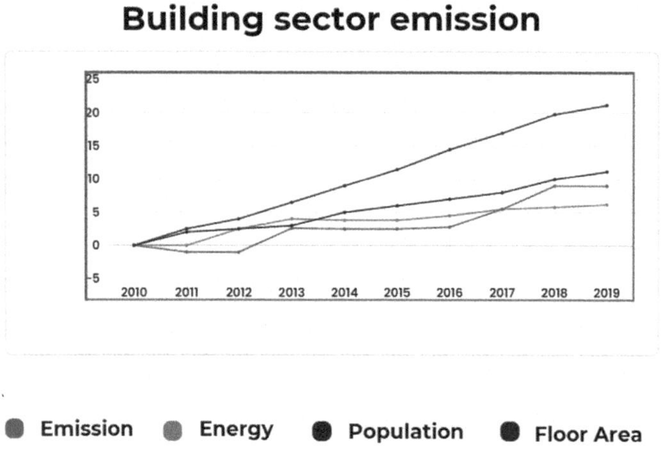

Fig. 3.2 Change in emission, energy, population and buildings from 2010 to 2019

issues. One tool by which AI can improve SHM is by training machine learning algorithms on historical data. These algorithms can detect changes in the behavior of a structure over time, authorizing early warning of potential problems before they turn into critical situations.

DL algorithms can also process data from sensors in real-time, enabling rapid detection and recognition of problems. By using AI, we can optimize the arrangement of sensors within a structure, the most critical positions can be monitored effectively without the need for trained staff, thus decreasing the costs and improving the accuracy of SHM systems.

DL has the potential to enhance the health of a structure in several ways, allowing early discovery of damage, predictive maintenance, upgraded protection, and optimized design. DL algorithms are able to analyze huge proportion of data produced by various sensors positioned on the structure, such as accelerometers, strain gauges, and emission sensors. By detecting even subtle changes in the data, DL algorithms

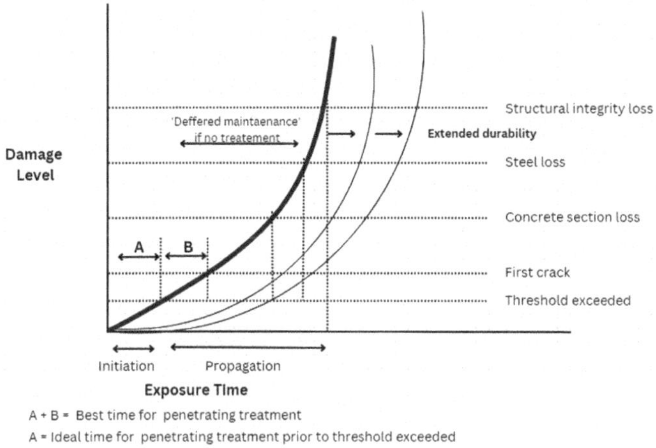

Fig. 3.3 Plot between the damage level of buildings versus time

can provide early warning of any potential damage or defects, allowing for timely maintenance or repair. Since these algorithms can train themselves from past data and make accurate predictions about the future behavior of the building. By accessing data from sensors, these algorithms can catch the pattern and trend that may indicate potential issues. This can help to schedule maintenance or repairs in advance, reducing downtime and maintenance costs. By detecting potential damage or defects early, the algorithm can help to prevent catastrophic failures that can pose a risk to human life or property. This can significantly improve the safety of structures, particularly those that are subject to continuous loads and stresses. DL Algorithms can also be used to elevate the outline of structures. By estimating data on the behavior of similar structures, it can provide insights into the most effective design parameters, such as material properties and structural configurations.

The process of damage identification in SHM involves six major fragments: (i) SHM, (ii) Statistical Process Control (SPC), (iii) Non-destructive Evaluation (NDE), (iv) Semi-Destructive Testing (SDE), (v) Condition Monitoring (CM), and (vi) Damage Prognosis (DP) [3–24] (Fig. 3.4).

Civil Engineers initially used SHM to keep track of dams [25]. The government of the United Kingdom was the first to begin this initiative by seeing the 254 Casualties due to the collapse of Dale Dyke Dam which is 29 m long, situated close to Sheffield in 1864 [26]. Reservoir Act was passed by the legislative body of the UK in the year 1975 which states that appointing an engineer to supervise reservoirs and dams is mandatory. His principal duties were to track records of the temperature of structures, level of water, meteorological circumstances, and many more.

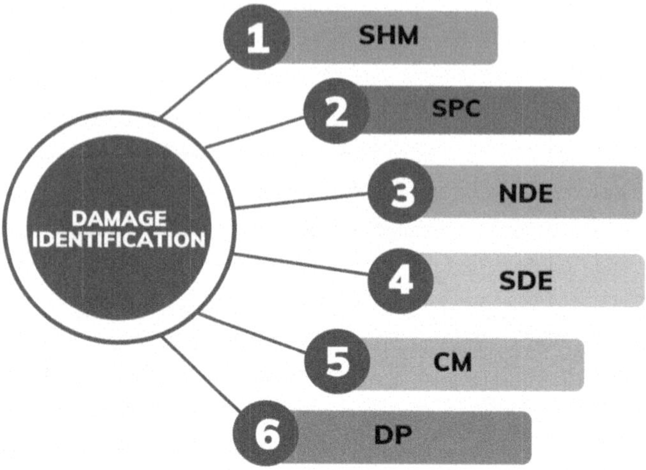

Fig. 3.4 Methods for identifying damage

3.2 Central Aim of Keeping Track of the Structure of Buildings

Preservation of buildings and assessing their resilience to threats, particularly earthquakes is a critical concern, in today's society. These structures, typically constructed using masonry materials that may have undergone modifications or additions over time pose challenges when it comes to evaluation due to their nature and limited information available. Of concern are masonry structures that bear heavy loads with towers, minarets and chimneys being especially susceptible to lateral forces.

Maintaining buildings and keeping a track of them will definitely put a stop to monetary loss accompanying increased health of people as well as buildings and ensuring their safety. These days, the extension of technological assistance can be employed for examining a structure's health and well-being soundly and skillfully. Rigorous rules and laws are enforced by the Governments of various countries with regard to the method acquired along with the time frame for periodical upkeeping of steel, concrete, and masonry buildings. SHM's main aim is to reduce the investment of working capital and fortuitous chances associated with conserving the health of buildings.

Extensive research has been conducted by the community in developing intrusive investigation technologies and reliable methodologies, for analyzing these structures structural integrity based on local or overall experimental responses.

Thoroughly it can be expressed as: (i) cost-efficient (economical) [27] (ii) identifying early risk [28] (iii) enhanced safety of citizens (iv) expanded life of buildings. The central aim of keeping track of the structures of buildings is shown in Fig. 3.5.

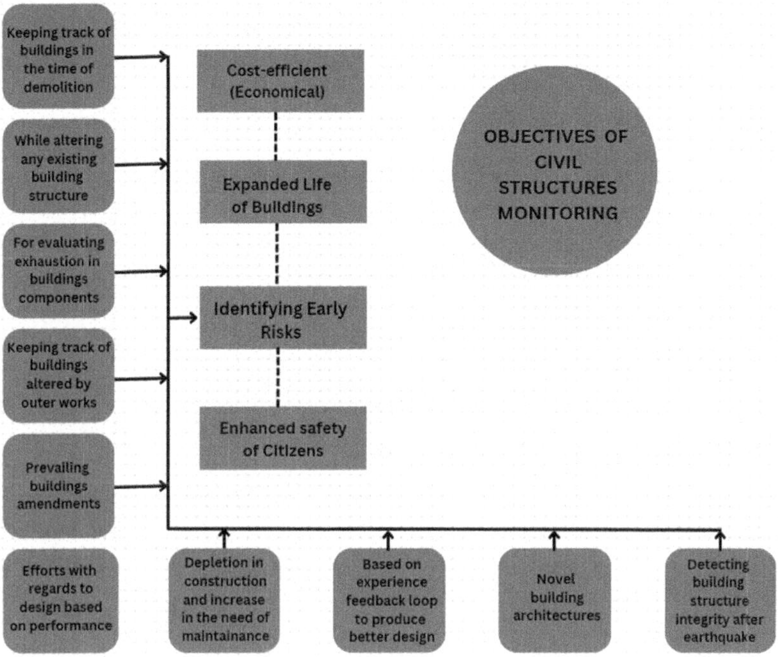

Fig. 3.5 Objectives of civil structures monitoring

3.2.1 Cost-Efficient (Economical)

Implementing appropriate structural health monitoring techniques can efficiently bring down structural expenses both for long as well as short periods of time. The industry is highly motivated to adopt SHM strategies as they can significantly enhance global profit. Researchers are consistently striving to minimize maintenance expenses through the implementation of condition-based monitoring techniques. The CBM technique involves the installation of sensors to constantly monitor structures and detect potential safety risks. When such risks are identified, the needed maintenance is promptly provided to confirm the well-being and reliability of structures [29–31]. Identifying defects within structures through SHM techniques can prevent costly failures and avoid the need for complete or partial demolition as well as rehabilitation of buildings. This can result in significant financial savings and minimize the risk of major losses [32, 33].

3.2.2 Identifying Early Risk

Through regular SHM, engineers are able to detect defects in buildings, assess their health status, monitor immediate impacts caused by Seismic Activities as well as Natural Disasters, and identify correlated risks in early stages. This permits for convenient and effective measures to be adopted to prevent potential structural collapses and guarantee the welfare of structure and its occupants. Early evaluation of buildings through SHM techniques can mitigate the threats correlated with natural calamities. Sensors are strategically placed in infrastructures such as Dams and Bridges to keep an eye on the faults and drifts. By collecting data and identifying movements in a timely manner, failures are avoided to a significant extent.

3.2.3 Enhanced Safety of Citizens

One of the most significant advantages of SHM is the increased level of civic protection it provides. Newly discovered SHM strategies utilize sensors to accumulate and gather data to provide accurate outcomes. Critical buildings are monitored in accordance with policies made by the government to look after the safety of citizens. As buildings age, they tend to develop delicate points as well as cracks. SHM techniques allow engineers to detect these cracks at an initial stage, permitting early identification of any problems attached to the buildings. In cases where situations are drastic and have no scope for amendment, the building may need to be isolated for public security concerns. By enhancing the security management of the citizens and addressing environmental issues, SHM techniques can guarantee the health of the buildings while prioritizing the safety of the public [34, 35].

3.2.4 Expanded Life of Buildings

By utilizing constant SHM methods, engineers can detect cracks and potential failures in the initial stages, thereby extending life and enhancing the productivity of buildings. Traditional techniques such as visual exploration and NDTs are often ineffective because of their accuracy. However, present-day SHM technologies such as Transient Thermography, Optical and Eddy Current Methods, and others, provide engineers with the ability to detect even the smallest failures. Addressing small cracks in a building at the early stage to prevent their future propagation and delay the failure of the structure [36, 37].

3.3 Artificial Intelligence

AI involves the replica of Intelligent Human practices by machines. It has the ability to integrate a wide range of diverse technological explanations and holds enormous potential for improving efficiency and addressing interdisciplinary challenges across fields such as medicine, science, engineering, as well as agriculture [37]. During a seminar at Dartmouth College in 1956, the label 'Artificial intelligence' was introduced by [38, 39]. Subsequently, in the 1960s, the US Department of Defense provided significant funding for AI research. Subsequent to its introduction, AI research faced several challenges, including condemnation and decreasing rate of success, leading to fluctuations in research financing and a period of decline during the period commonly referred to as the 'AI Winter' [40, 41]. AI research regained momentum and started to be implemented in real-time situations across various sectors such as Logistics, Data Mining, Medicine, Engineering, and others in the late 1990s [38, 42, 43]. With advancements in computational abilities, AI was able to yield impressive outcomes when integrated with fields such as Mathematics, Statistics, and Economics [44–46] AI's expansion into various sectors of engineering has continued since the late 1990s and early twenty-first century [37]. This expansion has brought significant potential benefits to various applications in Civil Engineering, including Structural Health Monitoring [47], construction management, building materials, and optimization, among others. The approaching decade is expected to witness numerous developments in AI and SHM because of the growing demand for the concept of intelligent cities at local as well as global levels [38, 48–50]. The integration of AI in SHM is poised to provide a significant boost to the structure as well as the construction field, enhancing safety and facilitating maintenance with greater ease.

Numerous studies are initiated by computer researchers, applied scientists and mathematicians who have made contributions, towards the realm of intelligence. Artificial neural networks (ANNs) belong to the field of intelligence. They draw inspiration from the system to assemble information similar to what human mind does. Key concept behind ANN is its arrangement of Information Processing System, which comprises interconnected processing elements collaborating to solve problems. In this system each neurons output is multiplied by weight coefficients. Fed into an input power through an excitation function.

The use of AI enables maintaining employees as well as end-users to analyze the current state and behavior of buildings with greater precision and ease throughout their entire lifecycle. With AI models, decision-making challenges such as abnormality identification, corrosion evaluation, and damage assessment are diagnosed and accurately solved, surpassing the intellectual capacity of human beings. AI in combination with IoT, data mining, smart buildings, machine learning, and deep learning is capable of performing SHM on different building structures. By leveraging past data and experiences, AI models can predict a set of patterns as well as understand the architecture of the data based on the belief hierarchy [51]. AI is a vast field that encompasses various subfields, including Machine Learning (ML). ML, in turn, has a subcategory known as DL, Soft computing methods (SCM) and Hard computing methods (HCM) are two groups in which ML is categorized based on the methods that are used. Soft Computing uses approximate reasoning and probabilistic logic to solve complex problems. It includes techniques such as neural networks, distorted logic, and probabilistic judging. In contrast, Hard Computing gives precise outcomes and utilizes techniques such as numerical examination as well as binary logic. SC is close to computational intelligence and is proficient in processing data acquired from experimental monitoring.

In recent years, the value of AI in Civil has grown significantly. SHM relies heavily on techniques such as DL, Pattern recognition (PR), ML and Image processing (IP). ML models used in SHM can be grouped into two classes based on their depth: shallow and deep architectures. In a model, the contrast between both Input and Output layer is termed 'model depth'. A model is considered "deep" if it has a large number of input layers compared to output layers. Conversely, a "shallow" model has little amount of input layers, and the contrast between them is minimal. DL technologies have a large number of layers that process the information. While these techniques were introduced in the eighties, they did not become popular for decades because of insufficient computer software and systems.

3.3.1 Machine Learning

Machine learning analyzes large input data sets and uses an algorithm to make predictions, without the need for explicit programming. In general, when creating a machine learning model there are some decisions take must be taken into consideration. Few being the specific type of learning approach to use, the desired level of performance for the learning goal, how to measure and assess target performance and which algorithm to employ for effective learning of models from instructional examples. ML algorithms recognize these patterns and detect threats before serious incidents occur. This can inform building owners and managers so they can act proactively to address such threats [44].

Based on learning, ML methods are divided into 3 groups: mainly Supervised Learning which includes learning with "Labeled Datasets," which means having instances with known outputs, forming input–output data pairs. With the help of these labeled datasets, algorithms create rules to sort out new sets of data. In contrast, Unsupervised Learning is about "Unlabeled Datasets" where there are no known outputs, but there are patterns that follow some rules, making it possible to group data. When it comes to SHM, Unsupervised Learning is helpful for spotting damage by organizing structural output data into clusters whereas supervised learning is useful for recognizing what type of damage is there and how serious it is, and the third group is Semi-Supervised Learning which combines elements of both learning methods and generally aims to classify data using both labeled and unlabeled data.

Various ML techniques, including Networks, Genetic-Programming, fuzzy logic, and Support Vector Machines (SVM) have been employed to produce models, for forecasting the machine-driven characteristics of concrete.

The tasks accomplished by machine learning arrangements can be outlined as follows; (i) Classification: the step aims to control the group to which the input belongs. (ii) Regression: In this stage, the output format focuses on quantifying the differences observed during classification. The objective is to model the relations among inputs and mathematical outputs. (iii) Prediction: The aim here is to forecast standards within a time frame. This is a form of Regression Analysis. (iv) Clustering: This task involves identifying data points across two or more datasets. Unlike classification, regression, and prediction, clustering employs an approach [57].

Before applying machine learning algorithms for data analysis, it is crucial to normalize the data. Data normalization is a practice in machine learning. Offers several advantages. For instance, it enhances descent performance when working with normalized data compared to normalized data.

For understanding it much better, let's take a sample model for detecting cracks in concrete structures. Consider that you have an extensive dataset at your disposal, which includes records of various concrete structures and whether or not they have cracks. The foremost step in modelling the arrangement would be to collaborate with structural engineers to identify the key indicators or characteristics that are typically linked with the presence of cracks in the concrete buildings.

Training Data Set is processed to extract info related to these key indicators from distinct structure's table, such as factors like age, environmental conditions, load history, and construction materials (i.e., Feature Extraction process). Subsequently, this treated dataset, containing info about these indicators from each structure along with the corresponding outcomes (presence or absence of cracks), is utilized to model AI system using a definite machine learning procedure. During the modelling phase, the arrangement autonomously learns the associations between these key indicators and its outcome, i.e., the presence or absence of cracks in the concrete structures. Once the system has been rigorously tested and verified for accuracy, it can be deployed for assessing the condition of any concrete structure. Given the relevant data about a structure, the trained system will be capable of determining whether it is likely to have cracks or not based on the identified key indicators.

Key benefits of machine learning include the ability to learn and adapt without the need for manual coding. ML is a method of Artificial Intelligence AI that can develop descriptive as well as predictive models based on the composed and observed data. In civil engineering field, ML is rapidly attaining popularity and its scope is expanding in various sub-branches. ML is not only contributing significantly to civil engineering but also to various other disciplines such as Computer Science Engineering, Information Technology, Finance, Statistics, Philosophy, and many more [50]. In machine learning, the key center of attention is image processing (IP), deep learning (DL), and pattern recognition (PR). IP as well as object recognition is a fragment of DL, while PR is used to identify patterns in information and group them correspondingly. Machine learning is further divided into subcategories such as supervised learning, unsupervised learning, and their branch categories in Fig. 3.6.

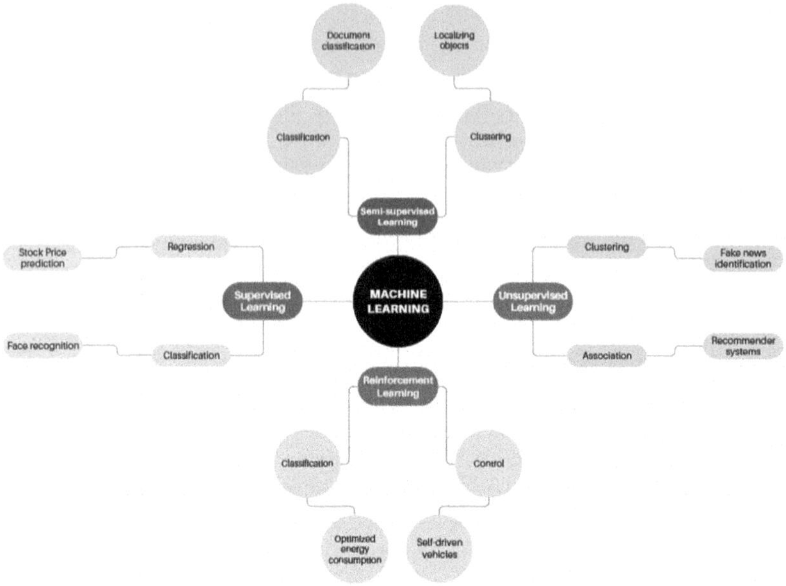

Fig. 3.6 Types of machine learning

3.3.2 Deep Learning

In today's world, DL techniques have become increasingly important in image recognition, natural language understanding, suggestion processing, and various other areas and these techniques have given jobs to us and have indeed worked well, saving time and reducing costs. DL is a special type of learning where the network learns highly imprecise features from raw data, automating its classification It is essentially a subset of machine learning, which is also an intelligent part of a wider project.

DL algorithms have the ability to assist in establishing connections, between features and the desired output as an aid in feature extraction. With training, a DL system can directly map raw inputs to target outputs without requiring explicit feature extraction. Additionally, it can identify features at a level of hierarchy that explain the simpler learned features, at a lower level. This unique ability lets DL Algorithms to split tasks into manageable problems and effectively resolve them [57].

DL has largely grown out of ANNs. In an ANN, the fundamental unit is known as a "neural cell." It consists of mainly three input parts named ×1, ×2, and ×3. These inputs further get multiplied by certain values referred to as weights, denoted as w1, w2, and w3. Thereafter, a number called "bias," symbolized as 'b,' is added to this sum to make some adjustments. Lastly, to produce an output, an activation function, denoted as 'f(x),' applies a non-linear transformation to this result.

It consists of an architecture that is built upon multiple masked layers. The predictions made by DL are built upon the learnings and outcomes acquired by these masked layers. DL and PR require a considerable amount of domain expertise to bring out relevant feature data from raw data consisting of images and convert it into a feature vector for classification. However, DL allows the training algorithm to work directly with the input data without the need for feature engineering. Physical coding is required for feature engineering, whereas DL automates this process. Hence, DL is more efficient than hand-coding as it can learn from the feature data automatically.

To understand it much better, let's suppose the goal is to develop a computer vision system for Crack Detection in Concrete Structures using DL. The available data for training the arrangement contains pairs of input as well as output models. For each and every instance, the structure's input is a picture of concrete structure, while output is precise location of any cracks in that picture. However, before modelling the system, it's essential to denote each input picture in relations to features that are likely to provide strong hints concerning the position of cracks. This feature extraction process is crucial because the appearance of cracks in concrete images can vary significantly due to factors such as lighting conditions, angle of the camera, surface texture, and the size and type of cracks. Deep learning techniques, similar to those used in computer vision for vehicle detection, can automatically learn and identify the complex patterns and characteristics associated with cracks in concrete.

Once, the DL model is skilled and thoroughly tested, then it is used to analyze images of concrete structures and accurately pinpoint the locations of cracks, despite variations in lighting, angles, and other conditions, ultimately aiding in efficient structural inspection and maintenance.

DL comprises networks that can learn without guidance, from unlabeled data. The prime goal of learning architecture is to comprehend and extract features from the input data. In essence, deep learning relies on networks, with hidden layers resulting in a network that becomes deeper as more layers are added. Convolutional neural networks (CNNs) and deep belief nets are all examples of deep learning architectures. Recently among DL algorithms CNNs have gained popularity. One of the reasons, behind their popularity is their ability to learn directly from signals in large scale datasets. This has led researchers to discover the use of CNNs to overcome the limitations of machine learning based Structural Damage Detection (SDD) methods. CNNs are a type of neural networks with multiple layers that are widely recognized as a standard approach for deep learning tasks. In engineering applications, they have consistently demonstrated state of the art performance [58].

Analyzing the DL-based approaches, encompassing their practical use and integral challenges, leads to several key observations: Innovation in the realm of deep learning, comprising the introduction of new architectural designs, efficient training and validation techniques, and the development of novel frameworks, is composed to simplify and enhance data processing capabilities. These advancements will empower scholars and engineers to address their professional challenges more effectively. Scientists and engineers in civil engineering have recognized the exciting possibilities and advanced technology of deep learning methods. They have made various efforts to use these techniques for SHM of civil infrastructure. The research mainly groups into two main areas: one related to detecting structural damage and the other focused on assessing the overall condition of structures. Tackling the challenges associated with implementing DL in SHM involves the collection of specialized datasets, the alteration of architectural designs to enhance performance, and the creation of innovative training strategies to mitigate issues like overfitting and the vanishing gradient problem. It is evident that DL approaches offer substantial value in addressing various SHM issues [52].

3.4 AI in Civil Engineering

The application of AI in Civil Engineering is vast and has an infinite reach. Every subdiscipline of civil engineering can benefit from AI. Concrete technology, Building Structure Analysis, Construction Management, Transportation Engineering, Green Structures, and Water Management are some of the sub-fields in civil engineering that are currently utilizing AI to a great extent. Concrete is indeed the most commonly utilized material for the construction of buildings globally, with approximately 3 tons for each person each year being used [53]. This material is a composite material, characterized by its hardness and strength, and its heterogeneous nature. Concrete is composed of thin and rough aggregate, water, and cement. The quality of concrete can be greatly influenced by variety of factors including the quality and properties of its ingredients, proportions in which they are mixed, the way it is placed and compacted, and the curing process. These aspects play a decisive role in defining the firmness, durability, and total status of the last product.

AI is playing a vital role in optimizing and predicting the feasibleness and condensing strength of concrete, as well as classifying concrete amounts, to improve concrete's properties such as robustness, airtight, resistance to erosion as well as fire, and great strength. The feasibility of concrete plays a crucial role in its strength as well as ease of placement on shuttering. However, concrete trials can be time-consuming, labor-intensive, and costly due to the wastage of manpower as well as material. To reduce these issues, Artificial neural networks (ANNs) are utilized to create feasibility models for concrete. These versions forecast the concrete crash value accurately, 1.41. helping to decrease wastage, design cost, and time. Concrete's condensing strength also leans on its composition, and its durability is reliant on its condensing strength.

AI has shown its capability to advance the learning process and simplify reasoning, which proves beneficial, in construction projects with characteristics. The fusion of technology and creativity has greatly enhanced our living conditions. Moreover, as we strive towards a future technological advancements and innovation will play a pivotal role. It has improved thruput that can aid in increasing efficiency lowering production costs and emissions optimizing resource utilization during production processes facilitating market alignment, and leveraging data to enhance the accessibility of public services.

Machine learning, a subset of AI has algorithms that are useful in various Vibration-Based Structural Damage Detection (SDD) methods. Similar to non-ML approaches, SDD methods using machine learning can be classified into parametric and nonparametric methods.

3.4.1 *Artificial Intelligence in SHM*

SHM is all about trying to understand how structures act under pressure conditions. You can go about this in couple different ways. The first is to use math models–come up with equations and stuff that show how the structure should work. But the problem is those models aren't perfect and they make assumptions about things like what the structure's made of and how it is supported and connected together. And those assumptions don't always match reality. So, then you need to test the real structure and compare it to the model predictions. That way you can twist the models until they match the real-world results better. This helps make sure the models work for different situations and can show if changes or damage are affecting things.

The other way is to skip models and just look directly at data from monitoring the structure. Advanced sensors can give you detailed info about what's going on with the structure right now. This data can show if there's any damage and even predict stuff about the future. You can use machine learning and neural networks to categorize patterns and trends in the information. That gives you a good idea of structural health without having to rely on theoretical models.

Artificial intelligence tools such as Image Processing and Data extraction, are extensively used in Structural Health Monitoring of buildings. Image processing tools play a crucial role in detecting and analyzing defects that include cracks, condensation, seepage, and other thermal substitutions. Recent observations have shown that image processing methods are enhancing productivity in identifying cracks in buildings. The process of crack identification in building infrastructures using image processing techniques such as 'Image binarization' is highly dependent on various factors including quality of the image, noise in the background, low image contrast, splitting of concrete, unequal illumination, and various other image-related specifications. The crack information obtained through these techniques helps decide the appropriate rehabilitation/repair techniques for the buildings.

Digital image correlation is applied in the technique of crack identification and their measurement by comparing healthy images of a building with images containing cracks that are deteriorated. The Image Binarization process is a technique that converts grayscale into the binary form of images containing only black-and-white information. This process changes the light as well as dark pixel elements of the original image into white-and-black information, respectively. IB is commonly used in Medical Image Processing, Text Pattern Recognition, and Crack Identification (Fig. 3.7).

Take a look at Fig. 3.8 and you'll see the basic setup of an image processing system. The process starts with a high-quality image of a component we want to

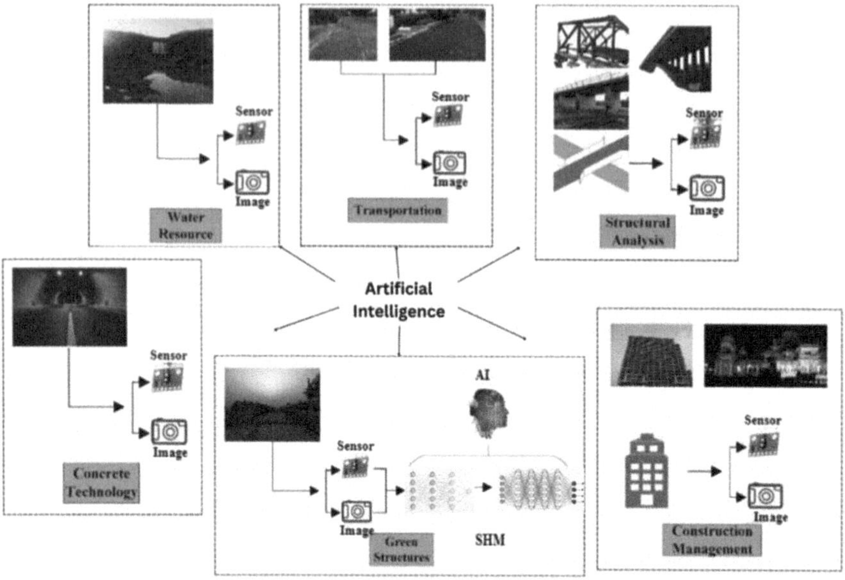

Fig. 3.7 Use of artificial intelligence in SHM

inspect taken by a camera or another imaging device. Before we can look for any cracks or other flaws, it needs to be cleaned up. If you've ever edited pictures you know there's always some sort of noise that comes with it. We have to somehow get rid of all those unnecessary additions and smooth out the photo. This is what our preprocessing step is dedicated to. It might also include simplifying the image in grayscale or binary forms. When we have a picture that's been cleaned up, it goes through techniques like edge detection, segmentation, and pixel analysis. These all allow us to determine where exactly a crack might be hiding in the image. Once we figure out this area, we move on to measuring how long, wide, deep, and dense it is in parameter estimation. We can't examine every little thing one by one so these measurements play an important role in determining if something's severe or not and how much concern it should bring us.

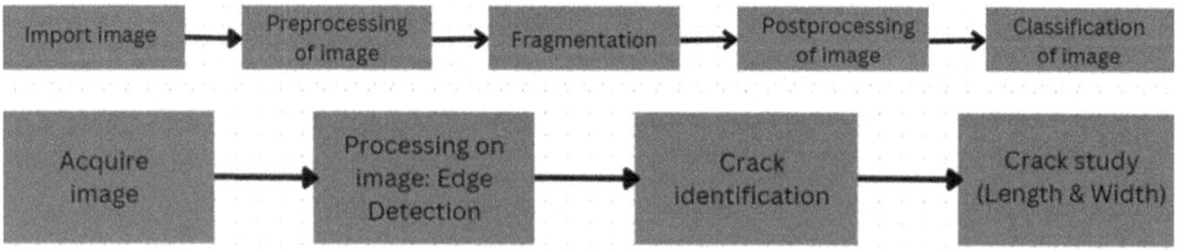

Fig. 3.8 Processing on a crack image

3.4.1.1 Image Binarization (IB)

The binarization process plays a crucial role in creating binary images from grayscale images to replace specific thresholds via 1 s and fixing all remaining values by 0 s. One popular method for binarization in MATLAB is Otsu's method which lowers the variance of intraclass for threshold black-and-white picture elements. In order to perform binarization on colored images, they must first be converted to grayscale images with thresholds ranging between 0 and 255 to represent black-and-white data, respectively. During this process, every picture element in the original picture is analyzed by associating its picture element value with its threshold. If the rate of pixel is higher than its Threshold Value, it is made to one that is white in binary image, while if it is lower, it is made zero that is black. Binarization variables such as size of window and sensitivity play a crucial part in the Threshold Calculation process. The window size determines the range of pixel values used in the calculation, while the sensitivity takes care of the allowance of statistical variables of the pixel intensities to calculate the threshold.

3.4.1.2 Threshold Values

Numerous researchers have compiled threshold values for obtaining the Binary form of images with much more accuracy and efficiency in the context of medical identification, by concerning the lowest and highest Intensities of a chosen window using various image binarization techniques which are arranged in Table 3.1. The method or technique used for image binarization can significantly impact the results obtained, leading to differences in the binary images produced. Additionally, the properties of the original images being processed can also affect the outcome of the binarization process. Therefore, it is important to consider both the method of image binarization and the images' properties when interpreting the IB outcomes. Otsu's Technique of Thresholding is a widely used clustering-based technique that can be applied to various sorts of Images [54]. This technique works by separating the picture elements of the image into 2 categories. The proportion of pixels and moderate gray level for each group is denoted as $\omega 0$ and $\mu 0$ for the foreground, and $\omega 1$ and $\mu 1$ for the background, respectively.

3.4.1.3 Considerations in Crack-Recognition in Images of Buildings

The primary purpose of image binarization is to accurately identify defects or anomalies within an image, such as the length of the crack and its width. To attain this, it is essential to cautiously select the sensitivity as well as the window size parameters used during the binarization process. By choosing appropriate values for these parameters, it is possible to point to the exact location of the defect pixels within the grayscale image, thereby facilitating accurate detection and characterization of the defect [58–60]. When the pixel of the background region in a grayscale picture

Table 3.1 Techniques to discover threshold numbers

Author	Equation	Application
Liu [53]	$\rho_n = sin(\alpha_r)$ $\rho_t = cos(\alpha_r)$	For automating detection of cracks on buildings using image processing and determines properly the kinematics of cracks
Khurshid et al. [54]	$T = m + k\sqrt{(B} + m2)$	Appending the square root of the variance to the thresholding process can help to minimize the effect of noisy backgrounds on the resulting binary image
Wolf et al. [55]	$T = (1 - k) \times m + k \times M + k \times s/R \times (m - M)$	Effective for images with a predominance of dark or dim colors
Sauvola and Pietikäinen [56]	$T = m \times \{1 - k \times (1 - s/R)\}$	Is able to work with images which has noise in the background
Niblack [57]	$T = m + k \times s$	Adaptive and works well when the background of the image contains less noise

is greater than that of crack element, increasing the window size during binarization can point to the false evaluation of the length of a crack in the resulting binary image. The optimization parameter for image binarization typically includes three main parameters: (i) precision of the dimension of the crack detected, (ii) Detection of the crack width, and (iii) computational processing period. To achieve these optimization criteria, color or original images are first changed to grayscale images, after which various binarization strategies can be applied to produce binary images that accurately capture the crack features. To measure the width of a crack, it is typically required to calculate the values of pixels at accurate locations along the length of the crack where the width is to be measured (Figs. 3.9 and 3.10).

3.5 Review and Discussion

See Tables 3.2, 3.3 and 3.4.

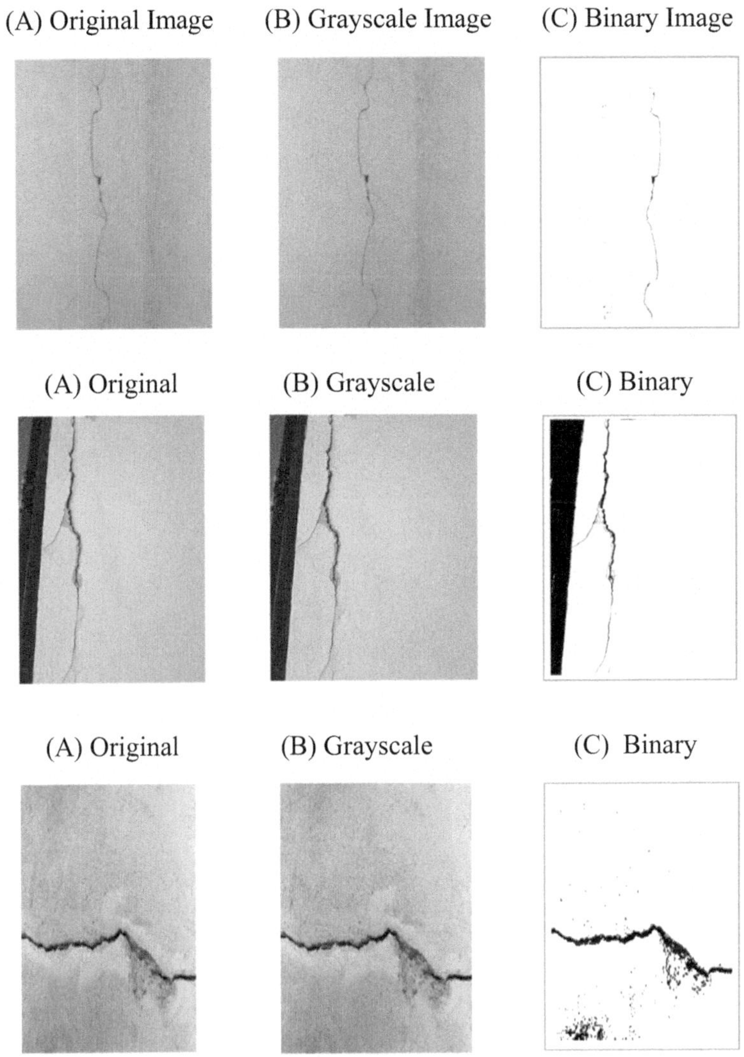

Fig. 3.9 Conversion of an original image to grayscale image and then to binary image

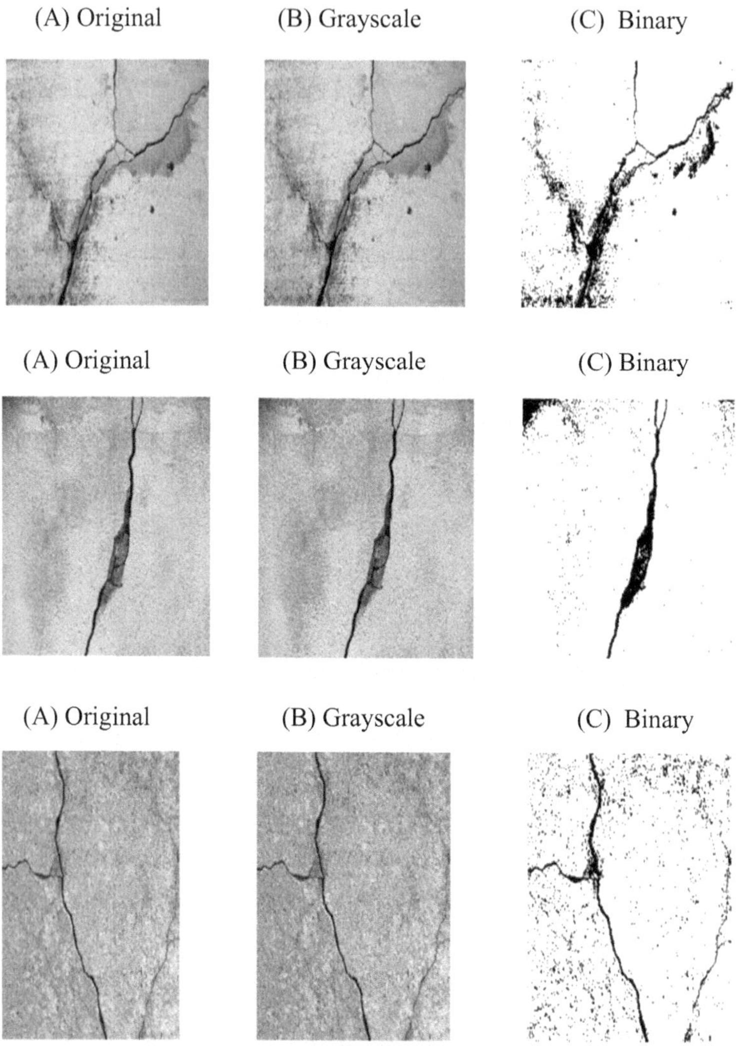

Fig. 3.9 (continued)

3.6 Conclusion

This Chapter presented a study focused on using image-processing strategies to detect and monitor surface cracks in buildings. The study aimed to develop an Image Processing based device that could be availed to identify cracks in concrete along with masonry buildings, with the ultimate goal of supporting structural health monitoring and maintenance efforts. Monitoring the integrity of buildings and other structures has always been an area of research. The well-being and safety of these structures

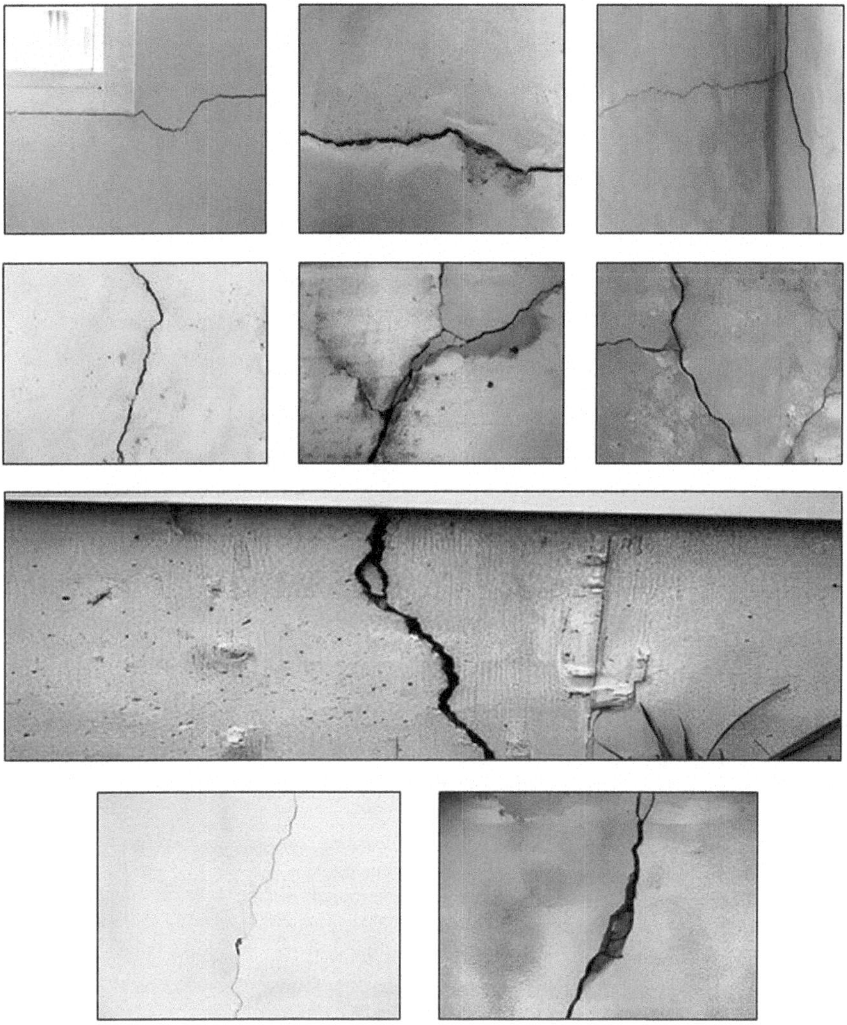

Fig. 3.10 Analysis of various kinds of cracks

rely heavily on monitoring any damage that may occur including its development and spread. There have been studies conducted on methods, for detecting and assessing structural damage using vibrations. These studies aim to identify and measure the extent of any harm, to the structure. The digital pictures used in this study to perform processing for crack detection and monitoring contained various issues that could impact the accuracy of the results, such as noise, up-down divergence, as well as irregular brightness. Our proposed framework, integrating CNNs and multi-modal data fusion, characterizes a substantial leap forward in the area of Structural Health Assessment. Through the analysis of visual data obtained from NDT methods and

Table 3.2 Applications of DL-based SDD

Application	Description	References
Civil structures	Detection of damage in bridges, buildings, and other civil infrastructure using deep learning models	Jia and Li [61], Ewald et al. [63]
Vibration-based damage detection	Utilizing deep learning for damage detection in civil structures through vibration analysis	Zhang and Li [62], Seventekidis et al. [65]
Smart building structures	Application of DL-based processes for damage identification in smart building constructions	Chen and Cui [64]
Bridge structures	Employment of DL practices for SHM and damage detection in bridges	Seventekidis et al. [65]
Non-destructive testing (NDT)	Using deep learning for non-destructive testing to identify structural damage and defects	Seventekidis et al. [65]
Mode shape analysis	Application of deep learning for damage detection through mode shape analysis of structures	Seventekidis et al. [65]
Acoustic emission testing	Utilizing deep learning for acoustic emission testing to detect and locate damage in structural components	Seventekidis et al. [65]
Finite element model (FEM) data analysis	Employing deep learning for damage detection using data from finite element models of structures	Seventekidis et al. [65]
Guided lamb wave technique	Application of DL for SHM based on the guided lamb wave method	Seventekidis et al. [65]
Feature fusion and neural network model	Using DL for feature mixture and neural network models for SDD	Seventekidis et al. [65]
Convolutional neural networks (CNNs)	Application of CNNs for structural damage detection, including feature extraction and mode shape analysis	Seventekidis et al. [65]
Optimal finite element model generated data	Utilizing deep learning for SHM using optimal finite element model generated data	Seventekidis et al. [65]

(continued)

Table 3.2 (continued)

Application	Description	References
Sensor data-driven damage detection	Using deep learning for sensor data-driven damage detection based on deep CNNs	Seventekidis et al. [65]
Structural damage features extraction	Application of deep learning for structural damage features extraction from mode shapes	Seventekidis et al. [65]
SHM systems	Employment of DL for SHM systems in buildings and bridges	Seventekidis et al. [65]

the incorporation of complementary information from various sensors, our system achieves a comprehensive understanding of concrete structure health. The advantages offered by our approach are numerous which includes real-time analysis, high accuracy in damage classification, and a reduced reliance on human interpretation distinguish our methodology from conventional inspection methods. Additionally, the proactive nature of our system enables the early detection of structural issues, paving the way for efficient and cost-effective maintenance strategies. By promoting a continuous monitoring framework, our research aligns with the principles of sustainability by minimizing the environmental impact associated with premature replacements and extensive repairs. The implementation of AI in structural health assessment not only enhances the resilience and longevity of concrete structures but also contributes to the broader goals of creating smart and sustainable urban environments. As we look to the future, our work lays the foundation for further advancements in the integration of deep learning techniques in infrastructure management. The potential for widespread adoption of such methodologies promises to reshape the landscape of urban development, fostering a new era of intelligent and sustainable infrastructure. It is our hope that this research stimulates further exploration, innovation, and collaboration in the pursuit of resilient and environmentally conscious urban structures.

Table 3.3 Applications of deep learning-based structural condition assessment

Application	Description	References
Bridge health monitoring	Utilizing deep learning for real-time assessment of bridge structural conditions and identification of potential damage	[67]
Civil infrastructure	Application of deep learning for continuous monitoring and assessment of structural health in civil infrastructure	[66]
Vibration-based damage identification	Employment of deep learning for identifying structural damage in civil structures through vibration analysis	[68]
Building health assessment	Using deep learning for real-time assessment of building structural conditions and identification of potential damage	[67]
Railway infrastructure	Application of deep learning for continuous monitoring and assessment of structural health in railway infrastructure	[66]
Roadway health monitoring	Utilizing deep learning for real-time assessment of roadway structural conditions and identification of potential damage	[67]
Pipeline integrity assessment	Application of deep learning for continuous monitoring and assessment of structural health in pipeline infrastructure	[66]
Offshore structure health monitoring	Using deep learning for real-time assessment of offshore structure conditions and identification of potential damage	[67]
Dam health assessment	Application of deep learning for continuous monitoring and assessment of structural health in dam infrastructure	[66]
Aircraft structural health monitoring	Utilizing deep learning for real-time assessment of aircraft structural conditions and identification of potential damage	[67]
Ship structural condition assessment	Application of deep learning for continuous monitoring and assessment of structural health in ship infrastructure	[66]

(continued)

Table 3.3 (continued)

Application	Description	References
Power plant structural health monitoring	Using deep learning for real-time assessment of power plant structural conditions and identification of potential damage	[67]
Industrial facility health assessment	Application of deep learning for continuous monitoring and assessment of structural health in industrial facility infrastructure	[66]
Seismic damage identification	Utilizing deep learning for real-time assessment of seismic damage in structures and identification of potential damage	[67]

Table 3.4 Different models made and used for SHM on the basis of technology used in them

Technology	Models used	Application	References
Smart sensors	AI, ML, DL	SHM of civil and mechanical structures, aerospace, naval, offshore fixed structures	Jia and Li [69]
Sensor integration	Strain gages, vibration/accelerometer sensors, displacement sensors, environmental/weather monitoring sensors	Tracking stresses, movement, environmental conditions	Zhang and Li [62]
Acoustic emission	Ultrasonic, thermal imaging	Non-destructive testing, damage detection	Jia and Li [61]
IoT and AI	Data-driven time series, predictive modeling, neural networks, genetic algorithms, support vector machines	Forecasting, portable and continuous monitoring, wireless communication, advanced sensor technologies	Zhang and Li [71]
Smartphone sensing	–	Cost-efficient, effective sensing system for SHM	Zhang and Li [70]
Ultrasonic testing	Pulse-echo, Non-destructive testing (NDT)	Detection of defects, cracks, and other damage in structures	Jia and Li [61]
Vibration monitoring	Accelerometers, Vibration data recorders (VDRs)	Monitoring structural vibrations, identifying damage and potential failure	Zhang and Li [62]
Infrared thermography	Infrared cameras, thermal imaging analysis	Detection of temperature differences, indicating damage or energy loss in structures	Jia and Li [61]

(continued)

Table 3.4 (continued)

Technology	Models used	Application	References
Fiber optic sensors	Optical fiber sensors (OFS), Brillouin fiber sensors (BFS)	Sensing strain, stress, and temperature in structures	Jia and Li [69]
Wireless sensor networks	Zigbee, Wi-Fi, Bluetooth low energy (BLE)	Remote monitoring and data collection from distributed sensors	Zhang and Li [71]
Data-driven models	ML, AI algorithms	Predicting structural performance and remaining life	Zhang and Li [71]
Finite element analysis (FEA)	Numerical simulation, FEA software	Analyzing structural behaviour and identifying potential failures	Zhang and Li [71]

References

1. Hussein AB, Abdi M (2021) Review paper on non-destructive testing and their accuracies to measure the mechanical properties of concrete. Int J Eng Appl Sci Technol 5(11):1–9
2. Malek J, Kaouther M (2014) Destructive and non-destructive testing of concrete structures. Jordan J Civil Eng 8(4):432–441
3. Han T, Wu G, Lu Y (2021) Crack monitoring using short-gauged Brillouin fiber optic sensor. Measurement 179:109461
4. Tan X et al (2021) Measurement and visualization of strains and cracks in CFRP post-tensioned fiber reinforced concrete beams using distributed fiber optic sensors. Autom Constr 124:103604
5. Floris I et al (2021) Fiber optic shape sensors: a comprehensive review. Opt Lasers Eng 139:106508
6. Kavitha S, Joseph Daniel R, Sumangala K (2016) Design and analysis of MEMS comb drive capacitive accelerometer for SHM and seismic applications. Measurement 93:327–339
7. Kavitha S, Joseph Daniel R, Sumangala K (2016) High performance MEMS accelerometers for concrete SHM applications and comparison with COTS accelerometers. Mech Syst Signal Process 66:410–424
8. Kavitha S, Joseph Daniel R, Sumangala K (2013) A simple analytical design approach based on computer aided analysis of bulk micromachined piezoresistive MEMS accelerometer for concrete SHM applications. Measurement 46(9):3372–3388.
9. Grillanda N et al (2021) SHM of a severely cracked masonry arch bridge in India: Experimental campaign and adaptive NURBS limit analysis numerical investigation. Constr Build Mater 280:122490
10. Miao H, Li F (2021) Shear horizontal wave transducers for structural health monitoring and nondestructive testing: a review. Ultrasonics 114:106355
11. Ribeiro D et al (2021) Non-contact structural displacement measurement using unmanned aerial vehicles and video-based systems. Mech Syst Signal Process 160:107869
12. Alamandala S, Prasad RS, Kumar PR (2020) Cost-effective load measurement system for health monitoring using long-period grating as an edge filter. Opt Fiber Technol 59:102328
13. Shahsavari V et al (2019) Diagnostic testing of a vertical lift truss bridge for model verification and decision-making support. Front Built Environ 5:92
14. Hou S et al (2018) Monitoring interstory drift in buildings under seismic loading using MEMS inclinometers. Constr Build Mater 185:453–467
15. Zhou Y et al (2020) Research on a novel inclinometer based on distributed optical fiber strain and conjugate beam method. Measurement 153:107404

16. Helmi K et al. Reference free method for real time monitoring of bridge deflections. Eng Struct 103:116–124
17. Marković MZ et al (2016) Application of fiber-optic curvature sensor in deformation measurement process. Measurement 92:50–57
18. Ma G, Du Q (2020) Optimization on the intellectual monitoring system for structures based on acoustic emission and data mining. Measurement 163:107937
19. Hesser DF et al (2021) Identification of acoustic emission sources for structural health monitoring applications based on convolutional neural networks and deep transfer learning. Neurocomputing 453:1–12
20. Mirgal P, Pal J, Banerjee S (2020) Online acoustic emission source localization in concrete structures using iterative and evolutionary algorithms. Ultrasonics 108:106211
21. Fu T et al (2009) Fiber optic acoustic emission sensor and its applications in the structural health monitoring of CFRP materials. Opt Lasers Eng 47(10):1056–1062
22. Xu D et al (2015) Temperature and loading effects of embedded smart piezoelectric sensor for health monitoring of concrete structures. Constr Build Mater 76:187–193
23. Rocha H, Semprimoschnig C, Nunes JP (2021) Sensors for process and structural health monitoring of aerospace composites: a review. Eng Struct 237:112231
24. Mieloszyk M, Majewska K, Ostachowicz W (2021) Application of embedded fibre Bragg grating sensors for structural health monitoring of complex composite structures for marine applications. Mar Struct 76:102903
25. Farrar CR, Worden K (2007) An introduction to structural health monitoring. Philos Trans Royal Soc A Math Phys Eng Sci 365(1851):303–315
26. Warren A (2011) News: new research published on lessons from historical dam incidents 145–146
27. Dong T, Kim NH (2018) Cost-effectiveness of structural health monitoring in fuselage maintenance of the civil aviation industry. Aerospace 5(3):87
28. Alokita S et al (2019) Recent advances and trends in structural health monitoring. Struct Health Monitor Biocomposites Fibre-Reinf Compos Hybrid Compos 53–73
29. Jardine AKS, Lin D, Banjevic D (2006) A review on machinery diagnostics and prognostics implementing condition-based maintenance. Mech Syst Signal Process 20(7):1483–1510
30. Boller C, Chang F-K, Fujino Y (eds) (2009) Encyclopedia of structural health monitoring, vol 1. Wiley, New York
31. Peil U (2009) Civil infrastructure load models for structural health monitoring. Encycl Struct Health Monit
32. Pines D, Emin Aktan A (2002) Status of structural health monitoring of long-span bridges in the United States. Progress Struct Eng Mater 4(4):372–380
33. Giurgiutiu V (2005) Tuned Lamb wave excitation and detection with piezoelectric wafer active sensors for structural health monitoring. J Intell Mater Syst Struct 16(4):291–305
34. Balageas D (2006) Claus-Peter FRITZEN a Alfredo GÜEMES. Struct Health Monit
35. Liu H et al (2021) Machine learning for glass science and engineering: a review. J Non-Cryst Solids 557:119419
36. Ko JM, Ni YQ (2005) Technology developments in structural health monitoring of large-scale bridges. Eng Struct 27(12):1715–1725
37. Glisic B (2013) Fiber optic sensors for subsea structural health monitoring. Subsea optics and imaging. Woodhead Publishing, pp 434–470
38. Krishnamoorthy CS (1996) Artificial intelligence and expert systems for engineers
39. Kumar A, Mor N (2021) An approach-driven: use of artificial intelligence and its applications in civil engineering. Artif Intell IoT Smart Converg Eco-friendly Topogr 201–221
40. Boobier T (2018) Advanced analytics and AI: impact, implementation, and the future of work. John Wiley & Sons
41. Thompson S (2008) Artificial intelligence comes of age. ICT Futures Delivering Pervasive Real-Time Secure Serv 153–164
42. Fisher D et al (1993) Applying AI clustering to engineering tasks. IEEE Expert 8(6):51–60
43. Faught WS (1986) Applications of AI in engineering. Computer 19(07):17–27

44. Marzban C (2009) Basic statistics and basic AI: neural networks. Artif Intell Methods Environ Sci 15–47
45. Lenat DB (1976) AM: an artificial intelligence approach to discovery in mathematics as heuristic search. Stanford University
46. Hoyles C, Noss R (2003) What can digital technologies take from and bring to research in mathematics education?. Second Int Handb Math Educ 323–349
47. Brownjohn JMW (2007) Structural health monitoring of civil infrastructure. Philos Trans Royal Soc A Math Phys Eng Sci 365(1851):589–622
48. Allam Z, Dhunny ZA (2019) On big data, artificial intelligence and smart cities. Cities 89:80–91
49. Chatterjee S, Kar AK, Gupta MP (2018) Success of IoT in smart cities of India: an empirical analysis. Govern Inf Quart 35(3):349–361
50. D'Alessandro A et al (2016) Towards smart concrete for smart cities: Recent results and future application of strain-sensing nanocomposites. J Smart Cities (Transferred) 1(1)
51. Kiani J, Camp C, Pezeshk S (2019) On the application of machine learning techniques to derive seismic fragility curves. Comput Struct 218:108–122
52. Ye XW, Jin T, Yun CB (2019) A review on deep learning-based structural health monitoring of civil infrastructures. Smart Struct Syst 24(5):567–585
53. Liu Y et al (2014) Automated assessment of cracks on concrete surfaces using adaptive digital image processing. Smart Struct Syst 14(4):719–741
54. Khurshid K, Siddiqi I, Faure C, Vincent N (2009) Comparison of Niblack inspired binarization methods for ancient documents. In: Document recognition and retrieval XVI (vol 7247. SPIE, pp 267–275
55. Wolf C, Jolion J-M (2004) Extraction and recognition of artificial text in multimedia documents. Form Pattern Anal Appl 6(4):309–326
56. Sauvola J, Pietikäinen M (2000) Adaptive document image binarization. Pattern Recogn 33(2):225–236
57. Niblack W (1985) An introduction to digital image processing. Strandberg Publishing Company
58. Gagg CR (2014) Cement and concrete as an engineering material: an historic appraisal and case study analysis. Eng Fail Anal 40:114–140
59. Kapoor NR et al (2022) Structural health monitoring of existing building structures for creating green smart cities using deep learning. Recurrent neural networks. CRC Press, pp 203–232
60. Li S, Zhao X (2020) Automatic crack detection and measurement of concrete structure using convolutional encoder-decoder network. IEEE Access 8:134602–134618
61. Jia J, Li Y (2023) Deep learning for structural health monitoring: data, algorithms, applications, challenges, and trends. Sensors 23(21):8824
62. Zhang Y, Li H (2021) Machine learning and structural health monitoring overview with emerging technology and high-dimensional data source highlights. J Intell Mater Syst Struct 32(16):2145–2162
63. Ewald V, Groves RM, Benedictus R (2019) Deep SHM: a deep learning approach for structural health monitoring based on guided Lamb wave technique. In: Proceedings of the SPIE, Denver, CO, USA, 27 March 2019, vol 10970, p 109700H-1-16
64. Chen G, Cui F (2020) Structural damage features extracted by convolutional neural networks from mode shapes. Appl Sci 10(12):4247
65. Seventekidis P, Giagopoulos D, Arailopoulos A, Markogiannaki O (2020) Structural Health Monitoring using deep learning with optimal finite element model generated data. Mech Syst Signal Process 145:106972
66. A review on deep learning-based structural health monitoring of civil infrastructures (2020) ResearchGate. https://www.researchgate.net/publication/342008901_A_review_on_deep_learning-based_structural_health_monitoring_of_civil_infrastructures
67. The application of deep learning in bridge health monitoring: a literature review (2022) Advances in bridge engineering. https://aben.springeropen.com/articles/https://doi.org/10.1186/s43251-022-00078-7
68. Deep learning based structural damage identification for the strain field of a subway bolster (2023) ScienceDirect. https://www.sciencedirect.com/science/article/pii/S1110016823008189

69. Zhang Y, Li H (2022) Deep learning-based damage detection and localization in concrete structures using vibration signals. J Sound Vib 536:116407
70. Kim S, Park J (2022) Smartphone sensing for structural health monitoring. Smart Mater Struct 25(6):065019
71. Chen X, Wang L (2023) IoT and AI for structural health monitoring. IEEE Sens J 23(8):3322–3331

Part II
Methods and Techniques for Damage Identification

Chapter 4
Methods of Assessing Concrete Structures and Identifying Damages Using Modal Data Processing

Hashem Jahangir, Farid Fazel Mojtahedi, Ali Golaghaei Darzi, and Moncef L. Nehdi

Abstract This study explores how to evaluate structures and detect damage using modal data processing methods. It introduces modal testing and the factors derived from it. The practical applications of modal tests for concrete structures are examined, along with the advantages and disadvantages of research on the vibration behavior of reinforced concrete beams. The study concludes that the results of modal testing can be used to identify damage patterns in reinforced concrete beams. Additionally, the main goals and categories of modal data processing methods for damage assessment in structures are reviewed, along with their advantages and disadvantages. Damage detection methods using modal data are divided into two categories based on changes in primary modal factors or modal calculation quantities. The study proposes a new algorithm based on the local energy of modal strains to identify the location and severity of damages in beam structures. The proposed solution can reveal single and multiple fictitious damages in the numerical model. Numerical analysis results show that the proposed method can determine the damage location and severity with a limited number of vibration modes. Applying the proposed algorithm to the modal strains can help identify the damage early and prevent the sudden failure of the structure.

H. Jahangir
Department of Civil Engineering, Research Group of Novel Technologies in Civil Engineering, University of Birjand, Birjand, Iran
e-mail: h.jahangir@birjand.ac.ir

F. F. Mojtahedi
Department of Civil Engineering, Sharif University of Technology, Tehran, Iran
e-mail: farid.fazel71@student.sharif.edu

A. G. Darzi
Department of Civil Engineering, Sharif University of Technology, Tehran, Iran
e-mail: ali.golaghaei@sharif.edu

M. L. Nehdi (✉)
Department of Civil Engineering, McMaster University, Hamilton, On, Canada
e-mail: nehdim@mcmaster.ca

H. Jahangir et al. (eds.), *Damage Detection and Structural Health Monitoring of Concrete and Masonry Structures*, Springer Tracts in Civil Engineering,
https://doi.org/10.1007/978-981-97-8975-7_4

Keywords Damage detection · Modal parameters · Damage location and intensity · Damage index

4.1 Introduction

Scientists and engineers are giving special attention to the health monitoring of structures and detecting damage, especially due to sudden and unpredictable failures that can result in irreversible financial and life losses. There are two types of damage detection methods: destructive and non-destructive. Thanks to recent advancements in sensors, computers, and electronic technologies, non-destructive methods are now easier and less expensive than destructive methods. There are two types of non-destructive methods: local and general. Local methods require access to the known damage area and test site, making it impossible to evaluate the entire structure. Though some local methods are useful for specific applications like detecting cracks in welded joints, general methods are more important. General methods are divided into static and vibration categories. Static methods use changes in locations and measured strains caused by known static loads and updated finite element models to determine changes in elasticity, stiffness, and load-bearing capacity. These methods are mostly used in bridges [1, 2].

Static methods are required to measure large amounts of data and accurately model finite elements with detailed material properties definitions. However, these methods can interfere with the use of the structure and are not ideal for timely damage detection. This highlights the importance of using general detection methods that can be applied to all complex structures without disrupting their operation. Vibration-based general damage detection methods have expanded significantly thanks to technological advancements, such as increased computer memory, sensor construction, and progress in finite element methods. These methods rely on the principle that damage in a structure alters its vibration characteristics and corresponding signals [3].

The characteristics and signals are recorded as time histories when testing for vibrations. However, analyzing this data in the time domain can be challenging to detect any damage. The common approach is to convert the data from the time domain to the frequency domain and then obtain modal domain data from the frequency domain data. Over the past thirty years, extensive research has been conducted in all three domains–time, frequency, and modal–and it continues because there is no single method to identify different damages in different structures. Modal factors, such as natural frequencies, modal damping, and mode shapes, have physical meanings and are easier to comprehend than purely mathematical concepts derived from the time and frequency domains. Consequently, this field has become more prevalent with the advancement of analytical methods and laboratory facilities and plays a greater role in recognizing structural damage. Evaluating the condition of structures can be done quickly and effectively through modal testing, which measures the vibrations of the structures. However, this process requires the processing of the collected modal data. To evaluate the extent of damage, experimental modal data must be processed to calculate damage indices. Numerous methods have been developed by researchers to assess the strength of structures and identify any damage. This research examines the findings of theoretical research and highlights the pros and cons of each damage assessment method [4].

Processing modal data is just as crucial as conducting experimental tests. Raw test data cannot accurately reflect changes in structural stiffness. When the amount of data received is substantial, modal data processing becomes even more essential. Large structures like bridges typically use between 15 and 50 sensors, while over 100 sensors may be required for investigating some of them [5]. Given the size of infrastructure structures and advancements in technology, engineers now have access to a significant amount of modal data. The next crucial step is to process this data to determine the structure's condition and develop a plan for maintenance and restoration. To accomplish this, modal data-based damage assessment techniques have been proposed. These methods are used to locate and evaluate any damages in the structure.

When selecting a method for assessing damage, it is important to consider the structure type, the type of damage, and the sensitivity of the utilized method. Not all damage assessment methods are suitable for every structure, and not all modal factors are equally sensitive to damage [6]. Failing to take these parameters into account can lead to unfavorable results. To address this, many studies have compared and evaluated damage determination methods. This chapter uses modal data to pinpoint the location and severity of damage in structures.

4.2 Modal Data Processing Methods for Structural Damage Assessment

There are two categories of damage assessment methods that use modal data. The first category involves examining changes in modal parameters, such as natural frequencies, mode shapes, and damping, that are directly obtained from experimental tests. However, these parameters may not always accurately determine the location and extent of damages. The second category involves methods that do not use experimental data directly, but instead rely on computational modal parameters such as modal strains, modal strain coefficients, and flexibility matrix. These parameters can help determine the location of damages and track their progress in the structure. In the following sections, we will examine the features, advantages, and disadvantages of each of these methods.

4.2.1 *Modal Data Processing Methods*

Over the past 30 years, extensive research has been conducted on vibration-based damage detection, resulting in significant advancements in this area. Numerous techniques, methods, and algorithms have been created to detect damage in different structures, from primary structural components like beams and plates to more intricate structures like bridges and buildings. Doebling et al. [7] have thoroughly reviewed the methods used for vibration-based damage detection. Several damage assessment methods have been proposed using modal data, which can be classified into the following categories:

- Techniques relying on specific frequencies
- Techniques relying on modes' shape
- Techniques relying on bending or strain mode shape
- Techniques relying on strain energy.

Additional techniques are based on modal factors, such as smoothness matrix and Ritz vectors. Most research on damage detection has focused on simple beam and plate structures. This is because many civil and mechanical engineering designs can be simplified to these basic shapes. Additionally, identifying damage in a beam or plate is an effective way to test the accuracy and efficiency of other identification methods. Below, each of these methods will be described in detail.

4.2.2 *Methods Based on Changes in Primary Modal Parameters*

The differences in a structure's mechanical properties can be observed through changes in its primary modal data, including frequencies, mode shape, and damping. These can be obtained by analyzing the structure's time or frequency domain response. In the time domain, statistical identification methods are used to analyze the time response signals. In the frequency domain, frequencies, damping, and mode shapes can be determined by fitting graphs to frequency response functions (FRF) [8]. FRF diagrams exhibit the fundamental qualities of primary modal parameters. Thus, some researchers utilize the alteration of frequency response functions to detect damages. The methods proposed based on primary modal parameters do not require complicated algorithms. Despite various drawbacks, these methods are still utilized for evaluating structures.

4.2.2.1 Natural Frequencies

It is challenging to locate and determine the severity of cracks using natural frequencies, as these frequencies are significantly affected by heavy damage and are a common feature of structural design. The natural frequencies decrease as the bending stiffness decreases due to loading, but this reduction varies across different vibration modes. Studying changes in the general characteristics of structures is more accessible through measuring specific frequencies compared to other modal factors [9]. The frequency of vibrations can be affected by environmental conditions like temperature and humidity [10]. This can make it challenging to analyze test results, especially when conducting modal testing, which can take a long time. During the experiment, changes in environmental conditions can occur and affect the results. These changes can create uncertainty in test results, and sometimes, they can be comparable to the development of damage [11]. Researchers have reported that temperature changes can cause specific frequencies to change within a day. For instance, after conducting long-term tests on the Alamos Canyon bridge, researchers at the Los Alamos laboratory found that the first three specific frequencies changed by approximately 4.7%, 6.6%, and 5% in 24 h, with the bridge deck temperature changing by 22 °C [12]. Peeters and De Roeck [13] also studied the Z24 bridge in Switzerland continuously for a year and reported that frequencies decrease as temperature increases. Similarly, Xia et al. [14] investigated the problem of reducing the specific frequencies due to an increase in temperature. They concluded that the mode shape is not sensitive to changes in temperature and humidity among the essential modal factors.

Determining the location of cracks in a structure solely through changes in specific frequencies is impossible, as it also depends on the stiffness reduction of the entire structure. Numerous studies have been conducted to identify the size and location of damage through frequency changes. The forward or reverse method is utilized to locate injuries. Alternatively, in an advanced approach, the location of the damage is estimated in a numerical model through trial and error and by comparing experimental and numerical results. The inverse method is employed to obtain the size and location of cracks by observing the frequencies. These methods require the use of a numerical model.

4.2.2.2 Vibrational Mode Shapes

By analyzing the vibrational mode shapes, damages can be located and evaluated without the use of a numerical model. Numerous damage indices have been proposed based on changes in mode shapes. To properly analyze mode shape, noise must be removed from experimental data. Depending on the desired evaluation method, the mode shape should be compared to the mass matrix or maximum range. Damage indicators are then used to determine the location and severity of damages. Shin et al. [15] introduced the Modal Confidence Criterion (MAC) to measure the correlation between two mode shape vectors. This relationship is widely used in research. For two shape vectors ϕ^A and ϕ^B, representing the undamaged and damaged states of a structure, the MAC criterion is defined below:

$$MAC_{A,B} = \frac{\left| \sum_{i=1}^{n} \phi_i^A \phi_i^B \right|^2}{\sum_{i=1}^{n} (\phi_i^A)^2 \sum_{i=1}^{n} (\phi_i^B)^2} \tag{4.1}$$

In Eq. (4.1), n is the number of degrees of freedom. The MAC value specifies the degree of dependence between two mode shapes, which varies between zero and one. Any deviation from one can be a sign of damage to the structure. Allemang [16] reviewed the applications of MAC criteria and other modal reliability criteria. According to [16], the MAC standard has undergone many changes during the twenty years since its first use, and solutions have been proposed to improve this standard. For example, Lieven and Ewins [17] introduced the coordinate-dependent modal reliability criterion (COMAC). The COMAC criterion is an enhanced form of the MAC criterion, and the degree of freedom i with m as the vibration mode is obtained from Eq. (4.2):

$$COMAC_{i,A,B} = \frac{\left| \sum_{j=1}^{m} \phi_{i,j}^A \phi_{i,j}^B \right|^2}{\sum_{j=1}^{m} (\phi_{i,j}^A)^2 \cdot \sum_{j=1}^{m} (\phi_{i,j}^B)^2} \tag{4.2}$$

The COMAC criterion presented in Eq. (4.2) is very similar to the MAC criterion which was presented in Eq. (4.1), with the only difference being that it shows strict changes in the degrees of freedom. The COMAC standard checks how much the stiffness is reduced in each degree of freedom, and its value ranges from zero to one, just like the MAC value. If the modal displacement in the degree of freedom i from two sets of measurements is the same, the COMAC value will be one. However, changes in the shape of the modes due to damage can cause the COMAC value to decrease, and it may even become close to zero.

4.2.2.3 Damping

Using a damping coefficient for damage assessment is still in the experimental stage. According to the research of Salawu and Williams [18], this issue is not only due to the lack of accuracy in determining damping ratios. In fact, obtaining a correct standard based on damping ratios is impossible. Nadambi et al. [19] used damping mechanisms and damping ratio as a sign of structural health monitoring for cracked reinforced concrete beams. They studied the strain energy propagation due to the transition stress between steel and concrete during cracking and the wasted energy caused by the friction of the crack corners. They considered the wasted work caused by friction by increasing the damage and increasing the damping coefficient. Casas and Aparicio [20] showed by testing on cracked concrete beams that there is no clear relationship between crack growth and damping increase. Salawu and Williams [18] also tested on a reinforced concrete bridge and did not obtain an explicit criterion for damping. For this reason, in most research, damping ratios are not used to evaluate damaged structures.

4.2.3 Methods Based on Changes in Computational Modal Parameters

In the previous methods discussed, only the initial modal parameters obtained directly from experimental tests are used. However, other methods evaluate the condition of structures using modal test calculation quantities. These quantities are obtained by processing primary modal data and include modal strain, modal curvatures, and flexibility matrix. Changes in these parameters can indicate changes in the structure's mechanical properties and are referred to as computational modal factors. Research has been conducted to compare primary and computational modal parameters in evaluating structures, and it has been found that computational modal factors better detect damages in structures [21]. This section will focus on discussing computational modal parameters.

4.2.3.1 Modal Strains

When a beam experiences a bending crack, its bending stiffness *EI* decreases, leading to a local decrease in bending. This phenomenon is more noticeable in flexural mode shape compared to modal displacements. To calculate the bending at a specific x loation in an Euler–Bernoulli beam, Eq. (4.3) as described by Alvandi and Ceremona [22] can be used.

$$\kappa(x) = \frac{M(x)}{EI} \tag{4.3}$$

For Eq. (4.3), E represents the elastic modulus, while I is the bare moment of inertia of the beam cross-section. Modal strains are bending strains that depend on each bending vibration mode. As damages increase, modal strains also increase. Therefore, changes in modal strains indicate the progression of damages. The concept of mode shape bending was first introduced by Pandey et al. [23]. They obtained curvatures from the displacement shape of the modes using central difference approximation. Pandey et al. [23] applied this method with the finite element method for a beam with simple supports and discovered that modal strains are much more sensitive to the MAC index. The relationship proposed by Pandey et al. [23] to calculate modal strains is expressed in the form of Eq. (4.4).

$$\kappa = \frac{\phi_{(j+1)r} - 2\phi_{jr} + \phi_{(j-1)r}}{h^2} \tag{4.4}$$

In Eq. (4.4), h represents the distance between the measurement locations, while ϕ_{jr} denoting the change in the j coordinate and the r mode. It is important to note that precise deflection measurements require a compact measuring network, which can be challenging to achieve for higher-frequency vibration modes [23].

4.2.3.2 Modal Strain Energy

The curvatures of each mode shape can be considered modal strains related to each vibration mode. By having modal strains and taking advantage of the resistance relations of materials, modal strain energy is obtained. Changes in modal strain energy during loading steps indicate the presence of damage. The evolution of modal strain efficiency (MSE) has been a common method for detecting damage in various structures. Stubbs and Park [24] proposed an index to determine damage based on reducing the modal strain function between two degrees of freedom, which was later applied to structural damage as below:

$$U = \frac{1}{2}\int_{0}^{L} EI\left(\frac{\partial^2 w}{\partial x^2}\right)dx \tag{4.5}$$

In Eq. (4.5), EI represents the bending stiffness of the beam, L is the length of the beam, and w is the deformation of the beam. Researchers have explored the effectiveness and enhancement of the modal strain energy to determine both the location and size of the damage, but it only works well in cases with very low noise [25–28].

4.2.3.3 Flexibility Matrix

To assess the damage, one method involves calculating the flexibility matrix from experimental data. This matrix can locate damages without the need to calculate modal strains. The flexibility matrix is determined by the displacement of the structure caused by incoming static force and is calculated using co-founded modes compared to the mass and natural frequencies matrix. All mode shapes and special frequencies must be measured to accurately calculate the flexibility matrix. However, it is not feasible to measure all of these in a laboratory setting, which can result in errors in the calculated matrix. The flexibility matrix is particularly sensitive to changes in lower frequencies. Therefore, an appropriate level of accuracy can be achieved by calculating the Softness matrix using a limited number of the first vibration modes of the structure. The flexibility matrix can be calculated by inversing the stiffness matrix of the structure. Many researchers have utilized the flexibility matrix method to detect damage in various structural elements [29–31].

4.3 The RC Beam FEM Models

4.3.1 Experimental-Based Verification of Models

In this chapter, numerical models of reinforced concrete (RC) beams were created using the finite element method (FEM) based on an experimental specimen tested in the study of Baghiee et al. [32]. The selected RC beam consisted of a cross-section with a width of 150 mm, height of 200 mm, the concrete cover of 25 mm, and length of 2200 mm. Two steel bars with a diameter of 12 mm were used at the top and bottom of the cross-section to withstand flexural forces, while rebars with a diameter of 8 mm and a center-to-center distance of 100 mm were used for standing the shear forces. The compressive strength of utilized concrete and the yield strength of steel bars were 20 and 240 MPa, respectively.

Baghiee et al. [32] conducted a modal test on the experimental RC beam in suspended condition to capture its natural frequencies and mode shapes. To achieve this goal, different degrees of freedom (DOFs) with 100 mm were considered alongside the longitudinal direction, resulting in 23 DOFs in the total length of the RC beam. An impact hammer was utilized to excite each DOF and a piezoelectric sensor was installed in the DOF number 7 to capture the vibrational responses. The frequency response functions (FRFs) then were calculated and the experimental natural frequencies and mode shapes of the RC beam were obtained.

In this chapter, as presented in Figs. 4.1 and 4.2, FEM models of the selected experimental RC beam were created using ABAQUS software [33]. In this FEM model, the modulus of elasticity, density, and Poisson ratio for the concrete and steel materials were considered 17 and 200 GPa, 2400 and 7850 kg/m^3, and 0.17 and 0.3, respectively. The resulting natural frequencies of the numerical FEM models and the selected experimental RC specimen showed a high degree of similarity, which are respectively 114.28 and 114.28 Hz for the first mode, 298.42 and 304.89 Hz for the second mode, and 560.24 and 563.19 Hz for the third mode.

A comparison was made between the first three mode shapes of the selected experimental RC beam and its corresponding FEM numerical model, which can be seen in Fig. 4.3. The results indicate a high degree of similarity between the two sets of data.

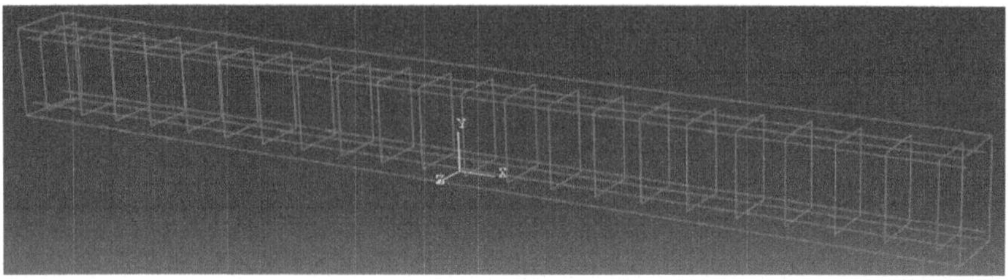

Fig. 4.1 Longitudinal and transverse rebars in the RC beam FEM models

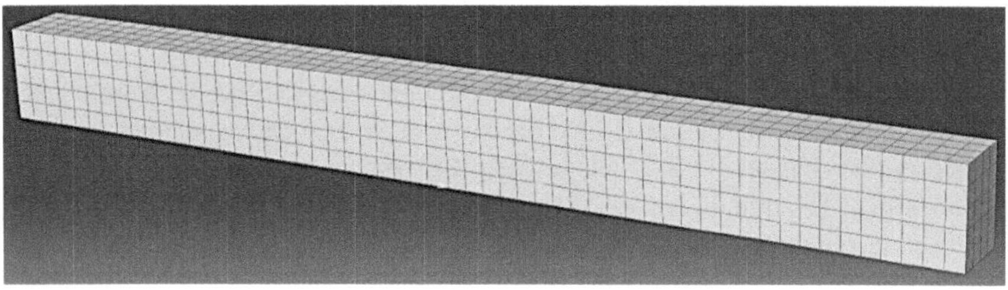

Fig. 4.2 Concrete mesh in the RC beam FEM models

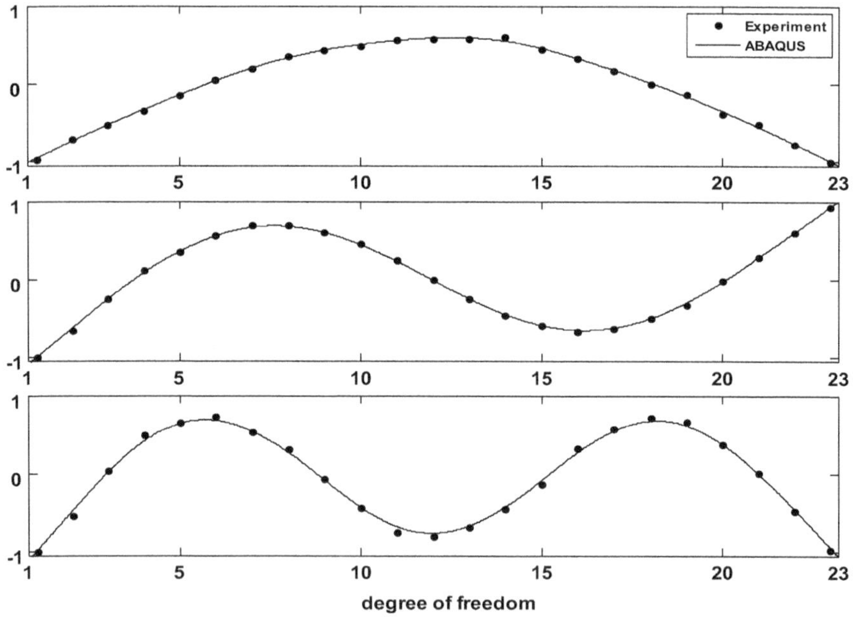

Fig. 4.3 The first three mode shape of experimental RC beam and its corresponding FEM numerical model

4.3.2 Damage Scenarios in FEM Models

In this chapter, to test the capabilities of damage detection methods, single, double, and triple damage scenarios were created in the numerical FEM models of RC beams. These damages were modeled as slots with a fixed width of 5 mm and with different depths of H_C, named h2 to h10, with values of different height values of 20 to 100 mm as reported in Table 4.1. As presented in Table 4.2, The single, double, and triple slots were located at different places alongside the RC beam. Figure 4.4 shows the first three mode shapes of the RC beam models in the undamaged stage (*C-0*) and the ultimate damage stage (h10) of the triple damage scenario (*C-3mLr*).

Figure 4.4 shows that the mode shapes of the undamaged and damaged RC beams in the final stage of damage are almost similar. As a result, no damage localization can be done utilizing these mode shapes. Therefore, the modal curvatures, the second derivative of mode shapes were calculated using the previously presented Eq. (4.4), suggested by Pandey et al. [23]. Figure 4.5, shows the process of calculating the first derivative of mode shape using the central difference method.

To obtain the second derivative of mode shapes using the central difference method, the first derivative must be used twice. However, the results obtained from this method may be irregular. To obtain precise values of mode shapes using this method, a compact measurement network is necessary. This becomes more complex when dealing with vibration modes with higher frequencies, as noted by Pandey et al. [23]. Curve fitting methods are commonly used to improve the accuracy of results obtained from the central difference method. These methods can eliminate irregularities resulting from the central difference method. One effective method for this purpose is using B-Spline functions. These functions create a polynomial and single-factor curve between several conductor points, establishing computational

Table 4.1 Damage stages and corresponding slot heights

Damage stages	No damage	h2	h4	h6	h8	h10
H_C (mm)	0	20	40	60	80	100

Table 4.2 Location of damage scenarios

Damage scenario	Type of damage	Location of slot (s)
C-0	No damage	–
C-1L	Single, left	500
C-1m	Single, middle	1100
C-1r	Single, right	1900
C-2Lr	Double, left and right	500 and 1900
C-2Lm	Double, left and middle	500 and 1100
C-2mr	Double, middle and right	1100 and 1900
C-3mLr	Triple, left, middle and right	500, 1100 and 1900

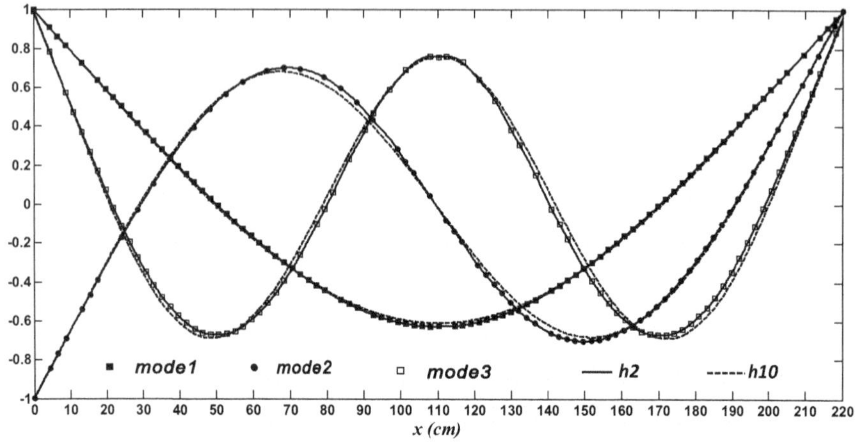

Fig. 4.4 The first three mode shapes of the RC beams in the undamaged state (*C-0*) and in the final damage stage (h10) of triple damage scenario (*C-3mLr*)

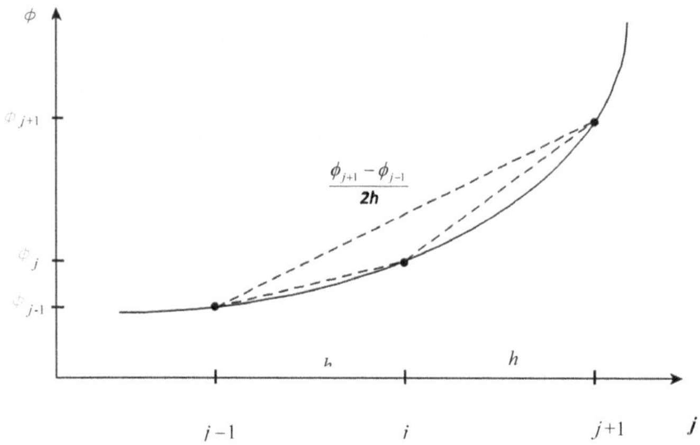

Fig. 4.5 The process of calculating the derivative using the central difference method

continuity. The third-order B-Spline functions can be expressed as the following relation, as reported by Goel and Anand [34]:

$$P_i(t) = N_{0,3}(i)V_i + N_{1,3}(i)V_{i+1} + N_{2,3}(i)V_{i+2} + N_{3,3}(i)V_{i+3} \quad (4.6)$$

Equation (4.6) involves V_i as the conductor points and N as the main b-Spline functions of the third order which were demonstrated in Fig. 4.6.

The third-order basic B-spline functions are defined by a factor that ranges from zero to one. When this factor is closer to one, the fitted curve closely resembles the original curve. This means that the vertices of the fitted curve are near the points

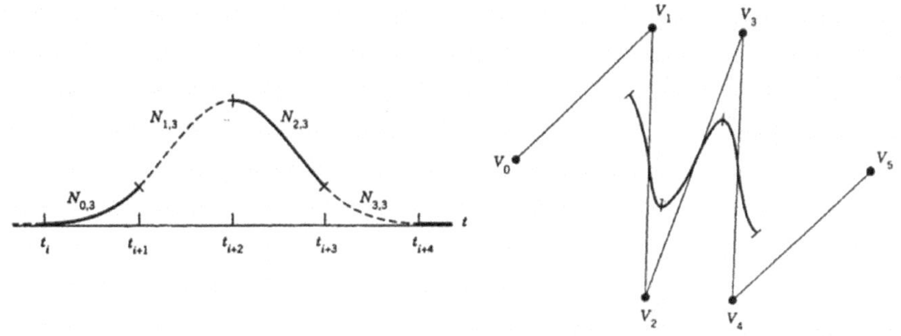

Main functions for third-order B-spline B-spline functions between 6 conductor points!

Fig. 4.6 The process of B-spline approximation method [34]

of the initial vector (V_i), resulting in a similar curve. On the other hand, when the factor approaches zero, the fitted curve deviates from the initial curve, resulting in a less irregular curve. The value of this factor can be determined based on the user's needs and the vector on which the curve fitting is to be performed. Once this factor is determined, the fitted curve can be used as a specified function.

This study utilized MATLAB software [35] to calculate the modal curvatures using the B-Spline curve-fitting method. In order to avoid creating further irregularities in the vectors, the factor for these functions was considered minor and equal to 0.01. As an instance, Fig. 4.7 shows the first three mode shapes of the final stage of the triple damage scenario after curve fitting with these functions. After fitting the curve using B-spline functions and obtaining the fitted mode shapes, the resulting functions are derived twice using the central difference method, and the modal curvatures are obtained, as presented in Fig. 4.8.

4.4 Proposed Damage Detection Method

In this study, modal curvatures were used to evaluate damage scenarios in the numerical model of the RC beams. A damage index was proposed based on the local energies obtained from the modal curvatures in the FEM numerical models. The first idea used in this chapter was based on the second-order norm of vector $f\ (t)$ which was defined as presented in Eq. (4.7).

$$\|f\|_2^2 = \int_R |f(x)|^2 dx \tag{4.7}$$

In computational mathematics, the second order of a vector represents its energy. This means that by utilizing the second-order soft function of a vector, the energy of that vector can be derived through Eq. (4.7). If a vector undergoes a change from its initial

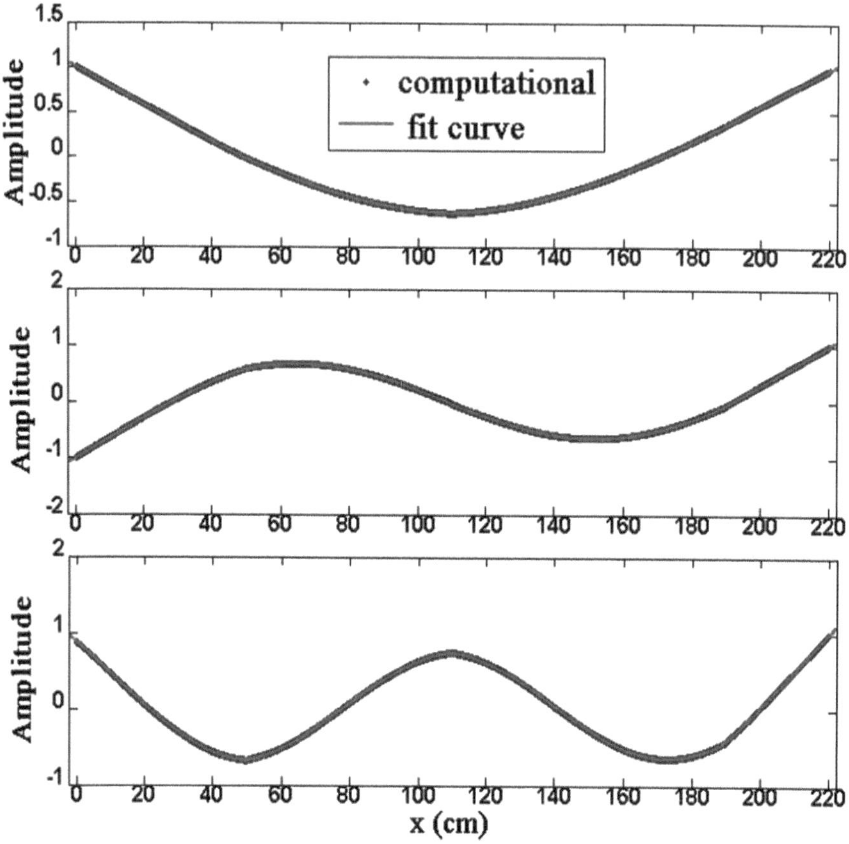

Fig. 4.7 The first three mode shapes and their fitted curve in the final damage stage (h10) in triple damage scenario (*C-3mLr*)

state, the energy obtained from that vector will also change accordingly. Essentially, the energy changes of a vector can be employed to assess any changes in the vector. If a vector is considered discretely, its energy is determined by calculating the sum of the squares of each of its vector values. For an arbitrary vector with components identified as {S(k), $k = 1, 2, \ldots, M$}, its energy is defined as below:

$$U = \sum_{k=1}^{M} |S(k)|^2 \tag{4.8}$$

In Eq. (4.8), the M shows the desired length, and U represents the energy of $S(k)$. An illustration of vector S with its components is shown in Fig. 4.9.

To determine the energy of a vector, its values at different points are squared and then summed. In the case of structural damage, irregularities will appear in the resulting mode shapes and corresponding modal curvatures, causing a change in

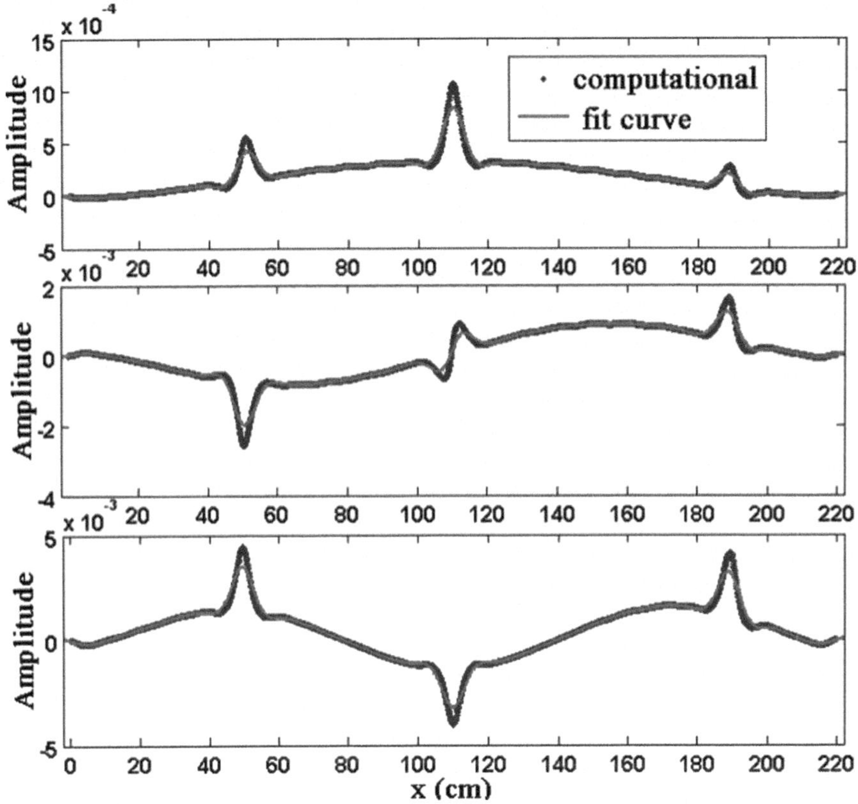

Fig. 4.8 The first three modal curvatures and their fitted curve in the final damage stage (h10) in triple damage scenario (*C-3mLr*)

the energy obtained from them compared to the undamaged status of the structure. This change in energy can be used as an indicator of the presence or absence of damage in the numerical model of the structure. However, the energy of a vector only provides a numerical value and cannot identify the location of the damage. Therefore, a solution is needed to address this limitation. To address the issue, this chapter proposed using local vector energy to take advantage of its benefits while also addressing its inability to identify the location of the damage. In this approach, the energy of a vector is calculated within a specific range, which can be considered as DOF in the numerical model. The location of damage can then be identified within these DOFs. To align with practical damage detection methods, the energy vector is cumulatively calculated across the DOFs. To calculate the energy of a vector in jth DOF using this method, the following equation is used:

$$U_j = \sum_{k=(j-1)+1/2}^{(j+1)-1/2} |S(k+1/2)|^2 \tag{4.9}$$

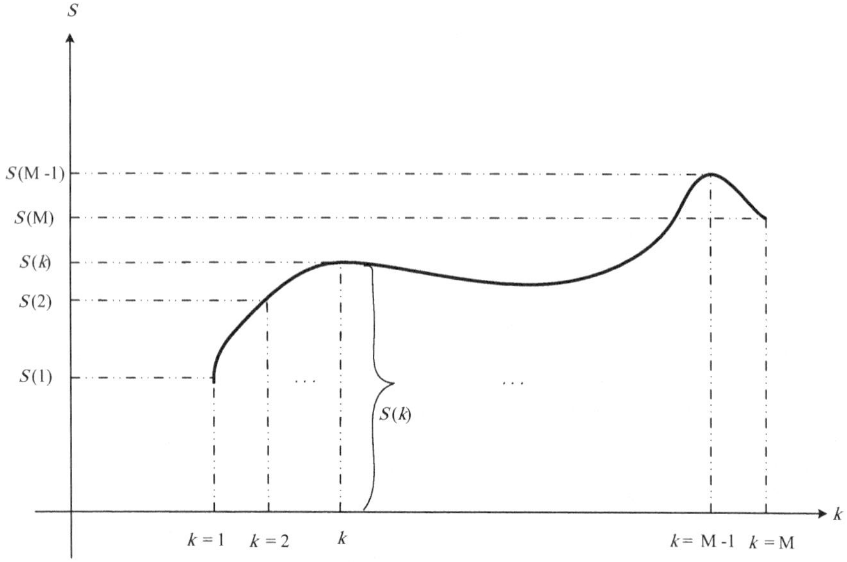

Fig. 4.9 An example of a vector and its components

In Eq. (4.9), the $S(k)$ is the values of the input vector and U_j is the energy vector in jth DOF. Figure 4.10 shows how to calculate local energy at point jth DOF. In Fig. 4.10, it is demonstrated that the local energy U_j can be obtained for each independent j DOF. This can be done using Eq. (4.9) in the middle region between the previous degree of freedom $(j-1)+1/2$ and the posterior $(j+1)-1/2$.

In this chapter, the modal curvature vectors are considered as local energy input vectors U_j, and the damage scenarios are identified with their help. The first three modal curvatures of the RC beam models were evaluated in single, double, and triple damage scenarios. The length of the input vectors, in this case, is equal to the length of the samples and equal to 2200. By considering the measured energies of damaged and undamaged status, the energy index EI_j is defined by Eq. (4.10). The modal curvatures are considered in the form of vectors as index inputs at different points of the sample $(j = 1, 2, \ldots, 2200)$.

$$EI_j = \left| \frac{U_j^d - U_j^0}{\max(U_j^0)} \right| \tag{4.10}$$

In Eq. (4.10), the dissimilarity between the local energies in the intact and impaired states (U_j^0 and U_j^d) is normalized to the maximum value in the intact state ($\max(U_j^0)$). This method enables a consistent framework for comparing damage index results across various scenarios. Additionally, dividing the local energy dissimilarities by the maximum local energy in the intact state, enhances the suitability of the damage index values, as the dissimilarities may otherwise be negligible.

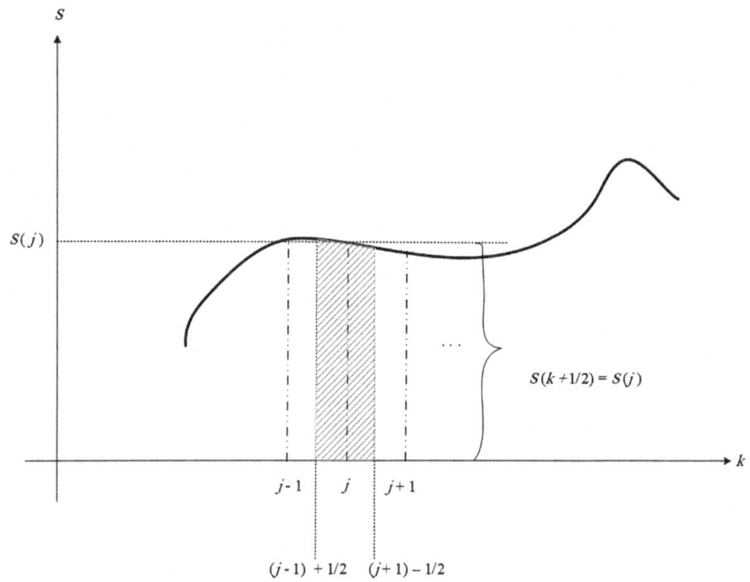

Fig. 4.10 The process of calculating local energy at *j*th DOF

Considering the first three modal curvatures as the inputs of energy index EI_j, the locations and severities of single, double, and triple damage scenarios at different damage stages (h2 to h10), were obtained. The results of applying energy index EI_j on the first, second, and third modal curvatures were illustrated in Figs. 4.11, 4.12, 4.13, 4.14, 4.15, 4.16 and 4.17, Figs. 4.18, 4.19, 4.20, 4.21, 4.22, 4.23 and 4.24, and Figs. 4.25, 4.26, 4.27, 4.28, 4.29, 4.30 and 4.31, respectively.

The results of Figs. 4.11, 4.12, 4.13, 4.14, 4.15, 4.16 and 4.17 showed that the proposed damage index EI_j is capable of identifying the location of damage and displaying the severity of damage in different damage scenarios. Applied to the first modal curvature, it is evident that the proposed damage index is more sensitive to middle damages rather than damages on the two sides. Among the side damages, the damage on the left exhibits a higher index value and is more easily identified than the damage on the right.

From the illustrated results in Figs. 4.18, 4.19, 4.20, 4.21, 4.22, 4.23 and 4.24, it can be concluded that the proposed damage index EI_j, using the second modal curvature, has a higher sensitivity to damage on the left side than other damages. Additionally, the damage index does not exhibit appropriate sensitivity to the middle damage.

As can be seen in Figs. 4.25, 4.26, 4.27, 4.28, 4.29, 4.30 and 4.31, the locations and severities of single, double, and triple damage scenarios were identified correctly by applying the proposed damage index EI_j on the third modal curvatures.

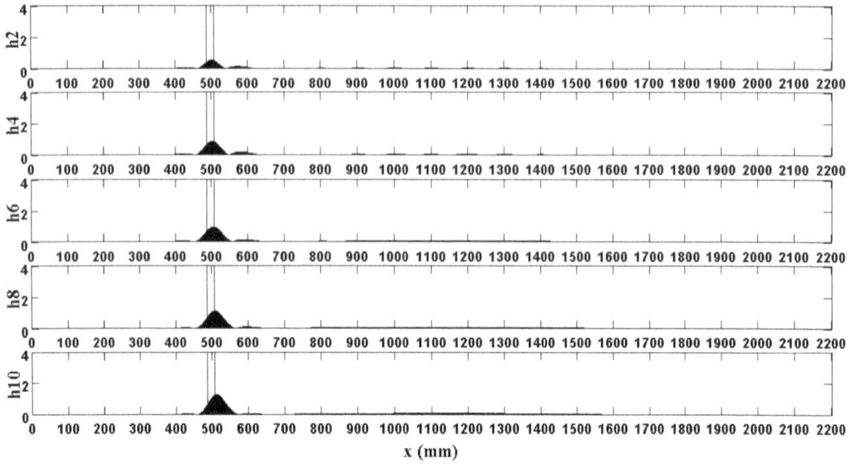

Fig. 4.11 The results of applying damage index EI_j on first modal curvatures in *C-1L* damage scenario

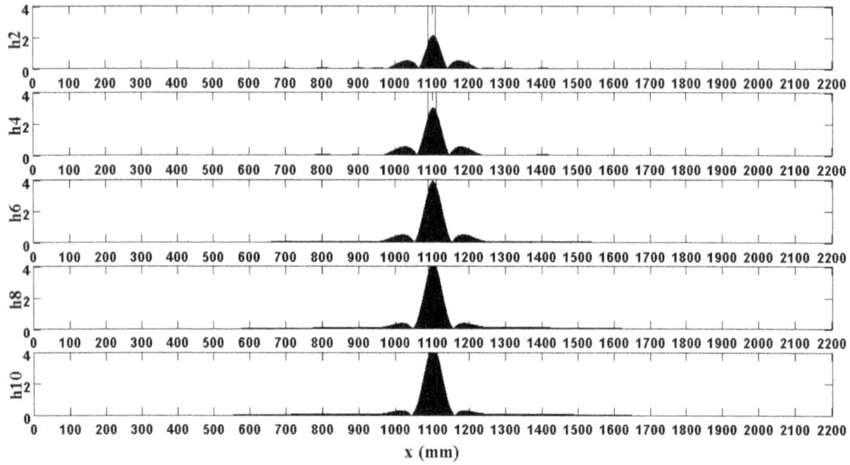

Fig. 4.12 The results of applying damage index EI_j on first modal curvatures in *C-1m* damage scenario

Comparing the energy damage index (EI_j) results in Figs. 4.11, 4.12, 4.13, 4.14, 4.15, 4.16, 4.17, 4.18, 4.19, 4.20, 4.21, 4.22, 4.23, 4.24, 4.25, 4.26, 4.27, 4.28, 4.29, 4.30 and 4.31, it was observed that each modal curvature is more sensitive to certain damage scenarios and locations than others. Therefore, it is not possible to pinpoint the location and severity of damage based on a single modal curvature. To properly assess damage locations and severities, multiple modal curvatures must be used simultaneously. In this chapter, a modified version of the energy damage index,

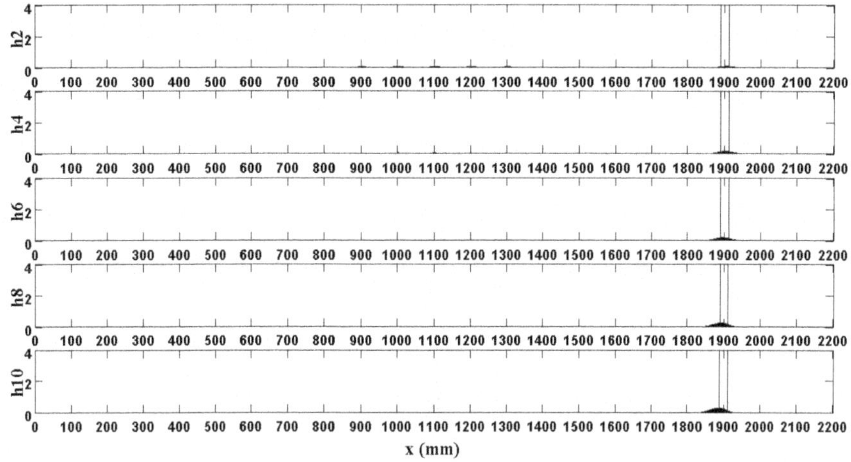

Fig. 4.13 The results of applying damage index EI_j on first modal curvatures in *C-1r* damage scenario

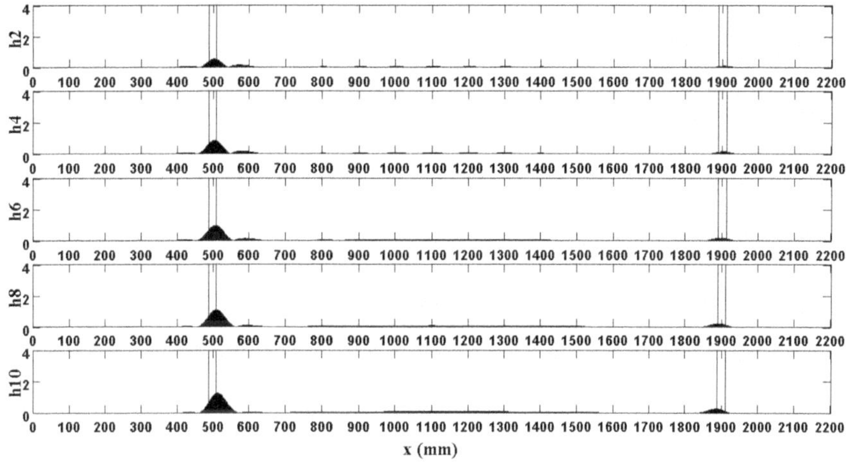

Fig. 4.14 The results of applying damage index EI_j on first modal curvatures in *C-2Lr* damage scenario

named MEI_j is proposed to identify damage scenarios. This index takes into account the combined effect of modal curvatures and sums up the EI_j values from each modal curvature as below:

$$MEI_j = \sum_{k=1}^{N} EI_j = \sum_{k=1}^{N} \left| \frac{U_j^d - U_j^0}{\max(U_j^0)} \right| \tag{4.11}$$

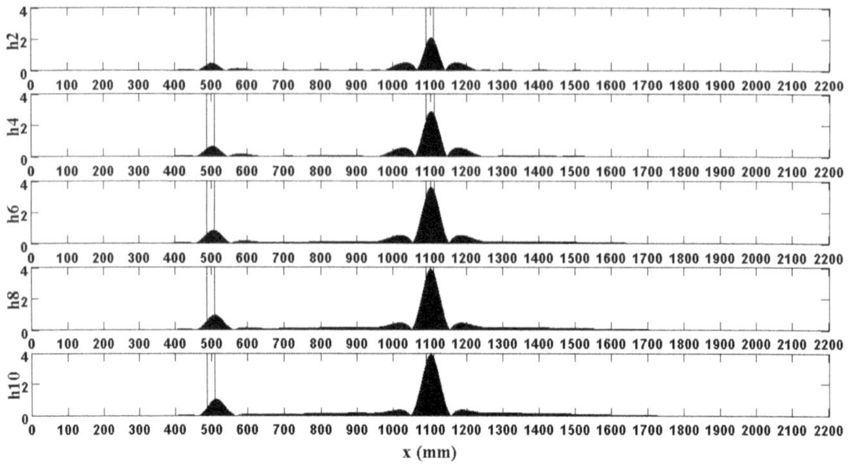

Fig. 4.15 The results of applying damage index EI_j on first modal curvatures in *C-2Lm* damage scenario

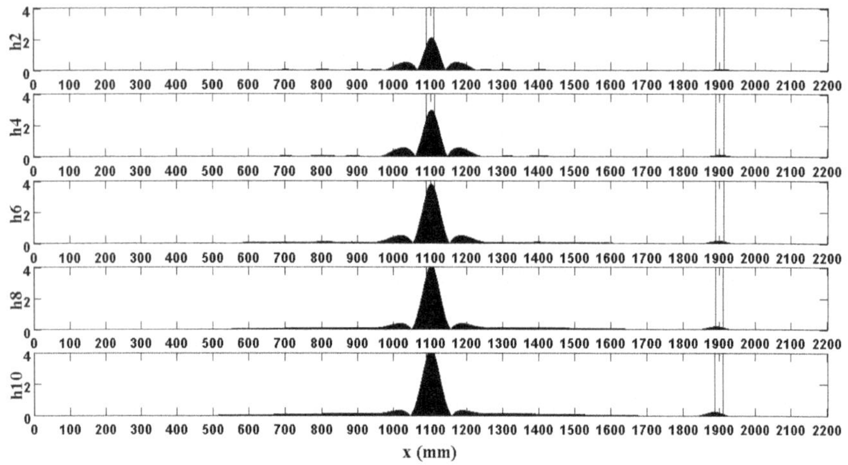

Fig. 4.16 The results of applying damage index EI_j on first modal curvatures in *C-2mr* damage scenario

In Eq. (4.11), the N is the number of modal curvatures, and therefore, in this chapter, the counter k will be equal to 1, 2, and 3. The modified energy damage index MEI_j was calculated for single, double, and triple damage scenarios and the obtained results are presented in Figs. 4.32, 4.33, 4.34, 4.35, 4.36, 4.37 and 4.38, respectively.

The presented results in Figs. 4.32, 4.33, 4.34, 4.35, 4.36, 4.37 and 4.38 showed that the modified energy damage index MEI_j values on the left and middle slots are almost the same, and are higher than the corresponding values on the right-side slots.

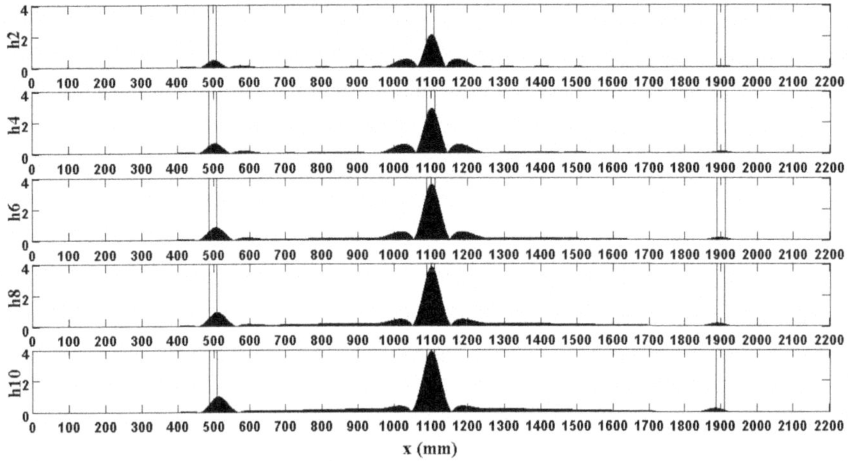

Fig. 4.17 The results of applying damage index EI_j on first modal curvatures in *C-3mLr* damage scenario

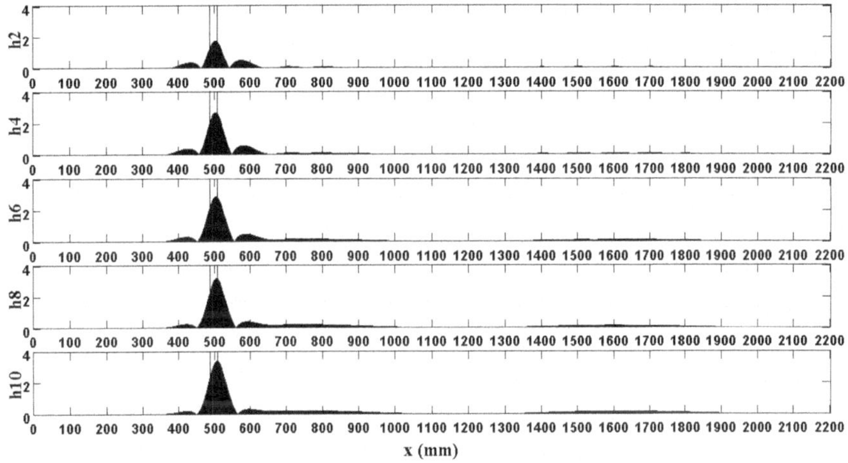

Fig. 4.18 The results of applying damage index EI_j on second modal curvatures in *C-1L* damage scenario

This indicates that the proposed modified energy damage index is less sensitive to the damage on the right side. The reason for this is due to the damage's proximity to the free end of the sample, which poses difficulties for damage identification methods when using modal data. In this chapter, the damage index value obtained for the damage near the end of the beam is lower than other points. However, the index has accurately estimated the location and increase of this damage.

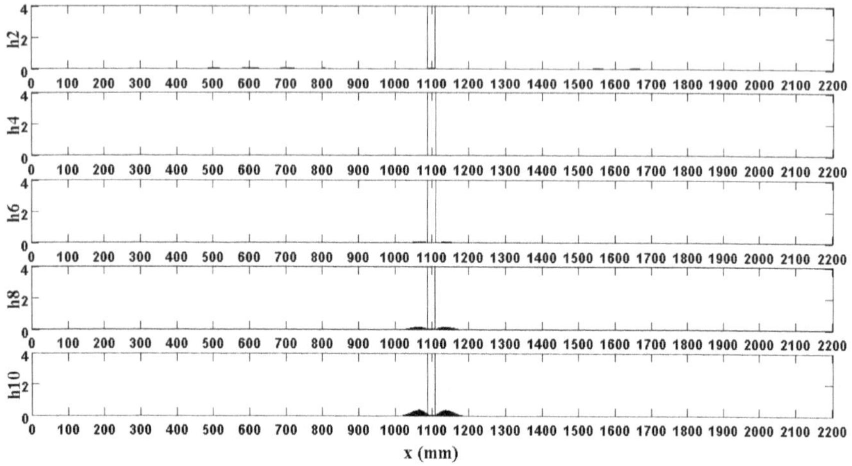

Fig. 4.19 The results of applying damage index EI_j on second modal curvatures in *C-1m* damage scenario

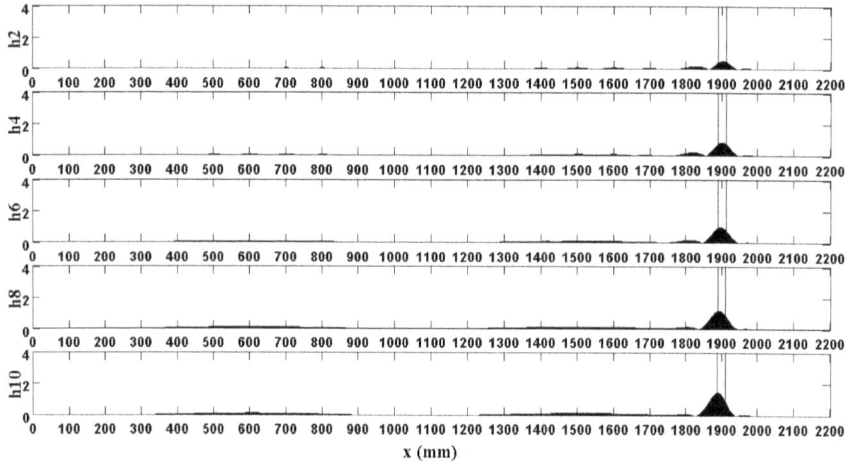

Fig. 4.20 The results of applying damage index EI_j on second modal curvatures in *C-1r* damage scenario

Overall, the proposed modified energy damage index has demonstrated the ability to identify the location and severity of damages simultaneously. This is in contrast to previous research, where most damage indices only identify the location or severity of damages separately. The proposed damage index can also estimate the severity of the damage and its increasing trend, making it a valuable tool for quickly analyzing the condition of structures.

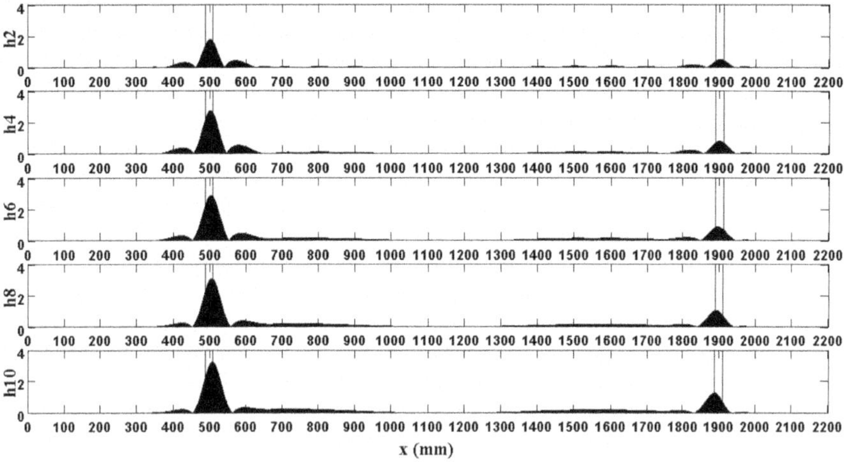

Fig. 4.21 The results of applying damage index EI_j on second modal curvatures in *C-2Lr* damage scenario

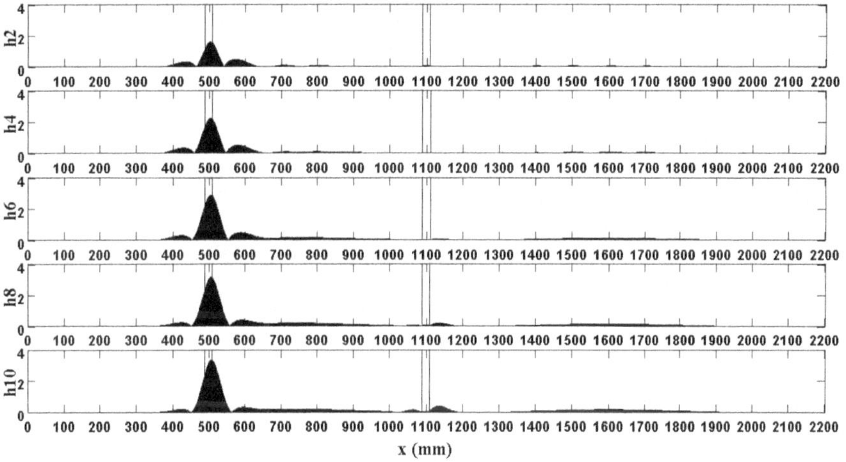

Fig. 4.22 The results of applying damage index EI_j on second modal curvatures in *C-2Lm* damage scenario

4.5 Conclusions

This chapter examines how to determine the location and severity of damage in FEM models of RC beams using modal data. Different single, double, and triple damage scenarios including slots at the left, middle, and right sides of the RC beams with various severities (heights) were created. Firstly, the finite difference method and B-spline curve fitting were used to obtain the modal curvatures of numerical models.

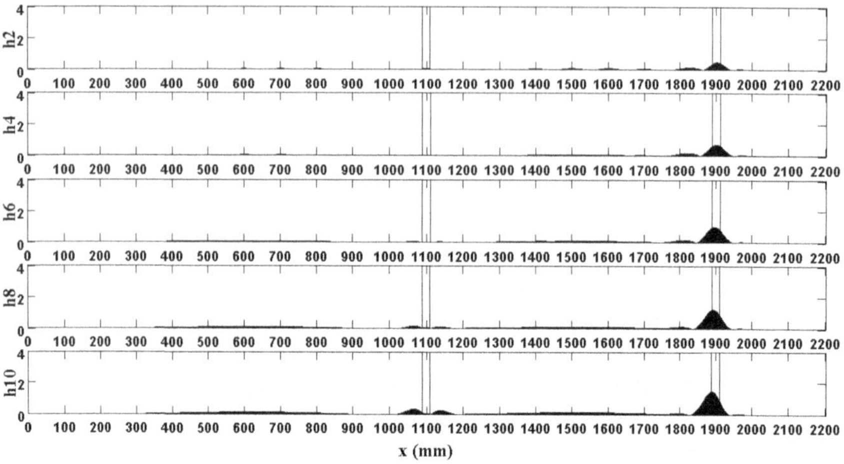

Fig. 4.23 The results of applying damage index EI_j on second modal curvatures in *C-2mr* damage scenario

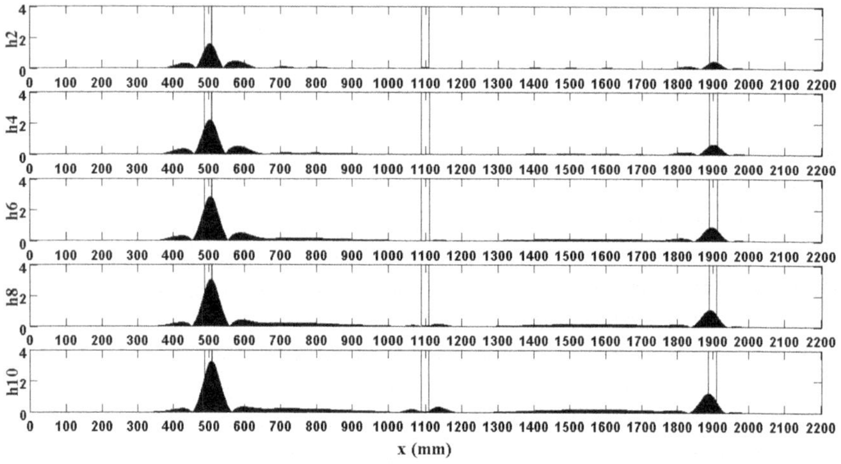

Fig. 4.24 The results of applying damage index EI_j on second modal curvatures in *C-3mLr* damage scenario

These modal curvatures were then used as input vectors for a damage index based on their local energy (EI_j). The results from this damage index showed that it can identify the location and severity of damages simultaneously. However, it was found to be more sensitive to certain damage scenarios for different modal strains. As a result, the index was modified by the superposition of modal curvatures (MEI_j). The following outcomes were driven in this chapter:

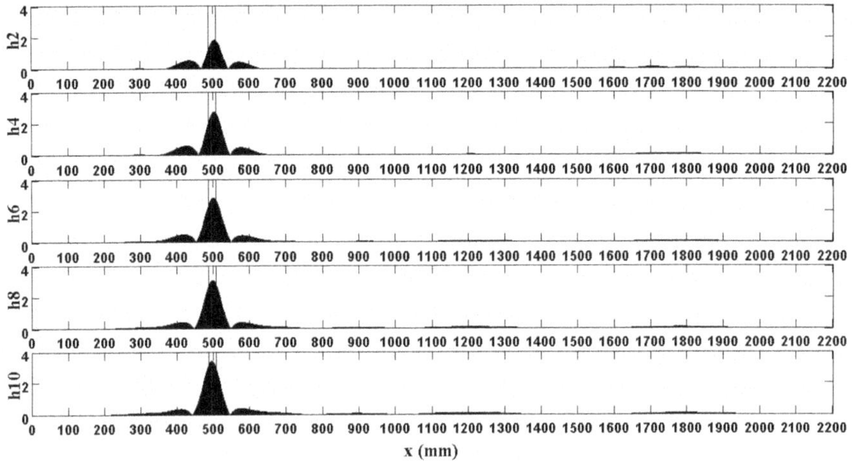

Fig. 4.25 The results of applying damage index EI_j on third modal curvatures in *C-1L* damage scenario

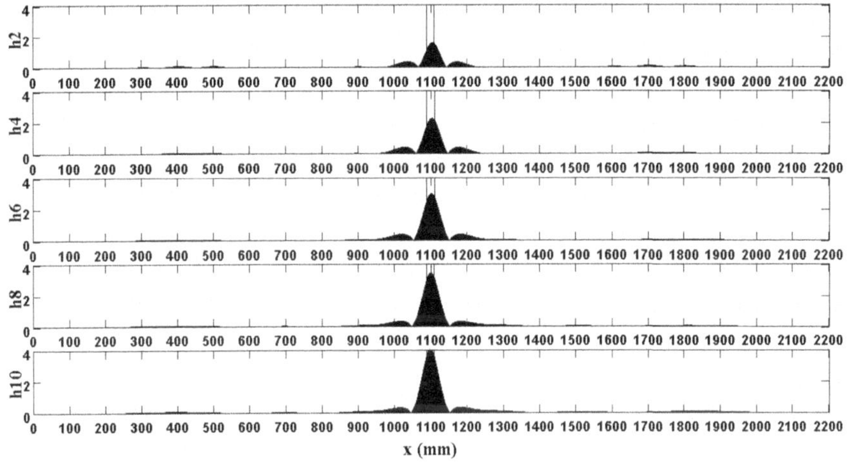

Fig. 4.26 The results of applying damage index EI_j on third modal curvatures in *C-1m* damage scenario

- The proposed damage index EI_j is capable of identifying the location of damage and displaying the severity of damage in different damage scenarios. Applied to the first modal curvature, it is evident that the proposed damage index is more sensitive to middle damages rather than damages on the two sides. Among the side damages, the damage on the left exhibits a higher index value and is more easily identified than the damage on the right.

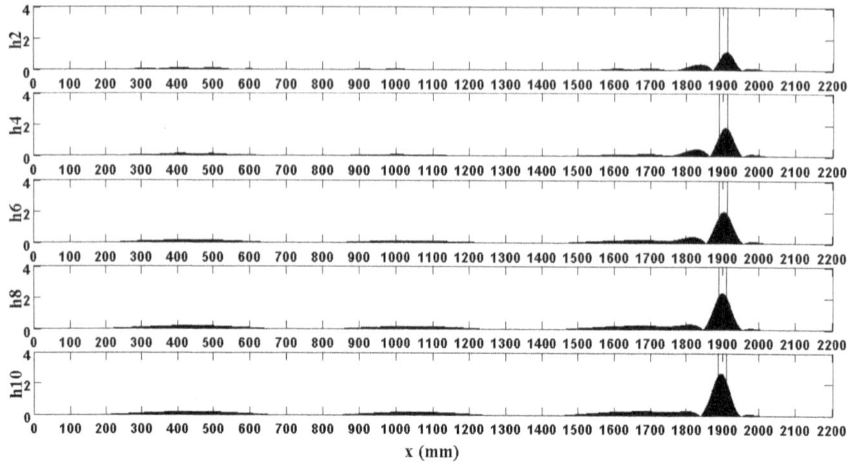

Fig. 4.27 The results of applying damage index EI_j on third modal curvatures in *C-1r* damage scenario

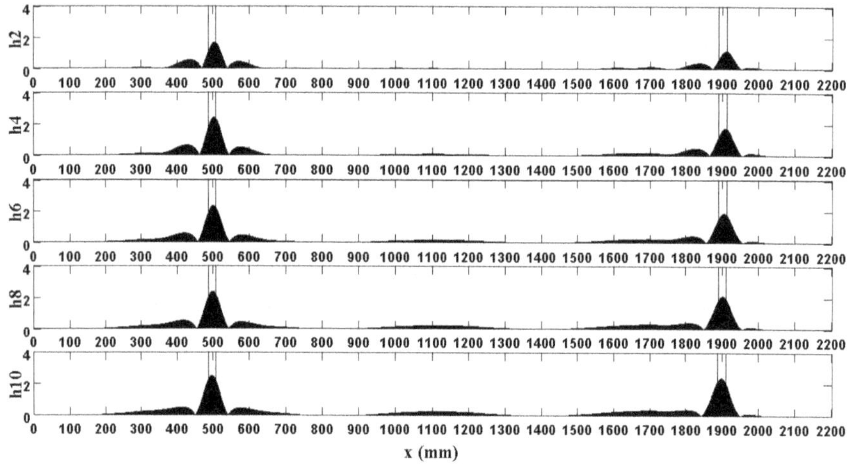

Fig. 4.28 The results of applying damage index EI_j on third modal curvatures in *C-2Lr* damage scenario

- It can be concluded that the proposed damage index EI_j, using the second modal curvature, has a higher sensitivity to damage on the left side than other damages. Additionally, the damage index does not exhibit appropriate sensitivity to the middle damage.
- The locations and severities of single, double, and triple damage scenarios were identified correctly by applying the proposed damage index EI_j on the third modal curvatures.

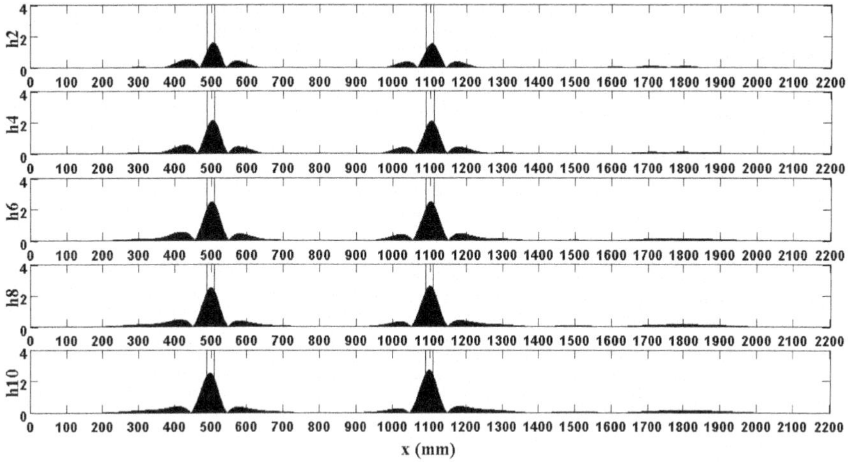

Fig. 4.29 The results of applying damage index EI_j on third modal curvatures in *C-2Lm* damage scenario

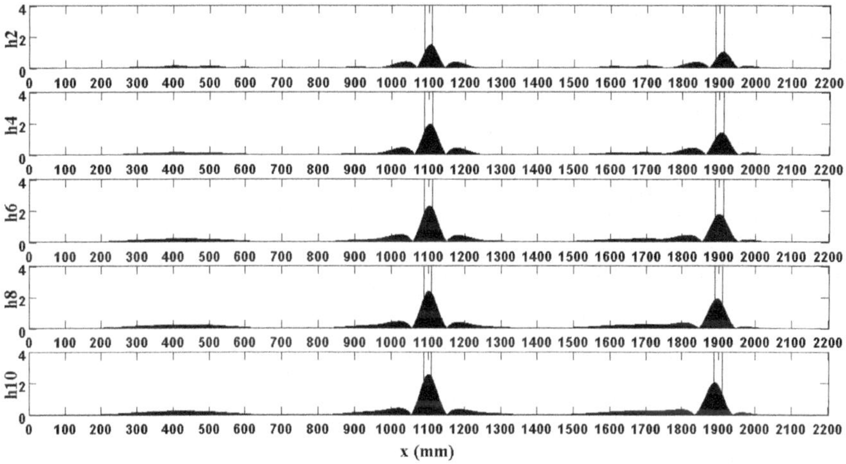

Fig. 4.30 The results of applying damage index EI_j on third modal curvatures in *C-2mr* damage scenario

- It was observed that each modal curvature is more sensitive to certain damage scenarios and locations than others. Therefore, it is not possible to pinpoint the location and severity of damage based on a single modal curvature. To properly assess damage locations and severities, multiple modal curvatures must be used simultaneously.

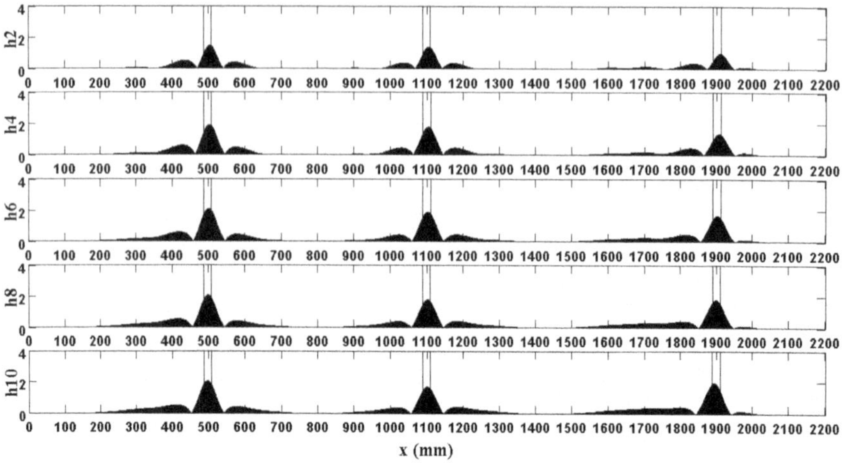

Fig. 4.31 The results of applying damage index EI_j on third modal curvatures in *C-3mLr* damage scenario

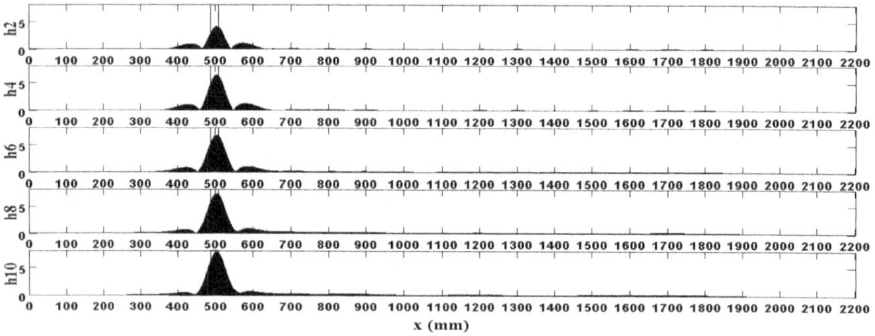

Fig. 4.32 The results of applying damage index MEI_j on first three modal curvatures in *C-1L* damage scenario

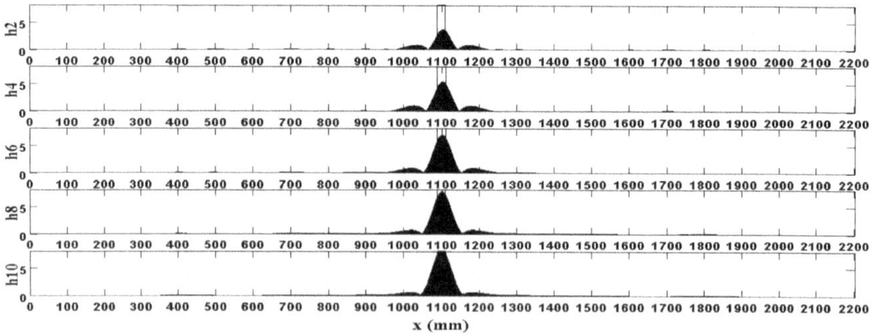

Fig. 4.33 The results of applying damage index MEI_j on first three modal curvatures in *C-1m* damage scenario

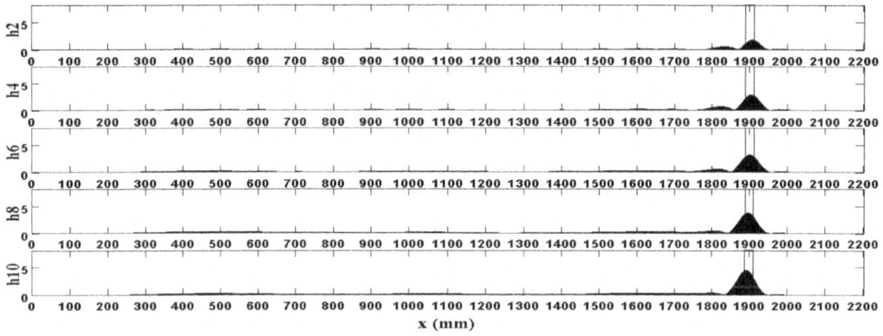

Fig. 4.34 The results of applying damage index MEI_j on first three modal curvatures in *C-1r* damage scenario

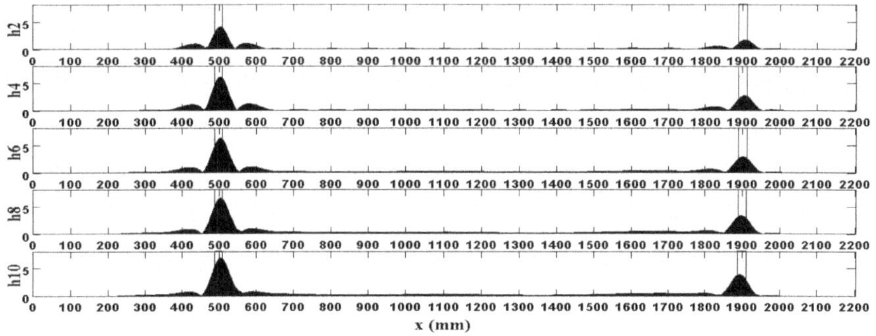

Fig. 4.35 The results of applying damage index MEI_j on first three modal curvatures in *C-2Lr* damage scenario

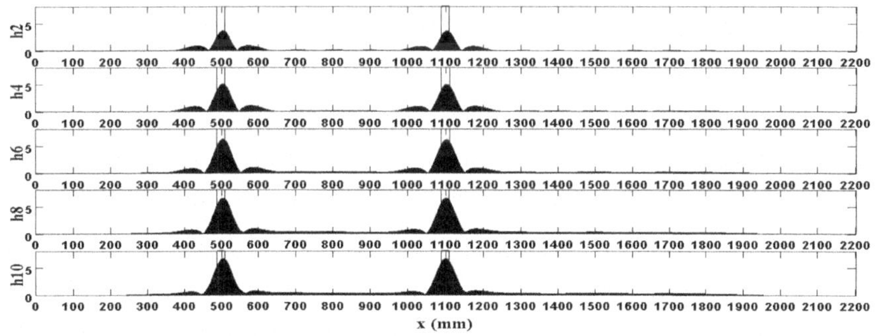

Fig. 4.36 The results of applying damage index MEI_j on first three modal curvatures in *C-2Lm* damage scenario

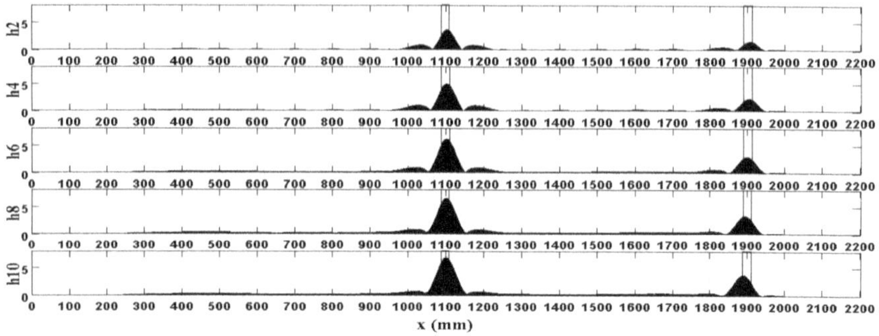

Fig. 4.37 The results of applying damage index MEI_j on first three modal curvatures in *C-2mr* damage scenario

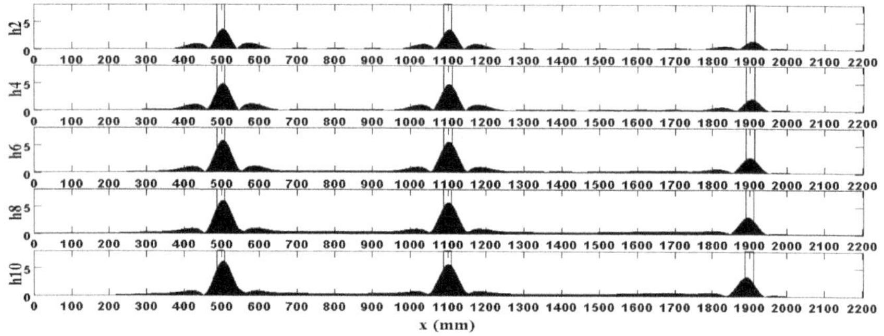

Fig. 4.38 The results of applying damage index MEI_j on first three modal curvatures in *C-3mLr* damage scenario

- The modified energy damage index MEI_j values on the left and middle slots are almost the same and are higher than the corresponding values on the right-side slots. This indicates that the proposed modified energy damage index is less sensitive to the damage on the right side. The reason for this is due to the damage's proximity to the free end of the sample, which poses difficulties for damage identification methods when using modal data.
- The proposed modified energy damage index has demonstrated the ability to identify the location and severity of damages simultaneously. This is in contrast to previous research, where most damage indices only identify the location or severity of damages separately.

References

1. Sohn H, Czarnecki JA, Farrar CR (2000) Structural health monitoring using statistical process control. J Struct Eng 126:1356–1363. https://doi.org/10.1061/(ASCE)0733-9445(2000)126:11(1356)
2. Worden K, Farrar CR, Haywood J, Todd M (2008) A review of nonlinear dynamics applications to structural health monitoring. Struct Control Health Monit 15:540–567. https://doi.org/10.1002/stc.215
3. Peng X, Yang Q (2023) Damage detection in beam-like structures using static shear energy redistribution. Front Struct Civ Eng. https://doi.org/10.1007/s11709-022-0903-4
4. Stepinac M, Gašparović M (2020) A review of emerging technologies for an assessment of safety and seismic vulnerability and damage detection of existing masonry structures. Appl Sci 10. https://doi.org/10.3390/app10155060
5. Entezami A, Sarmadi H, Behkamal B, Mariani S (2021) Health monitoring of large-scale civil structures: an approach based on data partitioning and classical multidimensional scaling. Sensors 21:1646. https://doi.org/10.3390/s21051646
6. Fan W, Qiao P (2011) Vibration-based damage identification methods: a review and comparative study. Struct Health Monit 10:83–111. https://doi.org/10.1177/1475921710365419
7. Doebling SW, Farrar CR, Prime MB (1998) A summary review of vibration-based damage identification methods. Shock Vibrat Dig 30:91–105. https://doi.org/10.1177/058310249803000201
8. Russo S (2013) On the monitoring of historic Anime Sante church damaged by earthquake in L'Aquila. Struct Control Health Monit 20:1226–39. https://doi.org/10.1002/stc.1531
9. Ubertini F, Comanducci G, Cavalagli N (2016) Vibration-based structural health monitoring of a historic bell-tower using output-only measurements and multivariate statistical analysis. Struct Health Monit 15:438–457. https://doi.org/10.1177/1475921716643948
10. Jahangiri M, Hadianfard MA (2019) Vibration-based structural health monitoring using symbiotic organism search based on an improved objective function. J Civ Struct Heal Monit 9:741–755. https://doi.org/10.1007/s13349-019-00364-5
11. Kessler SS, Spearing SM, Atalla MJ, Cesnik CES, Soutis C (2002) Damage detection in composite materials using frequency response methods. Compos B Eng 33:87–95. https://doi.org/10.1016/S1359-8368(01)00050-6
12. Farrar CR, James GH III (1997) System identification from ambient vibration measurements on a bridge. J Sound Vib 205:1–18. https://doi.org/10.1006/jsvi.1997.0977
13. Maeck J, Peeters B, De Roeck G (2001) Damage identification on the Z24 bridge using vibration monitoring. Smart Mater Struct 10:512–517. https://doi.org/10.1088/0964-1726/10/3/313
14. Xia Y, Hao H, Zanardo G, Deeks A (2006) Long term vibration monitoring of an RC slab: temperature and humidity effect. Eng Struct 28:441–452. https://doi.org/10.1016/j.engstruct.2005.09.001
15. Shih CY, Tsuei YG, Allemang RJ, Brown DL (1988) A frequency domain global parameter estimation method for multiple reference frequency response measurements. Mech Syst Signal Process 2:349–365. https://doi.org/10.1016/0888-3270(88)90059-3
16. Allemang RJ (2003) The modal assurance criterion–twenty years of use and abuse. Sound Vibrat 37:14–23
17. Lieven NAJ, Ewins DJ (1988) Spatial correlation of mode shapes, the coordinate modal assurance criterion (COMAC). In: 6th International modal analysis conference, Kissimmee Florida, USA, pp 690–5
18. Salawu OS, Williams C (1995) Review of full-scale dynamic testing of bridge structures. Eng Struct 17:113–121. https://doi.org/10.1016/0141-0296(95)92642-L
19. Ndambi JM, Peeters B, Maeck J, De Visscher J, Wahab MA, Vantomme J et al (2000) Comparison of techniques for modal analysis of concrete structures. Eng Struct 22:1159–1166. https://doi.org/10.1016/S0141-0296(99)00054-1
20. Casas JR, Aparicio AC (1994) Structural damage identification from dynamic-test data. J Struct Eng 120:2437–2450. https://doi.org/10.1061/(ASCE)0733-9445(1994)120:8(2437)

21. Huth O, Feltrin G, Maeck J, Kilic N, Motavalli M (2005) Damage identification using modal data: experiences on a prestressed concrete bridge. J Struct Eng 131:1898–1910. https://doi.org/10.1061/(asce)0733-9445(2005)131:12(1898)
22. Alvandi A, Cremona C (2006) Assessment of vibration-based damage identification techniques. J Sound Vib 292:179–202. https://doi.org/10.1016/j.jsv.2005.07.036
23. Pandey AK, Biswas M, Samman MM (1991) Damage detection from changes in curvature mode shapes. J Sound Vib 145:321–332. https://doi.org/10.1016/0022-460X(91)90595-B
24. Stubbs N, Park S (1996) Optimal sensor placement for mode shapes via Shannon's sampling theorem. Comput Aided Civil Infrastruct Eng 11:411–419. https://doi.org/10.1111/j.1467-8667.1996.tb00353.x
25. Entezami A, Shariatmadar H, Sarmadi H (2017) Structural damage detection by a new iterative regularization method and an improved sensitivity function. J Sound Vib 399:285–307. https://doi.org/10.1016/j.jsv.2017.02.038
26. Ooijevaar TH, Loendersloot R, Warnet LL, de Boer A, Akkerman R (2010) Vibration based structural health monitoring of a composite T-beam. Compos Struct 92:2007–2015. https://doi.org/10.1016/j.compstruct.2009.12.007
27. Hu H, Wang B-T, Lee C-H, Su J-S (2006) Damage detection of surface cracks in composite laminates using modal analysis and strain energy method. Compos Struct 74:399–405. https://doi.org/10.1016/j.compstruct.2005.04.020
28. Loendersloot R, Ooijevaar TH, Warnet L, de Boer A, Akkerman R (2011) Vibration based structural health monitoring and the modal strain energy damage index algorithm applied to a composite T-beam. Vibration and structural acoustics analysis. Dordrecht: Springer Netherlands, pp 121–50. https://doi.org/10.1007/978-94-007-1703-9_6
29. Zonta D, Lanaro A, Zanon P (2003) A strain-flexibility-based approach to damage location. Key Eng Mater 245–246:87–96. https://doi.org/10.4028/www.scientific.net/KEM.245-246.87
30. Choi FC, Li J, Samali B, Crews K (2008) Application of the modified damage index method to timber beams. Eng Struct 30:1124–1145. https://doi.org/10.1016/j.engstruct.2007.07.014
31. Hanumanthappa S (2022) A new structural damage detection method for cantilever beam using generalized flexibility quotient difference method. J Vibrat Eng Technol. https://doi.org/10.1007/s42417-022-00655-0
32. Baghiee N, Reza Esfahani M, Moslem K (2009) Studies on damage and FRP strengthening of reinforced concrete beams by vibration monitoring. Eng Struct 31:875–893. https://doi.org/10.1016/j.engstruct.2008.12.009
33. Smith M (2009) ABAQUS/standard user's manual, Version 6.9. Simulia
34. Goel A, Anand S (2019) Design of functionally graded lattice structures using b-splines for additive manufacturing. Proced Manuf 34:655–665. https://doi.org/10.1016/j.promfg.2019.06.193
35. The MathWorks Inc (2022) MATLAB version: 9.13.0 (R2022b)

Chapter 5
Cutting-Edge Network Based Concrete Crack Detection and Analysis for Structural Health Monitoring

S. Gandhimathi Alias Usha

Abstract Structural Health Monitoring (SHM) plays a crucial role in ensuring the safety and longevity of infrastructure, with concrete structures being integral components of our built environment. This chapter introduces DeepCrack, an innovative approach utilizing deep learning techniques for the detection and analysis of concrete cracks in the context of structural health monitoring. DeepCrack influences state-of-the-art Convolutional Neural Networks (CNNs) to automatically identify and characterize cracks in concrete surfaces. The proposed system exhibits high accuracy and efficiency in detecting both visible and subtle cracks, providing a comprehensive solution for structural integrity assessment. By employing a large dataset of annotated concrete images, DeepCrack learns intricate patterns and features associated with different types and stages of cracks. The detection process is complemented by a detailed analysis module that evaluates cross entropy, training and validation accuracy. This analysis not only aids in quantifying the severity of cracks but also contributes valuable insights into potential structural issues. DeepCrack's adaptability enables it to handle various environmental conditions, lighting scenarios, and surface textures commonly encountered in real-world applications. Furthermore, the integration of DeepCrack into existing structural health monitoring systems enhances the overall efficiency of maintenance strategies. Real-time monitoring and continuous analysis of concrete surfaces enable timely identification of structural vulnerabilities, facilitating proactive interventions and preventing potential disasters. The results of extensive experiments demonstrate the superior performance of DeepCrack compared to traditional methods, showcasing its potential as a reliable and accurate tool for concrete crack detection and analysis. The scalability and versatility of DeepCrack make it suitable for a wide range of applications, from routine inspections of buildings and bridges to large-scale infrastructure projects.

Keywords Structural strength · Image processing · Machine learning · Segmentation · Feature extraction · Strength assessment

---

S. G. A. Usha (✉)
Velammal College of Engineering and Technology, Viraganoor, Madurai, TamilNadu, India
e-mail: ushadears@gmail.com

H. Jahangir et al. (eds.), *Damage Detection and Structural Health Monitoring of Concrete and Masonry Structures*, Springer Tracts in Civil Engineering,
https://doi.org/10.1007/978-981-97-8975-7_5

5.1 Introduction

The assessment of building strength is a critical aspect of ensuring safety, stability, and resilience of built structures. Traditional methods of building strength assessment have relied heavily on manual inspection by trained engineers, which can be time-consuming, subjective, and prone to human errors. As buildings age and face various environmental factors, the need for more efficient and accurate assessment techniques has become increasingly evident [1]. The motivation behind exploring image processing methods for building strength assessment stems from the limitations of conventional approaches. Image processing offers the potential to automate and enhance the evaluation process, allowing for a more objective, data-driven analysis of building components and structural conditions. By leveraging high-resolution imaging technologies, such as drones, LiDAR, or infrared sensors, detailed visual information about building structures can be captured, enabling a more comprehensive understanding of their integrity [2].

The advent of machine learning and artificial intelligence further fuels the motivation to employ image processing methods for building strength assessment. Machine learning algorithms can be trained to recognize patterns, detect anomalies, and make predictions based on large datasets of labeled building images. This opens up possibilities for predictive maintenance strategies, where building weaknesses and potential failures can be anticipated, allowing for timely repairs and maintenance actions. Additionally, image processing methods offer non-destructive evaluation capabilities, eliminating the need for invasive testing that might compromise the structural integrity of the building [3]. This is particularly valuable for historic and heritage structures where preservation is paramount.

One of the primary objectives of using image processing methods for building strength assessment is to automate the evaluation process. By employing computer vision algorithms, the assessment can be performed efficiently and without the need for labor-intensive manual inspections. This automation can lead to faster and more consistent results, saving time and resources in the evaluation process [4]. Image processing techniques can enhance the clarity and resolution of building images, enabling the identification of even subtle cracks, material deterioration, or other anomalies that may not be easily noticeable during manual inspections. This level of precision can lead to more effective decision-making in maintenance and repair strategies. It offers the advantage of non-destructive evaluation, meaning that building strength can be assessed without causing any physical alterations to the structure [5]. This non-invasive approach is particularly important for historic buildings or structures where preservation is critical. By utilizing image data, engineers can gain insights into the building's condition without compromising its integrity [6].

An important objective is to enable predictive maintenance strategies based on image processing results. By adopting machine learning algorithms, the assessment can provide insights into the future health of the building, predicting potential failures or deterioration trends. This proactive approach allows for timely intervention and preventive maintenance, reducing the risk of catastrophic failures and extending the lifespan of the structure [7]. The use of image processing and machine learning allows for data-driven decision-making in building strength assessment. By analyzing large datasets of building images, patterns and trends can be identified, leading to more informed and evidence-based decisions regarding maintenance, repair, and rehabilitation.

Image processing methods can be integrated with structural health monitoring systems to continuously monitor the building's condition over time. Real-time image data can provide ongoing feedback on the structural integrity, enabling early detection of potential issues and facilitating a proactive approach to maintenance. By accurately identifying weaknesses and defects, engineers can implement targeted interventions to ensure the long-term safety and stability of the building, mitigating risks to occupants and the surrounding environment.

The motivation behind the development of DeepCrack and its application in concrete crack detection within SHM arises from the critical importance of ensuring the safety and longevity of infrastructure. Concrete structures form the backbone of our built environment, and their integrity is paramount. Traditional methods of crack detection are often time-consuming, prone to human error, and may lack the sensitivity to identify subtle yet potentially hazardous cracks. The advent of deep learning techniques, particularly CNNs, presents an opportunity to revolutionize the field by automating and significantly improving the accuracy of crack detection. The motivation is to fill the gap in current practices with a comprehensive solution like DeepCrack, capable of learning complex patterns, adapting to diverse conditions, and providing real-time insights. By enhancing the efficiency of SHM systems, DeepCrack aims to facilitate proactive interventions, ultimately contributing to the overall safety, reliability, and sustainability of vital infrastructure worldwide.

The primary goal of the presented work on DeepCrack is to advance the capabilities of SHM by introducing an innovative and highly accurate solution for concrete crack detection and analysis. The utilization of state-of-the-art CNNs in DeepCrack aims to address the critical need for a robust system that can automatically identify and characterize both visible and subtle cracks in concrete structures. By using a large dataset of annotated concrete images, DeepCrack not only achieves high detection accuracy but also learns intricate patterns associated with different types and stages of cracks. The detailed analysis module further contributes to the goal by evaluating cross-entropy, training, and validation accuracy, providing insights into crack severity and potential structural issues. Ultimately, the overarching objective is to enhance the efficiency of structural health monitoring systems, enabling real-time detection of vulnerabilities and facilitating proactive interventions to ensure the safety and longevity of critical infrastructure.

5.2 Literature Survey

Structural integrity is of paramount importance in civil engineering and infrastructure development. Cracks in structures, such as buildings, bridges, and dams, can lead to catastrophic failures if left undetected and untreated. Traditionally, crack detection and analysis have relied on manual inspection techniques, which are time-consuming, labor-intensive, and subject to human error. With the rapid advancements in artificial intelligence and deep learning, there is an emerging opportunity to revolutionize crack detection and analysis in structural engineering.

The conventional methods for crack detection include visual inspection, ultrasound, dye penetration, and magnetic particle testing. While these methods have been effective to some extent, they often require direct access to the structure and may miss subtle or internal cracks [8]. They are also labor-intensive and lack the efficiency and accuracy that modern technologies offer [9]. In recent years, image processing techniques have been explored for crack detection [10]. Researchers have utilized edge detection, morphological operations, and image segmentation to identify cracks in images captured by various imaging technologies, such as cameras and drones. However, these methods can struggle with complex crack patterns, lighting variations, and noise [11]. The unsupervised Deep Learning for SHM revolves around using the power of advanced machine learning techniques, specifically unsupervised learning, to automatically extract meaningful patterns and insights from sensor data without the need for labeled training examples. In the context of SHM, where monitoring the health and integrity of structures are critical, unsupervised deep learning methods, such as auto encoders or clustering algorithms, offer a promising approach. These algorithms can autonomously identify anomalies, patterns, or deviations in the structural data, such as vibrations, strain measurements, or acoustic signals, without relying on pre-existing knowledge of normal or faulty conditions. This concept is particularly advantageous in scenarios where obtaining labeled training data is challenging or impractical. By allowing the model to learn and adapt to the inherent structure of the data, unsupervised deep learning contributes to a more flexible and scalable approach for detecting structural abnormalities, ultimately enhancing the efficiency and effectiveness of structural health monitoring systems [12].

Deep learning has shown remarkable success in various fields, including computer vision [13]. In the realm of structural health monitoring, deep learning models, particularly CNNs, have been applied to detect cracks from images and videos captured by sensors or cameras. These models can automatically learn intricate features, adapt to various lighting conditions, and provide high accuracy in crack detection [14]. To train deep learning models effectively, researchers have created large-scale datasets containing images of cracked and non-cracked structures. Annotating these datasets with accurate crack location and severity information is crucial for training robust models [15]. Various techniques, such as manual annotation, crowd sourcing, and semi-automatic methods, have been employed to create these datasets.

Transfer learning in the context of SHM involves leveraging knowledge gained from one task or dataset to enhance the learning and performance of models in a different but related SHM task [16]. This approach acknowledges that data for structural health monitoring can be limited and expensive to acquire. By pre-training a neural network on a source dataset with abundant data, such as general structural images or vibration patterns, the model learns generalized features and representations. These learned features are then transferred and fine-tuned on a smaller target dataset specific to the SHM task, like crack detection or anomaly identification in a particular structure. This strategy accelerates the learning process and improves the performance of SHM models by capitalizing on the knowledge acquired from the source domain. Transfer Learning in SHM enables more efficient model training, mitigates the need for extensive labeled data in the target domain, and enhances the adaptability of AI systems to various structural monitoring scenarios, ultimately contributing to more accurate and robust structural health assessment.

Expertise-Based Dataset Construction and Learning Algorithms [17] involve a strategic approach to dataset creation that capitalizes on domain-specific knowledge and expertise to enhance the performance of machine learning algorithms. Here, constructing a dataset involves not only the gathering of diverse structural health data but also uses expert insights to curate and annotate the dataset effectively. Experts contribute their knowledge to identify critical features, anomalies, and nuanced patterns in the data that may be indicative of structural health issues. The created dataset is then used to train machine learning algorithms, with a focus on optimizing their performance for SHM tasks. By integrating expert domain knowledge into the dataset construction process, the algorithms can better capture and understand the intricacies of structural health data. This expertise-driven approach can improve the generalization and interpretability of the models, making them more effective in real-world applications. This iterative process of dataset refinement and algorithm training enhances the accuracy, reliability, and applicability of machine learning models in the domain of SHM, ultimately contributing to more effective and informed decision-making in the assessment of structural integrity.

Attention-based deep networks represent a significant advancement in the field of Automation in Construction, offering a sophisticated approach to information processing and feature extraction [18]. These networks incorporate attention mechanisms that enable them to selectively focus on relevant parts of input data, allowing for more effective and targeted learning. In the context of construction automation, attention mechanisms play a crucial role in handling complex and large-scale datasets, prioritizing key information, and improving the efficiency of various tasks. One notable application of attention-based deep networks in construction automation is in image analysis and computer vision tasks. For instance, these networks can be employed for automatic defect detection in construction materials or monitoring construction site safety. The attention mechanism enables the network to concentrate on specific regions of interest, enhancing the accuracy of detection and reducing false positives. It proves beneficial in tasks related to project scheduling and management. By focusing on critical factors or potential bottlenecks, these networks can optimize project timelines, resource allocation, and risk assessment. Also, it contributes to more informed decision-making and, ultimately, improved project outcomes.

Autonomous Crack Inspection System based on Unmanned Aerial Vehicles (UAVs) represents a transformative approach to infrastructure monitoring and maintenance [19]. The UAVs, equipped with high-resolution cameras or specialized sensors, autonomously navigate and capture detailed images of infrastructure surfaces. These images are then processed using computer vision algorithms to detect and analyze cracks, providing a comprehensive and efficient assessment of structural health. It eliminates the need for manual inspection, reducing the associated risks and costs. The UAVs can access difficult-to-reach areas, providing a more thorough and accurate inspection. The automation of data collection and analysis enhances the speed of the inspection process, allowing for timely identification of potential structural issues.

Deep learning in the frequency domain engineering signifies a cutting-edge approach to signal processing applications [20]. By using advanced neural network architectures, such as CNNs and Recurrent Neural Networks (RNNs), engineers can harness the power of deep learning to analyze and extract meaningful features from signals represented in the frequency domain. This approach is particularly advantageous in tasks like denoising, enhancement, anomaly detection, and predictive maintenance, where understanding complex patterns in frequency representations is crucial. Deep learning models excel at learning hierarchical features, enabling them to discern subtle nuances that may be challenging for traditional signal processing methods. The integration of deep learning in frequency domain engineering represents a paradigm shift towards more adaptive and data-driven signal processing, offering the potential for improved accuracy and efficiency in various engineering applications, ranging from communication systems to structural health monitoring and beyond.

Despite the progress made, challenges such as limited annotated data, model interpretability, and robustness to varying environmental conditions, and generalization across different structural types remain. Future research directions may involve semi-supervised learning, active learning strategies, and domain adaptation techniques to address these challenges. The integration of deep learning techniques into crack detection and analysis in structural engineering holds significant promise. This cutting-edge approach has the potential to enhance the safety, efficiency, and accuracy of structural health monitoring, contributing to the overall sustainability of critical infrastructure.

5.3 Traditional Building Strength Assessment Methods

Building strength assessment has been a fundamental aspect of ensuring the safety and reliability of structures. Traditionally, the primary approach to assess building strength involves manual inspection conducted by experienced engineers or experts. During manual inspection, trained professionals physically examine the building's structural components, such as beams, columns, walls, and foundations, to identify any signs of damage, deterioration, or weaknesses. They rely on visual observations, tactile examinations, and the use of basic tools to assess the structural condition.

5.3.1 Manual Inspection and Limitations

While manual inspection has been the standard practice for many years, it is not without its limitations. One of the main challenges of manual inspection is its time-consuming nature. Large and complex structures may require significant amounts of time and resources to be thoroughly examined, leading to potential delays in identifying critical issues. Moreover, manual inspection is subjective to some degree. Different inspectors may interpret structural conditions differently, potentially leading to variations in assessment results. This subjectivity introduces an element of uncertainty in the evaluation process. Furthermore, human errors may occur during manual inspection, especially when dealing with extensive datasets or complex structural geometries. Mistakes in data recording, measurements, or judgment can impact the accuracy and reliability of the assessment.

5.3.2 Challenges in Structural Assessment

Structural assessment poses several challenges that are not easily addressed by conventional methods. The identification of early signs of structural weaknesses, such as micro-cracks or subtle material degradation, is particularly challenging. These issues may remain undetected during manual inspections, leading to potential safety hazards if not addressed promptly. Additionally, the integration of multiple data sources, such as sensor data, blueprints, and historical records, presents a significant challenge in manual inspection. Data interpretation and integration become complex, particularly for large-scale and aging infrastructures.

5.3.3 Need for Innovative Approaches

Given the limitations and challenges associated with traditional methods, there is a pressing need for more innovative and advanced approaches to building strength assessment. Innovative image processing techniques offer promising solutions to overcome these limitations [21]. Image processing techniques leverage high-resolution imaging technologies to capture detailed visual information about the building's structural components. By employing computer vision algorithms, these techniques can automatically analyze and interpret the collected images, enabling rapid and objective assessment of building strength [22].

Moreover, image processing can facilitate non-destructive evaluation, as it does not require physical alterations to the structure during the assessment. This feature is particularly advantageous for historic buildings and heritage structures, where preservation is a key concern [23]. The integration of image processing with machine learning algorithms further enhances the assessment process [24]. Machine learning can recognize patterns, detect anomalies, and make predictions based on large datasets, contributing to more accurate and data- driven decision-making in building strength assessment [25].

5.4 Proposed Fundamentals of Image Processing in Building Assessment

This section focuses on the foundational principles and methodologies of image processing in the context of building strength assessment.

5.4.1 Image Acquisition Techniques

Various imaging technologies, such as digital cameras, drones, LIDAR, and infrared sensors, are explored, emphasizing their specific applications and advantages. The selection of the most suitable image acquisition technique based on the building type, size, and environmental conditions is elaborated upon. The subsection also addresses potential challenges in image acquisition, such as lighting conditions, camera calibration, and image resolution, and proposes strategies to mitigate these challenges.

Image acquisition techniques play a crucial role in building assessment using image processing methods. These techniques involve capturing high-quality images of building structures to enable accurate and detailed analysis. Various image acquisition methods are employed to obtain the necessary visual data for strength evaluation. Buildings are photographed from the air using unmanned aerial vehicles (UAVs) or drones that have high-resolution cameras. Drones offer the advantage of accessing hard-to-reach areas and providing a holistic view of the building's exterior, enabling efficient inspections. Drones can capture high-quality aerial images and videos of buildings, providing a comprehensive visual inspection of the entire structure, including roofs, facades, and other elevated areas. This enables inspectors and engineers to identify potential issues and assess the overall condition of the building.

5.4.2 Image Enhancement and Preprocessing

Image enhancement and preprocessing are crucial steps in image processing to improve the quality and clarity of the acquired images. There are different enhancement techniques, such as contrast stretching, histogram equalization, and noise reduction, to enhance the visibility of structural features and defects. Preprocessing methods, including image denoising, image registration and geometric correction, can be adopted to ensure data consistency and accuracy. The role of these techniques in optimizing the images for subsequent analysis is emphasized.

5.4.3 Feature Extraction Methods

Feature extraction is a key step in image processing, where relevant information and characteristics are extracted from the enhanced images. Various methods are edge detection algorithms, texture analysis, and shape recognition techniques to identify critical structural elements, cracks, and other anomalies.

5.4.4 Pattern Recognition Algorithms

Pattern recognition algorithms play a central role in analyzing the extracted features and identifying patterns related to building strength and defects. Different pattern recognition algorithms, such as neural networks, decision trees and support vector machines (SVM) are available. The suitability of each algorithm for different building assessment tasks is explained, along with considerations for training datasets and model validation. Additionally, the potential for hybrid approaches that combine multiple algorithms for improved accuracy and generalization is explored.

5.5 Innovative Image Processing Techniques for Building Strength Visualization

This section delves into the innovative image processing techniques employed for visualizing and analyzing building strength. These advanced methodologies outline the high-resolution imaging and cutting-edge algorithms to provide detailed insights into structural conditions.

5.5.1 Crack Detection and Analysis

Crack detection is a critical aspect of building strength assessment, as cracks can indicate structural weaknesses or potential failure points. This subsection focuses on innovative crack detection techniques using image processing algorithms. Methods like edge detection, thresholding, and morphology-based approaches are explored for accurate crack identification. Machine learning models play a vital role for automating crack detection and classification in buildings.

5.5.2 Material Deterioration Assessment

Material deterioration, such as corrosion, erosion, or spalling, can significantly impact building strength and longevity. This subsection discusses innovative techniques for assessing material deterioration using image processing. Image analysis algorithms, such as texture analysis and color-based methods, are examined for identifying signs of material degradation. The integration of spectroscopy and infrared imaging for detecting hidden material defects is also explored. Furthermore, the potential for real-time monitoring of material deterioration using image processing technologies is highlighted.

5.5.3 Image Segmentation for Structural Analysis

Image segmentation plays a vital role in extracting meaningful regions or regions of interest from building images. This subsection focuses on innovative image segmentation techniques for structural analysis. Algorithms like region- growing, watershed segmentation and deep learning-based segmentation are discussed for segmenting building components accurately. The utilization of segmented regions for further analysis, such as quantifying structural dimensions, calculating area and volume, and assessing building integrity, is explored.

Innovative image processing techniques offer a comprehensive approach to visualize and assess building strength. High-resolution imaging facilitates detailed analysis, while crack detection, material deterioration assessment, and image segmentation contribute to a more thorough understanding of building conditions. These advanced methods not only enhance the accuracy and efficiency of building strength assessment but also contribute to proactive maintenance and risk mitigation, ultimately ensuring the safety and resilience of built structures.

5.6 Machine Learning for Building Strength Evaluation

This section focuses on the application of machine learning techniques to evaluate building strength. Machine learning algorithms enable data-driven analysis and prediction, enhancing the accuracy and efficiency of building strength assessment.

5.6.1 Supervised Learning for Strength Prediction

In this subsection, supervised learning methods for building strength prediction are explored. The process involves training the machine learning model on labeled datasets, where the building strength is known. Various supervised learning algorithms, such as linear regression, support vector regression (SVR), and random forests, are discussed for predicting building strength based on input features extracted from image data. The selection of relevant features and strategies for feature engineering to improve prediction accuracy are also addressed. Additionally, the advantages of supervised learning in handling regression tasks for continuous strength values are highlighted.

5.6.2 Unsupervised Learning for Anomaly Detection

Unsupervised learning techniques are valuable for identifying anomalies or deviations from normal behavior in building strength assessment. In this subsection, unsupervised learning methods, such as clustering algorithms (e.g., k- means, DBSCAN) and autoencoders, are explored for anomaly detection in building structures. Anomaly detection helps identify structural defects or weaknesses that may not be evident during manual inspection. The application of unsupervised learning for identifying outlier building components or regions in the images is discussed, along with the potential for early detection of critical issues.

5.6.3 Hybrid Approaches for Enhanced Accuracy

Hybrid approaches combine both supervised and unsupervised learning techniques to capitalize on their respective strengths. This subsection focuses on the use of hybrid approaches for enhanced accuracy in building strength evaluation. Ensemble methods, such as Random Forests and Gradient Boosting, are an examples of hybrid approaches that combine the outputs of multiple models to make more robust predictions. Transfer learning, where pre- trained models are adapted to building strength assessment tasks, is explored as another form of hybrid approach. The advantages of combining different learning paradigms to handle complex building assessment scenarios are emphasized.

5.6.4 Deep Learning for Crack Detection

The role of deep learning in crack detection is pivotal, as it revolutionizes the traditional methods of identifying cracks in structures and infrastructure. Deep learning plays a transformative role in crack detection by enabling automated and efficient analysis, multi-scale feature learning, and accurate localization of cracks. The Deep Crack architecture [26] shown in Fig. 5.1 is a novel approach for crack detection that utilizes CNNs with a unique design. Instead of using fully connected layers, it incorporates side-output layers right after the convolutional layers. At each side-output layer, deep supervision is applied, enabling the network to learn multi-scale and multi-level features. These side-output layers capture information at different scales, as the input size becomes smaller and the receptive field size becomes larger. The final fused output is obtained by concatenating the predictions from all side-output layers.

The proposed architecture consists of:

- **Input Image**: The crack detection process starts with an input image, representing a part of a structure or infrastructure to be analyzed for cracks.
- **Convolutional Layers**: The input image undergoes a series of convolutional layers, where filters learn to extract relevant features from the image. These layers work on different receptive fields, capturing local and global information.
- **Side-Output Layers**: Instead of fully connected layers, side-output layers are introduced right after the convolutional layers. These side-output layers generate intermediate predictions based on the features extracted by the convolutional layers.

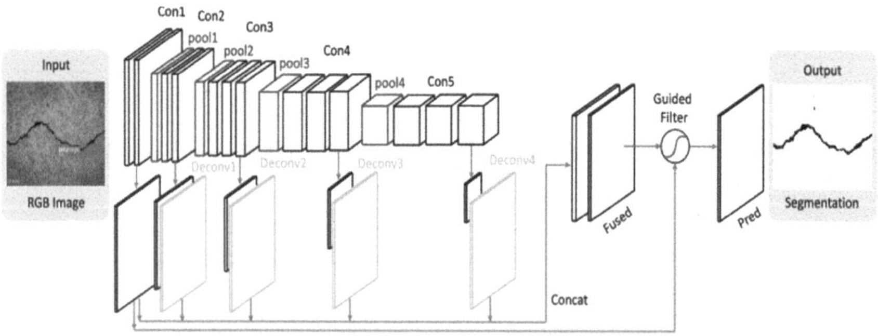

Fig. 5.1 DeepCrack architecture [26]

- **Deep Supervision**: At each side-output layer, deep supervision is applied. This means that the network is guided and trained not only to predict the final output but also to generate intermediate predictions at various scales.
- **Multi-Scale and Multi-Level Features**: The side-output layers capture features at different scales, corresponding to the varying sizes of input and receptive fields. This enables the model to gather information from both local and global perspectives.
- **Concatenation**: The predictions from all side-output layers are concatenated to form the final fused output. This fused prediction benefits from the multi-scale and multi-level features extracted by the side-output layers.
- **Guided Filtering**: To further refine the fused prediction, guided filtering is applied, specifically leveraging the first side-output layer. Guided filtering enhances the quality of the final prediction, making it more accurate and reliable.
- **Final Crack Detection**: The refined and fused output is then used to identify the presence of cracks in the input image. By incorporating multi-scale and multi-level features through deep supervision and guided filtering, the DeepCrack architecture achieves enhanced crack detection performance.

The above DeepCrack architecture is designed to exploit the benefits of side-output layers, deep supervision, and guided filtering to effectively detect cracks in structures and infrastructure. The multi-scale and multi-level features learned throughout the network enable it to discern cracks of various sizes and types, making it a powerful tool for accurate and robust crack detection and analysis.

5.6.4.1 Training Phase

Figures 5.2 and 5.3 shows the training and testing phases of the DeepCrack structure for crack detection and analysis. It follows the standard procedures used in deep learning tasks

(a) **Dataset Preparation**: The first step in the training phase is to prepare the dataset. This dataset consists of a large number of images containing examples of cracked and non-cracked structures. The images are labeled accordingly to indicate the presence or absence of cracks.
(b) **Data Augmentation**: Data augmentation techniques are applied to increase the diversity of the training data. Augmentation includes random transformations such as rotations, flips, and shifts of the images. This process helps prevent over fitting and improves the model's ability to generalize to new, unseen data.
(c) **Network Architecture**: The DeepCrack architecture, as described earlier, is selected for the crack detection task. The model comprises convolutional layers with side-output layers for deep supervision, replacing fully connected layers.
(d) **Loss Function**: The loss function used during training is typically binary cross-entropy since crack detection is a binary classification task. The loss function measures the dissimilarity between the predicted crack presence and the ground-truth labels.

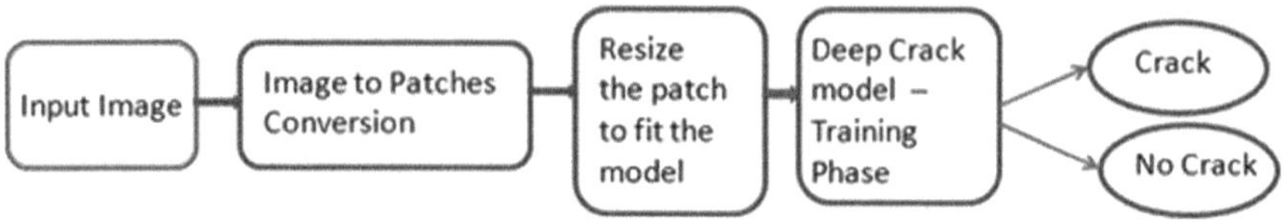

Fig. 5.2 Training phase

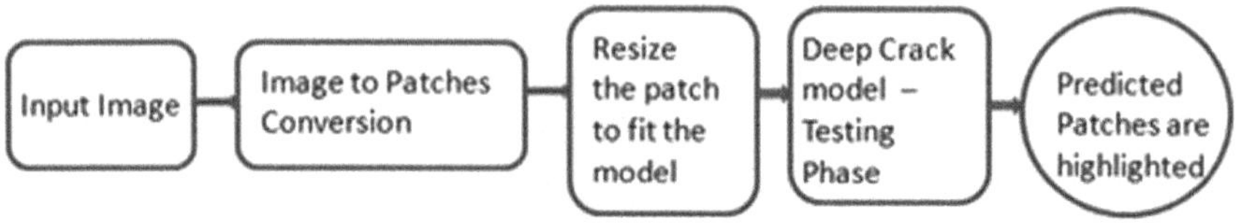

Fig. 5.3 Testing phase

(e) **Optimization**: The network is trained to minimize the loss function using an optimization algorithm, such as Stochastic Gradient Descent (SGD) with momentum or Adam. The optimization process adjusts the weights of the network to improve its performance.
(f) **Deep Supervision**: During training, deep supervision is applied at each side-output layer. This means that not only the final fused output but also the intermediate predictions from the side-output layers are used to compute the loss. Deep supervision helps in better learning of multi-scale and multi-level features.
(g) **Guided Filtering**: The fused prediction is refined by guided filtering using the information from the first side-output layer. Guided filtering enhances the quality of the final prediction and ensures more accurate crack detection.
(h) **Validation**: The model's performance is periodically evaluated on a validation set. This helps in monitoring the model's progress during training and avoiding over fitting. Early stopping may be applied if the validation performance does not improve.
(i) **Epochs and Batch Size**: The training process involves iterating over the entire dataset multiple times (epochs). The data is divided into batches to make the training more efficient, and the model's weights are updated after processing each batch.

5.6.4.2 Testing Phase

(a) **Dataset Preparation**: Similar to the training phase, a separate dataset is prepared for testing. This dataset contains images with ground-truth labels, but the model has not seen these images during training.
(b) **Inference**: The trained DeepCrack model is now used for inference on the testing dataset. The model predicts the presence or absence of cracks in each test image.

(c) **Evaluation**: The model's performance is evaluated on the testing dataset using various metrics such as accuracy, precision, recall, F1 score, and ROC-AUC. These metrics provide insights into how well the model generalizes to new, unseen data.
(d) **Visualization**: The predictions of the model can be visualized to gain a better understanding of its performance and to identify any potential false positives or false negatives.
(e) **Refinement**: If the model's performance is not satisfactory, further refinements or adjustments to the architecture or training process may be applied, followed by retesting and evaluation.

The training and testing phases are iterative processes, with adjustments made based on the evaluation results. The goal is to develop a DeepCrack model that accurately detects cracks in structures and provides reliable results in real-world applications.

5.7 Results and Analysis

The results of the DeepCrack model training can be evaluated in terms of cross-entropy loss, training accuracy, and validation accuracy across different epochs. These metrics provide valuable insights into how well the model is learning and generalizing during the training process.

5.7.1 Cross-Entropy Loss Versus Epoch

Cross-entropy loss is a common metric used in binary classification tasks, such as crack detection. It measures the dissimilarity between the predicted probabilities and the actual binary labels (0 for non-cracked and 1 for cracked). As the model undergoes training, the cross-entropy loss should generally decrease with each epoch. This indicates that the model is becoming more confident in its predictions and learning to distinguish cracked and non-cracked structures better. A smooth downward trend in the cross-entropy loss graph signifies effective learning.

5.7.2 Training Accuracy Versus Epoch

Training accuracy is the percentage of correctly classified examples from the training dataset. It is an essential metric to monitor how well the model is performing on the data it has seen during training. As the model learns, the training accuracy should increase with each epoch. However, it is crucial to keep an eye on over fitting. If the training accuracy continues to rise significantly while the validation accuracy stagnates or decreases, it may indicate over fitting, where the model is memorizing the training data rather than generalizing to new data.

5.7.3 *Validation Accuracy Versus Epoch*

Validation accuracy is the percentage of correctly classified examples from the validation dataset. The validation dataset contains data that the model has not seen during training and serves as an approximation of the model's performance on unseen data. The validation accuracy provides a measure of how well the model generalizes. Ideally, the validation accuracy should increase as the model learns and then stabilize or slightly improve, indicating that the model is learning to generalize effectively. A substantial gap between training and validation accuracy could be a sign of over fitting.

From the Figs. 5.4 and 5.5, it is observed that the cross-entropy loss decreases steadily over epochs, it indicates that the model is effectively learning from the training data. There is a steady increase in training accuracy indicates that the model is learning from the training data and becoming more confident in its predictions.

Monitoring the gap between training and validation accuracy is small and both accuracy metrics increase steadily, the model is likely generalizing well. The best performance is achieved by avoiding over fitting, hyper parameter tuning and regularization strategies during training. This proposed model reached a satisfactory

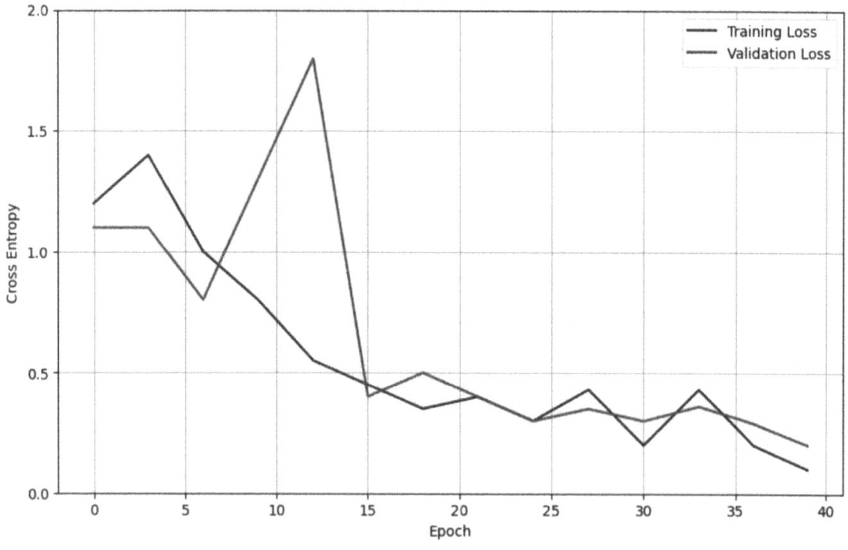

Fig. 5.4 Cross entropy versus Epoch during training of proposed model

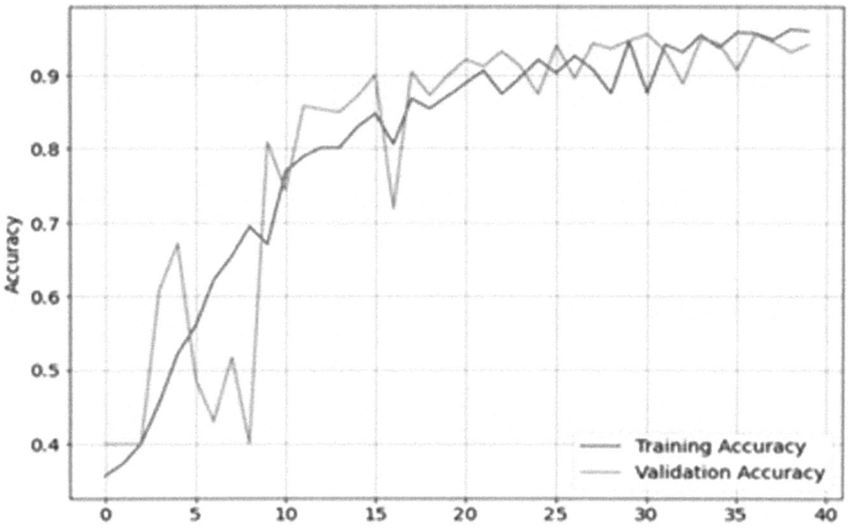

Fig. 5.5 Accuracy plots of the proposed model during testing

performance level; it can be deployed for crack detection on new, unseen data with confidence. The visualizations of building strength serve as compelling evidence of the success of the innovative techniques in enhancing building assessment accuracy and contributing to the overall safety and resilience of built structures.

5.8 Conclusion

The DeepCrack architecture offers a promising approach for crack detection and analysis using deep learning techniques. By using Convolutional Neural Networks with side-output layers, deep supervision, and guided filtering, the model gains the ability to learn multi-scale and multi-level features from the input images, which is critical for accurate crack detection.

During the training phase, the model undergoes an iterative process, optimizing its weights using an optimization algorithm to minimize the cross- entropy loss. With deep supervision at each side-output layer, the model learns intermediate predictions, enabling it to capture features at different scales and enhance its ability to distinguish cracks from non-cracked structures. The training accuracy increases as the model learns from the training data, while the cross- entropy loss decreases, indicating effective learning. However, it is essential to monitor the validation accuracy to ensure the model generalizes well to unseen data. A small gap between training and validation accuracy indicates good generalization, while a significant gap may signal over fitting, necessitating further refinements to the model.

By following these steps, the DeepCrack model can be fine-tuned and optimized to achieve accurate and reliable crack detection results. The ability to detect cracks in structures and infrastructure efficiently can significantly contribute to public safety and structural integrity. While the DeepCrack architecture shows promise, continuous research and refinement are essential to improve its performance further and address potential challenges. Additionally, the architecture's application can be extended to other related tasks, such as crack severity assessment or detecting cracks in various types of infrastructure, to enhance its versatility and applicability in real-world scenarios.

DeepCrack architecture represents a significant advancement in crack detection and analysis using deep learning, offering a powerful and effective tool to ensure the safety and longevity of our built environment. As technology continues to evolve, the future of crack detection using deep learning holds great potential for creating safer and more resilient infrastructures.

References

1. Dinesh A, Anitha Selvasofia SD, Datcheen KS, Rakhesh Varshan D (2023) Machine learning for strength evaluation of concrete structures–critical review. Mater Today Proceed
2. Liu J, Yang X, Lau S, Wang X, Luo S, Lee VCS, Ding L (2020) Automated pavement crack detection and segmentation based on two-step convolutional neural network. Comput Aided Civ Infrastruct Eng 35:1291–1305
3. Moein MM, Saradar A, Rahmati K, Mousavinejad SHG, Bristow J, Aramali V, Karakouzian M (2023) Predictive models for concrete properties using machine learning and deep learning approaches: a review. J Build Eng 63
4. Lee BY, Kim YY, Kim J-K (2013) Automated image processing technique for detecting and analyzing concrete surface cracks. Struct Infrastruct Eng 9:567–577
5. Mishra M (2021) Machine learning techniques for structural health monitoring of heritage buildings: a state-of-the-art review and case studies. J Cult Herit 47:227–245
6. Sun H, Burton HV, Huang H (2021) Machine learning applications for building structural design and performance assessment: state-of-the-art review. J Build Eng 33:101816
7. Mangalathu U, Burton HV (2019) Deep learning-based classification of earthquake-impacted buildings using textual damage descriptions. Int J Disaster Risk Reduct 36
8. Munawar HS, Hammad AWA, Haddad A, Soares CAP, Waller ST (2021) Image-based crack detection methods: a review. Infrastructures 6:115
9. Ahanshahi MR, Masri SF (2012) Adaptive vision-based crack detection using 3d scene reconstruction for condition assessment of structures. Autom Constr 22:567–576
10. Khan MA-M, Kee S-H, Pathan A-SK, Nahid A-A (2023) Image processing techniques for concrete crack detection: a scientometrics literature review. Remote Sens 15:2400
11. Mohan A, Poobal S (2018) Crack detection using image processing: a critical review and analysis. Alex Eng J 57(2):787–798
12. Boccagna R, Bottini M, Petracca M, Amelio A, Camata G (2023) Unsupervised deep learning for structural health monitoring. Big Data Cogn Comput 7:99
13. Yoon J, Lee J, Kim G et al (2022) Deep neural network-based structural health monitoring technique for real-time crack detection and localization using strain gauge sensors. Sci Rep 12:20204
14. Sarkar S, Reddy KK, Giering M, Giering M (2016) Deep learning for structural health monitoring: a damage characterization application. Annu Conf PHM Soc 8:1

15. Rasmussen CB, Kirk K, Moeslund TB (2022) The challenge of data annotation in deep learning—a case study on whole plant corn silage. Sensors 22:1596
16. Philip RE, Andrushia AD, Nammalvar A, Gurupatham BGA, Roy K (2023) A comparative study on crack detection in concrete walls using transfer learning techniques. J Compos Sci 7:169
17. Wibowo AP, Adha A, Kurniawan IF, Laory I (2022) Wall crack multiclass classification: expertise-based dataset construction and learning algorithms performance comparison. Buildings 12(12):2135
18. Chen W, He Z, Zhang J (2023) Online monitoring of crack dynamic development using attention-based deep networks. Autom Constr 154:105022
19. Tse KW, Pi R, Sun Y, Wen CY, Feng Y (2023) A novel real-time autonomous crack inspection system based on unmanned aerial vehicles. Sensors (Basel) 23(7):3418
20. Zhang Q, Barri K, Babanajad SK, Alavi AH (2021) Real-time detection of cracks on concrete bridge decks using deep learning in the frequency domain. Engineering 7(12):1786–1796
21. Hong T, Wang Z, Luo X, Zhang W (2020) State-of-the-art on research and applications of machine learning in the building life cycle. Energy Build 212
22. Sha H, Xu P, Yang Z, Chen Y, Tang J (2019) Overview of computational intelligence for building energy system design. Renew Sustain Energy Rev 108:76–90
23. Liu L, Silva EA, Wu C, Wang H (2017) A machine learning- based method for the large-scale evaluation of the qualities of the urban environment. Comput Environ and Urban Syst 65:113–125
24. Mirnaghi MS, Haghighat F (2020) Fault detection and diagnosis of large-scale HVAC systems in buildings using data-driven methods: a comprehensive review. Energy Build 229
25. Liu S-W, Huang JH, Sung J-C, Lee C (2002) Detection of cracks using neural networks and computational mechanics. Comput Methods Appl Mech Eng 191:2831–2845
26. Liu Y et al (2019) DeepCrack: a deep hierarchical feature learning architecture for crack segmentation. Neurocomputing 338:139–153

Chapter 6
Short-Term Health Monitoring and Damage Assessment of a Concrete Cable-Stayed Bridge by an Integrated Unsupervised Learning Approach

Hassan Sarmadi and Alireza Entezami

Abstract Civil structures are valuable assets of every society with significant influences on human life, economics, transportation networks, and energy supply. These assets are always susceptible to natural and man-induced hazards, aging, and material deterioration. The adverse consequences of these events are the occurrence and growth of different patterns of structural damage, failure, and even collapse. Vibration-based structural health monitoring (SHM) supported by sensory data and artificial intelligence is an emerging and innovative technology for assessing the safety and functionality of various civil structures, especially bridges. Although this technology can be implemented in short- and long-term programs, some limitations do not allow civil engineers to benefit from the Dunn long-term SHM, in which case an automated short-term monitoring program is an effective and practical alternative. However, some challenges make this program problematic. The most significant challenge relates to encountering limited vibration data during a short period of monitoring. The other important issue is profound effects of operational and environmental variability on vibration data such as modal frequencies, which cause false alarm and mis-detection errors. To address these challenges, this chapter proposes an innovative SHM approach suitable for short-term monitoring programs based on the concept of integrated unsupervised learning. Using limited modal frequencies as the main vibration features, this approach comprises three steps of local feature segmentation, unsupervised feature selection, and unsupervised anomaly detection. First, agglomerative hierarchical clustering in conjunction with the Dunn's index is employed to segment the limited set of modal frequencies into local subsets (clusters). Second, a filter-based unsupervised feature selector is proposed to find the most appropriate cluster with the features insensitive to environmental/operational

H. Sarmadi (✉)
IPESFP Startup Company, Mashhad, Iran
e-mail: hassan.sarmadi@alumni.um.ac.ir

A. Entezami
Department of Civil and Environmental Engineering, Politecnico di Milano, Milan, Italy
e-mail: alireza.entezami@polimi.it

H. Jahangir et al. (eds.), *Damage Detection and Structural Health Monitoring of Concrete and Masonry Structures*, Springer Tracts in Civil Engineering,
https://doi.org/10.1007/978-981-97-8975-7_6

conditions. Third, these features are used to develop an anomaly detector based on the local outlier factor and determine anomaly indices for damage assessment. A concrete cable-stayed bridge is considered to testify the proposed approach. It is observed that this approach succeeds in alarming damage with limited data under a short-term monitoring program.

Keywords Structural health monitoring · Damage assessment · Short-term monitoring · Environmental variability · Unsupervised learning · Clustering · Feature selection · Anomaly detection

6.1 Introduction

Bridge structures are important structural systems in any society due to their utilities. In contrast to other types of civil structures such as dams and buildings, long-span bridges are slender, flexible, low vibration, and sensitive to environmental and operational conditions [1]. Compared to short- and medium-span bridges, on the other hand, long-span bridges are further prone to particular environmental variations such as air temperature, weak and strong winds, and traffic loads [2]. Earthquakes, floods, typhoons, landslides, etc., as well as man-made hazards such as terrorist attacks, fires and blasts, human mistakes in design, construction, and maintenance, the crossing of unexceptional and heavy vehicles, etc. may threaten the safety and health of bridge structures and their main elements. Under such circumstances, the emergence of different patterns of structural damage in terms of concrete cracks and spalling, rebar detachment, fatigue, and corrosion is plausible.

Due to recent development in sensor technologies [3, 4], structural health monitoring (SHM) has become essential to every society due to its great benefits; that is, saving human lives and money [5]. The primary purpose of SHM is to provide an automated inspection strategy for assessing the functionality and safety of various structural systems and intelligently alarming the occurrence of any damage [6–9]. Despite various model-driven techniques for damage assessment [10–12], data-driven SHM methods have recently received considerable attention as a result of remarkable progress in artificial intelligence, machine learning, and pattern recognition. Such methods fully rely upon measured data without any numerical modeling and updating procedures [13]. A data-driven method is often composed of three steps of (i) sensor selection and installation as well as data measurement and transmission, (ii) structural feature extraction and selection, and (iii) feature analysis and discrimination. Alongside the first step, feature extraction is a methodology for discovering significant information called *feature* from sensory data that is more pertinent to structural attributes or damage patterns [14]. One of the most important and widely-used features in vibration-based SHM is the structural modal frequency. This is because it pertains to inherent structural properties; that is, mass and stiffness. Since the emergence of damage decreases structural stiffness, a reduction in modal frequencies is also indicative of damage occurrence provided that no mass increase occurs. For this reason, the monitoring of modal frequency changes under a feature analysis/classification framework is a well-accepted approach to detecting early damage in civil structures [15, 16].

Upon benefiting from data-driven SHM, machine learning can bring various solutions to this strategy. It is the science of enabling machines to learn algorithms that are not explicitly programmed [17]. This methodology is a product of artificial intelligence that intends to simulate the human brain for solving different problems based on sensory information [18]. Supervised learning, semi-supervised learning, and unsupervised learning are the three primary algorithms of machine learning. Though these algorithms have their pros and cons, unsupervised learning is a reasonable and practical approach to SHM, especially for early damage assessment [19]. This is due to the fact that the preparation of fully labeled data for a civil structure covering undamaged and all possible damaged cases is intractable and sometimes illogical [20]. Unlike to laboratory-scale models, full-scale civil structures are complex and expensive. In this case, it may be unreasonable to induce different intentional damage patterns in such structures for providing labeled data. Thus, it is beneficial for civil engineers to concentrate on unsupervised learning methods for early damage assessment [21–29].

Regarding variability in structural modal frequencies, damage is not the only influential cause. In real-world SHM of bridge structures, environmental and/or operational factors seriously affect modal frequencies. Ambient temperature, relative humidity, and wind characteristics are the key environmental parameters, while traffic is the main operational factor influencing bridges [30]. However, daily and seasonal temperature fluctuations are among the most critical reasons for variability in structural modal frequencies [31]. This is because temperature directly impacts on material and mechanical properties, particularly the elastic modulus, and also deforms the geometric dimensions of structures by thermal expansion and contraction. Under such circumstances, structural stiffness changes significantly, especially for bridge structures. In most cases, temperature-induced variability in bridge modal frequencies is about 5–10% that might be more important than moderate damage or normal operational loads [31]. When air temperature increases (i.e., warm weather), the elastic modulus decreases leading to a reduction in the structural stiffness (i.e., the structure becomes more flexible). Conversely, in cold weather, while the elastic modulus of common structural materials like steel and concrete does not significantly increase, the materials can become more brittle, which may make the structure seem stiffer.

Humidity increases the mass of concrete structures and makes slow variations with an increase in moisture content. In this case, there exists a tendency to decrease structural modal frequencies. However, humidity-induced frequency variability does not have serious influences on the stiffness of long-span bridge [30]. The effects of wind characteristics such as speed and direction on bridge structures depend on their types, sizes, geographical locations, and weather conditions. Generally, wind loads affect structural response amplitudes and aerodynamics. In this regard, wind is an important environmental factor for long-span cable-stayed and suspension bridges [32]. Under daily wind conditions, structural modal frequencies fluctuate instantaneously, while under strong wind or typhoon loads, modal frequencies reduce by increasing wind speed. However, this condition is not constant under weak winds. Traffic is the key parameter affecting the mass variability in bridge structures. Although air temperature and humidity directly influence the structural stiffness, the traffic volume on a bridge directly affects its mass properties. On this basis, traffic or overloading cause reductions in structural frequencies due to adding masses [30]. However, non-stationary excitation sources attributable to traffic produce random vibrations with daily and weekly fluctuations in modal frequencies. The removal of temperature-, humidity-, wind-, and traffic-induced variability conditions from structural modal frequencies is a serious challenge in both short- and long-term monitoring schemes [30]. In most cases, the environmental and operational impacts mask the features (e.g., modal frequencies) and cause deceptive interpretations. These interpretations appear as false positive and false negative errors [22]. Thus, it is imperative to remove such effects to achieve reliable SHM.

The other important issue regarding SHM is concerned with the duration of the monitoring process. Generally, this issue can be implemented in short- or long-term programs [24, 33]. A short-term monitoring scheme, as its name indicates, may be performed over a short period, such as hours of a day, some days, weeks, or months, while a long-term monitoring program is carried out over a long duration. In most cases of SHM applications, a one-year monitoring program is the minimum long-term plan and this program can be continued for several years or decades by exploiting permanent sensor systems [28]. An important note is that high variability in structural responses (i.e., modal frequencies) of civil structures monitored in a short period may be a warning of a sudden change or catastrophic event such as partial or entire collapse [34]. In other words, even though the removal of environmental and operational variations from structural responses/features in long-term monitoring is pertinent, it is also important to implement data normalization for features in short-term monitoring, particularly when the rate of variability is high.

The main objective of this chapter is to propose an innovative SHM approach to overcome the critical challenges of limited vibration features stemming from short-term monitoring programs and the effects of environmental and/or operational variations on such features. The proposed method is composed of three steps of local feature segmentation, unsupervised feature selection, and anomaly detection. First, agglomerative hierarchical clustering (AHC) with the aid of the Dunn's index is presented to divide the small set of modal frequencies associated with the structural undamaged state into clusters. Second, after segmenting the undamaged modal frequencies, a new unsupervised feature selection technique is proposed to choose the most appropriate cluster that consists of relevant features to structural stiffness (i.e., the damage factor) and not environmental/operational conditions. Third, an anomaly detector based on the local outlier factor (LOF) is developed to apply the selected features in the relevant cluster regarding the undamaged state and then give anomaly indices for damage assessment. A limited collection of structural natural frequencies of a concrete cable-stayed bridge monitored during a few days over several months in 2008 is utilized to validate the proposed method. Results show that the method presented here can address confounding variability conditions in the bridge modal frequencies and correctly detect damage in a short period of the health monitoring program.

6.2 Proposed Approach

The proposed integrated unsupervised learning approach is implemented in two parts; that is, training and monitoring phases. During the training period, there is a fixed training points concerning the undamaged condition (UC). In this phase, the three steps of local feature segmentation, unsupervised feature selection, and anomaly detection are carried out by training data. Once the anomaly detector based on a LOF model is constructed, the anomaly indices of the training points can be determined, thereby estimating a decision threshold for final decision-making. In the monitoring phase, it is assumed that and all test samples are collected sequentially. The test point is then applied to the developed anomaly detector to compute a new anomaly index. Utilizing the estimated threshold, the developed approach in this chapter warns of potential damage if the anomaly index of the test point rises above the threshold, i.e., indicating a damaged condition (DC); otherwise, it announces that the structure is in its undamaged state, as observed during the training phase. Figure 6.1 shows the process of the suggested integrated technique for convenience.

6.2.1 Local Feature Segmentation by Agglomerative Hierarchical Clustering

In machine learning, clustering is an approach that groups data points based on their similarities [35]. One of the great advantages and key properties of data clustering is its compatibility with the unsupervised learning framework. In essence, clustering is an inherent unsupervised learning technique for separating unlabeled data. For these reasons, it is among the most widely-used approaches to SHM [19, 26, 28, 36–41]. Hierarchical clustering is a connectivity-based method for cluster analysis that seeks to build a hierarchy of clusters in the form of a tree called *dendrogram.* This tree-like structure depicts the hierarchical relationships between clusters by iteratively separating clusters through a distance measure between data points. AHC is a type of hierarchical clustering in a bottom-up framework. This technique treats each data point as a single cluster as the bottom or lowest part of the dendrogram and successively agglomerates pairs of clusters until all clusters are combined into a single cluster that consists of all data points. Figure 6.2 shows the graphical schematic of the AHC by considering some samples (vectors) of modal frequencies, i.e., $\mathbf{x}_a$, $\mathbf{x}_b$, $\mathbf{x}_c$, $\mathbf{x}_d$, $\mathbf{x}_e$, and $\mathbf{x}_f$.

More precisely, each data point makes an individual cluster or a node (leaf) of a tree (i.e., Step 1), in which case one needs to calculate the distance of any cluster from

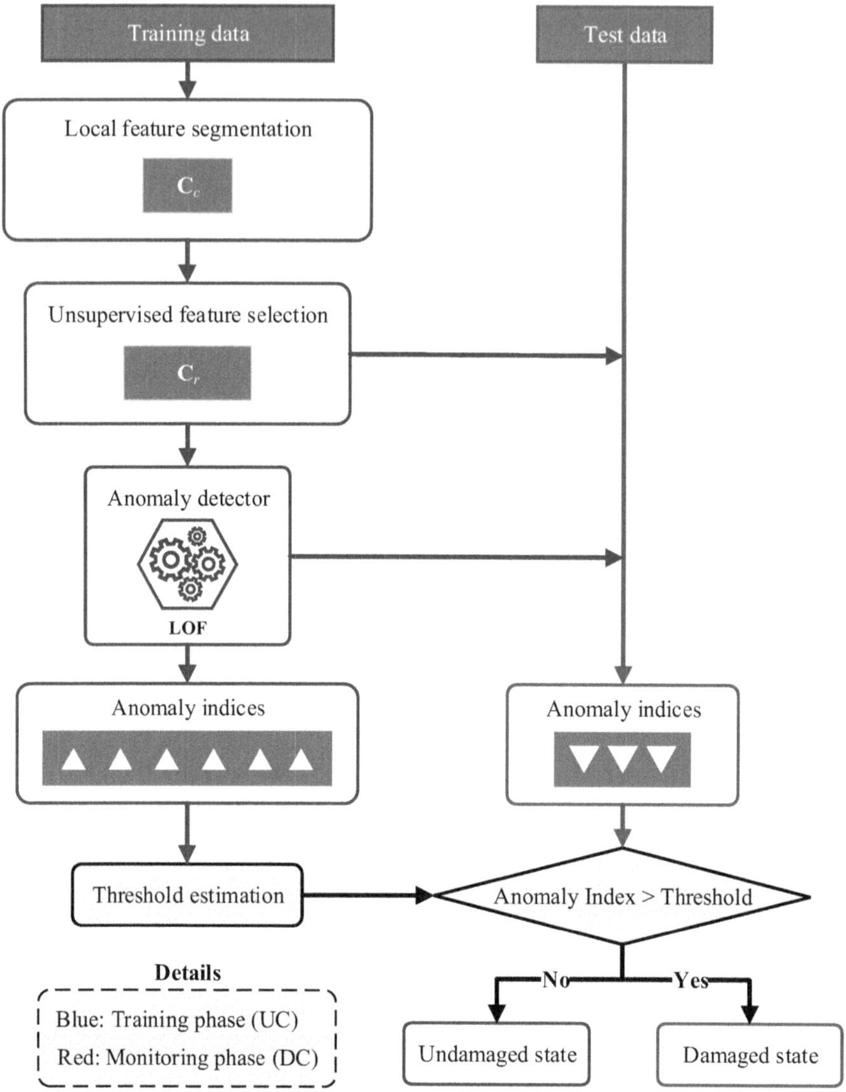

Fig. 6.1 Flowchart of the proposed integrated unsupervised learning method for short-term health monitoring and damage assessment

the other clusters. With the aid of a distance measure (e.g., the Euclidean distance), similar data points are merged together to form a new cluster (node). Given the two features (vectors) $\mathbf{x}_a$ and $\mathbf{x}_b$, the Euclidean distance between them, i.e., $d_E(\mathbf{x}_a, \mathbf{x}_b)$, is determined in the following form:

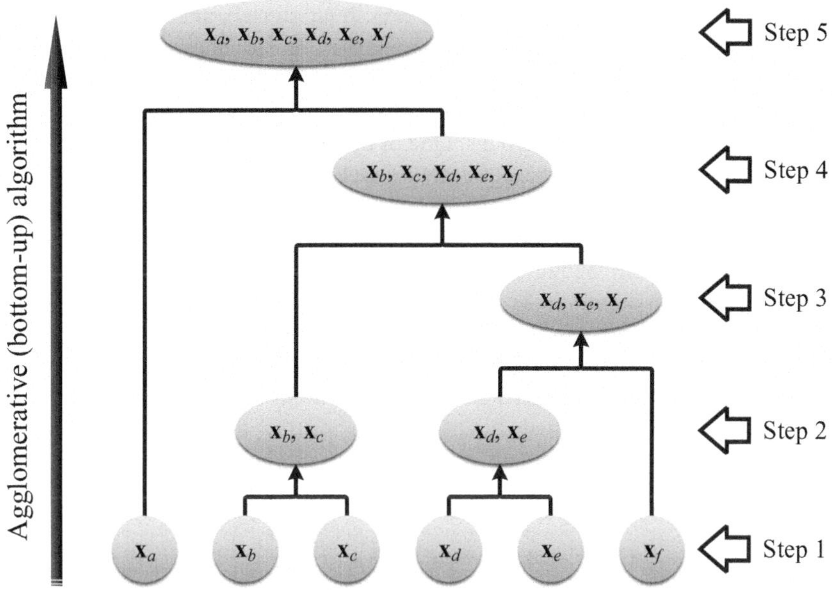

Fig. 6.2 The graphical schematic of the AHC

$$d_E(\mathbf{x}_a, \mathbf{x}_b) = \sqrt{\sum_{i=1}^{n_a} (x_a(i) - x_b(i))^2} \tag{6.1}$$

where, n_a is the number of samples of the feature vectors. In the following, in Step 2 of Fig. 6.2, similar data points (clusters); that is, $\mathbf{x}_b$ and $\mathbf{x}_c$ as well as $\mathbf{x}_d$ and $\mathbf{x}_e$, are combined to make new clusters $[\mathbf{x}_b, \mathbf{x}_c]$ and $[\mathbf{x}_d, \mathbf{x}_e]$. Until the first two steps, one can derive four clusters $[\mathbf{x}_a]$, $[\mathbf{x}_b, \mathbf{x}_c]$, $[\mathbf{x}_d, \mathbf{x}_e]$, and $[\mathbf{x}_f]$. On this basis, the dendrogram is created by connecting these clusters with rectangular shapes so that the heights of these shapes refer to the distances between the data points. By computing the distances between the new clusters, the cluster generation process is continued. In Fig. 6.2, the new clusters of Step 3 are $[\mathbf{x}_a]$, $[\mathbf{x}_b, \mathbf{x}_c]$, and $[\mathbf{x}_d, \mathbf{x}_e, \mathbf{x}_f]$ forming new rectangular shapes with higher heights compared to the previous steps. This procedure continues to merge similar pair clusters, i.e., the clusters $[\mathbf{x}_b, \mathbf{x}_c]$, and $[\mathbf{x}_d, \mathbf{x}_e, \mathbf{x}_f]$ are merged together to form a new cluster (Step 4), in which case there are two clusters $[\mathbf{x}_a]$ and $[\mathbf{x}_b, \mathbf{x}_c, \mathbf{x}_d, \mathbf{x}_e, \mathbf{x}_f]$. Finally, the two remaining clusters are combined to form a single cluster (Step 5). From Fig. 6.2, it can be realized that the AHC gives a dendrogram plot of the hierarchical binary cluster tree called *linkage*, which is the distance between two clusters. To generate the dendrogram, one can use some popular distance-oriented linkages such as the shortest distance (single linkage), furthest distance (complete linkage), unweighted average distance (average linkage), centroid distance (centroid linkage), and inner squared distance (Ward's linkage). Assuming that $c_i(k)$ and $c_j(l)$ are the kth and lth samples of the clusters $\mathbf{c}_i$ and $\mathbf{c}_j$ with n_i and n_j samples (i.e., both clusters are vectors), where $k = 1,\ldots, n_i$ and $l = 1,\ldots, n_j$, Table 6.1 summarizes the equations of these linkage functions.

Once the dendrogram of the AHC has been created, it is necessary to divide clustered data into well-structured groups. One of the effective and efficient approaches is to pre-determine the number of clusters. In this case, one can simply draw a horizontal line that intersects some vertical lines of the dendrogram. In other words, the number of vertical lines of the dendrogram intersected by the pre-determined horizontal line refers to the number of clusters. For this purpose, it suffices to apply a cluster evaluation metric and confirm the choice of the optimal cluster number. The Dunn's index is one of the well-known and useful metrics for cluster number evaluation [35], which is incorporated into this research. In this index, the maximum diameter across all clusters is used to represent intra-cluster compactness, while the lowest pairwise distance between data points in distinct clusters is considered to represent inter-cluster separation. The Dunn' index is defined as:

$$DI = \min_{1 \le i \le c_s}\left(\min_{1 \le j \le c_s, i \neq j}\left(\frac{\delta(\mathbf{C}_i, \mathbf{C}_j)}{\max\limits_{1 \le q \le c_s} \Delta(\mathbf{C}_q)}\right)\right) \tag{6.2}$$

where, $\delta(\mathbf{C}_i,\mathbf{C}_j)$ denotes the inter-cluster distance between the clusters $\mathbf{C}_i$ and $\mathbf{C}_j$; $\Delta(\mathbf{C}_q)$ represents the intra-cluster distance of the cluster $\mathbf{C}_q$, i.e., the distance within this cluster; and c_s is the number of sample clusters. Based on the c_s sample clusters, one of them with the maximum DI value is adopted as the optimal cluster number. Given the entire training set, which reflects the feature matrix concerning the structural undamaged status, i.e., $\mathbf{X} \in \Re^{p \times n}$, where, p and n signify the amounts of vibrating modes and observations (frequency samples or features), the output of the first step of the proposed method is c clusters $\{\mathbf{C}_1,\ldots,\mathbf{C}_c\}$, each of which is a matrix of p variables and n_s samples, where $s = 1,\ldots,c$ and $n_s < n$.

Table 6.1 Different linkage functions for generating the dendrogram plot of the AHC

Linkage function	Equations
Shortest distance	$d(\mathbf{c}_i, \mathbf{c}_j) = \min(\lvert c_i(k) - c_j(l)\rvert)$
Furthest distance	$d(\mathbf{c}_i, \mathbf{c}_j) = \max(\lvert c_i(k) - c_j(l)\rvert)$
Unweighted average distance	$d(\mathbf{c}_i, \mathbf{c}_j) = \frac{1}{n_i n_j}\sum_{k=1}^{n_i}\sum_{l=1}^{n_j}\lvert c_i(k) - c_j(l)\rvert$
Centroid distance	$d(\mathbf{c}_i, \mathbf{c}_j) = \left\lvert\frac{1}{n_i}\sum_{k=1}^{n_i} c_i(k) - \frac{1}{n_j}\sum_{l=1}^{n_j} c_j(l)\right\rvert$
Inner squared distance	$d(\mathbf{c}_i, \mathbf{c}_j) = \sqrt{\frac{2n_i n_j}{(n_i+n_j)}}\left\lvert\frac{1}{n_i}\sum_{k=1}^{n_i} c_i(k) - \frac{1}{n_j}\sum_{l=1}^{n_j} c_j(l)\right\rvert$

6.2.2 Unsupervised Feature Selection of Clustered Data

In machine learning, feature selection is a technique for choosing relevant features and disregarding irrelevant ones from the original set. Unsupervised feature selection is a branch of this strategy, for which the labels of features are unknown and the selection procedure is thoroughly based on the intrinsic properties of features [42]. Generally, an unsupervised feature selection can be performed using filter, wrapper, and embedding algorithms [43]. Here, we use the filter approach to choose one of the c clusters since it is straightforward and computationally efficient. The selected cluster is called the *relevant cluster* $\mathbf{C}_r$, which is obtained by the first step of the integrated approach to modeling the LOF-based anomaly detector (see Sect. 6.2.3).

Because the proposed approach does not consider any learning process, it is necessary to use a criterion for choosing the relevant cluster. With the aid of the theory of small distances within inter-clusters (i.e., all features in a cluster have small distances between them), $\mathbf{C}_r$ has the minimum distance value or contribution relative to other clusters. Since the clusters are matrices, the most suitable and effective distance measure is the Mahalanobis-squared distance (MSD) [44, 45]. Mathematically speaking, for a given matrix $\mathbf{X} \in \Re^{p\times n}$, the MSD of the ith column vector ($\mathbf{x}_i$) of this matrix is expressed as follows:

$$d_{MSD}(\mathbf{x}_i) = (\mathbf{x}_i - \overline{\mathbf{x}})^T \boldsymbol{\Sigma}_{\mathbf{x}}^{-1}(\mathbf{x}_i - \overline{\mathbf{x}}), \forall i = 1, \ldots, n \tag{6.3}$$

where, $\overline{\mathbf{x}} \in \Re^p$ and $\boldsymbol{\Sigma}_{\mathbf{X}} \in \Re^{p\times p}$ are the mean vector and covariance matrix of $\mathbf{X}$, respectively. Inspired by this statistical distance, the relevant cluster is selected when the following equation to be minimized:

$$d_{MSD}(\mathbf{C}_i) = \min_i \sum_{j=1}^{c-1} \sum_{h=1}^{p} \left(\mathbf{c}_h^{(j)} - \overline{\mathbf{c}}_i\right)^T \boldsymbol{\Sigma}_i^{-1}\left(\mathbf{c}_h^{(j)} - \overline{\mathbf{c}}_i\right) \tag{6.4}$$

where, $\mathbf{c}_h^{(j)}$ denotes the hth column vector of the jth cluster $\mathbf{C}_j$; $\overline{\mathbf{c}}_i$ and $\boldsymbol{\Sigma}_i$ is the mean vector and covariance matrix of the ith cluster $\mathbf{C}_i$; and $d_{MSD}(\mathbf{C}_i)$ is the distance value of the ith cluster. Accordingly, it can be determined c distance values $\{d_{MSD}(\mathbf{C}_1),\ldots,d_{MSD}(\mathbf{C}_c)\}$ in such a way that the relevant cluster $\mathbf{C}_r \in \Re^{p\times k}$ is one that has the minimum distance quantity.

6.2.3 Anomaly Detection by Local Outlier Factor

The LOF presents an unsupervised anomaly detector based on distance and density of data points [46, 47]. This method generates an anomaly index for every data point by dividing the local density of the region around a data point by using the local densities of the data points that surround it. This theory aids us to leverage the

LOF to detect anomalies (i.e., damage) by determining how dense a data point is proportionate to its surroundings. First, one needs to find some nearest neighbors of the data of interest and subsequently determine the local reachability density values of that data and its nearest neighbors. While choosing suitable nearest neighbors is a requirement for the LOF technique, this issue is not necessary in the proposed integrated method. This is because the second step of the proposed method chooses the relevant cluster, where its features are the nearest samples to all training data. In other words, the features in $\mathbf{C}_r$ can be selected as the main nearest neighbors of each training point. Therefore, the number of features (k) of the relevant cluster $\mathbf{C}_r \in \Re^{p\times k}$ refers to the number of nearest neighbors.

Given the training matrix $\mathbf{X}$, the LOF-aided anomaly detection is carried out in four steps. In the first step, the distances between the ith feature $\mathbf{x}_i$, where $i = 1,\ldots,n$, and its k nearest neighbors are computed by a distance measure, e.g., the Euclidean distance is a prime suggestion. The next step involves ascertaining the reachability distance between $\mathbf{x}_i$ and $\hat{\mathbf{x}}_o$, which is a new data sample and represents a characteristic of the pertinent cluster, where o represents one of the features of the relevant cluster ($o = 1,\ldots,k$):

$$d_R(\mathbf{x}_i, \hat{\mathbf{x}}_o) = \max\left(d_E(\mathbf{x}_i, \hat{\mathbf{x}}_o), d_{min}(\hat{\mathbf{x}}_o)\right) \tag{6.5}$$

In this equation, $d_E(\mathbf{x}_i, \hat{\mathbf{x}}_o)$ signifies the distance between $\mathbf{x}_i$ and $\hat{\mathbf{x}}_o$ gained by the Euclidean metric and $d_{\min}(\hat{\mathbf{x}}_o)$ is the minimum distance among all k nearest neighbors of $\mathbf{x}_i$. For the third step, the local reachability density of $\mathbf{x}_i$ and $\hat{\mathbf{x}}_o$ are expressed as:

$$\rho_R(\mathbf{x}_i) = \frac{k}{\sum_{\hat{\mathbf{x}}_o \in \mathbf{C}_r} d_R(\mathbf{x}_i, \hat{\mathbf{x}}_o)} \tag{6.6}$$

$$\rho_R(\hat{\mathbf{x}}_o) = \frac{k}{\sum_{\hat{\mathbf{x}}_v \in \mathbf{C}_r} d_R(\hat{\mathbf{x}}_o, \hat{\mathbf{x}}_v)} \tag{6.7}$$

Eventually, the fourth step determines the LOF value of $\mathbf{x}_i$ as follows:

$$LOF(\mathbf{x}_i) = \frac{1}{k} \sum_{\hat{\mathbf{x}}_o \in \mathbf{C}_r} \frac{\rho_R(\hat{\mathbf{x}}_o)}{\rho_R(\mathbf{x}_i)} \tag{6.8}$$

Having considered n training points, it can be obtained n anomaly indices of the undamaged structure; that is, $\{LOF(\mathbf{x}_1), \ldots, LOF(\mathbf{x}_n)\}$. Assume that $\mathbf{z}$ is a test point regarding the unknown structural state during the monitoring period. To obtain the anomaly index of this point, the four steps of the LOF-based anomaly detection should be performed by using the test point $\mathbf{z}$ and the selected features in the relevant cluster $\mathbf{C}_r$. In other words, it is necessary to determine the reachability distance between $\mathbf{z}$ and $\hat{\mathbf{x}}_o$ as follows:

$$d_R(\mathbf{z}, \hat{\mathbf{x}}_o) = \max\left(d_E(\mathbf{z}, \hat{\mathbf{x}}_o), d_{\min}(\hat{\mathbf{x}}_o)\right) \tag{6.9}$$

where, $\hat{\mathbf{x}}_o$ comes from the training data. Subsequently, the local reachability density of $\mathbf{z}$ is given by:

$$\rho_R(\mathbf{z}) = \frac{k}{\sum_{\hat{\mathbf{x}}_o \in \mathbf{C}_r} d_R(\mathbf{z}, \hat{\mathbf{x}}_o)} \tag{6.10}$$

Finally, the LOF value of the test data point $\mathbf{z}$ is obtained as follows:

$$LOF(z) = \frac{1}{k} \sum_{\hat{\mathbf{x}}_o \in \mathbf{C}_r} \frac{\rho_R(\hat{\mathbf{x}}_o)}{\rho_R(\mathbf{z})} \tag{6.11}$$

To make sure of a verifiable feature analysis, one should contrast all available values of anomalies with a threshold limit. This threshold is gained by using the undamaged anomaly scores $\{LOF(\mathbf{x}_1), \ldots, LOF(\mathbf{x}_n)\}$. On this basis, since these indices belong to the undamaged state, it is anticipated that they will drop below the threshold. Regarding the anomaly index of the current structural state $LOF(\mathbf{z})$, the proposed approach will declare that the structure is healthy if the index of interest falls below the threshold limit; otherwise, the approach alerts that the structure has undergone damage. Despite proposing various threshold estimators based on the central limit theory (CLT) [19, 28] and extreme value theory (EVT) [15, 22, 24, 26], we leverage a recently published estimator under the EVT proposed by Sarmadi et al. [43] to estimate the decision threshold.

6.3 Performance Evaluation

6.3.1 Description of the Bridge Structure

This structure is a concrete cable-stayed bridge with five spans, which was constructed in China. Figure 6.3a depicts a real picture of this bridge structure. The primary components of this structure emanate from two transversal beams attached to the piers and two double H-type towers, seventy-four prestressed concrete girder segments, and double cable planes. The bridge has a total length of 510 m, with the main span of 260 m and two side spans of 22.15 + 99.85 m as shown in Fig. 6.3b. The height of both towers corresponds to 60.5 m. The standard segment has the length of 5.8 m, and its cross section is shown in Fig. 6.3c. There existed eighty-eight couples of steel-wire cables with a diameter of 5 mm, with the minimum and maximum numers of wires ranging from 69 to 199. The main piers that supported the towers were built on caisson foundations, whereas secondary piers and abutments were formed on a number of tube piles equipped with pot-type rubber compression bearings.

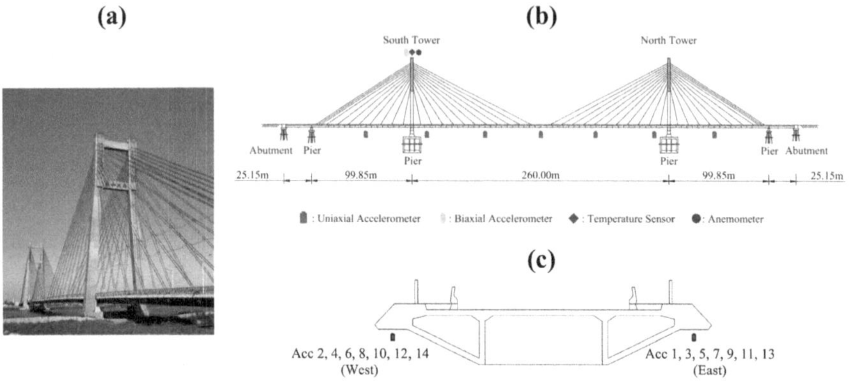

Fig. 6.3 **a** The cable-stayed bridge [48], **b** the elevation outlook and the key sizes along with the sensor systems, **c** the cross-section and the distribution of the accelerometers (i.e., Acc)

At the bottom of a girder segment spanning the mid-span, cracks up to 20 mm wide were identified after 19 years of the bridge service. Because the weights and numbers of vehicles on the bridge were significantly higher than those predicted in the original design, the cracks may have been caused by overweight vehicles. Figure 6.4 illustrates the crack occurred at the mid-span girder segment. Furthermore, the stay cables were highly corroded, particularly those close to the anchors. Repairs were carried out between 2005 and 2007, including the replacement of all stay cables and the recasting of the mid-span girder.

During the period of restoration and maintenance, an SHM system operated by researchers from the Harbin Institute of Technology (i.e., China) was incorporated to

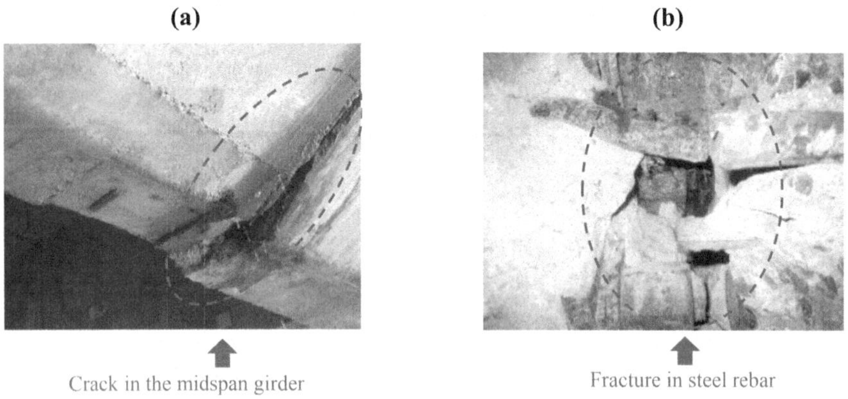

Fig. 6.4 Detailed pictures of the damage patterns in the cable-stayed bridge before installing the sensor systems: **a** the crack at the midspan girder segment, **b** the close view of the crack (i.e., the original pictures were taken by Li [49])

Table 6.2 Frequency samples in the short-term monitoring program

Date	Bridge status	Frequency samples
01 January 2008	Undamaged	1–24
17 January 2008	Undamaged	25–48
03 February 2008	Undamaged	49–72
19 March 2008	Undamaged	73–96
30 March 2008	Undamaged	97–120
19 April 2008	Undamaged	121–144
05 May 2008	Undamaged	145–168
18 May 2008	Undamaged	169–192
31 July 2008	Damaged	193–216

measure and collect structural responses and environmental/operational data. From Fig. 6.3b, the main sensor systems included fourteen uniaxial accelerometers permanently installed on the bridge deck, one biaxial accelerometer located on the south tower for monitoring the horizontal vibration of the tower, a temperature sensor and an anemometer mounted on top of the south tower for recording the air temperature and wind speed, respectively. Short-term monitoring data from some days in 2008 has been released to examine different SHM methods, while this research only considers the measured vibration (acceleration) data over nine days, i.e., see Table 6.2. The acceleration time histories were recorded hourly per day at a sampling frequency of 100 Hz. Among the fourteen uniaxial accelerometers, Acc #10 provided meaningless data points [22]; hence, the responses of Acc #1–9 and #11–14 are only considered to extract short-term modal frequencies of the bridge [24, 26].

6.3.2 *Investigation of the Bridge Modal Frequencies*

Frequency domain decomposition was previously utilized by the authors [26] to identify the bridge modal frequencies. Since the recording of acceleration responses was performed hourly, each day of the monitoring program contained 24 samples of the modal frequencies. Totally, four stable modes were extracted leading to a matrix of modal frequencies consisting of four vibrating modes (p) and 216 samples. Figure 6.5 shows the evolution of the bridge natural frequencies under four modes over nine days of the short-term monitoring. Moreover, Table 6.2 presents more details on the identified frequencies and their sample distributions. In this regard, the first eight days of the monitoring program relate to the undamaged state, while the final day is associated with the bridge damaged condition [48].

From Fig. 6.5 it is observed that the modal frequencies of the undamaged state vary significantly. Most of these variations follow upward and downward trends. Since there was no damage during the period from 01 January 2008 to 18 May 2008, such variability is most likely due to environmental and operational conditions such

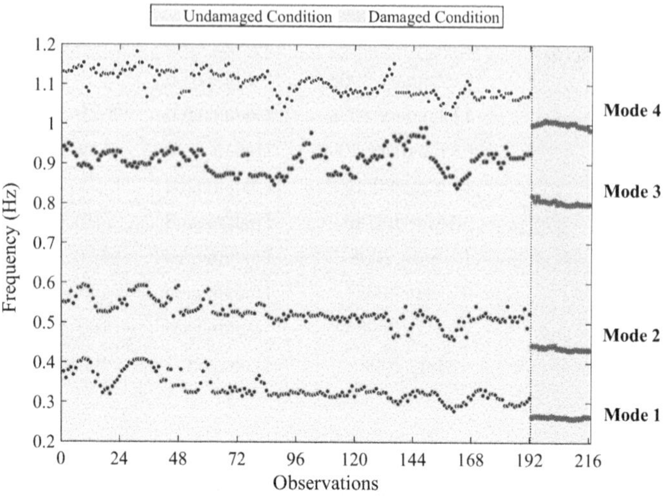

Fig. 6.5 Natural frequencies of the cable-stayed bridge

as daily ambient temperature fluctuations, humidity and wind changes, and traffic [13]. Each of these factors has the potential to alter (raise or decrease) the bridge modal frequencies in distinct ways [30]. The ambient temperature and wind speed were mentioned as two influential factors for changes in the bridge natural frequencies in the study by Li et al. [50]. Importantly, the positive aerodynamic stiffness had an impact on the bridge and raised the modal frequencies when weak winds were predominant. This event was likely the main reason for increasing the undamaged modal frequencies. In addition, the downward shifts in the bridge modal frequencies might be stemmed from relative increases in daily temperature or traffic volume. Generally, such unpredictable variability with uncertainties is a challenging issue in short-term monitoring processes and makes it problematic for correct decision-making. Thus, one can infer that the assessment of damage occurrence and investigation of the environmental and operational effects are not only important for long-term monitoring but also necessary for short-term monitoring.

6.3.3 Short-Term Assessment of Damage Occurrence

The limited collection of the bridge natural frequencies with hidden and uncertain environmental/operational variations is split into training and test datasets. Since the number of feature samples is limited, the training data consist of 172 samples (n) out of 192 undamaged features with the training ratio of 90% leading to the matrix $\mathbf{X} \in \Re^{4\times172}$. The test data comprise the remaining 20 samples of the undamaged features (i.e., known test points, which serve as validation data), and 24 damaged features generating the matrix $\mathbf{Z} \in \Re^{4\times44}$. Using the training matrix, the initial stage of the proposed approach begins by grouping 172 training points into clusters. Figure 6.6 displays the dendrogram of these points obtained from the Ward's linkage method and the Euclidean distance metric. Using the Dunn's index under 10 sample clusters ($c_s = 10$), Fig. 6.7a illustrates the evolution of the DI values, for which the optimum cluster number (c) coincided with the maximum Dunn's index is identical to 4. It should be clarified that the fourth-sixth clusters have the same maximum values of the Dunn's index; however, the fourth cluster is taken as the optimum choice in order to avoid oversampling in clusters. This phenomenon occurs when a large number of clusters is considered, some clusters may contain insufficient data samples. In such cases, the best choice is to use the minimum number of appropriate clusters. Figure 7b shows the distribution of the cluster labels over the training points. Therefore, the training matrix $\mathbf{X}$ is divided into four clusters $\{\mathbf{C}_1,\ldots,\mathbf{C}_4\}$, whose feature numbers are observable in Fig. 6.7b.

Once the training points have been clustered, the second stage of the proposed approach should choose the relevant cluster through the proposed unsupervised feature selection. In this regard, Fig. 6.8a shows the distance (MSD) contributions of all four clusters. As can be seen, the first cluster yields the minimum distance contribution (i.e., 4% out of 100%) and it is the best choice for the relevant cluster. Accordingly, Fig. 6.8b displays the 20 relevant features or modal frequencies ($k = 20$), which treat as the nearest neighbors of all training data. A simple comparison between the selected features and the original features presented in Fig. 6.5 reveals that the idea of choosing the relevant cluster can significantly reduce the influences of environmental variability. To further evaluation, Fig. 6.9 investigates the variability

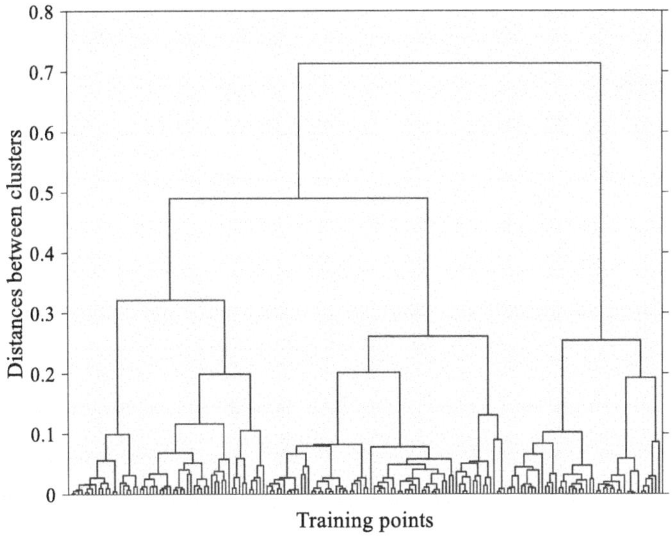

Fig. 6.6 The dendrogram of the training points based on the Ward's linkage function

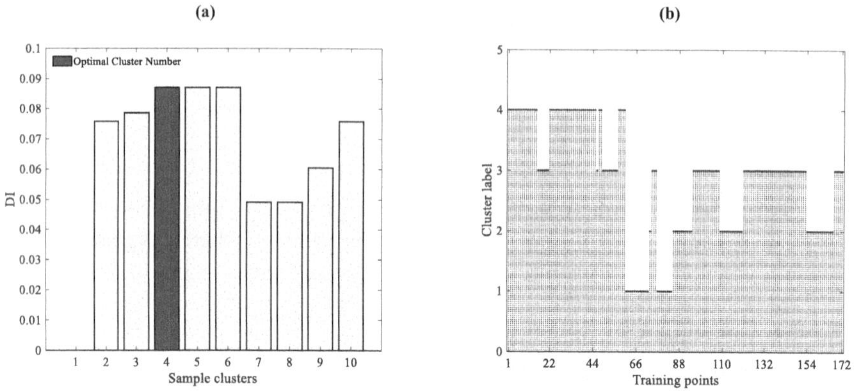

Fig. 6.7 The main outputs of the AHC: **a** the cluster number selection via the Dunn's index, **b** the cluster label distribution over the training points

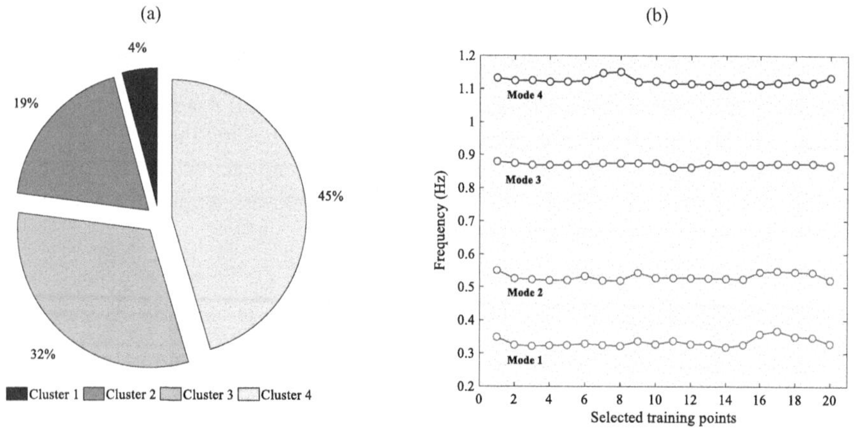

Fig. 6.8 The main outputs of the feature selection step: **a** the distance contributions over the clusters, **b** the selected training points in the optimal set (Cluster 1)

rates of the original and clustered features by the variance and median absolute deviation (MAD). As can be observed, the relevant cluster $\mathbf{C}_1$ not only possesses the minimum distance contribution but also yields the lower rate of variability compared to the original feature set ($\mathbf{X}$) and other clustered datasets except for the first mode. Hence, one can deduce that the features within $\mathbf{C}_1$ are suitable for developing the LOF model.

The outcome of the bridge damage assessment is presented in Fig. 6.10, where the first 172 LOF values labeled "Training Points (UC)" are used to estimate a decision threshold through an EVT-based technique, as discussed in detail in Sarmadi et al. [43]. In Fig. 6.10, the LOF quantities of the samples labeled "Validation Points (UC)" and "Testing Points (DC)" refer to the known and unknown test points considered in

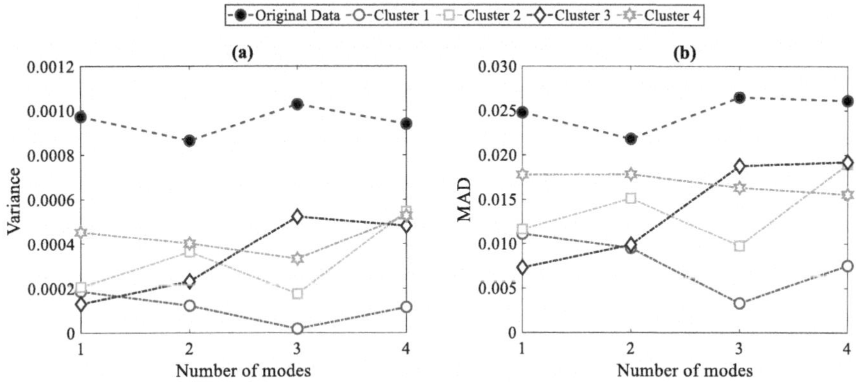

Fig. 6.9 Evaluation of variability rates of the original and clustered datasets: **a** variance, **b** MAD

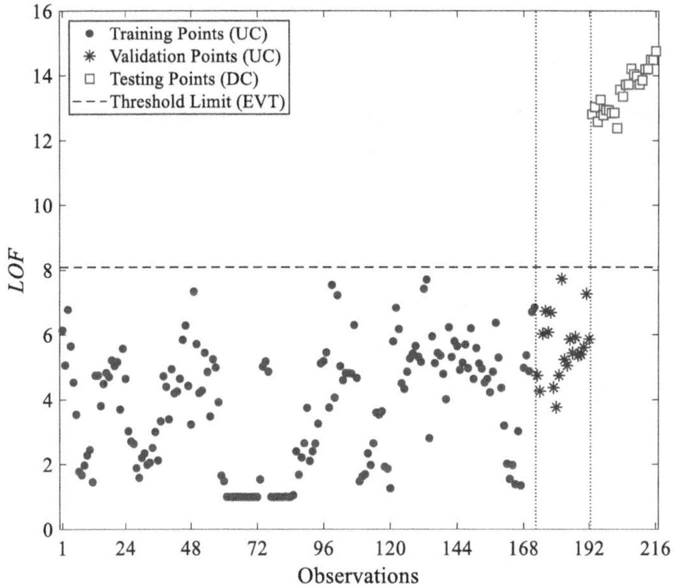

Fig. 6.10 Damage assessment of the cable-stayed bridge by the proposed integrated method

the monitoring phase. It is seen that the anomaly indices related to the undamaged state put below the horizontal line (i.e., threshold), which correctly indicate that no damage occurred during 01 January 2008 to 18 May 2008. In contrast, all LOF values of the samples 193–216 concerning 31 July 2008 exceed the threshold limit accurately implying the damage occurrence detected by the developed integrated approach.

Despite reliable and satisfactory result of damage assessment achieved by the approach developed in this research, it is compared to a popular anomaly detection technique based on the MSD [44]. This popularity is due to its simplicity, computational efficiency, and non-parametric nature. In order to have a fair comparison, the anomaly indices (i.e., MSD values) of this technique are compared with the EVT- and CLT-based threshold estimators, where the latter is a well-known technique under a 95% standard confidence interval. The EVT-aided threshold is obtained from the same technique [43]. The outcome of the bridge damage assessment via the MSD-based anomaly detection is presented in Fig. 6.11. In case of applying the EVT for threshold estimation, as Fig. 6.11a shows, all MSD values of the undamaged state lie below the threshold correctly indicating that no damage occurred. Despite such a reasonable result, it can be seen that all anomalous quantity in the damaged condition is below the threshold, proving the existence of a complete false negative. On the other hand, although the result in Fig. 6.11a demonstrates that the MSD quantities of the damaged state through the CLT-aided threshold estimator have been detected and labeled correctly, there are plenty of erroneous warnings (i.e., false positives) associated with the anomaly values associated with the undamaged condition. Without considering the thresholds, it is evident that for some samples, the anomaly indices pertaining to the undamaged state are larger than or equal to the corresponding indices of the damaged condition. Moreover, sudden increases and significant variability in the MSD values of the undamaged state are observable. These findings support the deficient performance of MSD-based anomaly detection in the short-term health monitoring of the cable-stayed bridge. Furthermore, when there is relatively high variability in dynamic features, the developed integrated technique is preferable to the well-known MSD.

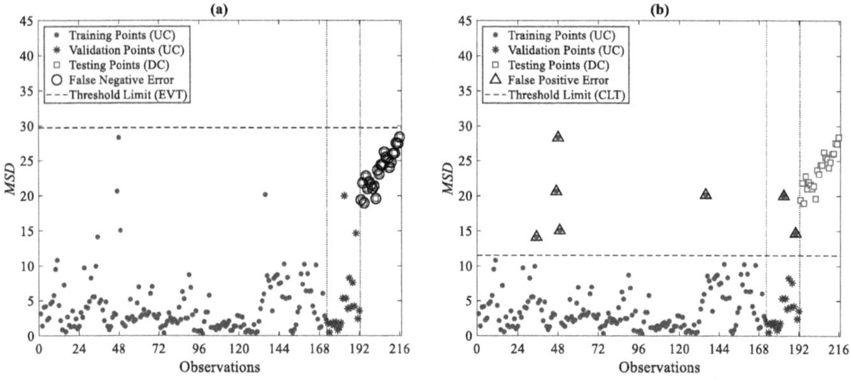

Fig. 6.11 Damage assessment of the cable-stayed bridge by the MSD: **a** the EVT-aided threshold estimator [43], **b** the CLT-aided threshold estimator under the 95% confidence interval

6.4 Conclusions

For the purposes of detecting damage to bridge structures over a short duration using modal frequencies, this chapter developed an integrated unsupervised learning method containing three steps of local feature segmentation, unsupervised feature selection, and anomaly detection. For the first step, the AHC combined with the Dunn's index for the cluster number selection was used to segment the entire training data into pre-determined clusters. Using the clustered data, a filter-based unsupervised feature selector via the MSD was proposed to choose the most appropriate features within one of the clusters. The LOF was suggested to establish an anomaly detection model that undertook computing the anomaly scores of the training and test samples for assessing damage employing features that were selected in the relevant cluster. A small set of modal frequencies of a cable-stayed bridge was considered to verify the proposed method.

The main conclusions of this research can be summarized as:

- The investigation of the bridge natural frequencies identified from a short period of a monitoring process indicated that the dynamic responses of cable-supported bridges can be influenced by a variety of environmental and operational factors including air temperature, humidity, wind direction and speed, and traffic even in a short-term monitoring program. Therefore, it is imperative to take notice to removing such conditions in not only long-term monitoring but also short-term monitoring programs.
- Having considered the environmental and operational variations in the small set of modal frequencies, the proposed approach could effectively mitigate such variations and improve the detectability of damage. Accordingly, the approach could properly discriminate the damaged state of the bridge from its undamaged one without any false alarm (false positive) and mis-detection (false negative) errors.
- When the problem of environmental and operational variability is solvable, the EVT-based threshold estimators are the best choices for unsupervised anomaly detection.
- The proposed unsupervised feature selector could choose the most proper features relevant to damage and not to the environmental and operational conditions. The selected features had the lower rates of variability compared to the original training data and also other clustered sets. This conclusion highlights the positive influence of unsupervised feature selection on early damage assessment.
- The proposed integrated unsupervised learning approach was superior to the state-of-the-art MSD-based anomaly detection, which used the entire training data. This conclusion also validates the positive effects of the first and second steps of the proposed approach to yielding more reliable results of damage assessment. Therefore, the utilization of local data makes better SHM outcomes compared to the entire data.

References

1. Xu Y-L, Xia Y (2012) Structural health monitoring of long-span suspension bridges. CRC Press, New York, United States
2. Mao J-X, Wang H, Feng D-M, Tao T-Y, Zheng W-Z (2018) Investigation of dynamic properties of long-span cable-stayed bridges based on one-year monitoring data under normal operating condition. Struct Contr Health Monit 25(5):e2146. https://doi.org/10.1002/stc.2146
3. Lynch JP, Sohn H, Wang ML (2022) Sensor technologies for civil infrastructures: volume 1: sensing hardware and data collection methods for performance assessment, 2nd edn. Elsevier, Cambridge, MA, United States. https://doi.org/10.1016/C2017-0-02662-9
4. Sarmadi H, Entezami A, Yuen K-V, Behkamal B (2023) Review on smartphone sensing technology for structural health monitoring. Meas 223:113716. https://doi.org/10.1016/j.measurement.2023.113716
5. Katam R, Pasupuleti VDK, Kalapatapu P (2023) A review on structural health monitoring: past to present. Innov Infrastruct Solutions 8(9):248. https://doi.org/10.1007/s41062-023-01217-3
6. Li H, Ou J (2016) The state of the art in structural health monitoring of cable-stayed bridges. J Civ Struct Health Monit 6(1):43–67
7. Soleymani A, Jahangir H, Nehdi ML (2023) Damage detection and monitoring in heritage masonry structures: systematic review. Constr Build Mater 397:132402. https://doi.org/10.1016/j.conbuildmat.2023.132402
8. Kontoni D-PN, Kumar A, Arora HC, Jahangir H, Kapoor NR (2024) Damage detection in reinforced concrete structures using advanced automatic systems: an overview. Autom Constr Toward Resilience 463–484. https://doi.org/10.1201/9781003325246-22
9. Tefera B, Zekaria A, Gebre A (2023) Challenges in applying vibration-based damage detection to highway bridge structures. Asian J Civ Eng 24(6):1875–1894. https://doi.org/10.1007/s42107-023-00594-5
10. Hou R, Xia Y (2021) Review on the new development of vibration-based damage identification for civil engineering structures: 2010–2019. J Sound Vib 491:115741. https://doi.org/10.1016/j.jsv.2020.115741
11. Rezaiee-Pajand M, Sarmadi H, Entezami A (2021) A hybrid sensitivity function and Lanczos bidiagonalization-Tikhonov method for structural model updating: application to a full-scale bridge structure. Appl Math Model 89:860–884. https://doi.org/10.1016/j.apm.2020.07.044
12. Daneshvar MH, Saffarian M, Jahangir H, Sarmadi H (2023) Damage identification of structural systems by modal strain energy and an optimization-based iterative regularization method. Eng Comput 39:2067–2087. https://doi.org/10.1007/s00366-021-01567-5
13. Entezami A, Sarmadi H, Salar M, De Michele C, Nadir Arslan A (2021) A novel data-driven method for structural health monitoring under ambient vibration and high dimensional features by robust multidimensional scaling. Struct Health Monit 20(5):2758–2777. https://doi.org/10.1177/1475921720973953
14. Zhang C, Mousavi AA, Masri SF, Gholipour G, Yan K, Li X (2022) Vibration feature extraction using signal processing techniques for structural health monitoring: a review. Mech Syst Sig Process 177:109175. https://doi.org/10.1016/j.ymssp.2022.109175
15. Sarmadi H, Yuen K-V (2021) Early damage detection by an innovative unsupervised learning method based on kernel null space and peak-over-threshold. Copmut Aided Civil Infrastruct Eng 36(9):1150–1167. https://doi.org/10.1111/mice.12635
16. Sarmadi H (2021) Investigation of machine learning methods for structural safety assessment under variability in data: Comparative studies and new approaches. J Perform Constr Facil 35(6):04021090. https://doi.org/10.1061/(ASCE)CF.1943-5509.0001664
17. Alpaydin E (2014) Introduction to machine learning. MIT Press
18. Harle SM (2023) Advancements and challenges in the application of artificial intelligence in civil engineering: a comprehensive review. Asian J Civ Eng. https://doi.org/10.1007/s42107-023-00760-9

19. Entezami A, Sarmadi H, Behkamal B, De Michele C (2023) On continuous health monitoring of bridges under serious environmental variability by an innovative multi-task unsupervised learning method. Struct Infrastruct Eng 20(12):1975–1993. https://doi.org/10.1080/15732479.2023.2166538
20. Sarmadi H, Entezami A (2021) Application of supervised learning to validation of damage detection. Arch Appl Mech 91(1):393–410. https://doi.org/10.1007/s00419-020-01779-z
21. Daneshvar MH, Gharighoran A, Zareei SA, Karamodin A (2021) Early damage detection under massive data via innovative hybrid methods: application to a large-scale cable-stayed bridge. Struct Infrastruct Eng 17(7):902–920. https://doi.org/10.1080/15732479.2020.1777572
22. Sarmadi H, Yuen K-V (2022) Structural health monitoring by a novel probabilistic machine learning method based on extreme value theory and mixture quantile modeling. Mech Syst Sig Process 173:109049. https://doi.org/10.1016/j.ymssp.2022.109049
23. Svendsen BT, Frøseth GT, Øiseth O, Rønnquist A (2022) A data-based structural health monitoring approach for damage detection in steel bridges using experimental data. J Civ Struct Health Monit 12(1):101–115. https://doi.org/10.1007/s13349-021-00530-8
24. Daneshvar MH, Sarmadi H (2022) Unsupervised learning-based damage assessment of full-scale civil structures under long-term and short-term monitoring. Eng Struct 256:114059. https://doi.org/10.1016/j.engstruct.2022.114059
25. Lydon D, Kromanis R, Lydon M, Early J, Taylor S (2022) Use of a roving computer vision system to compare anomaly detection techniques for health monitoring of bridges. J Civ Struct Health Monit 12(6):1299–1316. https://doi.org/10.1007/s13349-022-00617-w
26. Sarmadi H, Entezami A, De Michele C (2023) Probabilistic data self-clustering based on semi-parametric extreme value theory for structural health monitoring. Mech Syst Sig Process 187:109976. https://doi.org/10.1016/j.ymssp.2022.109976
27. Wang Z, Yi T-H, Yang D-H, Zhou P, Sun L (2023) Early warning method of structural damage using localized frequency cointegration under changing environments. J Struct Eng 149(2):04022230. https://doi.org/10.1061/(ASCE)ST.1943-541X.0003480
28. Entezami A, Sarmadi H, Behkamal B (2023) Long-term health monitoring of concrete and steel bridges under large and missing data by unsupervised meta learning. Eng Struct 279:115616. https://doi.org/10.1016/j.engstruct.2023.115616
29. Sarmadi H, Entezami A, Magalhães F (2023) Unsupervised data normalization for continuous dynamic monitoring by an innovative hybrid feature weighting-selection algorithm and natural nearest neighbor searching. Struct Health Monit 22(6):4005–4026. https://doi.org/10.1177/14759217231166116
30. Wang Z, Yang D-H, Yi T-H, Zhang G-H, Han J-G (2022) Eliminating environmental and operational effects on structural modal frequency: a comprehensive review. Struct Contr Health Monit 29(11):e3073. https://doi.org/10.1002/stc.3073
31. Han Q, Ma Q, Xu J, Liu M (2021) Structural health monitoring research under varying temperature condition: a review. J Civ Struct Health Monit 11(1):149–173. https://doi.org/10.1007/s13349-020-00444-x
32. Larsen A, Larose GL (2015) Dynamic wind effects on suspension and cable-stayed bridges. J Sound Vib 334:2–28. https://doi.org/10.1016/j.jsv.2014.06.009
33. Entezami A, Shariatmadar H, De Michele C (2022) Non-parametric empirical machine learning for short-term and long-term structural health monitoring. Struct Health Monit 21(6):2700–2718. https://doi.org/10.1177/14759217211069842
34. Entezami A, De Michele C, Arslan AN, Behkamal B (2022) Detection of partially structural collapse using long-term small displacement data from satellite images. Sensors 22(13):4964. https://doi.org/10.3390/s22134964
35. Aggarwal CC, Reddy CK (2016) Data clustering: algorithms and applications. CRC Press
36. Zonno G, Aguilar R, Boroschek R, Lourenço PB (2018) Automated long-term dynamic monitoring using hierarchical clustering and adaptive modal tracking: validation and applications. J Civ Struct Health Monit 8(5):791–808. https://doi.org/10.1007/s13349-018-0306-3
37. Chen S, Zhou S, Chen C, Li Y, Zhai S (2019) Detection of double defects for plate-like structures based on a fuzzy c-means clustering algorithm. Struct Health Monit 18(3):757–766

38. Sarmadi H, Entezami A, Salar M, De Michele C (2021) Bridge health monitoring in environmental variability by new clustering and threshold estimation methods. J Civ Struct Health Monit 11(3):629–644. https://doi.org/10.1007/s13349-021-00472-1
39. Wang S, Xu C, Gu C, Su H, Hu K, Xia Q (2020) Displacement monitoring model of concrete dams using the shape feature clustering-based temperature principal component factor. Struct Contr Health Monit 27(10):e2603. https://doi.org/10.1002/stc.2603
40. Salar M, Entezami A, Sarmadi H, De Michele C, Martinelli L (2022) On the data-driven damage detection of offshore structures using statistical and clustering techniques under various uncertainty sources: an experimental study. In: Current perspectives and new directions in mechanics, modelling and design of structural systems. CRC Press, pp 1797–1802. https://doi.org/10.1201/9781003348443-295
41. Daneshvar MH, Sarmadi H, Yuen K-V (2023) A locally unsupervised hybrid learning method for removing environmental effects under different measurement periods. Meas 208:112465. https://doi.org/10.1016/j.measurement.2023.112465
42. Solorio-Fernández S, Carrasco-Ochoa JA, Martínez-Trinidad JF (2020) A review of unsupervised feature selection methods. Artif Intell Rev 53(2):907–948
43. Sarmadi H, Entezami A, Behkamal B, De Michele C (2022) Partially online damage detection using long-term modal data under severe environmental effects by unsupervised feature selection and local metric learning. J Civ Struct Health Monit 12:1043–1066. https://doi.org/10.1007/s13349-022-00596-y
44. Sarmadi H, Entezami A, Saeedi Razavi B, Yuen K-V (2021) Ensemble learning-based structural health monitoring by Mahalanobis distance metrics. Struct Contr Health Monit 28(2):e2663. https://doi.org/10.1002/stc.2663
45. Sarmadi H, Karamodin A (2020) A novel anomaly detection method based on adaptive Mahalanobis-squared distance and one-class kNN rule for structural health monitoring under environmental effects. Mech Syst Sig Process 140:106495. https://doi.org/10.1016/j.ymssp.2019.106495
46. Breunig MM, Kriegel H-P, Ng RT, Sander J (2000) LOF: identifying density-based local outliers. In: Paper presented at the proceedings of the 2000 ACM SIGMOD international conference on management of data, Dallas, Texas, USA
47. Entezami A, Sarmadi H, Behkamal B (2023) A novel double-hybrid learning method for modal frequency-based damage assessment of bridge structures under different environmental variation patterns. Mech Syst Sig Process 201:110676. https://doi.org/10.1016/j.ymssp.2023.110676
48. Li S, Li H, Liu Y, Lan C, Zhou W, Ou J (2014) SMC structural health monitoring benchmark problem using monitored data from an actual cable-stayed bridge. Struct Contr Health Monit 21(2):156–172
49. Li H (2020) Structural assessment of concrete cable-stayed bridge after replacement of closure segment: the service stage. Pract Period Struct Des Constr 25(3):04020023. https://doi.org/10.1061/(ASCE)SC.1943-5576.0000508
50. Li H, Li S, Ou J, Li H (2010) Modal identification of bridges under varying environmental conditions: temperature and wind effects. Struct Contr Health Monit 17(5):495–512. https://doi.org/10.1002/stc.319

Chapter 7
Utilizing Rayleigh–Ritz Technique to Identify Damages of Multi Span Beams

Mohammad Hassan Daneshvar, Ali Reza Gharighoran, Seyed Reza Zareei, Hashem Jahangir, and Maria Rashidi

Abstract In this paper, a novel technique for identifying the location and the severity of damage in multi-spans beams and "beam and column" is proposed, which has an ability to reveal the structural damage with any type of support conditions. The proposed method is an expanded version of the finite element method (FE) by the assistance of the Ritz method which is called Ritz Damage Detection Method (RDDM). The innovation of the proposed method in this paper is the omission of some limitations of the RDDM method, including unique support conditions, and applying the effects of "beam and column". In this paper, by using the shape function, the definition of support conditions as a spring, and entering the interaction effect of "beam and column", the equations of RDDM method are developed to provide possible damage detection of the bridges with all support conditions. Also, singular value decomposition (SVD) method is used to determine the quantity and the severity of the damage, which is sensitive to dynamic characteristics changes, originated from the damage. The efficiency and capability of the proposed method for damage detection are evaluated by a numerical sample of "a multi-span beam" and

M. H. Daneshvar
Department of Civil Engineering, Ferdowsi University of Mashhad, Mashhad, Iran
e-mail: hasan.daneshvar@mail.um.ac.ir

A. R. Gharighoran
Department of Civil Engineering and Transportation, University of Isfahan, Isfahan, Iran
e-mail: a.ghari@trn.ui.ac.ir

S. R. Zareei
Department of Civil Engineering, Isfahan (Khorasgan) Branch, Islamic Azad University, Isfahan, Iran
e-mail: a.r.zareei@khuisf.ac.ir

H. Jahangir
Department of Civil Engineering, Research Group of Novel Technologies in Civil Engineering, University of Birjand, Birjand, Iran
e-mail: h.jahangir@birjand.ac.ir

M. Rashidi (✉)
Centre for Infrastructure Engineering, Western Sydney University, Sydney, NSW, Australia
e-mail: m.rashidi@westernsydney.edu.au

H. Jahangir et al. (eds.), *Damage Detection and Structural Health Monitoring of Concrete and Masonry Structures*, Springer Tracts in Civil Engineering,
https://doi.org/10.1007/978-981-97-8975-7_7

"beam to column connections. The investigation results showed that the proposed damage detection method has the ability to identify the location and the severity of the damages.

Keywords Rayleigh–Ritz technique · Damage indentification · Multi span beams · Finite element method

7.1 Introduction

Monitoring the structures to keep up and control their damages is an essential issue. The investigators have carried out a lot of research on the damage detection methods; most of which were based on finite element methods [1]. Some damage detection methods take benefits from utilizing static and vibrational based responses data, simultaneously. Eraky et al. (2015) focused on flexural structural elements and proposed damage localization indices to identify the location of its damage scenarios [2]. Limongell et al. (2016) tried to find the possible signs of defects in a post-tensioned concrete beam by conducted some static and dynamic experimental tests [3]. The literature review shows that there are a lot of finite element method (DEM) based damage detection techniques but limited research works focused on utilizing the basis of the well-known Rayleigh–Ritz technique to identify damages in structures.

The damage localization process in plate-like structures using the Ritz technique were studied by Li et al. (2002) [4]. Moreover, Sarker et al. (2011) tried to find the damage locations in shells of circular cylinders [5]. To Identify damages in laminated composite plates, Gracia et al. (2014) proposed a novel technique by utilizing derivatives of the vibrational mode shapes [6]. Maghsoodi et al. (2013) tried to propose a novel technique for crack localization and severity estimation of a Euler–Bernoulli beam with multiple spans [7]. They used rotational springs to model the cracks. Gharighoran et al. (2009) developed the Ritz techniques to detect the crack locations in simply supported one span reinforced and post-tensioned concrete beams [8].

This chapter utilized the Ritz technique to propose a novel and robust damage detection method to detect the location and severity of cracks in beam structures with any boundary condition. Actually, this investigation is assigned to improve the technique of Gharighoran et al. (2009) research work [8]. In this study, the damages are modelled by a reduction in flexural rigidity. In addition, the proposed damage detection technique utilizes the natural frequencies and vibrational mode shapes of the damaged structure.

7.2 Needful Equations

The free vibration behavior of a system can be shown by a differential equation of the mode shapes and structure frequencies are acquired from the system intrinsic features, such as mass (**M**), the damping value (**C**) and stiffness amount (**K**) as mentioned in Chopra [9]:

$$\mathbf{M\ddot{Z}} + \mathbf{C\dot{Z}} + \mathbf{KZ} = \mathbf{0} \tag{7.1}$$

In Eq. (7.1), $\mathbf{\ddot{Z}}$, $\mathbf{\dot{Z}}$ and $\mathbf{Z}$ respectively represent the vectors acceleration, velocity and displacement. To define damping value, Rayleigh's damping is used. On the other hand, the strain energy of a Euler–Bernoulli beam can be calculated by Eq. (7.2) Ilanko et al. [11]:

$$\Pi = \frac{1}{2}\int_0^L EI(x)\left(\frac{\mathrm{d}^2 w(x)}{\mathrm{d}x^2}\right)^2 \mathrm{d}x + \frac{1}{2}\int_0^L \rho A W^2 \mathrm{d}x + \sum_i^{nt} \frac{1}{2} K_i^t w^2(x_i) + \sum_i^{nr} \frac{1}{2} K_i^r w'^2(x_i) \tag{7.2}$$

In order to apply the boundary conditions, rotational and translational springs are utilized. In Eq. (7.2), K_i^t and K_i^r are rotational stiffness and translational stiffness, respectively. n^t and n^r are indicatives of the number of rotational and translational springs; the springs location is shown with x_i.

The multi-span bridges are sometimes made up of several connected beams and columns; thus, to model and detect the damages of the bridges with interaction effects between the pier and deck, their impact should be seen. Accordingly, the proposed equations of Lee and Ng (1994) are used to deploy the RDDM [11]. In the following, by using Lee and Ng's proposed equation, the Eq. (7.2) can be rewritten in the form of Eq. (7.3). In this equation, ns index represents the number of structure's members (beams and columns):

$$\begin{aligned}\Pi = &\sum_i^{ns} \frac{1}{2}\int_0^{L_1} EI_1(x)\left(\frac{\mathrm{d}^2 w_i(x)}{\mathrm{d}x_i^2}\right)^2 \mathrm{d}x_i + \sum_i^{nt1} \frac{1}{2} K_i^t w_1^2(x_i) + \sum_i^{nr1} \frac{1}{2} K_i^r w_1'^2(x_i) \\ &+ \sum_i^{nt2} \frac{1}{2} K_j^t w_2^2(x_j) + \sum_i^{nr2} \frac{1}{2} K_j^r w_2'^2(x_j) + \frac{K_i^t}{2}(W_1(L_1) - W_2(L_2))^2 \\ &+ \frac{K_i^r}{2}(\frac{dw_1}{dx_1}(L_1) + \frac{dw_2}{dx_2}(L_2))^2 + \sum_i^{ns} \frac{1}{2}\int_0^{L_1} \rho A_i W_i^2 \mathrm{d}x\end{aligned} \tag{7.3}$$

In the modified RDDM equation, the components of the structure are modelled separately, and then through some springs are connected to each other at the specified locations. The selected shape function should be according to the characteristics introduced by Ilanko et al. [10]. Equation (7.4) is representative of the function used in the improved RDDM, which is slightly different from the proposed equation:

$$N_i(x) = \begin{cases} \left(\frac{x}{L}\right)^{i-1} & i = 1, 2, 3 \\ \sin\left(\frac{(i-3)\pi x}{L}\right) + \cos\left(\frac{(i-3)\pi x}{L}\right) & i = 4, 5, ..., n \end{cases} \tag{7.4}$$

In Eq. (7.4), the n represents the number of vector entries. In consequence, Craig and Kurdila [12] proposed the follwong equation to solve the problem with Ritz technique:

$$w(x) = \mathbf{N}(x)\mathbf{B} \tag{7.5}$$

The generalized type of stiffness matrix and also the mass Ritz matrix can be calculated based on the principle of virtual displacement as proposed by Craig and Kurdila [12] as below:

$$\mathbf{K}= \int_0^L EI(x)\left(\frac{\mathrm{d}^2\mathbf{N}(x)}{\mathrm{d}x^2}\right)^{\mathrm{T}}\left(\frac{\mathrm{d}^2\mathbf{N}(x)}{\mathrm{d}x^2}\right)\mathrm{d}x \tag{7.6}$$

$$\mathbf{M}= \int_0^L \rho A(x)\mathbf{N}(x)^{\mathrm{T}}\mathbf{N}(x)\mathrm{d}x \tag{7.7}$$

To calculate the damage degree of each element, the structure Ritz stiffness matrix should be calculated based on the unknown coefficients in each element (Gharighoran et al. [8]):

$$\hat{\mathbf{K}}= \int_0^L \alpha_D EI(x)\left(\frac{\mathrm{d}^2\mathbf{N}(x)}{\mathrm{d}x^2}\right)^{\mathrm{T}}\left(\frac{\mathrm{d}^2\mathbf{N}(x)}{\mathrm{d}x^2}\right)\mathrm{d}x \tag{7.8}$$

The equation mentioned above is used to determine the stiffness and mass Ritz matrices. In the following, the method used by Gharighoran et al. (2009) was applied to determine the quantity and place of the damages [8]:

$$\hat{\mathbf{K}} = \sum_{e=1}^{NE} \alpha_{De}\mathbf{k}_e \tag{7.9}$$

In Eq. (7.9), **NE** is elements number. Correspondingly, the unknow damage indices can be extracted from the arrays of stiffness matrix and the system of equations can be obtained as presented below:

$$\mathbf{A}\alpha = \mathbf{R} \tag{7.10}$$

7.3 Numerical Examples

In this section, the performance and the accuracy of the proposed damage detection technique is evaluated by applying it on two numerical examples. The procedures begin with selecting a damage scenario and calculating the vibrational modal shapes and the structure's circular natural frequencies. To find the locations and severities of damages using the proposed damage detection technique, just a small number of vibrational mode shapes and circular natural frequencies are needed to be utilized.

7.3.1 A Two-Span Beam

In this part, as shown in Fig. 7.1, the beam similar to the provided by Kokot and Zembaty [13] with two damage patterns is considered. First, mode shapes calculation and the Eigen values are validated with accurate ones. As it is reported in Table 7.1, a good correspondence between the experimental specimen and the numerical model is achieved.

Table 7.2 reported the location and severity values obtained from the proposed damage detection for damage scenarios presented in the study of Kokot and Zembaty [13], using different number of eigen values and mode shapes. As it can be observed, the precision and damage detection location have an acceptable error, considering four eigen values and the mode shapes which indicates the good performance of the proposed method.

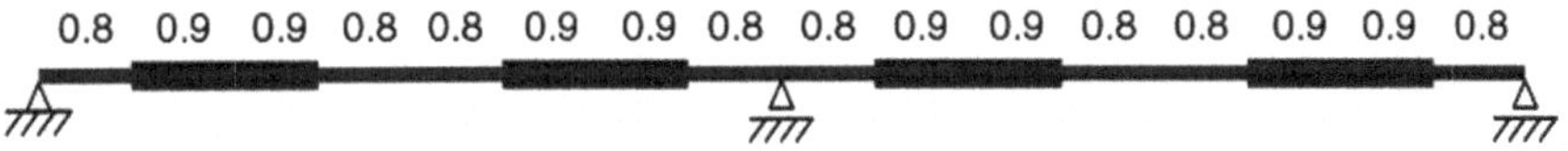

Fig. 7.1 Two-span beam with 10–20% decrease stiffness

Table 7.1 The comparison of natural frequencies in intact and damaged numerical model of two span beam

Description of the damaged pattern		f_1 (Hz)	f_2 (Hz)	f_3 (Hz)	f_4 (Hz)	f_5 (Hz)	f_6 (Hz)
Intact	Kokot and Zembaty (2009)	4.48	7.00	17.93	22.70	40.38	47.41
	Current study	4.48	7.00	17.94	22.75	40.39	47.50
10 and 20% stiffness loss	Kokot and Zembaty (2009)	4.12	6.40	16.82	21.05	38.83	45.57
	Current study	4.13	6.40	16.84	21.10	37.22	43.74

Table 7.2 The values of proposed damage indices

Element number	10 and 20% damage scenario (current study)			
	Damage scenario (Kokot and Zembaty 2009)	Two eigen-pairs	Three eigen-pairs	Four eigen-pairs
1	0.8	0.39	0.89	0.79
2	0.9	0.62	1.09	0.92
3	0.9	0.95	0.76	0.88
4	0.8	1.21	0.68	0.80
5	0.8	1.09	1.00	0.82
6	0.9	0.65	1.05	0.89
7	0.9	0.35	0.68	0.88
8	0.8	0.48	0.55	0.81
9	0.8	0.70	0.80	0.81
10	0.9	0.67	0.88	0.88
11	0.9	0.62	0.68	0.90
12	0.8	0.89	0.73	0.82
13	0.8	1.19	1.05	0.79
14	0.9	0.99	1.01	0.88
15	0.9	0.41	0.54	0.92
16	0.8	0.07	0.21	0.79

7.3.2 A Beam—Column Joint

A beam-column model, similar to the bridge and the deck, is considered as shown in Fig. 7.2. By applying the connection springs and penalty matrices, the stiffness and mass Ritz matrices is acquired. The validated Eigen values and mode shapes are demonstrated in Table 7.3.

In order to analyze the model, the beam and the column are divided into eight elements and two elements are assigned for the connection area. The analysis results are presented in Table 7.4.

Table 7.3 The comparison of natural frequencies in intact and damaged numerical model of beam-column

Damaged status	f_1 (Hz)	f_2 (Hz)	f_3 (Hz)
FEM	0.380	0.635	0.755
RDDM (healthy)	0.390	0.663	0.786
RDDM (senior damage)	0.375	0.579	0.722

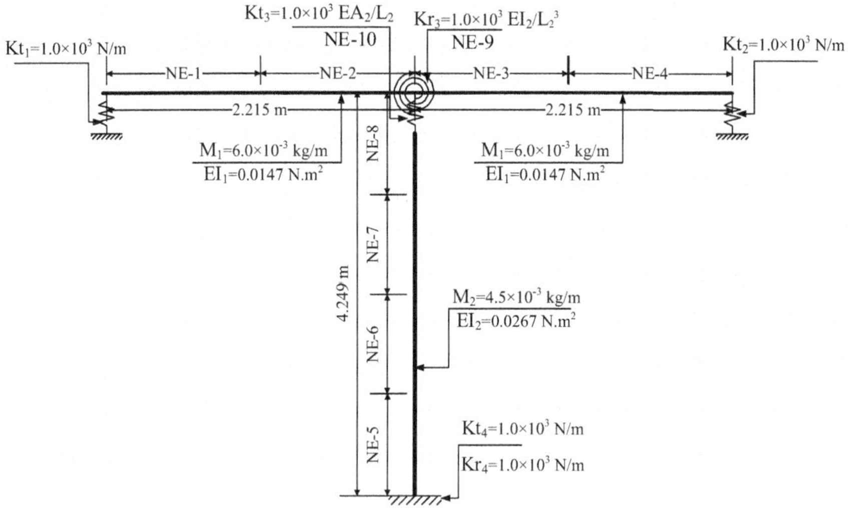

Fig. 7.2 The beam-column model of twin span beam with column supports

Table 7.4 The values of proposed damage detection indices in different damage scenarios

Element number	Damage senior	Three eigen-pairs	Four eigen-pairs	Five eigen-pairs
1	0.80	0.852	0.799	0.800
2	0.90	0.864	0.900	0.899
3	1.00	0.964	1.005	1.001
4	0.70	0.737	0.699	0.700
5	1.00	1.175	1.001	1.000
6	0.90	0.977	0.900	0.900
7	1.00	0.825	0.997	1.000
8	0.60	0.559	0.600	0.600
9	1.00	0.060	1.036	1.002
10	1.00	0.065	0.939	1.003

7.4 Conclusions

In this chapter, a novel method for damage localization and severity estimation is proposed based on improved RDDM method, in which some limitations such as support conditions and beams to column connection are eliminated. In the proposed method, rotational and translational springs have the role of connecting elements and fulfill boundary conditions. The efficiency of the proposed damage detection technique was investigated and evaluated by two models of multi-span beam and beam—column joint.

References

1. Farrar CR, Baker WE, Bell TM, Cone KM, Darling TW, Duffey TA, Migliori A (1994) Dynamic characterization and damage detection in the I-40 bridge over the Rio Grande
2. Eraky A, Anwar AM, Saad A, Abdo A (2015) Damage detection of flexural structural systems using damage index method—experimental approach. Alex Eng J 54(3):497–507. https://doi.org/10.1016/j.aej.2015.05.015
3. Limongelli MP, Siegert D, Merliot E, Waeytens J, Bourquin F, Vidal R, Cottineau LM (2016) Damage detection in a post tensioned concrete beam—experimental investigation. Eng Struct 128:15–25. https://doi.org/10.1016/j.engstruct.2016.09.017
4. Li YY, Cheng L, Yam LH, Wong WO (2002) Identification of damage locations for plate-like structures using damage sensitive indices: strain modal approach. Comput Struct 80(25):1881–1894. https://doi.org/10.1016/S0045-7949(02)00209-2
5. Sarker L, Xiang Y, Uy B, Zhu XQ (2011) Damage detection of circular cylindrical shells by Ritz method. In: Paper presented at the 9th international conference on damage assessment of structures
6. García PM, Araújo dos Santos JV, Lopes H (2014) A new technique to optimize the use of mode shape derivatives to localize damage in laminated composite plates. Compos Struct 108:548–554. https://doi.org/10.1016/j.compstruct.2013.09.050
7. Maghsoodi A, Ghadami A, Mirdamadi HR (2013) Multiple-crack damage detection in multi-step beams by a novel local flexibility-based damage index. J Sound Vib 332(2):294–305. https://doi.org/10.1016/j.jsv.2012.09.002
8. Gharighoran A, Daneshjoo F, Khaji N (2009) Use of Ritz method for damage detection of reinforced and post-tensioned concrete beams. Constr Build Mater 23(6):2167–2176. https://doi.org/10.1016/j.conbuildmat.2008.12.017
9. Chopra AK (2001) Dynamics of structures: theory and applications to earthquake engineering. Prentice-Hall
10. Ilanko S, Monterrubio L, Mochida Y (2015) The Rayleigh-Ritz method for structural analysis. John Wiley & Sons
11. Lee H, Ng T (1994) Natural frequencies and modes for the flexural vibration of a cracked beam. Appl Acoust 42(2):151–163
12. Craig RR, Kurdila AJ (2006) Fundamentals of structural dynamics. John Wiley & Sons
13. Kokot S, Zembaty Z (2009) Vibration based stiffness reconstruction of beams and frames by observing their rotations under harmonic excitations—numerical analysis. Eng Struct 31(7):1581–1588

Chapter 8
Identification of Damages in Concrete and Steel Structures: A Comprehensive Review

Abolfazl Yosefi, Farid Fazel Mojtahedi, and Michael Bahrami

Abstract This chapter examines various methods and techniques for detecting, assessing, and monitoring damage in concrete and steel structures. It covers a wide range of damage types, including corrosion, fatigue, impact, fire, earthquake, burst, cracks, fractures, anomalous shape change, abnormal deflections, abnormal bending, abnormal vibration, and unbalance. The importance of identifying and addressing structural damage to ensure safety and integrity is emphasized. The non-destructive testing (NDT) methods such as visual inspection, ultrasonic inspection, magnetic particle inspection, radiographic testing, and eddy current inspection is discussed. It also explores imaging techniques, quasi-static loading, structural monitoring, and artificial intelligence (AI) techniques for damage detection. Additionally, structural health monitoring (SHM) methods, including Acoustic Emission Testing (AET), the PZT-Based Active Wave Method, noise emission control, vibration-based methods, strain-based methods, and the utilization of machine learning algorithms, were investigated. The advantages, limitations, and applications of each method are highlighted, emphasizing the need for a comprehensive approach that combines multiple techniques for effective structural health assessment. The conclusion included discussing data collection and analysis methods, such as sensor data, seismic observations, visual data, and satellite data, along with imaging techniques like thermography, magnetic imaging, radar imaging, and electrical imaging. By utilizing these methods and techniques, engineers can enhance their understanding of structural health and make informed decisions to ensure the safety and longevity of critical infrastructure.

A. Yosefi
Department of Civil Engineering, University of Birjand, Birjand, Iran
e-mail: abolfazl_yosefi@birjand.ac.ir

F. F. Mojtahedi (✉) · M. Bahrami
Department of Civil Engineering, Sharif University of Technology, Tehran, Iran
e-mail: farid.fazel71@student.sharif.edu

M. Bahrami
e-mail: Meikaeil.bahrami@sharif.edu

H. Jahangir et al. (eds.), *Damage Detection and Structural Health Monitoring of Concrete and Masonry Structures*, Springer Tracts in Civil Engineering,
https://doi.org/10.1007/978-981-97-8975-7_8

Keywords Structural damage detection · Non-destructive testing (NDT) · Structural health monitoring (SHM) · Artificial intelligence (AI) techniques · Comprehensive structural health assessment

8.1 Introduction

Structural damage in concrete and steel structures poses significant challenges and risks to the safety and performance of these vital engineering elements. Identifying and understanding different types of damage is crucial for ensuring the durability and integrity of structures. This chapter aims to explore the various types of damage that can occur in concrete and steel structures, including corrosion[1–14], fatigue [15], impact, fire [8, 9, 15, 16], earthquake [8, 9, 15, 17–19], burst [15], cracks and fractures, anomalous shape change, abnormal deflections [1], abnormal bending [1], abnormal vibration [1], and unbalance [15]. It also examines the methods used to detect and assess these damages, emphasizing the importance of early identification and appropriate remedial measures [20, 21]. By comprehensively understanding and addressing structural damage, engineers and professionals can enhance the resilience and safety of concrete and steel structures [8, 9, 22]. Non-destructive testing (NDT) methods play a vital role in the field of structural engineering by providing techniques to detect and evaluate damage in concrete and steel structures without causing any harm [8, 9]. The current research work aims to explore and discuss the advantages and limitations of various NDT methods employed in structural inspections [8, 9, 23]. The methods covered include visual inspection, ultrasonic inspection, magnetic particle inspection, radiographic testing, and eddy current inspection [5, 8, 9]. Each method is evaluated based on its ability to detect different types of damage, its cost-effectiveness, and its practicality in different scenarios [8, 9]. The importance of early detection of damage in concrete and steel structures is highlighted [20, 21]. By employing NDT methods, structural engineers can identify damage in its early stages, enabling timely interventions and preventing catastrophic failures [20, 21]. The current study also emphasizes the significance of accurate identification of damage, as different types of damage require specific repair strategies [8, 9].

The proposed vibration-based health monitoring approach is highlighted as a valuable method for continuous monitoring of structural health [15, 24, 25]. This approach allows for the detection of changes in modal parameters, such as natural frequencies and damping ratios, which can indicate the presence of damage [26, 27]. By continuously monitoring structures, engineers can analyze data and implement appropriate maintenance and repair actions to ensure structural safety [28, 29]. Furthermore, the research works explored the use of artificial intelligence (AI) techniques in detecting structural damage [24, 26, 27]. By training neural networks with data from healthy and damaged structures, AI algorithms can learn to recognize patterns of damage in concrete and steel structures [30]. This enables accurate detection of damage through data analysis, providing valuable information for construction professionals and civil engineers [24, 31]. Structural health monitoring (SHM) is a critical aspect of maintaining the safety and performance of concrete and steel structures. This study provides an overview of various methods employed in SHM to detect damage and evaluate the condition of structures [8, 9, 15, 21, 23, 29, 32, 33]. The first method discussed is Acoustic Emission Testing (AET), which utilizes the analysis of emitted noise to detect and locate active damage, such as cracking and delamination [8, 9, 34]. AET offers real-time monitoring capabilities and is suitable for both internal and surface damage. Another method explored is the PZT-Based Active Wave Method, which uses lead zirconate titanate (PZT) to generate and detect waves within a structure. This method allows for real-time monitoring and is particularly useful for complex geometries and identifying potential areas of damage [8, 9]. Noise emission control is also examined as a method to manage and regulate the generation of high-frequency sound waves emitted by structures. By identifying and analyzing noise emissions, measures can be taken to minimize their negative impact on human health and the environment [34]. Vibration-based methods in SHM utilize the inherent vibrations of structures to extract information about their health. Output-only systems focus on modal parameters, while input/output systems aim to differentiate between environmental influences and actual damage events. These methods enable early detection of damage and are widely applicable in civil, mechanical, and aerospace engineering [8, 9]. Strain-based methods involve the measurement of strain to assess the extent of damage. By monitoring strain levels, deviations from expected behavior can be identified early, allowing for timely intervention and prevention of further damage [8, 9]. Lastly, the utilization of machine learning algorithms, including artificial neural networks, is discussed. Machine learning enhances the analysis of structural data and aids in the detection of damage patterns. By recognizing patterns and rules in the data, machine learning enables accurate predictions of damage location and severity [15, 23, 33]. The accurate assessment of structural integrity is essential in the fields of structural analysis and engineering. Gathering reliable information about the condition of structures enables engineers to evaluate their stability, detect potential damage, and implement appropriate maintenance and repair strategies [15]. Various data sources and imaging techniques are employed to acquire a comprehensive understanding of a structure's health. Sensors provide real-time data on parameters such as temperature, pressure, acceleration, deformation, and weight, enabling engineers to detect deviations and potential issues [32]. Seismic observations capture ground motion during earthquakes, aiding in the evaluation of structural response and retrofitting measures [1]. Visual data, including images and videos, helps identify visible signs of deterioration and facilitates the creation of accurate 3D models [23]. Satellite data offers a broad-scale perspective on structures and their surroundings, enabling the identification of changes and potential risks [33]. Imaging techniques such as thermography, magnetic imaging, radar imaging, and electrical imaging provide valuable insights into temperature variations, internal structures, and environmental changes [2–5]. The research work explored the significance of these data sources and imaging techniques in data collection and analysis for structural assessment. Detecting and addressing structural damage in concrete and steel structures is of utmost importance to ensure safety and prevent accidents. Various methods and techniques are employed to identify damage accurately and efficiently. Visual inspection allows for the examination of surface damage, such as cracks and corrosion [1]. Leak testing involves subjecting the structure to liquid pressure to detect internal damage [12, 35]. Ultrasound procedures utilize ultrasonic waves to detect internal issues like deep cracks and hidden corrosion [4, 13]. Thermography uses thermal imaging to analyze surface temperature changes and identify damage [36, 37]. X-ray methods provide insights into internal damage [1]. Imaging methods using UAVs capture images of structures for analysis [38]. Wireless sensor networks collect data to identify damage [35, 39]. Vibration analysis detects damage based on structural vibrations [1, 40]. Thermal image analysis utilizes thermal imaging for damage identification [41]. Acoustic analysis detects damage based on emitted noise [1, 40]. Shear strength analysis evaluates the strength of structures to identify damage [39, 42]. Artificial neural networks and deep learning utilize data and images for automatic damage detection [1, 40]. Image processing and signal processing techniques are used for structural damage identification [42, 43]. Multitasking networks and shear resistance generator networks offer automated and optimized recognition of structural damage [40]. Clustering sensor data and analysis of sensor data provide insights into damage localization and identification [1, 42, 43]. By employing these methods, early detection of structural damage can be achieved, leading to prompt repairs and maintenance. Damage to concrete and steel structures can lead to a range of common damages with significant consequences. One prevalent damage is a waterfall accident, which occurs when individuals lose their footing on top of a damaged structure. The instability of the structure can cause them to slip, resulting in a fall that can lead to broken bones, head damages, and other severe trauma. These falls can be particularly dangerous if the height is considerable or if there are sharp objects or debris present.

8.2 Different Kinds of Damages

In the realm of engineering and structural systems, safeguarding the integrity and stability of components necessitates a comprehensive understanding of the diverse forms of damage that can compromise their functionality. Table 8.1 presents an inclusive compilation of damage types, categorized based on their underlying causes and consequences. "Material Degradation" includes corrosion, a gradual material deterioration due to chemical reactions, and fatigue, the weakening of materials over time from repeated loading cycles. "External Forces" encompass impact damage from external objects and the peril of fire weakening materials and structures. "Earthquake" damage arises from seismic effects, causing intense shaking and potential collapse. "Structural Failure" encompasses burst, sudden ruptures, and cracks and fractures resulting from stress concentrations.

Deformation and instability are covered by "Anomalous Shape Change," "Abnormal Deviation," and "Abnormal Bending," representing unexpected structural alterations. Lastly, "Abnormal Vibration" and "Unbalance" involve dynamic response damage, such as excessive stress and erratic behavior in rotating machinery.

Table 8.1 Damage type

Damage type	Category
Corrosion	Material Degradation
Fatigue	Material Degradation
Impact	External Forces
Fire	External Forces
Earthquake	Seismic Effects
Burst	Structural Failure
Cracks and fractures	Structural Failure
Anomalous shape change	Deformation and Instability
Abnormal deviation	Deformation and Instability
Abnormal bending	Deformation and Instability
Abnormal vibration	Dynamic Response
Unbalance	Dynamic Response

By recognizing these damage types, engineers and professionals can implement proactive measures to bolster safety, reliability, and longevity in engineering and architectural endeavors.

8.2.1 Corrosion

Corrosion is a form of damage that affects concrete and steel structures when the steel components start to rust. It is characterized by the gradual degradation and deterioration of the material due to chemical reactions with elements such as moisture, salt, and chemicals. Corrosion can lead to various types of visible and hidden damage, including cracking, pitting, loss of structural integrity, spalling of concrete, and deformation. Visual inspection is a commonly used method to identify corrosion and other forms of damage in structures. This method involves visually examining the structure for signs of corrosion, such as surface damage, cracks, and fissures. However, it may not be effective in detecting hidden or internal corrosion. To detect deeper damage, NDT techniques are employed. NDT methods, such as ultrasonic inspection, X-ray inspection, magnetic particle inspection, and eddy current inspection, are more accurate than visual inspection and can identify internal corrosion, cracks, and pitting. These techniques are capable of detecting hidden or internal damage, but they can be costly and time-consuming. Another approach to identifying corrosion involves analyzing the geometry and surface characteristics of the structure using specialized software. By examining the height difference and shape of the structure surface, surface damage like cracks, plaque, and corrosion can be identified. Corrosion is a significant concern for concrete and steel structures as it can compromise their structural integrity. Detecting corrosion and assessing its extent is crucial for ensuring the safety and durability of these structures. Therefore, a combination of visual inspection, NDT, and analytical methods is employed to accurately detect and evaluate the presence of corrosion in order to initiate appropriate repair and maintenance measures [1–14].

8.2.2 Fatigue

Fatigue, in the context of structural engineering, refers to the phenomenon of material degradation and failure that occurs due to repeated loading and unloading cycles. It is a major concern in both concrete and steel structures and is primarily caused by cyclic loading, temperature variations, and other environmental changes. In the case of steel structures, fatigue manifests as cracking, deformation, and failure of the steel members. The repeated application of loads induces stress concentrations, which can lead to the initiation and propagation of cracks within the material. Over time, these cracks can grow and eventually result in structural failure if not addressed. Similarly, in concrete structures, fatigue is characterized by cracking, delamination, and spalling. The repeated application of loads causes the concrete to undergo gradual deterioration, leading to the formation of cracks and separation of layers. Environmental exposure, such as temperature changes and moisture, can exacerbate the fatigue process in concrete. To assess and detect fatigue in structures, various analytical and computational techniques are employed. These methods involve applying fixed and semi-fixed loads to structures and utilizing gauges and instruments to monitor and measure damage. By analyzing the behavior of structures under these loads, engineers can identify signs of fatigue, such as abnormal deformations and fractures. Quasi-static loads, which mimic the characteristics of cyclic loading, are commonly employed in fatigue testing of steel structures. By subjecting steel members to these loads, engineers can observe the response of the structure and identify any indications of fatigue, such as crack initiation or propagation. In summary, fatigue is a critical consideration in structural engineering, as it can lead to the deterioration and failure of concrete and steel structures. Understanding the causes and effects of fatigue is essential for ensuring the safety and longevity of these structures. Through the use of analytical techniques and load testing, engineers can detect and mitigate the risks associated with fatigue, thereby enhancing the structural integrity and performance of the built environment [8–11, 15]. Static and fatigue tests are performed using an electro-hydraulic servo static and dynamic universal testing machine as presented in Fig. 8.1.

8.2.3 Impact

In academic terms, the concept of impact refers to the effects or consequences resulting from a specific action or event. Within the context of structural engineering,

(a) View

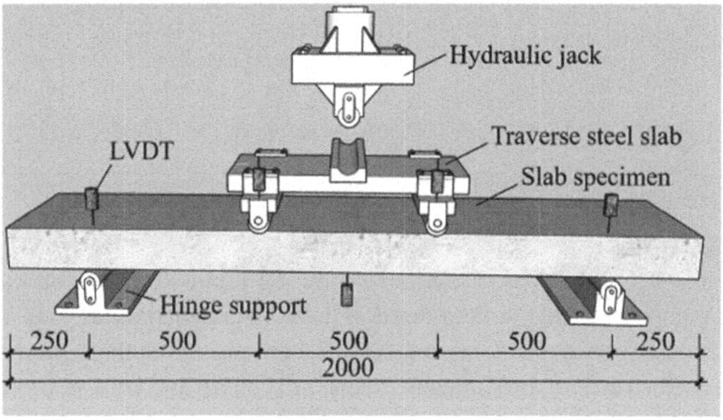

(b) Scheme (unit: mm)

Fig. 8.1 Electro-hydraulic servo static and dynamic universal testing machine for Static and fatigue tests [44]

impact is often employed as a method to assess the damage and behavior of concrete and steel structures. Dynamic loading, which involves subjecting a structure to high-frequency and sudden strength loads, is a widely used approach for investigating structural integrity. By applying dynamic loads through impactors, heavy equipment, vibrating tools, or other means, researchers and engineers can evaluate how the structure responds and behaves under these extreme conditions. The impact exerted on concrete structures can lead to various detrimental effects. Cracking, spalling, and deformation are among the primary issues that can arise as a result of impact. When subjected to high-frequency and sudden strength loads, the concrete may develop cracks, compromising its structural integrity. Additionally, the impact can cause spalling, where fragments of the concrete surface break off, further compromising its durability. Moreover, deformation, such as bending or warping, may occur, leading to structural instabilities in the concrete elements. In the case of steel structures, impact can result in distinct forms of damage. The application of dynamic loads through impact can cause plastic deformation, altering the shape and structure of steel members. The excessive forces experienced during impact can induce bending or warping of the steel components, compromising their load-bearing capacity. Moreover, impact can also lead to shear failure, where the steel members fail due to the applied forces exceeding their shear strength. These consequences highlight the significance of understanding the impact behavior of steel structures, enabling engineers to design and implement appropriate measures to ensure their structural stability and safety. In summary, impact within the academic context of structural engineering refers to the effects of subjecting a structure to high-frequency and sudden strength loads. This method allows researchers and engineers to investigate the behavior and damage incurred by concrete and steel structures. The impact on concrete structures can result in cracking, spalling, and deformation, while steel structures may experience plastic deformation, bending, and shear failure. By comprehensively studying the impact response of structures, professionals can enhance their understanding and develop strategies to improve the safety and performance of these vital engineering elements [8, 9].

8.2.4 Fire

Fire is a natural phenomenon characterized by the rapid oxidation of combustible material, typically resulting in the release of heat, light, and flames. It is a chemical reaction that occurs when fuel, oxygen, and heat are present in the right proportions. Fire can have destructive consequences, especially when it comes into contact with various materials, such as concrete and steel. When exposed to high temperatures, concrete can undergo spalling, cracking, and a loss of strength. The intense heat causes the moisture trapped within the concrete to turn into steam, exerting pressure and leading to the formation of cracks and fragments. This process weakens the structural integrity of the concrete, making it more susceptible to collapse or failure. Similarly, fire can have detrimental effects on steel members. The high temperatures cause the steel to soften and deform, compromising its structural integrity and load-bearing capacity. This loss of strength can result in structural failure and collapse, posing significant risks to buildings and other structures. Furthermore, fire damage occurs when structures are engulfed in flames. The intense heat of the fire causes the materials within the structure to burn and deteriorate. This can lead to structural collapse, destruction of property, and the loss of valuable assets. Fire damage can have severe consequences, not only in terms of physical destruction but also in terms of economic and personal losses.

In summary, fire is a powerful and potentially destructive force that can cause significant damage to various materials, including concrete and steel. It is essential to understand the risks associated with fire and take appropriate measures to prevent and mitigate its impact to ensure the safety and integrity of structures and the protection of lives and property [8, 9, 15, 16].

8.2.5 Earthquake

An earthquake is a natural phenomenon that can have devastating effects on the built environment. It can cause various forms of damage to both concrete and steel structures, including cracking, deformation, and collapse as shows in Fig. 8.2. These destructive consequences are often a result of unbalance and excessive stress exerted on the structures during seismic activity. One factor that contributes to structural damage during an earthquake is unbalance. Asymmetric loading or unfavorable conditions can lead to imbalances in concrete and steel structures. These imbalances can be detected through quasi-static loading, which is commonly used to check for damage. However, it is recommended to employ dynamic loading and other methods to assess earthquake damage comprehensively.

To effectively monitor and detect earthquake-induced damage, an earthquake monitoring system is implemented. This system involves placing sensitive devices at critical points on the structure to measure its vibrations during seismic events. By utilizing sensors that measure deflection and acceleration, these systems provide valuable information about the behavior of structures during earthquakes. The collected data is then analyzed using specialized algorithms to accurately identify any structural damage without causing harm to the building itself. The advantages of employing an earthquake monitoring system are numerous. Firstly, it offers high accuracy in detecting damage caused by earthquakes, allowing for prompt action and assessment. When an earthquake occurs, this system provides timely information to relevant authorities and rescue workers, enabling quick response and efficient allocation of resources. For instance, in the event of an earthquake, the system can promptly notify authorities of the specific areas within a building that have been damaged, facilitating targeted rescue efforts. Seismic data collected from a seismic observation network also plays a crucial role in detecting earthquake damage.

This data is valuable in assessing the effects of seismic activity and can be used in conjunction with other monitoring methods to identify structural vulnerabilities and potential damage. Overall, the detection and monitoring of earthquake damage in concrete and steel structures are of utmost importance for ensuring their safety. By employing a combination of monitoring techniques, such as dynamic loading, seismic data analysis, and the implementation of earthquake monitoring systems, it becomes possible to assess and respond to structural damage caused by earthquakes effectively. This proactive approach can potentially save lives, minimize repair costs, and contribute to the overall resilience of built environments [8, 9, 15, 17–19].

8.2.6 Burst

A burst refers to the sudden and often catastrophic failure or rupture of concrete and steel structures. It occurs as a result of excessive loads or abnormal and unwanted deformation of the structure. When subjected to high stress or pressure beyond their

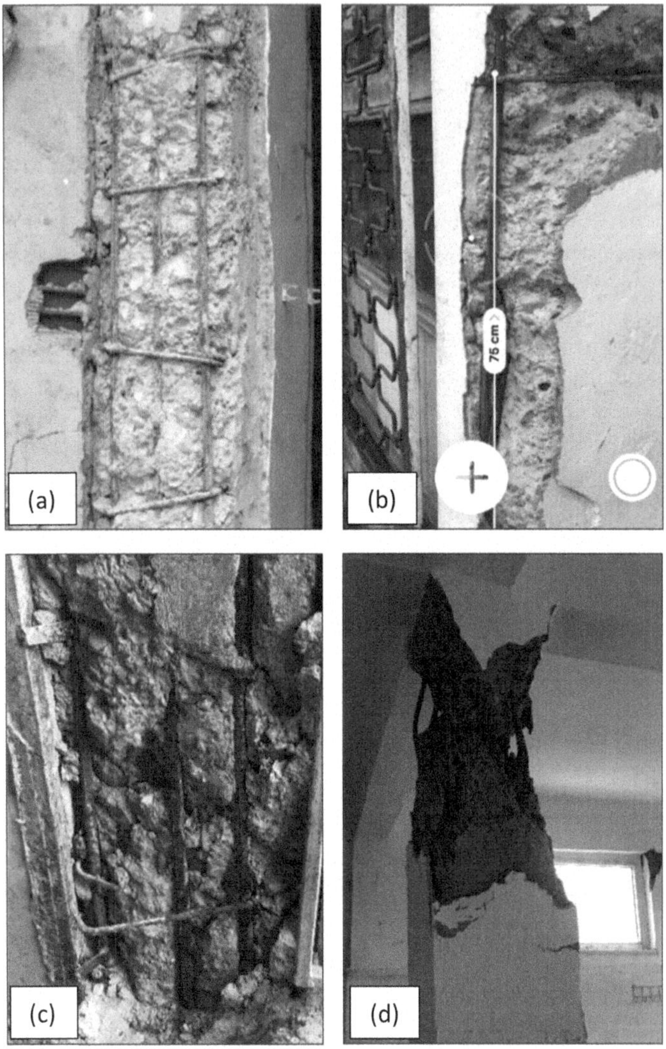

Fig. 8.2 Earthquake damages caused by the stirrup application errors [45]: **a** Inclined stirrups; **b** No stirrups in confinement area; **c** Opening of hooled stirrups and **d** The column without stirrups

design limits, these structures can experience a sudden collapse or rupture, leading to significant damage or even loss of life. The term "burst" is commonly used to describe such structural failures, which can have severe consequences in various fields such as civil engineering, construction, and infrastructure. Detecting and predicting bursts is crucial for ensuring the safety and integrity of these structures, and techniques such as quasi-static loading can be employed to identify potential signs of structural rupture before they occur. By understanding the causes and implications of bursts, engineers and professionals can work towards designing and maintaining structures that are more resilient and less prone to such failures [15].

8.2.7 *Cracks and Fractures*

Cracks and fractures are types of damage that can occur in concrete and steel structures when subjected to dynamic loads. Cracks are visible openings or fissures that can be observed on the surface of structures, while fractures refer to internal damage within the structure. Visual inspection is a common and straightforward method used to detect cracks and other visible damage. This involves physically examining the structure to identify any visible cracks, peeling, or deformation. However, visual inspection may not be effective in detecting hidden or internal damage. NDT methods are employed to identify internal damage, including cracks. Ultrasonic inspection, X-ray inspection, magnetic particle inspection, and eddy current inspection are some examples of NDT techniques. Ultrasonic waves are transmitted through the structure, and their reflections are used to detect deep cracks and other structural damage. X-ray inspection utilizes X-rays to examine the structure and identify internal damage, making it suitable for detecting hidden cracks. In addition to visual inspection and NDT methods, other techniques such as imaging can be employed. Thermography, magnetic imaging, radar imaging, and electrical imaging are imaging techniques that use specialized equipment to create images of the interior of structures. These methods aid in identifying hidden damage, including small cracks. When conducting tests to detect damage in concrete and steel structures, cracks are one of the most common types of damage to look for. Factors such as stress, fatigue, and environmental exposure can contribute to the formation of cracks in concrete structures. For concrete, visual inspection, surface unevenness tests, and destructive tests can help identify cracks and other damages. Similarly, for steel structures, visual inspection and the application of static and semi-static loads can help detect cracks and other mechanical damage. In summary, cracks and fractures are forms of damage that can occur in concrete and steel structures. Visual inspection, NDT methods, and imaging techniques are used to identify cracks and other types of damage. These methods play a crucial role in assessing the integrity of structures and determining the extent of damage for appropriate repairs or maintenance [1–7].

8.2.8 *Anomalous Shape Change*

Anomalous shape change refers to the phenomenon where concrete and steel structures experience abnormal deformation that deviates from the expected or typical behavior. This deviation occurs due to various factors, including the application of strong forces or exposure to adverse conditions. Unlike regular or anticipated deformations, anomalous shape changes exhibit distinctive characteristics that can be identified through the process of quasi-static loading. Concrete and steel structures are designed to withstand certain loads and environmental conditions, and their behavior under normal circumstances follows predictable patterns. However, when subjected to unusually high forces or unfavorable conditions, these structures may undergo unexpected deformations. Anomalous shape changes can manifest as alterations in the structural geometry, such as bending, twisting, or warping, that exceed the anticipated limits or patterns of deformation. The causes of anomalous shape change can vary. It may result from extreme weather events like earthquakes, hurricanes, or floods that subject the structures to forces beyond their design specifications. Additionally, human-made factors such as accidents, improper construction techniques, or material deficiencies can also contribute to anomalous shape change. Identifying and understanding anomalous shape change is crucial for assessing the structural integrity of buildings and infrastructure. By conducting quasi-static loading tests, engineers can simulate the forces acting on the structures and observe their responses. Through this process, any deviations from expected behavior can be detected, allowing for further investigation and appropriate remedial measures to be taken. Overall, anomalous shape change signifies the atypical deformation experienced by concrete and steel structures when exposed to strong forces or adverse conditions. Recognizing and addressing these abnormal deformations is essential to ensure the safety and reliability of structures and prevent potential structural failures [15].

8.2.9 Abnormal Deflections

Abnormal deflections refer to atypical or excessive movements or displacements exhibited by structures under the influence of dynamic loads. When a structure is subjected to dynamic forces, such as vibrations, impact, or sudden loads, it may experience deflections that are beyond the expected or normal range. These abnormal deflections often serve as indicators of structural damage. Structures, whether they are buildings, bridges, or other engineered systems, are designed to withstand certain loads and maintain their stability and integrity. However, when subjected to excessive or unexpected dynamic forces, the structural elements may deform or displace more than anticipated. These abnormal deflections can manifest as bending, twisting, sagging, or any other form of movement that deviates significantly from the expected behavior. Abnormal deflections can be a result of various factors, including material fatigue, overloading, poor construction quality, or the presence of cracks, fractures, or other forms of structural damage. These deflections can compromise the safety, functionality, and durability of the structure, potentially leading to structural failure if not addressed. To identify abnormal deflections, engineers and inspectors often employ techniques such as structural monitoring, visual inspections, and NDT methods. By analyzing the deflection patterns and comparing them to the expected behavior, they can assess the extent of structural damage and determine the necessary repairs or reinforcements. Addressing abnormal deflections is crucial to ensure the safety and reliability of structures. Timely detection and appropriate remedial measures can prevent further deterioration, mitigate the risk of collapse, and extend the service life of the structure. By understanding the causes and implications of abnormal deflections, engineers can design structures with improved resilience and develop maintenance strategies to detect and address such issues promptly [1].

8.2.9.1 Abnormal Bending

Abnormal bending refers to the phenomenon where structures experience an unusual or excessive bending deformation due to dynamic loading. When subjected to external forces or loads, structures are expected to undergo a certain degree of bending or deformation in order to accommodate the applied loads. However, in cases of abnormal bending, the amount or manner of bending deviates from what is considered normal or expected. This abnormal bending can serve as an indication of structural failure. Dynamic loading refers to the application of forces or loads that change over time, such as those resulting from vibrations, impacts, or sudden changes in the environment. These dynamic loads can place significant stress on structures, causing them to bend or deform. In most cases, structures are designed to withstand these loads within acceptable limits, ensuring their stability and integrity. However, when the applied forces exceed the structural capacity or if there are underlying weaknesses or defects in the material or design, abnormal bending can occur. Abnormal bending can manifest in various ways depending on the type of structure and the nature of the dynamic loading. It may involve excessive bending, where the structure bends beyond its intended or safe limits. This can lead to visible deformations, such as sagging, buckling, or twisting. Additionally, abnormal bending can also involve irregular or uneven bending patterns, where the structure bends in an unpredictable or asymmetric manner. Identifying abnormal bending is crucial in assessing the structural health and integrity of a system. It serves as a warning sign that the structure may be compromised and at risk of failure. Monitoring and detecting abnormal bending can be done through various means, including visual inspections, strain measurements, and structural analysis techniques. By recognizing and addressing abnormal bending early on, appropriate measures can be taken to mitigate further damage or failure, ensuring the safety and longevity of the structure [1].

8.2.9.2 Abnormal Vibration

Abnormal vibration refers to the occurrence of unusual or excessive vibrations in structures, typically caused by dynamic loads. Vibrations are a natural phenomenon that can be observed in various systems, including buildings, bridges, machinery, and vehicles. However, when these vibrations deviate from their expected or normal patterns, it often indicates the presence of structural damage or underlying issues.

Dynamic loads, such as heavy machinery operations, seismic activity, wind forces, or even human activities, can induce vibrations in structures. Under normal circumstances, structures are designed to withstand and dissipate these vibrations without compromising their integrity. However, when the vibrations exceed the anticipated levels or exhibit irregular patterns, they are classified as abnormal. Abnormal vibrations can arise due to several reasons, including material fatigue, mechanical failure, inadequate structural design, or unexpected external forces. These vibrations can manifest in different forms, such as oscillations, shaking, or resonance, depending on the nature of the structure and the underlying causes. Detecting and addressing abnormal vibrations is crucial to ensure the safety and functionality of structures. Continuous or repeated abnormal vibrations can lead to progressive structural damage, compromising the integrity and stability of the system. Therefore, it is important to conduct regular inspections, monitoring, and maintenance to identify and rectify any abnormalities in vibrations. Various techniques and tools are available for the assessment and mitigation of abnormal vibrations. These may include vibration sensors, accelerometers, modal analysis, and computer simulations to analyze the vibrational behavior of structures. By pinpointing the source and characteristics of abnormal vibrations, engineers and technicians can develop effective strategies to repair, reinforce, or redesign the affected components or systems. In summary, abnormal vibrations refer to unusual or excessive oscillations experienced by structures due to dynamic loads. They serve as indicators of potential structural damage or underlying issues. Timely detection and appropriate action are necessary to prevent further deterioration and ensure the safe and reliable operation of the affected structures [1].

8.2.9.3 Imbalance

Imbalance refers to the condition where there are asymmetries or unfavorable conditions present in concrete and steel structures. This can result in imbalances in the distribution of loads within the structures. Quasi-static loading is commonly used to identify and assess such imbalances. It involves applying slow and gradual loads to the structures to evaluate their response and potential damage. While quasi-static loading is effective for detecting imbalances in concrete and steel structures, it may not be sufficient to identify other types of damage, such as that caused by earthquakes. Therefore, it is recommended to complement the assessment with other methods like dynamic loading when checking for earthquake damage. Wooden structures have different characteristics compared to concrete and steel structures. Although quasi-static loading can be used for damage testing in wooden structures, there are variations in the loading methods due to these structural differences. It is important to consider these distinctions when evaluating the unbalance and potential damage in wooden structures [46].

8.3 Methods

In Table 8.2, various methods are listed under different categories for inspection and testing purposes. These categories include NDT, Imaging Techniques, Surface Testing, Destructive Testing, Environmental Monitoring, Structural Health Monitoring, and Machine Learning and AI techniques. Each method serves a specific function in assessing the integrity and health of structures and materials.

8.3.1 Non-Destructive Testing

NDT methods are crucial in structural engineering as they can identify and evaluate damage in concrete and steel structures without causing harm. These methods offer cost-effective and efficient ways to assess structural integrity and ensure the safety of buildings and bridges. One widely used method is vibration-based health monitoring, which continuously monitors structures to detect changes and identify necessary maintenance. Vibration analysis helps uncover damage and condition changes by analyzing dynamic characteristics like frequencies and mode shapes. Quasi-static loading applies a constant load to the structure to identify damage within its elements. Observing the response helps identify signs of distress such as cracks or deformations. Dynamic loading subjects the structure to dynamic loads to analyze its behavior and detect various types of damage, such as cracks. Visual inspection,

Table 8.2 Methods

Category	Methods
NDT	Visual inspection, Ultrasonic Testing (UT), X-ray method, Ultrasound procedure, Imaging method using UAV, Wireless sensor network method, Acoustic analysis method, Method of shear strength analysis, Magnetic Particle Inspection (MPI), Radiographic Testing (RT), Eddy Current Testing (ECT), Acoustic Emission Testing (AET), PZT-Based Active Wave Method, leak test, ultrasound
Imaging techniques	Thermography, Magnetic Imaging, Radar Imaging, Electrical Imaging, Image processing and signal processing
Surface testing	Surface unevenness tests
Destructive testing	Destructive tests
Environmental monitoring	Noise emission control
Structural health Monitoring	Vibration-based methods, Strain-based methods
Machine learning and AI	Machine learning techniques, Artificial Neural Networks, Deep learning, Multitasking network, Shear resistance generator network, Clustering sensor data

though limited to surface-level damage, provides an initial assessment. Complementary methods like ultrasonic testing, X-ray inspection, magnetic particle inspection, and eddy current inspection detect internal damage more accurately. Magnetic particle inspection detects surface-breaking defects in ferromagnetic materials by applying magnetic fields and particles that gather at defect locations. Radiographic testing uses X-rays or gamma rays to examine the internal structure and identify flaws like cracks and voids. Eddy current inspection detects surface and near-surface defects in conductive materials by inducing eddy currents and detecting variations caused by defects. NDT also includes methods like acoustic emission testing, PZT-based active wave method, and surface roughness inspection, which monitor noise, perform real-time monitoring, and measure surface roughness, respectively.

NDT saves time and costs, detects both superficial and deep damage, and allows for timely repairs. However, some methods require specialized equipment and expertise, and certain types of damage may require a combination of non-destructive and destructive testing. In summary, NDT methods are invaluable in detecting and evaluating damage in concrete and steel structures. They offer numerous benefits and ensure structural safety and longevity. By incorporating these methods into regular inspections, professionals can maintain the integrity of structures effectively [8, 9, 15, 20, 21, 23, 47–53].

8.3.1.1 Visual Inspection

Visual inspection is a crucial method used for detecting damage in concrete and steel structures. It involves a thorough examination of the structure's surface to visually identify any visible signs of damage such as cracks, corrosion, peeling, or deformation. This approach is widely employed due to its simplicity, affordability, and ease of implementation. Visual inspection serves as a primary tool in the initial assessment of structures, allowing for the identification of surface defects and damage. During a visual inspection, a team of experts with specialized knowledge and experience in concrete and steel construction carefully examines the structure. They systematically inspect various components, including walls, columns, beams, and surfaces, to identify any indications of damage. The experts rely on their expertise to recognize and analyze visual cues that suggest structural deterioration or vulnerabilities. To aid in the inspection process, inspectors may utilize tools such as hammers or magnets. By tapping on surfaces or using magnetic devices, they can identify hollow sounds, unevenness, or anomalies that may indicate underlying issues. This hands-on approach helps detect cracks, fissures, or areas of corrosion that may be visible on the surface. However, it is important to note that visual inspection has its limitations. While it can effectively detect surface damage, it may not be able to identify hidden or internal damage that is not visible to the naked eye. Some structural defects, such as delamination or subsurface cracks, may remain undetectable through visual inspection alone. Consequently, relying solely on visual inspection may result in the oversight of critical damage that requires immediate attention. To complement the limitations of visual inspection, NDT methods are often employed. These techniques, such as ultrasonic inspection, X-ray inspection, magnetic particle inspection, and eddy current inspection, provide a more accurate and comprehensive assessment of the structure. NDT methods can penetrate the surface and identify hidden or internal damage that visual inspection alone may not reveal. However, it is important to note that NDT methods can be more expensive and time-consuming than visual inspection. The early detection of damage in concrete and steel structures is vital for ensuring structural integrity and mitigating potential risks. By combining visual inspection with complementary testing methods, engineers and inspectors can obtain a more comprehensive understanding of the structure's condition. This integrated approach allows for the identification of both surface damage visible through visual inspection and hidden or internal damage detected through NDT methods. By leveraging the strengths of each technique, stakeholders can make informed decisions regarding maintenance, repairs, and the overall safety of the structure [2–10, 12, 54].

8.3.1.2 Ultrasonic Inspection

Ultrasonic inspection, also known as ultrasonic testing (UT), is a NDT method used to detect internal damage and flaws in various structures. It is one of several NDT techniques, including X-ray inspection, magnetic particle inspection, and eddy current inspection, that offer higher accuracy compared to visual inspection. Ultrasonic inspection utilizes ultrasonic waves transmitted into the structures being examined. The waves then reflect or scatter, providing valuable information about the presence of structural damage such as cracks, voids, pitting, and corrosion.

This procedure is particularly effective in identifying internal flaws and hidden damage that may not be visible to the naked eye. By analyzing the reflections of the ultrasonic waves, trained technicians can pinpoint deep cracks and detect corrosion within the structure. However, it is important to note that ultrasonic inspection has limited penetration depth, meaning it may not be able to detect flaws located too far below the surface. Ultrasonic inspection requires specialized equipment to transmit and receive ultrasonic waves and interpret the resulting signals. It is capable of identifying both internal and surface damage, making it suitable for a wide range of applications. Unlike some other inspection methods, ultrasonic inspection does not require direct contact with the surface of the structure, allowing for non-intrusive examination. However, it is essential to consider certain limitations of ultrasonic inspection. This method may not be as effective in enclosed spaces or areas obstructed by mines or other obstacles, as the waves need a clear path for accurate detection. Additionally, while ultrasonic inspection offers valuable insights into internal damage, it may not be as proficient in identifying surface damage compared to other inspection techniques. In summary, ultrasonic inspection is a NDT method that utilizes ultrasonic waves to detect internal damage and flaws in structures. It is a precise and reliable technique for identifying deep cracks, voids, and hidden corrosion. Although it requires specialized equipment, it offers the advantage of non-intrusive examination and can be applied to various industries where structural integrity is crucial [5, 10, 54].

8.3.1.3 X-ray Method

X-rays are used in this method to examine structures and determine the presence of damage within them [1]. It is particularly useful for detecting internal damage, including hidden cracks and corrosion. However, it necessitates specialized equipment and is not as proficient in detecting surface damage.

8.3.1.4 Ultrasound Procedure

Similar to the previous mention, this method involves transmitting ultrasonic waves into a structure and detecting damage based on their reflection and scattering patterns [4]. It can identify both internal and surface damage without requiring direct contact with the structure's surface. However, it requires special equipment and may not be suitable for enclosed spaces or areas with obstacles like mines.

8.3.1.5 Imaging Method Using UAV

Unmanned aerial vehicles (UAVs) are utilized to capture images of structures, which are then analyzed to identify structural damage [55]. This method is particularly beneficial for structures in challenging environments or difficult-to-access areas. However, it does require access to a drone and associated equipment [1].

8.3.1.6 Ultrasonic Testing

Ultrasonic testing is widely used to detect internal flaws in concrete and steel structures, such as cracks, voids, and delamination. This method relies on the transmission of high-frequency sound waves into the material, and the analysis of the reflected waves provides information about the presence and characteristics of internal defects. Ultrasonic testing is known for its ability to penetrate into the material and detect flaws at various depths, although its effectiveness may be limited by the material's thickness and properties.

8.3.1.7 Wireless Sensor Network Method

Wireless sensor network method This method involves installing a network of wireless sensors on the structure to collect information that can be used to identify structural damage [56, 57]. It is suitable for detecting both internal and surface damage and offers the advantage of reducing the cost and time required for damage detection. However, it necessitates the installation of wireless sensors on the structure.

8.3.1.8 Acoustic Analysis Method

Acoustic analysis method Structural damage is identified by analyzing the noise emitted from the structure [1, 40]. This method is suitable for detecting both internal and surface damage and can be performed remotely and non-destructively. However, access to a speech analyzer is required [58]. Method of shear strength analysis: This technique involves analyzing the shear strength of the structure to identify damage [39, 42]. It is suitable for detecting both internal and surface damage and can be performed remotely and non-destructively. However, access to shear strength data for structures and associated equipment is required.

8.3.1.9 Magnetic Particle

Magnetic Particle Inspection (MPI) is a NDT method utilized to detect surface-breaking defects like cracks and corrosion in ferromagnetic materials. As part of the broader field of NDT, MPI is recognized for its accuracy in identifying internal damage that may not be visible through visual inspection alone. It is one of several imaging techniques used in NDT, along with thermography, radar imaging, and electrical imaging. These imaging technologies employ specialized equipment and algorithms to create images or videos that aid in the extraction of information and identification of various patterns within structures. However, it is important to note that MPI is limited to ferromagnetic materials, meaning it cannot be applied to non-ferromagnetic substances [5, 8, 9].

8.3.1.10 Radiographic Testing (RT)

Radiographic Testing (RT) is a non-destructive method that uses X-rays or gamma rays to detect internal flaws and measure wall thickness in various materials. It provides detailed information about an object's internal condition without causing damage. RT ensures the safety and quality of components in industries like aerospace and manufacturing. However, radiation safety precautions are crucial due to the potential harm caused by X-rays and gamma rays. Proper shielding, monitoring, and training are necessary to protect personnel and maintain safety during the testing process [8, 9].

8.3.1.11 Eddy Current Testing

Eddy Current Testing (ECT) is an NDT method used to detect surface and near-surface defects in conductive materials. It is one of several NDT techniques, including ultrasonic inspection, X-ray inspection, and magnetic particle inspection, that are employed to identify internal damage such as cracks, pitting, and corrosion. Compared to visual inspection, these NDT methods offer greater accuracy and the ability to detect hidden or internal damage that may not be visible to the naked eye. ECT utilizes the principle of electromagnetic induction to detect flaws in conductive materials. A probe containing a coil is placed in close proximity to the test material, and an alternating current is passed through the coil, creating a changing magnetic field. This magnetic field induces eddy currents in the conductive material, which in turn produce their own magnetic field. Any disruptions in the material's conductivity or variations in its surface can alter the eddy currents, resulting in measurable changes in the probe's electrical properties. One of the key advantages of ECT is its ability to detect defects on or just beneath the material's surface. It is particularly effective in identifying cracks and corrosion, which are common types of damage that can compromise the structural integrity of conductive materials. However, it should be noted that ECT is limited to materials that are conductive in nature, as non-conductive materials do not allow for the formation of eddy currents. While ECT offers accurate and reliable defect detection, it does come with certain limitations. The process can be time-consuming and costly, requiring specialized equipment and trained personnel to perform the inspections. Additionally, the technique is not suitable for non-conductive materials, which may require alternative NDT methods to assess their integrity. Nonetheless, ECT remains a valuable tool in various industries where the detection of surface and near-surface defects is crucial for ensuring the safety and reliability of conductive materials [8, 9].

8.3.1.12 Acoustic Emission Testing

Acoustic Emission Testing (AET) is a technique used in Structural Health Monitoring (SHM) to continuously monitor the health of structures in real-time. It is designed to detect and accurately locate active damage such as cracking and delamination. AET relies on the analysis of the noise emitted from the structure, making it suitable for detecting both internal and surface damage. This method can be performed remotely and non-destructively, allowing for efficient and convenient monitoring. However, access to a speech analyzer is necessary for conducting the analysis. It is important to note that AET requires the damage to be actively propagating, meaning it may not be effective in detecting dormant or inactive damage. By utilizing AET, early detection of damage is made possible, enabling timely intervention and maintenance to ensure the structural integrity and safety of the monitored system [8, 9, 34].

8.3.1.13 PZT-Based Active Wave Method

The PZT-Based Active Wave Method is a non-destructive technique used for monitoring damage in real-time. It employs the use of a material called lead zirconate titanate (PZT) to generate and detect waves within a structure. This method is particularly useful for assessing the integrity of complex geometries and identifying potential areas of damage. The primary advantage of the PZT-Based Active Wave Method is its ability to provide accurate and real-time monitoring of structural damage. By utilizing

PZT transducers, which can both emit and receive waves, it becomes possible to detect and analyze changes in the material properties of a structure. This allows for the early identification of damage, such as cracks or fractures, which may not be visible to the naked eye. However, it is important to note that the accuracy and reliability of the PZT-Based Active Wave Method can be influenced by external factors and complex geometries. External factors, such as temperature variations or environmental noise, can affect the quality of the wave signals and introduce uncertainties in the monitoring process. Additionally, the presence of intricate geometries in the structure can create wave reflections and diffractions, making it more challenging to interpret the data accurately. To mitigate the impact of these external factors and complex geometries, researchers and practitioners employ various signal processing and data analysis techniques. These techniques aim to filter out noise, enhance the signal-to-noise ratio, and develop algorithms for reliable damage detection and localization. Ongoing advancements in technology and computational methods continue to improve the accuracy and applicability of the PZT-Based Active Wave Method, making it a valuable tool for structural health monitoring in a wide range of applications [8, 9].

8.3.1.14 Leak Test

A leak test is a procedure used to identify potential structural damage by applying liquid pressure to a system or structure. It involves subjecting the structure to a certain amount of liquid pressure for a specific duration. Subsequently, the liquid level within the structure is examined to locate any points of damage or leakage. This testing method is particularly effective for detecting internal damage within structures. However, conducting a leak test requires specialized equipment and can be a time-consuming process. The equipment used is designed to generate and maintain the necessary liquid pressure while ensuring the safety of the testing environment. Additionally, the test must be performed for a sufficient duration to allow any leaks or damage to become apparent. Despite the advantages of the leak test in detecting internal damage, its implementation can be resource-intensive and may require skilled personnel. The need for specialized equipment and the time-consuming nature of the test can pose challenges in terms of cost and efficiency. Nevertheless, the leak test remains a valuable method for identifying structural damage, particularly when internal issues need to be addressed [59].

8.3.1.15 Ultrasound Procedure

The ultrasound procedure is a diagnostic technique that involves the transmission of ultrasonic waves into structures, with the purpose of detecting any potential structural damage. By analyzing the reflections of these waves, it becomes possible to identify and assess internal issues, such as deep cracks or hidden corrosion. This method serves as a valuable tool for inspecting the integrity of various objects and materials. While the ultrasound procedure proves highly effective in detecting internal damage, it is important to note that its capabilities are limited when it comes to identifying surface-level issues. This means that surface damage, such as scratches or minor cracks, may not be readily detected through this method alone. Instead, other inspection techniques or complementary procedures may be necessary to thoroughly assess the condition of an object or structure. Due to the specialized nature of ultrasound equipment, this procedure requires the use of specific tools and technologies designed for transmitting and receiving ultrasonic waves. Trained professionals typically carry out the ultrasound procedure, utilizing the equipment to scan and examine the target area. The collected data is then analyzed and interpreted to determine the presence and severity of any underlying structural damage. Overall, the ultrasound procedure offers a non-invasive and efficient means of detecting internal damage within objects and structures. By harnessing the power of ultrasonic waves and their reflections, this method provides valuable insights into the integrity and health of materials, allowing for timely repairs and maintenance to be carried out as needed [13].

8.3.2 *Imaging Techniques*

8.3.2.1 Thermography

Thermography, also known as infrared imaging, is a non-destructive imaging technique that detects and visualizes temperature variations on the surface of an object. It relies on the principle that all objects emit infrared radiation based on their temperature. By using an infrared camera, heat patterns can be captured and translated into visible images, known as thermograms. Thermography finds applications in various fields, including building inspections, electrical systems analysis, and medical diagnostics. In building inspections, it helps detect heat loss, water leaks, and insulation defects. In electrical systems analysis, it identifies overheating components and potential failures. In the medical field, thermography aids in identifying abnormal temperature patterns that may indicate diseases or damages [10, 23, 33, 36, 37, 54].

8.3.2.2 Magnetic Imaging

Magnetic imaging, commonly referred to as magnetic resonance imaging (MRI), utilizes a powerful magnetic field and radio waves to generate detailed images of internal body structures. This technique provides exceptional soft tissue contrast and is particularly valuable in medical diagnostics. MRI works by aligning the hydrogen atoms within the body's tissues using the magnetic field. When radio waves are applied, the atoms emit signals that are captured by specialized detectors. These signals are then processed to create highly detailed cross-sectional images of the body, aiding in the diagnosis of various conditions, including tumors, damages, and neurological disorders [2, 3, 23, 33].

8.3.2.3 Radar Imaging

Radar imaging, also known as synthetic aperture radar (SAR) imaging, employs radio waves to create images of objects or landscapes. It is commonly used in remote sensing applications, such as mapping terrains, monitoring vegetation, and assessing environmental changes. Radar imaging works by transmitting radio wave pulses toward the target and then capturing the reflected signals. By analyzing the time delay and amplitude of these signals, a radar system can reconstruct a detailed image of the object or landscape. This technique is particularly useful in situations where visibility is limited, such as during bad weather conditions or in dense vegetation [6, 23, 24, 28].

8.3.2.4 Electrical Imaging

Electrical imaging, also known as electrical impedance tomography (EIT), is an imaging technique that measures the electrical conductivity and impedance distribution within an object or subject. It involves applying small electrical currents through electrodes placed on the surface and measuring the resulting voltage changes. By analyzing the distribution of electrical impedance, electrical imaging provides valuable information about the internal structure or composition of an object. It has applications in medical imaging, geophysics, and industrial process monitoring. In medical imaging, EIT can be used to monitor lung function, detect tumors, and assess cerebral blood flow [23, 24].

8.3.2.5 Image Processing and Signal Processing

Image processing refers to the manipulation and analysis of digital images and videos to identify, calculate, and interpret various features. It utilizes complex algorithms, including deep neural networks and deep learning, to carefully analyze and inspect images in order to find common patterns and extract valuable insights. Image processing has a wide range of applications, such as object detection, face detection, motion detection, security and video surveillance, medical imaging, and aesthetics. For example, in the field of medical imaging, images are used to identify internal structures, diagnose cancer, and detect brain damages and cardiovascular diseases. Furthermore, image processing techniques can be applied to detect and locate surface and deep damage in steel structures. By utilizing various imaging technologies like digital images, thermographic images, laser images, and vibration images, image analysis software can be used to identify and locate damage in steel structures. This non-destructive method provides a fast and accurate way to assess the condition of steel structures, enabling the determination of necessary repairs and reinforcements. Signal processing, on the other hand, involves the manipulation and analysis of signals using techniques such as image and signal processing algorithms. In the context of detecting structural damage, this method utilizes sensor data and structural images along with image processing and signal processing algorithms. By combining these techniques, it becomes possible to automatically identify and assess structural damage in a non-destructive and remote manner. The use of signal processing techniques allows for the extraction of valuable information from signals, enabling the detection and interpretation of damage within structures. In summary, image processing and signal processing techniques are utilized to detect and interpret structural damage. Image processing focuses on the analysis and interpretation of digital images and videos, extracting information from visual data to identify features and patterns. Signal processing, on the other hand, involves the analysis and manipulation of signals, including sensor data and structural images, to detect and assess structural damage. Both methods provide non-destructive and effective approaches for identifying and interpreting damage within structures [23, 24, 42, 43, 60, 61].

8.3.3 Destructive Testing

Destructive testing is a method employed in the field of structural engineering to assess the condition of concrete and steel structures. It involves intentionally separating specific parts of the structure and subjecting them to various forms of stress, such as compression, bending, and tension. The primary objective of destructive testing is to evaluate the structural integrity, detect hidden damages, and assess the vulnerability of the components. Unlike NDT methods, which focus on surface inspections and measurements, destructive testing provides a higher level of accuracy and allows for a more detailed analysis of the structural damage and weaknesses that may not be readily visible. By subjecting the separated parts to controlled stress, engineers can gauge their response and measure factors such as resistance, deformation, and failure points. This information is crucial in understanding the structural behavior, identifying potential flaws, and determining the necessary repairs and reinforcements. When conducting destructive tests on concrete and steel structures, the initial step is to carefully select the location and type of the test. Concrete samples can be obtained by separating a portion of the structure using nuts and bolts or machining techniques. These samples are then sent to a specialized laboratory where they undergo thorough analysis. By subjecting the concrete samples to tests such as compression, bending, and permeability, engineers can assess the extent of damage, determine the compressive and bending strength of the concrete, and evaluate its permeability properties. Similarly, for steel structures, samples may need to be cut or bent to isolate sections for testing. These steel samples are transferred to the laboratory where tests such as tensile, compression, rebound, and destructive tests are performed. These tests allow engineers to measure the tensile strength, compression resistance, rebound properties, and overall integrity of the steel components. Once the destructive tests are conducted, the results are carefully analyzed to identify any damage present within the structure. Engineers examine the data to understand the severity and extent of the damage, identify potential weaknesses or defects, and determine the appropriate measures for repair and strengthening. The analysis also helps in estimating the cost and resources required for the necessary repairs. Although destructive testing is a powerful tool for detecting deeper damage in concrete and steel structures, it should only be utilized when non-destructive methods fail to provide sufficient information or when it is absolutely necessary. It is crucial to consider the associated costs and consequences, as destructive testing requires the intentional destruction of part of the structure. This means that subsequent repairs or replacements will be required, adding to the overall project expenses. In some cases, non-destructive methods such as visual inspections and surface roughness tests may provide quick and accurate results for identifying surface damages in concrete and steel structures. However, for detecting hidden or internal damages, destructive testing and imaging techniques become essential. By employing advanced imaging technologies and digital imaging methods, engineers can investigate both surface defects and deeper flaws, such as fractures, enabling a comprehensive assessment of the structure's condition. To execute visual techniques effectively, expertise and experience in structural design and fault detection are crucial. However, when it comes to identifying deeper damages that cannot be detected through surface inspections alone, more sophisticated methods, such as destructive testing and imaging techniques, may be necessary. These advanced techniques allow engineers to gain a comprehensive understanding of the structural damage, guiding them in making informed decisions regarding the direction of repairs and strengthening efforts. In conclusion, destructive testing plays a vital role in the evaluation of concrete and steel structures. By intentionally separating and subjecting parts of the structure to stress, engineers can obtain accurate and detailed information about the structural integrity, damage, and vulnerabilities. Although destructive testing should be used judiciously due to the associated repair costs, it is an effective method for detecting hidden damages and guiding repair and reinforcement efforts. When combined with non-destructive methods and imaging techniques, destructive testing enables a thorough examination of the structure, facilitating the implementation of appropriate measures to ensure the safety and longevity of the infrastructure [1, 2, 6, 7, 23, 33, 56, 62–67].

8.3.3.1 Surface Unevenness Tests

Surface unevenness tests are a set of assessments conducted to identify and measure damages, such as pavements and cracks, present on a given surface. These tests serve as valuable tools for evaluating the overall condition and quality of various structures, including roads, pavements, and other surfaces. By examining the unevenness and irregularities on a surface, engineers and inspectors can gather crucial information about its structural integrity and determine the necessary repairs or maintenance actions. One of the primary objectives of surface unevenness tests is to identify any form of damage or deterioration. These tests involve the use of specialized equipment and techniques that allow for precise measurements of irregularities on a surface. By carefully assessing the presence and severity of pavements and cracks, professionals can determine the extent of the damage and develop appropriate strategies for repair or remediation. Surface unevenness tests also play a vital role in ensuring the safety and functionality of various infrastructures. Uneven surfaces can pose risks to pedestrians, cyclists, and motorists, leading to accidents or damages. Through these tests, engineers can identify potential hazards and implement necessary corrective measures to enhance safety. Moreover, the results obtained from surface unevenness tests can inform decisions regarding road maintenance schedules, pavement resurfacing, or even the design of new structures. To perform surface unevenness tests, various methods and technologies are employed. These may include laser-based devices, profilometers, or even visual inspections. The chosen method depends on the nature of the surface and the level of precision required. Regardless of the approach, the goal remains consistent: to detect and quantify any irregularities present and determine their impact on the structural integrity and usability of the surface. In summary, surface unevenness tests are critical assessments that help identify and measure damages such as pavements and cracks on different surfaces. By conducting these tests, professionals can evaluate the condition of structures, ensure safety, and plan appropriate repair or maintenance actions. Through the use of specialized equipment and techniques, accurate measurements are obtained, aiding in decision-making processes related to infrastructure management and improvement [2].

8.3.4 Environmental Monitoring

8.3.4.1 Noise Emission Control

Noise emission control refers to the process of managing and regulating the generation of high-frequency sound waves emitted by various structures when they are subjected to loads or external stimuli. It involves identifying and analyzing these sound waves in order to understand their source and the extent of damage they may cause. By utilizing this method, it becomes possible to detect and monitor the real-time effects of noise emissions, helping to pinpoint the underlying causes of damage. The primary objective of noise emission control is to mitigate the negative impact of excessive noise on both human health and the surrounding environment. Excessive noise can lead to various detrimental effects, including hearing impairment, sleep disturbance, and increased stress levels. Moreover, it can disrupt ecosystems, harm wildlife, and interfere with communication systems. By identifying the sources of noise emissions and assessing their intensity, measures can be taken to minimize or eliminate the noise at its origin, thereby reducing the overall noise pollution. To achieve effective noise emission control, several approaches can be employed. These may include implementing engineering solutions, such as soundproofing materials or vibration damping techniques, to reduce the generation and transmission of noise. Additionally, regulations and standards can be established to limit noise emissions from specific industries or activities. Regular monitoring and analysis of noise levels can help assess compliance and identify areas that require further control measures. Furthermore, noise emission control plays a crucial role in identifying and preventing potential structural damage. By monitoring the sound waves produced by structures, it becomes possible to detect abnormalities or anomalies that could indicate underlying issues. Real-time analysis of noise emissions can provide valuable insights into the structural integrity of buildings, machinery, or infrastructure, enabling proactive maintenance and repairs to prevent further damage or failure. In summary, noise emission control encompasses the identification, analysis, and management of high-frequency sound waves emitted by structures under various conditions. By understanding the source and extent of noise emissions, it becomes possible to minimize their negative impact on human health, the environment, and structural integrity. Through a combination of engineering solutions, regulations, and proactive monitoring, noise emission control aims to create a quieter and safer living and working environment for all [34].

8.3.5 Structural Health Monitoring

Monitoring the health of a structure is crucial for its safety and efficiency. One effective method is vibration-based health monitoring, which involves continuously monitoring a structure's vibrations to detect potential damage and changes in condition [8]. This approach has two components: identification of output-only systems and identification of input/output systems [9]. Output-only systems extract modal parameters from vibration measurements to detect damage, while input/output systems isolate environmental and damage-related influences [21]. Vibration-based health monitoring is non-destructive, cost-effective, and provides a comprehensive overview of a structure's health [29]. It is used in civil engineering for infrastructure like bridges, tunnels, and buildings. It is also applied in mechanical engineering for monitoring rotating machinery [8, 55]. Aerospace engineering benefits from vibration-based health monitoring as well, particularly in monitoring aircraft for potential damage or changes in condition [9]. Another method of monitoring structure health is earthquake monitoring systems [21]. These systems utilize seismic sensors to capture structural behavior during earthquakes, enabling accurate detection of damage without destructive testing [29]. This information can aid authorities and rescue workers in effective response and recovery [32]. By employing these monitoring techniques, engineers can proactively maintain and repair structures, ensuring their safety and longevity [8, 9, 21, 29, 32].

8.3.5.1 Vibration-Based Methods

Vibration-based methods in SHM have gained significant attention due to their non-destructive nature and cost-effectiveness in detecting damage and assessing the condition of concrete and steel structures. These methods utilize the inherent vibrations present in structures to extract valuable information about their health and performance. One of the primary aspects of vibration-based methods is the identification of output-only systems. This technique involves extracting modal parameters from vibration measurements without requiring detailed knowledge of the specific excitations that cause the structural vibration. By analyzing the vibration data collected from sensors placed on the structure, key features such as natural frequencies, mode shapes, and damping ratios can be determined. These modal parameters provide important insights into the dynamic characteristics of the structure and serve as indicators of its condition. Changes in these parameters can signify the presence of damage or structural degradation. In addition to output-only systems, another approach in vibration-based health monitoring is the identification of input/output systems. This technique aims to differentiate between environmental influences, such as wind or temperature variations, and actual damage events within the structure. By employing statistical pattern recognition and machine learning techniques, vibration data can be analyzed to identify patterns associated with damage. This process involves comparing the current vibration signature with baseline measurements or established damage signatures to determine the presence and extent of damage. To validate the effectiveness of vibration-based damage detection methods, quasi-static stress tests are often conducted. These tests involve applying controlled loads or displacements to the structure while monitoring the corresponding response. By comparing the structural behavior during these tests with the vibration-based damage detection results, the accuracy and reliability of the method can be assessed. It is worth noting that the success of vibration-based methods relies on the installation of appropriate sensors and data analysis systems. These sensors are strategically placed on the structure to capture vibrations across different locations. The collected data is then processed and analyzed using specialized algorithms and techniques. This step often requires expertise in signal processing, statistical analysis, and machine learning to interpret the data effectively and extract meaningful information regarding the structural health. While vibration-based methods have proven to be valuable in detecting various types of damage, they may not be equally effective in identifying certain forms of damage, such as corrosion, which may not significantly affect the vibration characteristics of the structure. Therefore, additional complementary techniques, such as the use of thermal sensors or numerical models, may be employed to enhance the accuracy and reliability of damage detection. This method uses two measures, RMSD and CCDM, to quantify damage. RMSD measures the difference between corrupted and uncorrupted wavelet coefficients, and CCDM measures the correlation between them. These criteria can be used to determine the location and extent of structural damage [68]. Overall, vibration-based methods provide a comprehensive and dynamic approach to structural health monitoring. By continuously monitoring the vibrations of concrete and steel structures, these methods enable early detection and accurate identification of damage, facilitating timely maintenance and repair actions. The ability to assess the condition of critical infrastructure and machinery in real time contributes to improved safety, prolonged structural longevity, and optimized resource allocation. With ongoing advancements in sensor technology, data analysis algorithms, and machine learning techniques, vibration-based methods continue to evolve and offer even greater potential for enhancing structural health monitoring practices in various engineering applications [8, 9, 34].

8.3.5.2 Strain-Based Methods

Strain-based methods refer to a set of techniques used to assess and quantify the extent of damage in various materials and structures. These methods provide valuable quantitative data that can help determine the severity and progression of damage. By measuring strain, which is the deformation experienced by a material under load, strain-based methods enable researchers and engineers to evaluate the structural integrity and identify areas of concern. One significant advantage of strain-based methods is their ability to detect damage in the early stages. By continuously monitoring strain levels, deviations from the expected behavior can be identified promptly, indicating the presence of damage even before it becomes visually apparent. This early detection allows for timely intervention and preventive measures to mitigate the progression of the damage and potential failure. However, the implementation of strain-based methods typically requires the installation of sensors. These sensors are strategically placed at critical locations within the structure to measure strain accurately. While this installation process may involve some effort and cost, it enables ongoing monitoring and data collection, providing valuable insights into the behavior of the structure under various loading conditions. Despite their effectiveness in many scenarios, strain-based methods may not be suitable for detecting certain types of damages, such as fatigue. Fatigue damage is often characterized by microscopic cracks that can propagate gradually over time, leading to sudden failure. As strain-based methods primarily focus on macroscopic deformation, they may not be as sensitive to the early stages of fatigue damage. In such cases, complementary techniques, such as NDT methods or material analysis, may be required to identify and evaluate fatigue-related damages accurately. In conclusion, strain-based methods offer a quantitative approach to assessing damage and evaluating the structural integrity of materials and structures. They enable early detection of damage, providing opportunities for timely intervention. However, their effectiveness may vary depending on the type of damage, and they require the installation of sensors to monitor strain levels accurately. By understanding the capabilities and limitations of strain-based methods, researchers and engineers can employ them as part of a comprehensive approach to ensuring the safety and reliability of various systems [8, 9].

8.3.6 Machine Learning

Machine learning is a branch of artificial intelligence (AI) that utilizes algorithms and techniques to enable systems and programs to learn, analyze data, recognize patterns, and make decisions autonomously. It is a subset of AI and forms one of its main components. Machine learning involves the automatic extraction of meaningful information and rules from input data or sets of information, which can be used for various purposes such as prediction, decision making, information retrieval, and pattern recognition. In the context of detecting damage to concrete and steel structures, machine learning techniques play a crucial role in structural health monitoring. Vibration-based damage detection methods have been developed to identify damage in these structures by analyzing changes in their vibration properties. Machine learning algorithms, such as neural networks, can be utilized to analyze the vibration data and apply statistical pattern recognition to identify the presence of damage. By automatically recognizing patterns and rules in the data, machine learning enables accurate predictions of the location and amount of damage. The use of artificial intelligence techniques, including machine learning, allows for the analysis of structural data to identify damage. These techniques involve the application of algorithms that enable systems and programs to learn, think, and make decisions similar to humans. Artificial intelligence encompasses machine learning and natural language processing, where the latter focuses on analyzing and understanding human language, including text, speech, and image data. The increasing adoption of artificial intelligence across various fields and industries is driven by its ability to enhance accuracy, speed, decision-making capabilities, and the analysis of big data. However, the successful implementation of artificial intelligence, including machine learning, requires qualitative and actionable data, appropriate algorithms, and expertise in analyzing the data. Training data from both healthy and damaged structures is necessary for machine learning algorithms to effectively detect damage. Additionally, specialized knowledge and skills may be required to analyze the data and extract meaningful insights. It is important to consider ethical considerations, ensure security, and protect privacy when utilizing artificial intelligence in any application. Moreover, one specific method within machine learning is the use of artificial neural networks. This technique employs neural models created from sensor data and structural images to detect structural damage. By leveraging artificial neural networks, damage identification can be conducted remotely and non-destructively, optimizing the detection and location of damage within structures. In summary, machine learning is a subset of artificial intelligence that enables systems and programs to learn, recognize patterns, and make decisions based on input data. Within the domain of structural health monitoring, machine learning techniques, including artificial neural networks, are employed to analyze vibration data and detect damage in concrete and steel structures. The use of artificial intelligence techniques, encompassing machine learning and natural language processing, offers enhanced accuracy, speed, decision-making capabilities, and the ability to analyze big data. However, successful implementation requires high-quality data, appropriate algorithms, expertise, and adherence to ethical considerations [7–9, 48].

8.3.6.1 Deep Learning

Deep learning is a branch of artificial intelligence that focuses on training algorithms to automatically learn and extract meaningful patterns from data. It is particularly effective in the field of image processing, where complex algorithms such as deep neural networks are used to analyze digital images and videos, and extract valuable information and interpret various features. One application of deep learning in image processing is the detection of damage in concrete and steel structures. By utilizing primary data and video images, deep learning techniques and neural networks are employed to learn the damage patterns of structures. Through the application of classification techniques and analysis of collected data, structural damage can be accurately detected. This method can incorporate various types of data, including satellite data, seismic data, sensor data, and video imagery, which are fed into a neural network trained with deep learning algorithms to recognize different types of damage. The detected damage and relevant information can then be shared with construction professionals and civil engineers, aiding in the assessment and maintenance of structures. In the context of damage detection, deep learning algorithms are specifically used to create a deep learning model by utilizing sensor data and structural images. This model is capable of automatically and optimally identifying defects in structures. While the training of this model requires a significant amount of data, its ability to accurately detect damage makes it a valuable tool in ensuring the structural integrity of buildings and infrastructure. Overall, deep learning plays a crucial role in image processing and damage detection. By leveraging the power of complex algorithms and neural networks, it enables the extraction of rich information from images and videos, aiding in the identification of patterns, features, and defects in structures. Through its ability to learn from data and make accurate predictions, deep learning contributes to the improvement of performance and the increase in the accuracy of structural damage detection [23, 24, 30, 31, 33, 42].

8.3.6.2 Artificial Neural Networks

Artificial Neural Networks (ANNs) are a computational approach inspired by the human brain's neural networks. They have proven to be an effective method for detecting and analyzing structural damage in various fields. In the context of structural damage detection, ANNs utilize sensor data and structural images to construct neural models that can accurately identify and locate any potential damage. The process begins by feeding the ANN with a vast amount of input data, consisting of sensor readings and images captured from the structure under examination. Through a series of interconnected layers, the neural network learns and extracts complex patterns and relationships within the data. These patterns can include subtle indicators of structural damage, such as changes in vibration patterns or irregularities in the captured images. As the ANN continues to train on the input data, it gradually develops the ability to recognize and differentiate between healthy and damaged structural conditions. This training phase is crucial in enabling the network to make accurate predictions and classifications when presented with new, unseen data. Once the ANN has been sufficiently trained, it can be deployed to remotely and non-destructively evaluate structures for potential damage. By inputting new sensor data and structural images into the trained network, it can swiftly analyze the information and generate predictions regarding the presence and location of any structural damage. The advantage of using ANNs for structural damage detection lies in their ability to capture and understand intricate patterns and features that may be challenging for conventional analytical methods. By leveraging the power of neural networks, this technique allows for automated and optimized damage identification, streamlining the inspection and maintenance processes for various structures, such as bridges, buildings, and pipelines. Ultimately, artificial neural networks offer a promising approach to efficiently and accurately detect structural damage, enhancing safety and facilitating timely repairs and maintenance efforts [69].

8.3.6.3 Multitasking Network

A multitasking network is a specialized approach used to identify structural damage in various contexts. This method involves the utilization of a multifunctional neural network that combines sensor data and structural images to accomplish multiple tasks simultaneously. Not only does this network possess the ability to detect damage in structures, but it also performs other functions related to the structural domain, such as predicting the service life of a building and identifying potential hazards. By integrating multiple tasks into a single network, the multitasking approach enables efficient and automated recognition of building damage. Traditional methods often require separate models or algorithms to address individual tasks, leading to increased complexity and reduced effectiveness. In contrast, the multitasking network streamlines the process by leveraging shared information and learned representations across multiple tasks. One significant advantage of the multitasking network is its ability to overcome the challenges associated with building prediction and analysis. By encompassing various structural-related functions within a single model, it offers a comprehensive solution that addresses multiple aspects of structural assessment. This integrated approach not only enhances the accuracy and reliability of damage detection but also contributes to the optimization of building maintenance and management processes. Overall, the use of a multitasking network for structural damage identification provides a powerful tool that combines sensor data, structural images, and neural network capabilities. This approach offers automated and optimized solutions for recognizing building damage, predicting service life, and identifying hazards, thereby improving the efficiency and effectiveness of structural analysis and maintenance procedures [4].

8.3.6.4 Shear Resistance Generator Network

The shear resistance generator network is a method employed to detect and assess structural damage in various infrastructure systems. This approach utilizes a specialized network known as the shear resistance generation network, which is designed to predict the shear strength of structures based on sensor data and structural images. By continuously updating the network with new data, it can generate accurate predictions of the shear strength, allowing for the identification of potential damage. To determine the damage status of structures, the shear resistance generator network compares the predicted shear strength values with the actual measured values. This comparison enables the detection of any deviations or anomalies, thereby indicating the presence of structural defects. The network's ability to automatically and optimally identify these defects makes it a valuable tool for structural engineers and maintenance teams. It is important to note that training the shear resistance generator network requires a substantial amount of data. The network needs to be exposed to a wide range of structural conditions, including both undamaged and damaged states, in order to learn and generalize effectively. This extensive training allows the network to accurately predict the shear strength of structures, even when faced with real-world scenarios. In summary, the shear resistance generator network is a powerful tool for assessing structural damage. By leveraging sensor data and structural images, it can predict the shear strength of structures and compare it with actual measurements, enabling the identification of defects in an automated and optimized manner. While requiring a significant amount of data for training, this method offers a reliable means of detecting and assessing structural damage [40].

8.3.6.5 Clustering Sensor Data

Clustering sensor data is a technique that involves the application of a clustering algorithm to organize and partition the sensor data collected from a structure. By employing this method, the data can be grouped into distinct clusters based on their similarities and patterns. The objective of clustering sensor data is to identify and isolate potential damage within each cluster. One of the key advantages of this approach is that it allows for remote and non-destructive analysis of structures. Instead of physically inspecting each component or area, the clustering algorithm enables the detection of damage automatically and optimally. This means that the algorithm can effectively pinpoint the location of damage within the structure without the need for extensive manual intervention. The clustering algorithm employed in this method operates by analyzing various parameters and features extracted from the sensor data. These parameters can include measurements such as vibration levels, temperature readings, or strain gauges. By considering these factors, the algorithm can identify similarities and dissimilarities among the data points and group them accordingly. Once the clustering algorithm has organized the sensor data into distinct clusters, it becomes easier to analyze and interpret the information. The presence of damage within a particular cluster can be inferred by observing anomalies or deviations from the expected patterns. By examining the characteristics of the data within each cluster, it is possible to assess the severity and extent of the identified damage. Overall, clustering sensor data offers a valuable approach for effectively detecting and localizing damage in structures. Its remote and non-destructive nature makes it particularly advantageous for large-scale monitoring and assessment. By leveraging this method, engineers and researchers can gain valuable insights into the structural health of various systems and take appropriate actions to address any identified issues promptly. The following Table 8.3 shows the advantages and limitations of some of the methods.

Table 8.3 Advantages and limitations

Methods	Advantages	Limitations
Visual Inspection	- Simple and affordable - Can detect visible damage	- May not be effective for detecting - hidden or internal damage - Limited to surface damage
NDT	- Can detect internal damage - Can provide quantitative data on the extent of damage	- Can be expensive and time consuming - May require specialized equipment and expertise - May not be effective for detecting certain types of damages such as fatigue
Monitoring the health of the structure	- Continuous and real-time monitoring of structure health - Can detect damage in the early stages - Can provide data on damage progression over time	- It requires the installation of sensors and data analysis systems - May require specialized expertise to analyze data - It may not be effective for detecting certain types of damage such as corrosion
Strain-based methods	- Can provide quantitative data on the extent of damage - Can detect damage in the early stages	- Requires sensor installation - It may not be effective for detecting certain types of damages such as fatigue
Vibration-based methods	- Non-destructive and cost-effective approach - Can detect damage in the early stages - Can provide data on damage progression over time	- It requires the installation of sensors and data analysis systems - May require specialized expertise to analyze data - It may not be effective for detecting certain types of damage such as corrosion
Machine learning techniques	- Can provide accurate predictions of the location and amount of damage - It can be applied to a wide range of structures and types of damage Limitations	- Requires training data from healthy and damaged structures - May require specialized expertise to analyze data - It may not be effective for detecting certain types of damages such as fatigue

8.4 Data Collection and Analysis

In the fields of structural analysis and engineering, gathering accurate information about the condition of a structure is of utmost importance. This information helps assess its integrity, identify potential damage, and implement appropriate maintenance and repair measures. Various data sources, including sensors, seismic observations, visual imagery, and satellite data, are employed to acquire a comprehensive understanding of the structure's health. Each source provides unique insights into the structure's behavior, enabling engineers to make informed decisions. Moreover, imaging techniques such as thermography, magnetic imaging, radar imaging, and electrical imaging play a crucial role in capturing and visualizing different aspects of objects or subjects of interest. In this chapter, we will delve into the details of these data sources and imaging techniques, exploring their significance in the process of information gathering, damage detection, and scientific advancements.

8.4.1 Data from Sensors

One of the primary sources of data for structural assessment is the information collected from sensors. Sensors are strategically placed at various locations on the structure to capture important parameters such as temperature, pressure, acceleration, deformation, and weight. By continuously monitoring these variables, engineers can identify any deviations or anomalies that may indicate damage or structural instability. The data gathered from sensors is crucial in detecting and assessing the severity of potential issues. Sensors provide real-time information, enabling engineers to promptly respond to any structural changes. They play a vital role in structural health monitoring systems, which are designed to ensure the safety and longevity of infrastructure. By integrating sensor data with advanced analytics, engineers can detect early signs of deterioration, enabling proactive maintenance and preventing catastrophic failures [38, 70–72].

8.4.2 Seismic Data

Seismic data is another valuable source used to gather information about structures, especially in regions prone to earthquakes. Seismic observations are obtained from a network of specialized sensors that record ground motion during seismic events. These sensors, known as seismometers, are strategically placed to capture data related to ground acceleration, velocity, and displacement. By analyzing seismic data, engineers can determine the dynamic behavior of a structure during an earthquake. This information helps in assessing the structural response, identifying vulnerable areas, and evaluating the effectiveness of existing design and retrofitting measures. Seismic data also aids in calibrating structural models and validating their accuracy, ensuring that buildings and infrastructure can withstand seismic forces [17, 18, 27, 31].

8.4.3 Visual Data

Visual data, including video images and photographs, serves as a valuable source of information for detecting structural damage. Through visual inspection, engineers can identify visible signs of deterioration, such as cracks, corrosion, or deformation. Visual data is often collected during regular inspections or when specific issues are suspected. Advanced imaging techniques, such as photogrammetry, enable engineers to create accurate 3D models of structures using images or videos. These models provide a detailed representation of the structure's current condition, facilitating the identification of potential problem areas. Visual data is particularly effective when combined with other data sources, providing a comprehensive understanding of the structure's health [31, 54].

8.4.4 Satellite Data

Satellite data, including satellite imagery and radar data, offers a broad-scale perspective on structures and their surroundings. Satellites equipped with high-resolution cameras capture images that can be analyzed to assess the condition of buildings, bridges, and other infrastructure. Radar data, on the other hand, provides information about structural deformation and movement. Satellite data is especially useful for large-scale infrastructure monitoring, such as pipelines, power grids, and transportation networks. By comparing historical and current satellite images, engineers can detect changes in the structure's appearance or identify potential risks. This data aids in identifying damage caused by natural disasters, land subsidence, or other environmental factors [24, 31, 73].

8.5 Conclusions

In conclusion, the chapter emphasizes the importance of employing a combination of methods and techniques for the detection and assessment of damage in concrete and steel structures. Various types of damage, such as corrosion, fatigue, impact, fire, earthquake, burst, cracks, fractures, anomalous shape change, abnormal deflections, abnormal bending, abnormal vibration, and unbalance, can compromise the structural integrity and longevity of these structures. Different methods for detecting and assessing damage, including visual inspection, NDT, imaging techniques, quasi-static loading, structural monitoring, acoustic emission testing, and artificial intelligence (AI) techniques, were discussed. NDT methods, such as ultrasonic inspection, magnetic particle inspection, radiographic testing, and eddy current inspection, serve as valuable tools for detecting and evaluating damage. The selection of a method depends on the specific application and type of damage being assessed. The significance of vibration-based health monitoring and AI techniques in providing continuous monitoring and accurate identification of damage patterns were highlighted. By implementing a comprehensive approach that combines NDT methods, advanced techniques, and data analysis, engineers can ensure the safety, reliability, and longevity of concrete and steel structures. Regular monitoring, early detection of damage, and timely maintenance and repair actions are essential for preventing accidents and optimizing structural performance. Moreover, the continuous improvement and advancement of these methods, contributing to the ongoing enhancement of structural health monitoring practices was emphasized. Additionally, the utilization of data collection and analysis, including sensors, seismic observations, visual data, and satellite data, along with advanced imaging techniques, allows for comprehensive assessment and informed decision-making. These methods facilitate proactive maintenance, timely repairs, and effective resource allocation, ensuring the safety and longevity of critical infrastructure. To prevent accidents resulting from damage to concrete and steel structures, the regular inspections, structural stability analysis, investigation of the building's history, inspection of sensitive areas, immediate repair and maintenance work, and continuous monitoring through sensor installations, were suggested. Public awareness, adherence to safety protocols, and prompt repairs also contribute to reducing the occurrence and severity of damages caused by damaged structures. Overall, by implementing a combination of methods, techniques, and preventive measures, engineers and professionals can enhance the resilience, safety, and longevity of concrete and steel structures, ensuring the well-being of individuals and the reliability of critical infrastructure.

References

1. De Sortis A, Antonacci E, Vestroni F (2005) Dynamic identification of a masonry building using forced vibration tests. Eng Struct 27:155–165. https://doi.org/10.1016/j.engstruct.2004.08.012
2. Zhu HP, He B, Chen XQ (2005) Detection of structural damage through changes in frequency. Wuhan Univ J Nat Sci 10:1069–1073. https://doi.org/10.1007/BF02832469
3. Jahangir H, Hasani H, Esfahani MR (2021) Wavelet-based damage localization and severity estimation of experimental RC beams subjected to gradual static bending tests. Structures 34:3055–3069. https://doi.org/10.1016/j.istruc.2021.09.059
4. Oyawa WO, Sugiura K, Watanabe E (2001) Polymer concrete-filled steel tubes under axial compression. Constr Build Mater 15:187–197. https://doi.org/10.1016/S0950-0618(00)00034-9
5. Shariati M, Mafipour MS, Haido JH, Yousif ST, Toghroli A, Trung NT, Shariati A (2020) Identification of the most influencing parameters on the properties of corroded concrete beams

using an Adaptive Neuro-Fuzzy Inference System (ANFIS). Steel Compos Struct. 34 155–170. https://doi.org/10.12989/scs.2020.34.1.155
6. Zhao XL, Grzebieta R (2002) Strength and ductility of concrete filled double skin (SHS inner and SHS outer) tubes. Thin-Walled Structures. 40:199–213. https://doi.org/10.1016/S0263-8231(01)00060-X
7. Aied H, González A, Cantero D (2016) Identification of sudden stiffness changes in the acceleration response of a bridge to moving loads using ensemble empirical mode decomposition. Mech Syst Signal Process 66–67:314–338. https://doi.org/10.1016/j.ymssp.2015.05.027
8. Xu D, Liu Q, Qin Y, Chen B (2021) Analytical approach for crack identification of glass fiber reinforced polymer–sea sand concrete composite structures based on strain dissipations. Struct Health Monit 20:2778–2790. https://doi.org/10.1177/1475921720974290
9. Of SE, Wang TBC (2000) R Esearch T Oward U Se of 18:918–920
10. Maeck J, Peeters B, De Roeck G (2001) Damage identification on the Z24 bridge using vibration monitoring, Smart. Mater Struct 10:512–517. https://doi.org/10.1088/0964-1726/10/3/313
11. Zeng L, Xiao Y, Chen Y, Jin S, Xie W, Li X (2017) Seismic damage evaluation of concrete-encased steel frame-reinforced concrete core tube buildings based on dynamic characteristics. Appl Sci (Switz). 7. https://doi.org/10.3390/app7040314
12. Maeck J, Abdel Wahab M, Peeters B, De Roeck G, De Visscher J, De Wilde WP, Ndambi JM, Vantomme J (2000) Damage identification in reinforced concrete structures by dynamic stiffness determination. Eng Struct. 22 1339–1349. https://doi.org/10.1016/S0141-0296(99)00074-7
13. Liu J, Li K (2022) Research on an improved SOM model for damage identification of concrete structures. Appl Sci (Switz). 12. https://doi.org/10.3390/app12094152
14. Skolnik D, Lei Y, Yu E, Wallace JW (2006) Identification, model updating, and response prediction of an instrumented 15-story steel-frame building. Earthq Spectra 22:781–802. https://doi.org/10.1193/1.2219487
15. Mahin SA (1998) Lessons from damage to steel buildings during the Northridge earthquake. Eng Struct 20:261–270. https://doi.org/10.1016/S0141-0296(97)00032-1
16. Gross JL, Phan LT (2010) Summary of best practice guidelines for structural fire resistance design of concrete and steel buildings. Struct Congr 2010:2369–2379. https://doi.org/10.1061/41130(369)215
17. Oladazimi A, Mansour S, Hosseinijou SA, Majdfaghihi MH (2021) Sustainability identification of steel and concrete construction frames with respect to triple bottom line, Buildings. 11. https://doi.org/10.3390/buildings11110565
18. Huth O, Feltrin G, Maeck J, Kilic N, Motavalli M (2005) Damage identification using modal data: experiences on a prestressed concrete bridge. J Struct Eng 131:1898–1910. https://doi.org/10.1061/(asce)0733-9445(2005)131:12(1898)
19. Wang S, Ren Q, Qiao P (2006) Structural damage detection using local damage factor. JVC/J Vib Control. 12:955–973. https://doi.org/10.1177/1077546306068286
20. Betti M, Facchini L, Biagini P (2015) Damage detection on a three-storey steel frame using artificial neural networks and genetic algorithms. Meccanica 50:875–886. https://doi.org/10.1007/s11012-014-0085-9
21. Ren W-X, De Roeck G (2002) Structural damage identification using modal data. II: Test Verification. J Struct Eng. 128 96–104. https://doi.org/10.1061/(asce)0733-9445(2002)128:1(96)
22. Soleymani A, Saffari H (2023) Seismic improvement and rehabilitation of steel concentric braced frames: a framework-based review. Rehabil Civ Eng. 11 1–23. https://doi.org/10.22075/JRCE.2022.26179.1611
23. Diederichsen SZ, Grønhøj MH, Mickley H, Gerke O, Steffensen FH, Lambrechtsen J, Rønnow Sand NP, Rasmussen LM, Olsen MH, Diederichsen A (2017) CT-Detected growth of coronary artery calcification in asymptomatic middle-aged subjects and association with 15 biomarkers. JACC: Cardiovasc Imaging. 10 858–866. https://doi.org/10.1016/j.jcmg.2017.05.010
24. Azenha M, Faria R, Ferreira D (2009) Identification of early-age concrete temperatures and strains: Monitoring and numerical simulation. Cement Concr Compos 31:369–378. https://doi.org/10.1016/j.cemconcomp.2009.03.004

25. Soleymani A, Jahangir H, Nehdi ML (2023) Damage detection and monitoring in heritage masonry structures: Systematic review. Constr Build Mater 397:132402. https://doi.org/10.1016/j.conbuildmat.2023.132402
26. Unger JF (2002) System identification and damage detection in civil engineering
27. Wijesundara LMG, Clubley SK (2016) Damage state identification for reinforced concrete columns in uplift due to internal building detonations. Eng Struct 127:804–823. https://doi.org/10.1016/j.engstruct.2016.07.051
28. Friswell MI (2007) Damage identification using inverse methods, philosophical transactions of the royal society a: mathematical, physical and engineering. Sci 365:393–410. https://doi.org/10.1098/rsta.2006.1930
29. Pathirage CSN, Li J, Li L, Hao H, Liu W, Wang R (2019) Development and application of a deep learning–based sparse autoencoder framework for structural damage identification. Struct Health Monit 18:103–122. https://doi.org/10.1177/1475921718800363
30. Mohan V, Parivallal S, Kesavan K, Arunsundaram B, Ahmed AKF, Ravisankar K (2014) Studies on damage detection using frequency change correlation approach for health assessment. Procedia Eng. 86:503–510. https://doi.org/10.1016/j.proeng.2014.11.074
31. Rackauskaite E, Hamel C, Law A, Rein G (2015) Improved formulation of travelling fires and application to concrete and steel structures. Struct 3:250–260. https://doi.org/10.1016/j.istruc.2015.06.001
32. Zong ZH, Jaishi B, Ge JP, Ren WX (2005) Dynamic analysis of a half-through concrete-filled steel tubular arch bridge. Eng Struct 27:3–15. https://doi.org/10.1016/j.engstruct.2004.08.007
33. Jahangir H, Esfahani MR (2013) Damage localization of structures using adaptive neuro-fuzzy inference system, In: 7th national congress on civil engineering, Zahedan, Iran, 2013.
34. Cook DC (2005) Spectroscopic identification of protective and non-protective corrosion coatings on steel structures in marine environments. Corros Sci 47:2550–2570. https://doi.org/10.1016/j.corsci.2004.10.018
35. Cantieni R (2005) Experimental methods used in system identification of civil engineering structures. In: Proceedings of the 1st international operational modal analysis conference, IOMAC 2005. 10–11
36. Ai L, Bayat M, Ziehl P (2023) Localizing damage on stainless steel structures using acoustic emission signals and weighted ensemble regression-based convolutional neural network. Meas: J Int Meas Confed. 211. https://doi.org/10.1016/j.measurement.2023.112659
37. Abbasi A, Khatibinia M, Jahangir H, Dos Santos JVA, Lopes HMR (2023) Shearography-Wavelet-Based damage detection methodology for aluminum beams. Rehabil Civ Eng. 11 22–43. https://doi.org/10.22075/jrce.2023.29435.1780
38. Donze V, Hentz S, Daudeville L, Donzé FV (2004) Identification and Validation of a Discrete Element Model for Concrete. J Eng Mech 130:709–719. https://doi.org/10.1061/(asce)0733-9399(2004)130:6(709)
39. Dahmani L, Khennane A, Kaci S (2010) Crack identification in reinforced concrete beams using ANSYS software. Strength Mater 42:232–240. https://doi.org/10.1007/s11223-010-9212-6
40. Kunnath SK, Kalkan E (2004) Evaluation of seismic deformation demands using non-linear procedures in multistory steel and concrete moment frames. ISET J Earthq Technol 41:159–181
41. Zhang W, Li J, Hao H, Ma H (2017) Damage detection in bridge structures under moving loads with phase trajectory change of multi-type vibration measurements. Mech Syst Signal Process 87:410–425. https://doi.org/10.1016/j.ymssp.2016.10.035
42. Sabet Divsholi B, Yang Y (2008) Application of PZT sensors for detection of damage severity and location in concrete. Smart Struct, Devices, Syst IV. 7268 726813. https://doi.org/10.1117/12.810696
43. Farrar CR, Doebling SW, Nix DA (2001) Vibration-based structural damage identification, philosophical transactions of the royal society a: mathematical, physical and engineering. Sci 359:131–149. https://doi.org/10.1098/rsta.2000.0717
44. Mai G, Li L, Chen X, Xiong Z, Liang J, Zou X, Qiu Y, Qiao S, Liang D, Liu F (2023) Fatigue performance of basalt fibre-reinforced polymer bar-reinforced sea sand concrete slabs. J Market Res 22:706–727. https://doi.org/10.1016/j.jmrt.2022.11.135

45. Caglar N, Vural I, Kirtel O, Saribiyik A, Sumer Y (2023) Structural damages observed in buildings after the January 24, 2020 Elazığ-Sivrice earthquake in Türkiye. Case Stud Constr Mater. 18:e01886. https://doi.org/10.1016/j.cscm.2023.e01886
46. Cole GL, Dhakal RP, Turner FM, Building pounding damage observed in the, (2011) Christchurch earthquake. Earthquake Eng Struct Dynam 41(2012):893–913. https://doi.org/10.1002/eqe.1164
47. Jahangir H, Hasani H, Esfahani MR (2021) Damage localization of rc beams via wavelet analysis of noise contaminated modal curvatures. J Soft Comput Civ Eng. 5 101–133. https://doi.org/10.22115/SCCE.2021.292279.1340
48. Huang M, Zhao W, Gu J, Lei Y (2020) Damage identification of a steel frame based on integration of time series and neural network under varying temperatures. Adv Civ Eng. 2020. https://doi.org/10.1155/2020/4284381
49. Entezami A, Shariatmadar H, Sarmadi H (2017) Structural damage detection by a new iterative regularization method and an improved sensitivity function. J Sound Vib 399:285–307. https://doi.org/10.1016/j.jsv.2017.02.038
50. Ko JM, Sun ZG, Ni YQ (2002) Multi-stage identification scheme for detecting damage in cable-stayed Kap Shui Mun Bridge. Eng Struct 24:857–868. https://doi.org/10.1016/S0141-0296(02)00024-X
51. Keyhani A, an Energy-Based Damage Detection, (n.d.) 1–8
52. Rent WX, Zong ZH (2004) Output-only modal parameter identification of civil engineering structures. Struct Eng Mech. 17 429–444. https://doi.org/10.12989/sem.2004.17.3_4.429
53. Sharma N, Sharma S, Sharma SK, Mehta R (2020) Evaluation of corrosion inhibition and self healing capabilities of nanoclay and tung oil microencapsulated epoxy coatings on rebars in concrete. Constr Build Mater 259:120278. https://doi.org/10.1016/j.conbuildmat.2020.120278
54. Assouli B, Simescu F, Debicki G, Idrissi H (2005) Detection and identification of concrete cracking during corrosion of reinforced concrete by acoustic emission coupled to the electrochemical techniques. NDT and E Int 38:682–689. https://doi.org/10.1016/j.ndteint.2005.04.007
55. Zaid O, Ahmad J, Siddique MS, Aslam F, Alabduljabbar H, Khedher KM (2021) A step towards sustainable glass fiber reinforced concrete utilizing silica fume and waste coconut shell aggregate. Sci Rep 11:1–14. https://doi.org/10.1038/s41598-021-92228-6
56. Morino S, Uchikoshi M, Yamaguchi I (2001) Concrete-filled steel tube column system-its advantages. Steel Struct. 1:33–44
57. Arora V, Wijnant YH, De Boer A (2014) Acoustic-based damage detection method. Appl Acoust 80:23–27. https://doi.org/10.1016/j.apacoust.2014.01.003
58. Sadkovyi V, Andronov V, Semkiv O, Kovalov A, Rybka E, Otrosh Y, Udianskyi M, Koloskov V, Danilin A, Kovalov P (2021) Fire resistance of reinforced concrete and steel structures, fire resistance of reinforced concrete and steel structures. 1–166. https://doi.org/10.15587/978-617-7319-43-5
59. Yu T, Chen X, Yan W, Xu Z, Ye M (2023) Leak detection in water distribution systems by classifying vibration signals. Mech Syst Signal Process 185:109810. https://doi.org/10.1016/j.ymssp.2022.109810
60. Yosefi A, Jahangir H (2023) The influence of natural additives on mechanical properties of concrete specimens : a review 5:11–20
61. Yan S, Dai Y, Zhao P, Liu W (2018) Interfacial damage identification of steel and concrete composite beams based on piezoceramic wave method. J Appl Biomater Funct Mater. 16:70–80. https://doi.org/10.1177/2280800017753051
62. Biswas AK, Aloke Kumar Datta, Pijush Topdar, Sanjay Sengupta (2023) Acoustic emission-based damage detection in steel framed structure-a review. Int J Comput Civ Struct Eng. 19 188–204. https://doi.org/10.22337/2587-9618-2023-19-2-188-204
63. Ahmad J, Arbili MM, Majdi A, Althoey F, Farouk Deifalla A, Rahmawati C (2022) Performance of concrete reinforced with jute fibers (natural fibers): A review. J Eng Fibers Fabr. 17. https://doi.org/10.1177/15589250221121871

64. di Prisco M, Felicetti R, Iorio F, Gettu R (2001) On the identification of SFRC tensile constitutive behaviour. Fracture Mechanics of Concrete Structures. AA Balkema Publishers, Rotterdam, pp 541–548
65. Kong L, Xie S, Wang C, Wang L (2023) Effect of Iron tailings as fine aggregate and mineral admixture on strength and microstructure of cement mortar. Int J Concr Struct Mater. 17. https://doi.org/10.1186/s40069-023-00584-6
66. Syed H, Nerella R, Madduru SRC (2020) Role of coconut coir fiber in concrete, Materials Today. Proceedings 27:1104–1110. https://doi.org/10.1016/j.matpr.2020.01.477
67. Solahuddin BA, Yahaya FM (2021) Properties of concrete containing shredded waste paper as an additive, Materials Today. Proceedings 51:1350–1354. https://doi.org/10.1016/j.matpr.2021.11.390
68. Campeiro LM, Budoya DE, Baptista FG (2022) Damage detection in noisy environments based on EMI and Lamb waves: A comparative study. J Intell Mater Syst Struct 34:1042–1056. https://doi.org/10.1177/1045389X221128583
69. Tashima MM, De Silva A, Akasaki JL, Barbosa MB (2004) The possibility of adding Rice Husk Ash (Rha) to concrete. International RILEM conference on the use of recycled materials in building structures. 778–786
70. Tatar A, Dowden DM (2022) Analytical and numerical investigation of a low-damage uplift friction damper for self-centering cross-laminated timber rocking walls. Eng Struct 254:113836. https://doi.org/10.1016/j.engstruct.2021.113836
71. Ayele A, Woldearegay K, Meten M (2021) A review on the multi-criteria seismic hazard analysis of Ethiopia: with implications of infrastructural development. Geoenvironmental Disasters. 8. https://doi.org/10.1186/s40677-020-00175-7
72. Soleymani A, Hasani H, Jahangir H, Yazdanpanah E (2022) A comprehensive review of FRP-Concrete bond behaviour. In: 13th national congress on civil engineering
73. Unger JF, Teughels A, De Roeck G (2006) System identification and damage detection of a prestressed concrete beam. J Struct Eng 132:1691–1698. https://doi.org/10.1061/(asce)0733-9445(2006)132:11(1691)

Chapter 9
Detection of Damages in Concrete Structures by Signal Processing and Image Processing Techniques: A Critical Review

S. M. Priok Rashid, Atefeh Soleymani, Hashem Jahangir, and Moncef L. Nehdi

Abstract Concrete cracks are among the primary indications of the structure' failure, which is crucial for maintenance and the resulting exposure will damage the environment. To support vibration-based structural health monitoring (SHM), this chapter provides a review on the assessment of current studies on signal processing methods and image processing. Data may be collected, displayed, analyzed, and then utilized to inform maintenance choices using various tools and procedures. Civil construction, such as buildings, columns, beams, and slabs, has been the focus of attention. SHM based on vibration relies heavily on signal processing. In signal processing, the goal is to find, quantify, and locate the structure's damage by extracting subtle changes in the vibration signals. There remains a gap in the exact measurement of the width of cracks in millimeters, despite the fact that significant research has been done on crack identification utilizing image processing. By highlighting and analyzing some of the most influential papers from around the world, this review article illuminates research from all over the world.

S. M. P. Rashid
Department of Civil Engineering, Faculty of Engineering and Built Environment, Universiti Kebangsaan Malaysia, UKM Bangi 43600, Malaysia
e-mail: p131825@siswa.ukm.edu.my

A. Soleymani
Department of Civil Engineering, Shahid Bahonar University of Kerman, Kerman, Iran
e-mail: Atefeh_Soleymani@eng.uk.ac.ir

H. Jahangir
Department of Civil Engineering, Research Group of Novel Technologies in Civil Engineering, University of Birjand, Birjand, Iran
e-mail: h.jahangir@birjand.ac.ir

M. L. Nehdi (✉)
Department of Civil Engineering, McMaster University, Hamilton, ON, Canada
e-mail: nehdim@mcmaster.ca

H. Jahangir et al. (eds.), *Damage Detection and Structural Health Monitoring of Concrete and Masonry Structures*, Springer Tracts in Civil Engineering,
https://doi.org/10.1007/978-981-97-8975-7_9

Keywords Concrete structures · Damage detection · Signal processing · Image processing · Wavelet transform · Contourlet transform

9.1 Introduction

Concrete is a widely employed artificial substance that is commonly utilized in various applications. It can be difficult to increase concrete performance and regulate its structural behavior using simple, cost-efficient methods. Predicting concrete construction performance has become crucial for economic and safety purposes, particularly in emerging nations [1, 2]. Mechanical overloading appears to be the leading cause of concrete deterioration. Mechanical stresses, which are frequently localized, can lead to micro cracks. These cracks have the potential to grow and spread throughout the building, resulting in catastrophic damage [3]. Monitoring cracks is important because of this. Operationally, this monitoring is often carried out by periodically assessing the onset of surface cracks using extensometers or optical methods. The increasing adoption of non-destructive testing (NDT) to identify cracks at an early stage is due to the fact that interior damage cannot be detected using these conventional methods.

Concrete surface cracking can result from various factors such as thermal shrinkage, overloading, deformation, and salt reactivity. This phenomenon is quite common and particularly relevant in South Korea where a large proportion of reinforced concrete structures are aged over 30 years. In 2014, this percentage was estimated to be 3.8%. However, it is projected to increase significantly to 13.8% and 33.7% in 2024 and 2029, respectively [4]. Therefore, early crack detection is essential, and this entails examining and evaluating the usage of concrete structures as well as their overall stability.

Over the course of the previous twenty years, the domain of SHM has arisen as a prominent and expeditiously burgeoning area of study across multiple disciplines, such as aeronautics, mechanical engineering, and civil engineering. SHM aims to evaluate the structural health and dynamic properties in real time. The SHM system comprises three key stages for detecting damages, namely signal monitoring, signal processing, and interpretation [5]. Since vibration measurements may be used to identify damage that is unseen or buried inside the interior parts of the building before it can be seen by visual examination, vibration-based monitoring has become a key area of attention for SHM in recent years. The fundamental concept behind SHM using vibration analysis is that structural damage often has an impact on the inherent properties of the structure, such as stiffness, which can be detected through the analysis of the structure's vibration response. So, it is important to identify the structural damage and assess its severity using a powerful signal processing system that can extract minute variations in the vibration signal.

In the last ten years, crack detection has become a key topic of research in computer vision, an image processing-based technology. Because of this, it is now a significant factor in automating the crack identification process. Image-based crack detection techniques are being developed and improved by researchers from all over the world. Overall, these approaches are currently of significant interest to both scientists and engineers [6].

Three distinct and innovative transformations, namely the wavelet, curvelet, and contourlet, have proven to be highly effective in capturing the fundamental geometric properties of various forms of data, particularly in the context of removing noise from remotely sensed images through the use of multiresolution analysis. These techniques essentially represent innovations for reducing noise on high resolution multispectral pictures [7]. In recent years, the multiscale directed contourlet transform, which can identify defects in 2D structures with complicated geometric forms, has been presented as an alternative to the wavelet transform.

Data gathering, data processing, locating properties at risk of harm, and decision-making for determining the extent of damage are all included in the SHM technique set. Numerous methodologies for assessing damage have been devised with regards to the direct time domain examination of vibration [8]. Since the raw sensor data is typically nonlinear and extremely complicated, its analysis requires effective tools like soft computing approaches [9–12] or signal processing techniques. The wavelet transform is a unique technique for signal processing that may analyze local signals at multiple times and frequencies [13–15]. A signal's wavelet transform can reveal some of its hidden characteristics, such as discontinuities and fractured spots. As a result, it is used in certain study to analyze vibration responses in the time domain and identify structural damage [16–20]. At times, it is common practice to utilize modal properties such as the natural frequency, mode shape, and modal damping ratio in order to analyze the information obtained from sensors affixed to a structure in the frequency domain. The structure's unique features are described by its modal parameters, which vary as a result of damage to the structure's mass, stiffness, or damping values and their related modal parameters. As a result, study of the modal parameters might be used to identify structural degradation [21–25]. The wavelet transform of modal characteristics is employed in various research studies to find structural deterioration [26–28]. Wavelets' primary benefit is its capacity for local analysis. Based on this, wavelets might find hidden features of signals like rupture sites and discontinuities. The proper utilization of wavelet analysis necessitates expertise and familiarity, as opting for an alternative family or mother wavelet may substantially impact the efficacy and results of the analysis. In certain cases, the wavelet transform of vibration data in the time domain is employed to assess structures and identify areas of damage.

9.2 Damage Detection Using Signal Processing Techniques

One of the key components and most demanding aspects of vibration-based machine position monitoring is the extraction of characteristics utilizing signal processing. The goal is to highlight characteristics of the detected signals, which are typically complicated and noisy and require additional processing to produce the characteristics that are of interest to us. To accurately detect and diagnose faults that point to probable machine failure, the choice of the signal processing technique is crucial. In light of the intricate nature of the structural response to dynamic loads, numerous methods, methodologies, and algorithms have been devised to oversee the health of buildings. The degree of identification demanded by various approaches determines the classification of the approach utilized in assessing the current health status of a structure [29, 30]: Level 1 pertains to the identification of structural damage, while Level 2 involves the determination of the precise location of the failure in geometric terms. Level 3, on the other hand, centers on the evaluation of the magnitude of the damage. Level 4 is concerned with the projection of the remaining useful life of the structure, and lastly, Level 5 focuses on the development of self-healing structures.

Shape memory alloys, particularly Cu-Al-Ni and Ni–Ti (NiTi) alloys, have been classified as 'level 5' in the realm of smart structures as posited by Park and colleagues [30]. Within the field of SHM, the principal objective is to determine the residual useful life of the structure. However, when excessive noise is present in vibration signals, signal processing algorithms become incapable of extracting any valuable information [31]. In this regard, adaptive noise cancellation represents a promising method of enhancing the signal-to-noise ratio. The raw data of the structure is measured in the time domain using a variety of vibration transducers, such as accelerometers, etc. The disadvantage of time domain analysis is that it does not clearly identify all of the frequencies present in the structure's response despite being straightforward and effective. Real-time analysis necessitates the employment of the rapid Fourier transform for the expeditious conversion of functions from the time domain to the frequency domain, or vice versa. Additional analysis is generally undertaken to procure modal domain data from frequency domain data. Time domain data may now be immediately converted into modal domain data using modern techniques [32]. Linear systems undergo a certain level of loss of information when transitioning from the time domain to the frequency domain, despite undergoing some data compression throughout the conversion process [33]. In recent research, signal processing techniques that are based on the modal domain have been extensively explored [34]. The reason for this is that mode shapes, modal damping factors, and natural frequencies are easier to comprehend than features derived from the time domain (such as residuals of the autoregressive model) and frequency domain (such as the distortion identification functions). Of course, there are no universally applicable physical constants or signal processing methods, and it is not feasible to establish a signal processing approach for every application [35].

Multiple components exhibit distinct responses at various levels that contribute to the detection of seismic damage in tall buildings. Multiple investigations have been carried out to effectively detect damage in diminutive constructions, such as a solitary or manifold span beam with diverse boundary circumstances. To the utmost of our awareness, none of the techniques reliant on signals have managed to accomplish the three goals of identifying damage. Using just the output data from the sensors inserted in the structure, signal processing techniques and neural networks were used in this study to fulfill the triple objectives [36].

Signal-processing techniques are useful for detecting structural degeneration. Several approaches hinge on the utilization of the Fourier transform. This mathematical method facilitates the relocation of the signal from the time domain to the frequency domain. Despite the advantages of Fourier transform, it is important to note that the conversion process leads to the loss of sometime information of the signal. To remedy this predicament, the wavelet transform emerged as a viable alternative to the Fourier transform. Wavelet transform, a signal processing method, offers the possibility of a historical opening with configurable dimensions. By highlighting the short time period for high-frequency components and wide time interval for low-frequency components, this transform improves the analysis of signals with fluctuations and local changes. Wavelets perform remarkably well with one-dimensional signals. However, the extended version of one-dimensional wavelets is frequently used to handle the two-dimensional signals. The processing of data with additional dimensions is beyond the capabilities of wavelet transformations, thereby creating a need for new signal processing methods. The contourlet transform, a modern generation of signal-processing techniques, has been developed to overcome the limitations of the wavelet approach for analyzing multidimensional data.

9.2.1 Wavelet Transform

Image processing was utilized by Tsao et al. [37] and Wang et al. [38] to identify asphalt defects. The asphalt cracks were categorized using the neural network model proposed by Kaseko et al. [39].

According to Sun and Chang's [40] investigation of the productivity of the wavelet transforms, the wavelet transform outperforms the traditional Fourier transform in the analysis of nonstationary signals. Although fracture identification is a local issue that is probably represented by high-frequency modes, the wavelet transform resolution is weak in the high-frequency area, which might affect the performance of this approach in assessing damage.

A neuro-fuzzy technique was developed by Chae and Abraham [41] to assess the conditions of wastewater pipelines. In this model, neural networks were given pre-processed photos, and the networks identified the presence of various properties including fractures and connections. Finally, the fuzzy logic model was combined with the photos and characteristics acquired from them to assess the overall health and conditions of the tubes.

In recent years, there has been a surge in the presentation of damage detection algorithms that utilize picture inspection as their basis [42–44]. A study conducted by Abdel-Qader et al. [45] examined the use of various techniques, such as wavelet transform, Fourier transform, Sobel filter, and Canny filters, to detect fractures in pictures of a bridge's surface. Their findings revealed that wavelet transformation produced more dependable outcomes than the other methods. Hutchinson and Chen [46] also implemented wavelet processing and a Canny filter to detect and estimate the necessary parameters required to analyze the receiver operating characteristics.

Ito et al. [47] proposed a digital camera-based automatic image inspection system that identified and analyzed concrete surface cracks by combining several image-processing methods, including wavelet transform, shade correction, and binarization. Kawamura et al. [48] introduced an approach that utilized a genetic algorithm to accurately recognize cracks through the semi-automated optimization of image processing settings. However, these techniques have not addressed certain crucial crack properties, such as connectivity. To enhance the quality and accuracy of pictures, global image-processing techniques, such as wavelet transformations, were also employed.

The time–frequency spectrum of a signal can be characterized by utilizing time and scale window functions along with the wavelet transform method [49]. Wavelet transform and its advanced version, wavelet packet transform (WPT), have recently gained popularity as the most widely used signal processing methods. By utilizing two discrete-time low- and high-pass filters, a wavelet transform can be created to provide approximations and details. The first-level approximation is then divided into new details and an approximation based on multiresolution analysis, which is then repeated. WPT creates a complete binary tree of band decomposition by decomposing both the detail and approximation coefficients. A variety of wavelet mother functions including the Mexican hat, Daubechies, Haar, Gauss, Morlet, Shannon, Symlets, Coiflets, Meyer, Spline, and Gabor have been proposed (Fig. 9.1). In order to obtain the approximations and detailed coefficients of the Wavelet transform and WPT, the mother wavelets need to be translated and dilated.

In recent times, the utilization of wavelet transform has become increasingly important as a signal processing technique for identifying systems and evaluating damages in civil projects. Quek et al. [50] conducted a comparison and analysis of the sensitivity of wavelet transform coefficients, employing both Haar and Gabor wavelets to identify fractures in a model beam. They utilized sudden changes in the coefficients as an indicator of deterioration. The researchers noted that the Haar wavelet method can more accurately determine the location and severity of the damage in the beam. Han et al. [51] also adopted WPT and Daubechies wavelets for the identification of degradation in a basic steel beam.

Adeli and Jiang [52] utilized the NARMAX model to construct a dynamic fuzzy wavelet neural network model, which is specifically designed for non-parametric identification of high-rise building structures. The model combines wavelets with the computational intelligence techniques of neural networks [53] and fuzzy logic [54–56]. In addition, the authors developed a backtracking inexact linear search technique with an adaptive Levenberg–Marquardt-least squares algorithm for training the model. The proposed model was applied to two high-rise moment-resisting building structures, a twenty-story steel frame, and a twelve-story steel frame, while considering their geometric nonlinearities. The study's findings indicate that the suggested technique is especially useful for modeling large-scale structures with complex nonlinear behaviors and noisy, incoherent, and incomplete structural responses to high loadings. The identification inaccuracy is seldom more than roughly 3%.

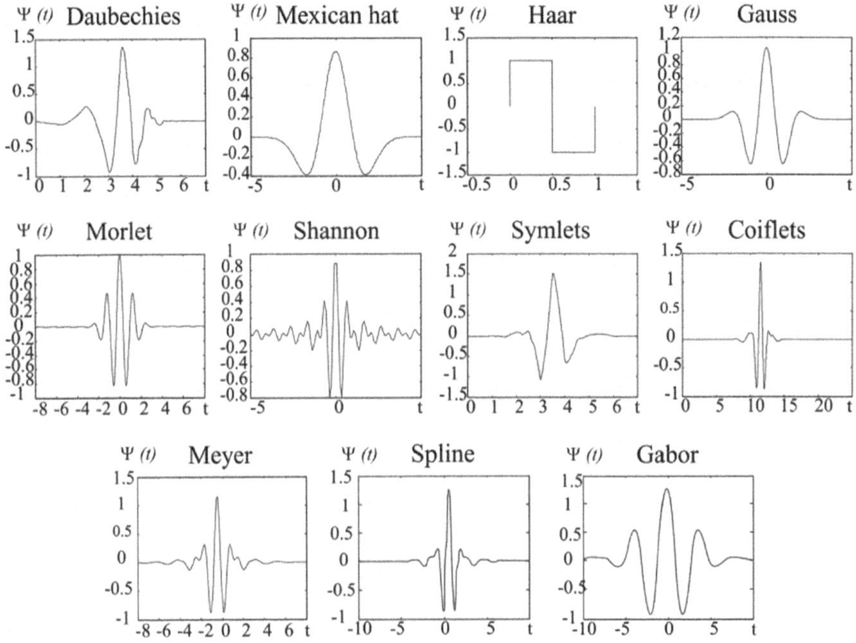

Fig. 9.1 Wavelets used in SHM [49]

The authors also draw the conclusion that this method will assist structural engineers in creating more effective earthquake-resistant buildings by improving their understanding of how structures behave during earthquakes [57].

9.2.1.1 Synchrosqueezed Wavelet Transform (SWT)

When utilizing wavelet transform to extract features from extremely noisy data, it may be necessary to employ numerous decomposition layers. To tackle this issue, Daubechies et al. [58] developed the SWT, which offers a more precise time–frequency representation in the presence of noise. The SWT technique operates by redistributing the coefficients of the continuous wavelet transform according to the frequency information to create a sharper representation in both the time and frequency domains.

9.2.1.2 Empirical Wavelet Transform

The invention of the adaptive wavelet transform is a recent breakthrough. Empirical wavelet transforms were first introduced by Gilles [59], and are capable of separating signals based on the information they contain. The proposed method comprises a set of wavelet filters that utilize Fourier supports derived from the processed signal spectrum. In order to construct the wavelet filter bank, the Fourier spectrum was utilized to ascertain the modes or frequencies that exist within the signal.

9.2.1.3 Multiresolution Analysis Based on Wavelets

The notion of multiresolution analysis [60, 61] pertaining to a signal is founded upon the signal's disintegration across an extensive spectrum of scales, a process that can be likened to cartography. The best approximation that can be drawn will be used to replace the signal at each scale. We may obtain ever-more-accurate representations of the provided signal by working our way up from the coarsest scales to the finest scales. Investigating what changes from scale to scale is made possible by computing the details at a certain resolution. This permits the production of a representation of greater quality by enhancing a still rather imprecise approximation.

The following goals should be attained by an appropriate representation of the data to be shown [60]:

- A global view at any resolution;
- An exact local view;
- Network-assisted progressive transmission;
- Compression.

The techniques deployed must possess an O(N) minimum average computational expense, both in terms of time and space, whereby N denotes the size of the signal under examination. The hierarchical data structure is based on multiresolution wavelet transform analysis, and all of the aforementioned conditions are fulfilled by it. The multiresolution analysis techniques employed in this structure enable the representation of a signal at numerous resolution levels by capturing the errors that exist between the different levels and the coarsest resolution level. Instead of retaining all resolution levels, the multiresolution analysis techniques' linear space cost is accounted for by the coding of the subsequent errors. The inaccuracies are contained in the detail coefficients. Due to its oscillating nature, each of these coefficients has a local basis function that is known as a wavelet. Applications that attract attention in the context of visualization are produced by choosing some detail coefficients and removing others after analysis but before synthesis.

The inaccuracy resulting from the suppression of detail coefficients is linked to the magnitude of these coefficients through their orthogonality characteristics. To minimize error, low magnitude coefficients should be suppressed. The region affected by the suppression of a detail coefficient can be identified using the locality attributes. To generate a global image at any resolution, researchers can select the required fraction of the most significant detail coefficients, i.e., those with the highest absolute value. Compression is based on the same principle. Progressive transmission is optimized when the detail coefficients are transmitted in decreasing order of magnitude. To create an accurate local image, all the detail coefficients whose sphere of influence intersects the region to be observed are selected. These properties have made multiresolution analysis an effective tool in visualization, especially in the visualization of volumetric medical or oceanographic data. In this study, multiresolution analysis is employed to visualize concrete cracks.

9.2.2 Wavelet Based Damage Detection Literature Review

Figure 9.2 provides an overview of the recommended method, which automatically detects inner cracks in a concrete specimen using NDT, multiresolution analysis, and classification stages [61]. The three essential steps of the procedure are depicted in the image. The initial phase entails conducting a NDT on the concrete specimen, thereby generating an ultrasonic signal that serves to identify the existence of an internal flaw. Subsequently, in the second step, the aforementioned signal is subject to processing through discrete wavelet transforms (DWT)-based multiresolution analysis.

The translated ultrasonic signal was used to generate a scalogram, a space-scale image that enables the localization of the damage in location and at each resolution. The third stage of the approach involves the employment of a convolutional neural network block, which receives the scalogram image and subsequently isolates the pertinent characteristics that enable the identification of the specific type of internal flaw, ultimately classifying it as either a crack or non-crack.

The wavelet Daubechies10, abbreviated Db10, was selected by the authors because it has the most vanishing moments—10 in this case—for a certain support width, or 20 for the number of coefficients of the related filters [60].

In other words, because there would be a high connection between the studied signal and this wavelet, the wavelet employed for multiresolution analysis (also

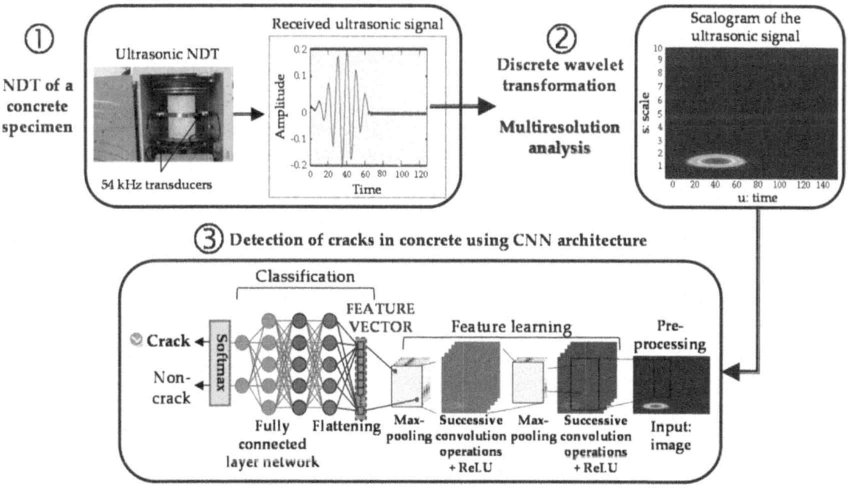

Fig. 9.2 Flowchart of cracks detection in concrete [61]

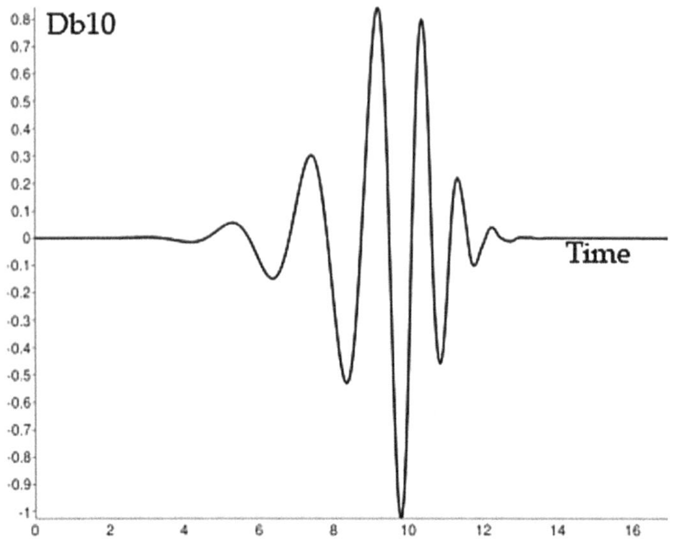

Fig. 9.3 Daubechies 10 or Db10 wavelet employed to analyze the received ultrasound signal [61]

known as the wavelet transform) correctly analyzes the ultrasonic signal. Additionally, the Db10 wavelet and the received ultrasound signal have a striking visual similarity (for context, see the ultrasound signal in Fig. 9.2).

The Daubechies wavelet, specifically the Db10 variant, is depicted in Fig. 9.3. As an orthogonal wavelet, it is extensively employed for the purpose of identifying anomalies such as signal discontinuities and structural cracks. This rationale supports its application in the present context.

Zeebaree et al. [62] presented a valuable conceptual model that is founded on the concealment of watermarks that are embedded within the coefficients of images that have been disintegrated by the contourlet transform. This model is characterized by an additional level of support for the potential and actual security of the hiding methods, which strengthens the distribution techniques of watermarking within the cover image. By combining these methods with additional image-processing techniques, it is possible to achieve this. In order to enable the algorithms of the suggested system (non-blind), a case study technique was employed that involved hiding the watermarking.

This approach led to the capability of reciprocally changing the watermarking dimensions with fixed cover measurements. Another significant practical implication is that the quartet attempted to divide the techniques into four sections before embedding them in the cover image, and after they had been divided, they distributed them among the watermark's cover image fragments. Using the three recommended methods and watermarks on cover images of different sizes, it was demonstrated that the correlation factor ratio before and after hiding the cover picture was more than

0.99%. The study's most striking result was the watermark measured before and after the process based on the PSNR, SNR, MSE, and NC.

Jahangir et al. [63] conducted an investigation into the modal data and identification of defects in prestressed concrete slab models. Their study involved the use of the contourlet transform. Interestingly, the proposed damage index was able to identify single, double, and triple damage scenarios in models with different geometric shapes, locations, and severity levels. Specifically, the suggested damage index was most sensitive to damages characterized by inclined slots (IS), curved slots (CS), transverse slots (TS), and longitudinal slots (LS). The maximum values of the recommended damage index confirmed its high sensitivity to damage.

Jahangir et al. [64] investigated the use of various geometric shapes (transverse, longitudinal, and inclined slots) and locations (middle and center) of prestressed concrete slab models in detecting damage scenarios by analyzing the vibration responses' time histories. The authors proposed a number of distinct damage indices using the acceleration time histories corresponding to the degrees of freedom for the damaged and undamaged models. These indices were obtained using the wavelet transform applied to the time histories of acceleration responses, which yielded the maximum detail coefficients (DI_MW), the area under the diagram of acceleration time histories (DI_SA), and the area under the diagram of detail coefficients (DI_SW). The proposed damage indices used rbio2.2, rbio2.4, and rbio2.8 from the Reverse Biorthogonal wavelet family and db2, db4, db6, and db8 mother wavelets from the Daubechies wavelet family to identify various damage situations. The study found that the suggested damage index DI_SA was not effective in locating damage situations in the numerical models of prestressed concrete slabs. Furthermore, the DI_MW damage index was unable to locate damage in prestressed concrete slabs regardless of the wavelet family used. In contrast, the results showed that the DI_SW damage index was more sensitive to inclined, longitudinal, and transverse slots in all numerical models of prestressed concrete slabs. The values of the suggested damage index DI_SW varied for different damage types.

Zhao et al. [65] reported that higher resolution remote sensing images have a greater amount of information, noise and complex feature geometry and texture information, which makes the task of building information extraction more challenging. To address this problem, a Markov model-based high-resolution remote sensing image building extraction method was developed in this study. This approach incorporated contourlet domain map clustering and a Markov model, in addition to collecting and enhancing contour and texture information of high-resolution remote sensing picture characteristics in various directions. Furthermore, a spectral feature index was developed to detect "pseudo-buildings" in the building area. By utilizing multiscale segmentation and feature extraction on the image, the building region can be accurately separated from the rest of the structure. Results have shown that this method outperforms conventional pixel-level image information extraction in terms of building extraction precision, accuracy, and completeness. It can also reduce noise in high-resolution remote sensing images, minimize interference of non-target ground texture information, and eliminate shadow, vegetation, and other fake building information.

Jahangir et al. [66] employed wavelet transform, a unique modal data processing technique, to identify faults in reinforced concrete beams. A threshold selection and hypothesis test on the detailed signal of modal curvatures were used to achieve damage detection using statistical indexes. The validity of the proposed method was examined and evaluated using computational and experimental data from reinforced-concrete beams. The concrete elastic modulus reduction was used in various areas of the numerical models to produce damage scenarios in reinforced concrete beams that were solitary, multiplex, and dispersed. Damage sites that were discovered after implementing the suggested technique were compared to what was taken those considered in situations of concentrated and spread damage. Therefore, it would be useful to discover and identify damaged reinforced concrete beams using responses obtained from the modified damage index on the superposition of various modal curvatures.

In their numerical models of RC beams, Mirkazemi et al. [67] created signal-processing techniques to identify distinct damage areas. To accomplish this task, the vibrational reactions of numerical simulations under modal testing were examined at varying levels of freedom in both damaged and undamaged conditions. The wavelet transform technique, a unique signal processing methodology, was employed for this analysis. By examining the region underneath the wavelet transform coefficients, with the mother wavelet being bior2.8, single, double, and triple damage scenarios were discovered. A new damage threshold based on statistical factors, such as the mean and standard deviation of the computed damage index, was developed to prevent any inaccuracies in damage localization. The results indicate that the proposed DI_SW damage index possesses the ability to effectively recognize all single, double, and triple damage sites.

Traditional methods for determining the breadth of cracks in concrete buildings depend mostly on the manual and random collection of data, as well as on subjective considerations and judgment, suggested by Shahrokhinasab et al. [68]. Since these procedures take time and frequently entail human error, more emphasis has recently been concentrated on creating fresh ways for crack monitoring and detection. One of the pivotal tactics involves the utilization of techniques centered around images. These particular systems rely on field photography and images produced by cameras to ascertain various characteristics such as the emergence of damage, its placement, degree of severity, extent of fissures, and depth of cracks. These devices' designers wanted to be able to use a series of timed images to track the crack evolution over time. Two common strategies, targeting and image processing, each have benefits and drawbacks. The findings showed that the image processing methodology's mistaken identification of surface noises as fractures is the most important weakness in this technique. The focused approach, however, presents issues with pinpointing fractures. Researchers have created more innovative techniques to get over these limitations, such as wavelet transform and digital picture correlation.

9.3 Crack Detection Employing Image Processing

This section presents a depiction of the fundamental framework for crack identification by means of image-processing methodology [69]. The image-based analysis is more precise, which is the primary advantage of utilizing image processing for crack detection in lieu of conventional manual techniques [70]. The magnitude of the image plays a significant role in the difficulty of fracture identification. Contemporary digital cameras boast an image resolution of over ten megapixels. This escalation in resolution facilitates the acquisition of comprehensive images of the concrete surfaces. With the aid of modern-day commercial cameras, a wide-ranging view of the concrete surface can be captured as a single image, which can be employed [71].

Figure 9.4 illustrates the all-encompassing structure for the detection of cracks via image processing. The stages involved in this image processing approach are as follows. Firstly, an image of the structure to be examined for fracture detection is obtained using a camera or another relevant means. Secondly, the acquired images are preprocessed to execute techniques such as segmentation and enhance the efficacy of the image processing system. Thirdly, specific image-processing techniques are applied to process the sample of the obtained images. Fourthly, the image processing output demonstrates the detection of fractures in the structure. Lastly, the detection of cracks involves the separation of the detected fractures based on their width, depth, and the direction of propagation of the crack.

Wang et al. [69] suggested a method to describe cracks based on their efficacy and to perform image-based crack identification. They divided the currently

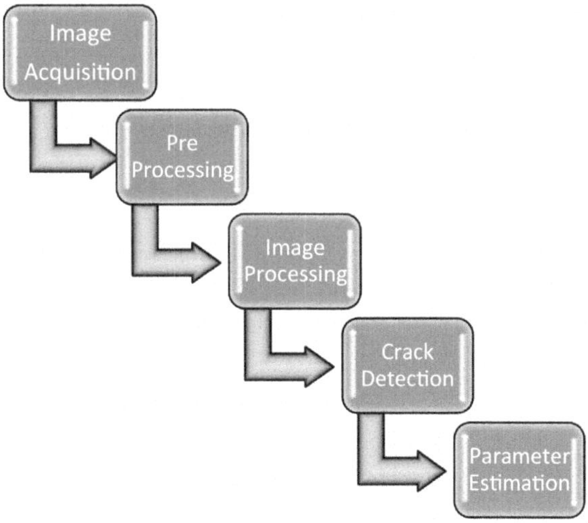

Fig. 9.4 Image processing based on crack detection

used image-based crack detection methods into four groups. These are morphological methods, percolation strategies, integrated algorithms, and practical techniques. An integrated algorithm was used to adjust the shading. Using the percolation approach, an ambiguous crack prediction was found. For microfracture identification, a morphological approach was used, using a practical method that offers high-performance feature extraction.

Jahanshahi and Masri [72] have developed a method as an alternative to the present monitoring approach. They suggested an approach that required less time. For the analysis of 2D pictures, they deployed an autonomous robotic system with a vision-based fracture detection algorithm. The autonomous system automatically changes the depth settings. Depth perception was then achieved using the 3D reconstruction method. Reconstruction of a 3D scene was used to determine the depth perception. Their approach is suitable because they separate the entire crack from its backdrop.

A detection method that combines digital picture correlation and acoustic emission was proposed by Alam et al. [73]. As a result of the earlier technique, extremely accurate assessments of the surface displacements, fracture openings, and crack spacing were identified. The acoustic emissions caused by interior damage were also examined to supplement this approach and examine the damage processes. To categorize the various classes of acoustic emission energy generated by beams of three different sizes, a manual grouping approach (similar to the K-means method) was employed. To ensure the efficacy of the results, they employed three alternative beam proportionalities in their research.

Zou et al. [74] established a fully automated technique for identifying pavement cracks. By maintaining the fracture, pavement shadows can be eliminated using a geodesic shadow removal technique. After removing shadows, a fracture probability map was constructed using tensor voting techniques. A graph model was then used to create the fracture probability map. Following the representation of the model, the generation of Minimum Spanning Trees was initiated, which facilitated the application of recursive tree-edge pruning to eliminate the extracted crack data.

An automatic crack detection method was established by Oliveira and Correia [75]. The basis for cracking detection is the sample paradigm. An image database subset was automatically selected to train system photographs in the unsupervised sample paradigm. Operations were established on the classification of non-overlapping image components. Subsequently, crack block-based detection was employed to estimate the crack width. They published their strategy in accordance with the suggestions of the Portuguese Distress Catalogue [76].

9.3.1 Contourlet Transform

In recent years, the multi-scale directed contourlet transform, which can identify defects in 2D structures with complicated geometric forms, has been developed as an alternative to wavelet transform [77]. For instance, Shahrokhinasab et al. [78] examined the effectiveness of concrete structure fracture-detection systems. The contourlet transform was used by Ma et al. [79] to determine the road pavement distress. Other research efforts have also examined the contourlet transform's capacity to identify road cracks [80–82]. The contourlet transform was utilized by Ai and Xu [83] to determine the damage to metal plates. The effectiveness of the wavelet transforms and contourlet transform with regard to the detection of damage in plate-like structures was contrasted in the study by Hajizadeh et al. [84]. The results show that in addition to damage localization, the contourlet transform can also identify the geometric shape of the damage.

Li et al. [85] conducted a study wherein they photographed the nickel alloy surfaces before and after damage to identify surface cracks. The images obtained were processed using a contourlet transform, which allowed for precise identification of surface problems with an accuracy rate of 88.9%. On the other hand, Vafaei and Salajegheh [86] employed wavelet and contourlet transformations to locate curved-shaped damage in plate structures with fixed supports. The findings of the study revealed that the contourlet transform yielded more accurate results in detecting curved-shaped damage than the wavelet transform.

The twin filter banks are used sequentially in the contourlet transform, which has a directed multi-scale structure. The Laplacian Pyramid (LP) in this structure initially identifies point discontinuities across the input signal [87]. Point discontinuities and linear structures were then connected using a directional filter bank (DFB) [88]. Figure 9.5 depicts the contourlet transform, which involves dividing the bandpass signal generated from the LP into multiple directions using the DFB. This procedure can be extended with a low-pass signal [89]. The resulting signal representation, called a contourlet, is composed of the initial signals as contour sections in various directions and scales.

The discrete contourlet transform separates multiscale and directional decomposition, resulting in a variable number of directions with different scales and a flexible multiscale and directional expansion. Furthermore, any tree structure can be used with the contourlet transform to decompose the DFB entirely into binary

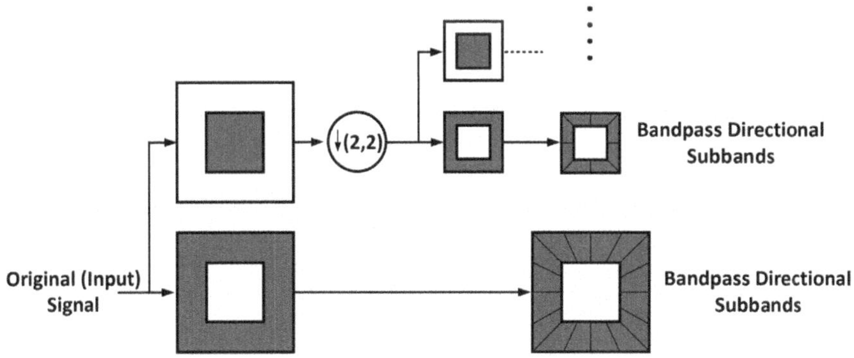

Fig. 9.5 Multiscale decomposition employing the Laplacian Pyramid (LP) and applying the Directional Filter Bank on each bandpass signal in the contourlet transform [81]

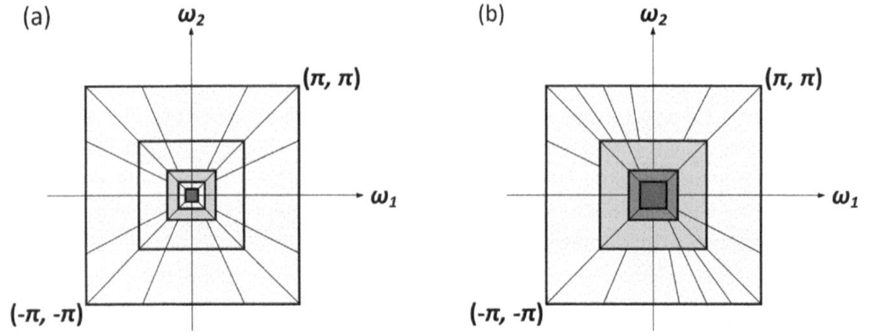

Fig. 9.6 Possible frequency decompositions example using: **a** contourlet transform; **b** contourlet packets [80]

trees. Consequently, wavelet-like packets known as contourlet packets, which are a collection of directional multiresolution expansions, are created. Figure 9.6 depicts the proposed contourlet transform and contourlet packet frequency decomposition [80].

In comparison to the contourlet transform, Fig. 9.3 demonstrates that the contourlet packets offer finer angular resolution breakdown at any direction or scale. Therefore, this transform might better smooth contours of diverse curvatures due to the flexibility of contourlet packets [80].

The contourlet transform's fundamental concept is to extract directional information with multiscale resolution. The contourlet transform may produce degeneration in any direction and at any scale, making it a true sense of two-dimensional performance of pictures. The problem of the wavelet transform's insufficient idiom in directed data is resolved by the contourlet transform, which has several excellent qualities including multiresolution, directionality, localization, and anisotropy [90, 91].

The DWT, also referred to as the contourlet transform, is acknowledged for its enhanced or updated characteristics. The DWT technique subdivides an image into four subbands: low-low (approximation coefficient), low–high (vertical details), high-low (horizontal details), and high-high (diagonal details). Furthermore, the contourlet transform has the potential to generate various subbands depending on the specific requirements [92, 93].

There are several types of filter banks in the context of contourlet transform theory, such as directional and quincunx sampling filters. Quincunx sampling filters were obtained for the segmentation of the two-dimensional spectrum. Various subbands are created using various frequency response categories. If a diamond filter is used, low-pass and high-pass sub-bands should be captured at a 45° angle. However, horizontal and vertical sub-bands must be obtained in a 45°orientation when using a fan filter [94, 95].

The contourlet transform reduces the wavelet [96]. A multiscale and multidirectional discrete image architecture is called contourlet transform. This approach uses many scales and separates them serially so that point abruption is considered by the LP, which is then combined with linear structures using a directional filter bank.

9.4 Conclusion

In light of recent significant advancements in innovative sensors, signal processing methodologies, and robust computing systems, SHM has become an area of substantial activity in the domain of structural engineering studies. This study presents an overview of the primary signal processing and image processing techniques employed in SHM. Conducted by the authors of this research, the investigation scrutinized various state-of-the-art techniques for detecting and tracking cracks in concrete structures. These techniques primarily target the initial stages of cracks or fractures that are not perceptible from the exterior.

This article also highlights fresh signal processing methods that have been proposed in recent years as viable options for SHM studies. The wavelet transforms and counterlet transform, as well as occasionally a combination of any two of them, are the most commonly used signal processing methods in SHM. Large-scale structural vibration measurements were noisy. Novel methods for online SHM should be accurate, scalable, portable, and computationally efficient while handling noisy data. The analysis was completed by the authors based on the image-processing methods employed in each system. In addition, we discuss research questions that may be pertinent to future work on image-processing-based crack-detection systems. According to their findings, more researchers have employed camera-type images for analysis using superior segmentation algorithms, such as threshold and reconstructable feature extraction techniques, for in-depth damage analysis. As this work provides a thorough analysis of noninvasive methods of crack detection, the authors want to conduct a survey on the many approaches available for invasive method-based fracture identification in the future.

References

1. Hassoum MNH, El-Manaseer A (2015) Structural Concrete: Theory and Design; John Wiley & Sons: Hoboken. NJ, USA
2. Mehta PK, Monteiro PJM (2006) Concrete: microstructure, properties and materials. McGraw-Hill: New York, NY, USA
3. Mehta PK (1997) Durability Critical Issues for the future. Concr Int 9:27–33
4. Peng T, Kavya TS, Jang Y-M, Kim B-W (2020) Concrete Crack Detection using Relative Standard Deviation for Image Thresholding. Int. J. Eng. Res. Technol. 13:2720
5. Adeli H, Jiang X (2009) Intelligent infrastructure: neural networks, wavelets, and chaos theory for intelligent transportation systems and smart structures. CRC Press, Boca Raton
6. Khan MA-M, Kee S-H, Pathan A-SK, Nahid A-A (2023) Image processing techniques for concrete crack detection: a scientometrics literature review. Remote Sens. 15, 2400. https://doi.org/10.3390/rs15092400

7. Ansari RA, Budhhiraju KM (2016) A comparative evaluation of denoising of remotely sensed images using wavelet, curvelet and contourlet transforms. J Indian Soc Remote Sens 44:843–853. https://doi.org/10.1007/s12524-016-0552-y
8. Seyedpoor SM, Ahmadi A, Pahnabi N (2019) Structural damage detection using time domain responses and an optimization method. Inverse Probl Sci Eng 27:669–688. https://doi.org/10.1080/17415977.2018.1505884
9. Rezazadeh Eidgahee D, Haddad A, Naderpour H (2019) Evaluation of shear strength parameters of granulated waste rubber using artificial neural networks and group method of data handling. Sci Iran 26:3233–3244. https://doi.org/10.24200/sci.2018.5663.1408
10. Naderpour H, Rezazadeh Eidgahee D, Fakharian P, Rafiean AH, Kalantari SM (2020) A new proposed approach for moment capacity estimation of ferrocement members using Group Method of Data Handling. Eng Sci Technol an Int J 23:382–391. https://doi.org/10.1016/j.jestch.2019.05.013
11. Rezazadeh Eidgahee D, Rafiean AH, Haddad A (2019) A novel formulation for the compressive strength of IBP-Based geopolymer stabilized clayey soils using ANN and GMDH-NN approaches. Iran J Sci Technol Trans Civ Eng. https://doi.org/10.1007/s40996-019-00263-1
12. Jahangir H, Rezazadeh ED (2021) A new and robust hybrid artificial bee colony algorithm—ANN model for FRP-concrete bond strength evaluation. Compos Struct 257:113160. https://doi.org/10.1016/j.compstruct.2020.113160
13. Heidari A, Raeisi J (2018) Optimum design of structures against earthquake by simulated annealing using wavelet transform. J Soft Comput Civ Eng 2:23–33. https://doi.org/10.22115/SCCE.2018.125682.1055
14. Naderpour H, Fakharian P (2016) A synthesis of peak picking method and wavelet packet transform for structural modal identification. KSCE J Civ Eng 20:2859–2867. https://doi.org/10.1007/s12205-016-0523-4
15. Jahangir H, Esfahani MR (2012) Structural damage identification based on modal data and wavelet analysis. 3rd Natl Conf Earthq Struct, Kerman, Iran
16. Shah M, Mevada S, of VP-IJ, 2016 undefined. Comparative study of diagrid structures with conventional frame structures. IngentaconnectCom n.d
17. Jiang T, Kong Q, Patil D, Luo Z, Huo L, Song G (2017) Detection of debonding between fiber reinforced polymer bar and concrete structure using piezoceramic transducers and wavelet packet analysis. IEEE Sens J 17:1992–1998. https://doi.org/10.1109/JSEN.2017.2660301
18. Qu H, Li T, Chen G (2018) Adaptive wavelet transform: Definition, parameter optimization algorithms, and application for concrete delamination detection from impact echo responses. Struct Heal Monit 147592171877620. https://doi.org/10.1177/1475921718776200.
19. Naito H, Bolander JE (2019) Damage detection method for RC members using local vibration testing. Eng Struct 178:361–374. https://doi.org/10.1016/j.engstruct.2018.10.031
20. Mokhtari Masinaei M, Jahangir H, Khatibinia M (2019) Damage detection in prestressed concrete slabs using vibrational responses in time domain. 5th Natl Conf Recent Adv Civ Eng Archit Urban Dev, Shahid Beheshti University, Tehran, Iran
21. Jahangir H, Esfahani MR (2012) Assessment of structural damage severity using the energy due to dynamic response (In Persian). 2nd Int Conf Acoust Vib, Tehran, Iran
22. Pahlevan Mosavari, A. Jahangir H, Esfahani MR (2016) The effect of sensor weight on obtained data from modal tests (In Persian). 9th Natl Congr Civ Eng, Mashhad, Iran: Ferdowsi University of Mashhad
23. Moughty JJ, Casas JR (2017) A state of the art review of modal-based damage detection in bridges: development, challenges, and solutions. Appl Sci 7:510. https://doi.org/10.3390/app7050510
24. Yang YB, Yang JP (2018) State-of-the-Art review on modal identification and damage detection of bridges by moving test vehicles. Int J Struct Stab Dyn 18:1850025. https://doi.org/10.1142/S0219455418500256
25. Wang S, Xu M (2019) Modal strain energy-based structural damage identification: a review and comparative study. Struct Eng Int 29:234–248. https://doi.org/10.1080/10168664.2018.1507607

26. Jahangir H, Daneshvar Khoram MH, Esfahani MR (2014) Application of vibration modal data in gradually detecting structural damage (In Persian). 4th Int Conf Acoust Vib, Tehran, Iran
27. Seyedi SR, Keyhani A, Jahangir H (2015) An energy-based damage detection algorithm based on modal data. 7th Int Conf Seismol Earthq Eng, Int Inst Earthq Eng Seism (IIEES). pp 335–6
28. Daneshvar MH, Gharighoran A, Zareei SA, Karamodin A (2020) Damage detection of bridge by Rayleigh-Ritz method. J Rehabil Civ Eng 8:111–120. https://doi.org/10.22075/JRCE.2019.17603.1337
29. Rytter A, Kirkegaard PH (1994) Vibration based inspection of civil engineering structures. Aalborg Universitetsforlag, Aalborg
30. Park G, Muntges DE, Inman DJ (2001) Self-monitoring and self-healing jointed structures. Key Eng Mater 204:75–84
31. Khemili I, Chouchane M (2005) Detection of rolling element bearing defects by adaptive filtering. Eur J Mech-A/Solids 24(2):293–303
32. Maeck J, Abdel Wahab M, Peeters B, De Roeck G, De Visscher J, De Wilde WP et al (2000) Damage identification in reinforced concrete structures by dynamic stiffness determination. Eng Struct 22(10):1339–1349
33. Friswell MI, Penny JET (1997) Is damage location using vibration measurements, practical. In: Proceedings of EUROMECH 365 international workshop: DAMAS, 97
34. Carden EP, Fanning P (2004) Vibration based condition monitoring: a review. Struct Health Monit 3(4):355–437
35. Li CJ (2006) Signal processing in manufacturing monitoring. In: Wang L, Gao RX (eds) Condition monitoring and control for intelligent manufacturing. Springer, London, pp 245–265
36. Beheshti Aval SB, Ahmadian V, Maldar M, Darvishan E (2020) Damage detection of structures using signal processing and artificial neural networks. Adv Struct Eng 23(5):884–897
37. Tsao S, Kehtarnavaz N, Chan P, Lytton R (1994) Image-Based expert-system approach to distress detection on CRC pavement. J Transp Eng 120:52–64. https://doi.org/10.1061/(ASCE)0733-947X(1994)120:1(52)
38. Wang KCP, Nallamothu S, Elliott RP (1998) Classification of pavement surface distress with an embedded neural net chip. ASCE, Artif. neural networks Civ. Eng. Adv. Featur. Appl., pp 131–161
39. Kaseko MS, Lo Z, Ritchie SG (1994) Comparison of traditional and neural classifiers for pavement—crack detection. J Transp Eng 120:552–569. https://doi.org/10.1061/(ASCE)0733-947X(1994)120:4(552)
40. Sun Z, Chang CC (2002) Structural damage assessment based on wavelet packet transform. J Struct Eng 128:1354–1361. https://doi.org/10.1061/(ASCE)0733-9445(2002)128:10(1354)
41. Chae MJ, Abraham DM (2001) Neuro-Fuzzy approaches for sanitary sewer pipeline condition assessment. J Comput Civ Eng 15:4–14. https://doi.org/10.1061/(ASCE)0887-3801(2001)15:1(4)
42. Japan AI of. Shrinkage Cracking in Reinforced Concrete Structures—Mechanisms and Practice of Crack Control 2003
43. Dare P, Hanley H, Fraser C, Riedel B, Niemeier W (2002) An operational application of automatic feature extraction: the measurement of cracks in concrete structures. Photogramm Rec 17:453–464. https://doi.org/10.1111/0031-868X.00198
44. Chen L-C, Shao Y-C, Jan H-H, Huang C-W, Tien Y-M (2006) Measuring system for cracks in concrete using multitemporal images. J Surv Eng 132:77–82. https://doi.org/10.1061/(ASCE)0733-9453(2006)132:2(77)
45. Abdel-Qader I, Abudayyeh O, Kelly ME (2003) Analysis of edge-detection techniques for crack identification in bridges. J Comput Civ Eng 17:255–263. https://doi.org/10.1061/(ASCE)0887-3801(2003)17:4(255)
46. Hutchinson TC, Chen Z (2006) Improved image analysis for evaluating concrete damage. J Comput Civ Eng 20:210–216. https://doi.org/10.1061/(ASCE)0887-3801(2006)20:3(210)
47. Ito A, Aoki Y, Hashimoto S, Accurate extraction and measurement of fine cracks from concrete block surface image. IEEE, (2002) 28th Annu Conf Ind Electron Soc IECON 02, 3. IEEE 2002:2202–2207

48. Kawamura K, Miyamoto A, Nakamura H, Sato R (2003) Proposal of a crack pattern extraction method from digital images using an interactive genetic algorithm. Doboku Gakkai Ronbunshu 2003:115–131
49. Amezquita-Sanchez JP, Adeli H (2016) Signal processing techniques for vibration-based health monitoring of smart structures. Arch Comput Methods Eng 23:1–15
50. Quek ST, Wang Q, Zhang L, Ang KK (2001) Sensitivity analysis of notch detection in beams by wavelet technique. Int J Mech Sci 43:2899–2910
51. Han JG, Ren WX, Sun ZS (2005) Wavelet packet based damage identification of beam structures. Int J Solids Struct 42(26):6610–6627
52. Adeli H, Jiang X (2006) Dynamic fuzzy wavelet neural network model for structural system identification. J Struct Eng ASCE 132(1):102–111
53. Liu F, Er MJ (2012) A novel efficient learning algorithm for self-generating fuzzy neural network with applications. Int J Neural Syst 22(1):21–35
54. Li H, Yi W, Yuan X (2013) Fuzzy-valued intensity measures for near-fault ground motions. Comput Aid Civ Infrastruct Eng 28(10):780–795
55. Yan L, Ma ZM (2012) Comparison of entity with fuzzy data types in fuzzy object-oriented databases. Integr Comput Aid Eng 19(2):199–212
56. Yan L, Ma ZM (2012) Incorporating fuzzy information into the formal mapping from web data model to extended entity– relationship model. Integr Comput Aid Eng 19(4):313–330
57. Jiang X, Adeli H (2005) Dynamic wavelet neural network for nonlinear identification of highrise buildings. Comput Aid Civ Infrastruct Eng 20(5):316–330
58. Daubechies I, Lu J, Wu HT (2011) Synchrosqueezed wavelet transforms: an empirical mode decomposition-like tool. Appl Comput Harmonic Anal 30:243–261
59. Gilles J (2013) Empirical wavelet transform. IEEE Trans Signal Process 61(16):3999–4010
60. Ouahabi, A (2013) Signal and image multiresolution analysis. ISTE-Wiley: London, UK; Hoboken, NJ, USA.
61. Arbaoui A, Ouahabi A, Jacques S, Hamiane M (2021) Concrete cracks detection and monitoring using deep learning-based multiresolution analysis. Electron 10(15):1772
62. Zeebaree DQ, Abdulazeez AM, Hassan OMS, Zebari DA, Saeed JN (2020) Hiding image by using contourlet transform. 83, 16979–16990
63. Jahangir H, Khatibinia M, Kavousi M (2021) Application of contourlet transform in damage localization and severity assessment of prestressed concrete slabs. J Soft Comput Civ Eng 5(2):39–67
64. Jahangir H, Khatibinia M, Mokhtari Masinaei M (2022) Damage detection in prestressed concrete slabs using wavelet analysis of vibration responses in the time domain. J Rehabil Civ Eng 10(3):37–63
65. Zhao W, Yan L, Chang Y, Gong L (2018) High-Resolution remote sensing image building extraction based on markov model. Int Arch Photogramm Remote Sens Spat Inf Sci 42:2411–2418
66. Jahangir H, Hasani H, Aval FS (2021) Identification of concentrated and distributed damages in reinforced concrete beams. no. December
67. Mirkazemi SB, Jahangir H, Zahmatkesh A (2021) Multiple damage identification in reinforced concrete beams via signal processing techniques. In 2nd conference on urban planning, architecture, civil engineering, environment, Oslo, Norway
68. Shahrokhinasab E, Hosseinzadeh N, Monirabbasi A, Torkaman S (2020) Shape optimization of gravity dams using a nature-inspired approach. J Soft Comput Civ Eng 4(1):127–139
69. Pingrang Wang, Hongwei Huang (2010) Comparison analysis on present image-based crack detection methods in concrete structures, in: Proceedings of 2010 3rd international congress on image and signal processing (CISP2010), 5, pp 2530–2533
70. Patrik (2013) Broberg, Surface crack detection in welds using thermography, NDT E Int 57 69–73
71. Rodrı´guez-Martı´na M, Lagu¨elaa S, Gonza´lez-Aguileraa D, Martı´nezb J (2016) Thermographic test for the geometric characterization of cracks in welding using IR image rectification, Autom. Constr. 61 58–65

72. Jahanshahi MR, Masri SF (2012) Adaptive vision-based crack detection using 3D scene reconstruction for condition assessment of structures. Autom Constr 22:567–576
73. Alam SY, Loukili A, Grondin F, Rozie're E (2015) Use of the digital image correlation and acoustic emission technique to study the effect of structural size on cracking of reinforced concrete, Eng Fract Mech 143 17–31
74. Qin Zou, Yu Cao, Qingquan Li, Qingzhou Mao, Song Wang (2012) CrackTree: automatic crack detection from pavement images, Pattern Recogn. Lett. 33(3), 227–238
75. Oliveira H (2012) Paulo Lobato Correia, Automatic road crack detection and characterization. IEEE Trans Intell Transp Syst 14(1):155–168
76. Cata´logo de Degradac¸o˜es dos PavimentosRodovia´riosFlexı´veis—2nd ver, JAE, Ex Junta Auto´noma das Estradas, Almada, Portugal, 1997
77. Do MN, Vetterli M (2002) Contourlets: a directional multiresolution image representation. Proceedings int conf image process, 1, IEEE, pp I-357-I–360. https://doi.org/10.1109/ICIP.2002.1038034
78. Shahrokhinasab E, Hosseinzadeh N, Monirabbasi A, Torkaman S (2020) Performance of image-based crack detection systems in concrete structures. J Soft Comput Civ Eng 4:127–139. https://doi.org/10.22115/SCCE.2020.218984.1174
79. Ma C-X, Zhao C-X, Pavement HY, Based DD, on Nonsubsampled Contourlet Transform. (2008) Int conf comput sci softw eng. IEEE 2008:28–31. https://doi.org/10.1109/CSSE.2008.1027
80. Zhibiao S, Yanqing G (2013) Algorithm on contourlet domain in detection of road cracks for pavement images. J Algorithm Comput Technol 7:15–25. https://doi.org/10.1260/1748-3018.7.1.15
81. Fan Y (2015) A pavement cracks detection algorithm based on NSCT Domain. J Inf Comput Sci 12:4791–4798. https://doi.org/10.12733/jics20106356
82. Qian Q, Dang Q, Zhao C, He J, Recognition of pavement damage types based on features fusion. (2018) Tenth int conf adv comput intell. IEEE 2018:239–242. https://doi.org/10.1109/ICACI.2018.8377613
83. Ai Y, Xu K (2012) Feature extraction based on contourlet transform and its application to surface inspection of metals. Opt Eng 51:113605. https://doi.org/10.1117/1.OE.51.11.113605
84. Hajizadeh AR, Salajegheh J, Salajegheh E (2016) Performance evaluation of wavelet and curvelet transforms based-damage detection of defect types in plate structures. Struct Eng Mech 60:667–691. https://doi.org/10.12989/sem.2016.60.4.667
85. Li J, Cao B, Nie F, Zhu M (2019) Feature extraction of foam nickel surface based on multi-scale texture analysis. J Adv Comput Intell Intell Informatics 23:175–182. https://doi.org/10.20965/jaciii.2019.p0175
86. Vafaie S, Salajegheh E (2019) Comparisons of wavelets and contourlets for vibration-based damage identification in the plate structures. Adv Struct Eng 22:1672–1684. https://doi.org/10.1177/1369433218824903
87. Burt P, Adelson E (1983) The laplacian pyramid as a compact image code. IEEE Trans Commun 31:532–540. https://doi.org/10.1109/TCOM.1983.1095851
88. Bamberger RH, Smith MJT (1992) A filter bank for the directional decomposition of images: theory and design. IEEE Trans Signal Process 40:882–893. https://doi.org/10.1109/78.127960
89. Po DD-Y, Do MN (2006) Directional multiscale modeling of images using the contourlet transform. IEEE Trans Image Process.15:1610–20. https://doi.org/10.1109/TIP.2006.873450
90. Zhang Y, Sun Y (2019) An image watermarking method based on visual saliency and contourlet transform. Optik 186:379–389
91. Zeebaree DQ, Haron H, Abdulazeez AM, Zeebaree SR (2017) Combination of K-means clustering with Genetic Algorithm: A review. Int J Appl Eng Res 12(24):14238–14245
92. Kalita M, Majumder S (2016) "A new steganographic method using Contourlet Transform", 2016 Int. Conf. Signal Process. Commun. ICSC 2016:274–278
93. "Advanced Image Steganography Technique using DWT and SVD with High PSNR," 7(5), pp 12234–12236, (2017)

94. Zhu J, Zhang H, Li X (2017) Information hiding based on gray relation and contourlet transform. pp 179–184
95. Zebari DA, Haron H, Zeebaree SR, Zeebaree DQ (2019) Enhance the mammogram images for both segmentation and feature extraction using wavelet transform. In 2019 International conference on advanced science and engineering (ICOASE), pp 100–105: IEEE
96. Hassan OMS, Abdulazeez AM, Tiryaki VM (2018) Gait-based human gender classification using lifting 5/3 wavelet and principal component analysis. In: 2018 International conference on advanced science and engineering (ICOASE), pp 173–178, IEEE

Part III
Signal Processing and Analysis

Chapter 10
Signal Processing in Real-Life Structural Health Monitoring: MATLAB and Python Implementation

Chandrabhan Patel, Naman Garg, S. K. Panigrahi, and Vikesh Kumar

Abstract Signal and data processing are necessary for analyzing experimental data. Signal processing techniques are well-defined and handy for real-life applications, but applying those techniques to experimental data is challenging for the researcher. In structural health monitoring, post-processing helps remove noise and residuals from the sensor data. Various software and programming can help achieve the best signal-processing algorithms. Each package has its solver to solve the signal processing algorithms. Due to the countless options available for signal processing, choosing the best and most reliable package to solve the problem is foremost. Nowadays, MATLAB and Python are two packages that are reliable, popular, and accurate for multiple applications. This article evaluates the signal processing capabilities of MATLAB and Python in the context of structural health monitoring applications. File handling, simulating the signals, filter realization, filter response, and adding and removing noise from the signal are discussed and implemented for structural health monitoring. Signal processing is implemented, keeping structural health monitoring as the focus area. The real-life demonstration used a supported beam instrumented with Fiber Bragg Grating strain sensors. It has been determined that both MATLAB and Python can be utilized for structural health monitoring based on cross-correlation techniques, as the absolute difference between their results is approximately 0.0001, which is negligible.

Keywords Signal processing · MATLAB · Python · Structural health monitoring · Damage identification

---

C. Patel (✉) · N. Garg · S. K. Panigrahi · V. Kumar
CSIR-Central Building Research Institute, Roorkee, India
e-mail: cpatel@cbri.res.in

N. Garg
e-mail: gargnaman712@gmail.com

S. K. Panigrahi
e-mail: skpanigrahi@cbri.res.in

V. Kumar
e-mail: vikeshparmar12@gmail.com

H. Jahangir et al. (eds.), *Damage Detection and Structural Health Monitoring of Concrete and Masonry Structures*, Springer Tracts in Civil Engineering,
https://doi.org/10.1007/978-981-97-8975-7_10

Abbreviation List

UI	User Interface
SHM	Structural Health Monitoring
LPF	Low Pass Filter
FBG	Fiber Bragg Grating
FTT	Fast Fourier Transform
DAQ	Data Acquisition

10.1 Introduction

Continuous measurement of structural behaviour enables damage detection induced in the structures. This measurement requires various sensors like strain gauges, accelerometers, etc. The sensor measurements are processed to extract the structure response [1]. Different sensing technologies are available to measure the structure response, like optical fiber, acoustic emission, and vibration sensors [2]. In structural health monitoring (SHM), signal processing is crucial in getting information from the measurement. Signal processing focuses on analyzing, modifying, and synthesizing signals. Representation, transformation, and manipulation of signals are also important. For example, if we wish to separate two combined signals, signal processing can help. The development and evolution of digital computers and microchips have taken signal processing to another level. For example, the Fast Fourier Transform (FFT) technique has evolved with digital computers, giving rise to the digital signal processing field [3]. Most real-life signals captured from sensors are analog and function of continuous quantities like time. Digital signal processing is an alternative method of processing analog signals. To perform digital processing, analog-to-digital (A/D) converters are required [4]. Steps involved in Analog and digital signal processing are discussed by Syed for measurement [5]. Signal processing is not limited to one or two fields but is necessary for almost all the engineering fields like communication, information processing, control systems, seismology, radar, and biomedical, etc.; the contribution of various scientists from 1948 to 1998 is discussed by Nebeker et al. [6].

The signal is a quantity that conveys information and can be transferred from one place to another. The signal may include errors due to measurement or environment. For example, an electrical strain gauge may record unwanted signals due to electromagnetic interference or temperature variation. This unwanted signal can be removed by computational approximation and requires mathematical knowledge of differential equations, time–frequency analysis, calculus, etc. [7]. The need for mathematical solutions leads us toward computer involvement, as the computer can help solve complicated equations. Paul A. et al. explained the scope of computers for the signal process [8]. As the new advancement in computer technology for fast processing evolves, we are at the stage where researchers are focusing on nano-computers and super-computers [9]. In 1978, Oppenheim listed various digital signal-processing applications [10]. Many signal processing techniques have been developed, and we will see more in the coming days as all industries move towards automation. The first step of automation is to understand and process the signal without any unwanted signal.

MATLAB, Octave, and Scipy are three software tools for signal processing listed by Wikibooks. MATLAB is a software package for simulating and manipulating the signal, but it is not limited. Octave is a free alternative for MATLAB with limitations. Scipy is a Python library that allows numeric calculation performance [11]. MATLAB is a programming platform for engineers and scientists to analyze and design systems and products. MATLAB can analyze the data, develop the algorithm, create models, and more [12]. Python is an open-source and popular programming language used for web development, mathematics, machine learning, etc. [13]. Various modules have been developed to perform different applications. Django is used for web development, and Numpy and Scipy are helpful for mathematical and numerical calculations. The evolution of Python modules attracted researchers to use Python for signal processing and data processing.

Measurement is critical for automation and monitoring systems involving sensors in real life. Various types of sensors are available for different applications. For example, structural health monitoring of buildings may use strain gauges, accelerometers, displacement, and tilt sensors [14].

Most of the sensor's output is electrical and is recorded by DAQ devices. Neither DAQ devices nor sensors are protected from unwanted signals. These unwanted signals are known as noise and should be removed before processing. Another real-life example is a communication system where signal processing is essential. Communication system devices first convert an audio signal into a sinusoidal signal to process and transfer from one device to another. On the other end, an original signal must be retrieved from the transmitted signal and signal processing is the only way to achieve it [15]. Amplification of signal, decoding of signal, and processing of an image are all examples of the real-life signal process.

This article presents the implementation and analysis of MATLAB and Python signal-processing techniques for SHM applications. The article discusses file management as an essential aspect of signal analysis, including importing different experimental files. Simulating waveforms for initial analysis and adding noise to signals are demonstrated. Frequency domain analysis is explored, specifically, power spectrum determination using Fast Fourier Transform (FFT). Spectral signal analysis is performed to examine signal characteristics. Furthermore, a digital Low-Pass Filter (LPF) Infinite Impulse Response (IIR) is implemented to remove high-frequency components from measurements. Cross-correlation analysis is employed to identify similarities in structural behaviours across different nodes. Statistical analysis, focused on feature-finding, is performed to extract information from recorded signals.

A comparison between MATLAB and Python regarding statistical analysis is made. Real-life applications of SHM are illustrated through experimental analysis of a supported beam. Frequency domain analysis is utilized to identify frequency modes. Cross-correlation coefficients are calculated using MATLAB and Python packages to assess the similarities among beam nodes. The article includes the developed MATLAB and Python codes for the tasks above, listed in the appendix.

10.2 Experimental Files Handling

In SHM, sensor data is captured using Data Acquisition (DAQ) devices, and each device stores data in a unique file format, depending on the manufacturer. Importing these files directly into software packages eliminates the need to convert the recorded files into formats such as Excel. Additionally, there may be instances where audio files, image files, or CDF (Common Data Format) files must be imported, requiring software that can seamlessly read various data types. MATLAB offers extensive support for importing and exporting a wide range of file formats, simplifying the data processing workflow in SHM. Table 10.1 shows some of the file formats MATLAB supports [16].

MATLAB provides the capability to convert file formats from one type to another. For instance, it can convert a text file into a spreadsheet format. The process involves importing the text file into MATLAB and utilizing a specific code to facilitate the conversion. The code '*txtToXls.m*' is listed in the appendix, demonstrating the conversion and export of a 'txt' file into a spreadsheet format. This code can be executed to perform the desired file format conversion.

Two types of files are directly supported in Python text and binary files [17]. But with the Pandas module, more files can be imported and exported. The file format supported by the pandas module is listed in Table 10.2 [18].

In Python, files can be opened using an in-built function. But for the spreadsheet, the pandas' module of Python is the best choice. The Pandas offer functions/methods

Table 10.1 File formats supported in MATLAB

S. No.	Content	Extension
1	MATLAB formatted data	MAT
2	Text	TXT, CSV
3	Spreadsheet	XLS, XLSX, XLSM, ODS, etc.
4	Extensible Markup Language	XML
5	Scientific data	CDF, FITS, HDF, H5, NC
6	Image data	PNG, JPEG, TIFF, BMP, etc.
7	Audio	MP4, WAV, AU, SND, AIFF, AIFC, FLAC, OGG etc.

Table 10.2 File format supported in Python

S. No.	Content	Extension
1	Text	TXT, CSV
2	MATLAB formatted data	MAT
3	Spreadsheet	XLS, XLSX
4	Extensible Markup Language	XML, HTML, JSON
5	Other	PDF, DOCX, SQL
6	Audio	MP3, MP4

to read and write data as data frames and are reasonable for further processing. A snip of code to convert text into a spreadsheet is listed in the appendix (*txtToXls.py*).

MATLAB and Python can easily handle the maximum number of files. But MATLAB has various options to import files when working with the user interface (UI), where it can directly use the import data options present on the home screen and save it into.mat format or as a variable without writing a script.

10.3 Signal Generation

Signal generation is crucial in signal analysis, allowing for a comprehensive assessment before proceeding with real-life signal processing. Researchers can predict outcomes based on experimental setups by simulating and analyzing signals. Simulating the vibration signature of a structure enables the analysis of its physical properties. MATLAB and Python offer capabilities for generating and resampling various types of signals. This section compares multiple generated signals and explores the statistical analysis of signals. In the first case, implementation is made between a particle's position and velocity curves. By examining these curves, valuable insights can be gained regarding the particle's behavior and characteristics. Suppose a particle is moving in the form of an infinity symbol '∞'. Position and velocity can be plotted with time as they are a function of time, and the velocity of X and Y directions are represented in Eqs. 10.1 and 10.2.

X-Position of particle—

$$x = \sin(t) \tag{10.1}$$

Y-Position of particle—

$$y = \sin(2 * t) \tag{10.2}$$

Velocity can be calculated by differentiating the X and Y positions of the particle. The velocity of the particle is calculated and plotted as shown in Fig. 10.1 with time for 10 s to compare outcomes from MATLAB and Python.

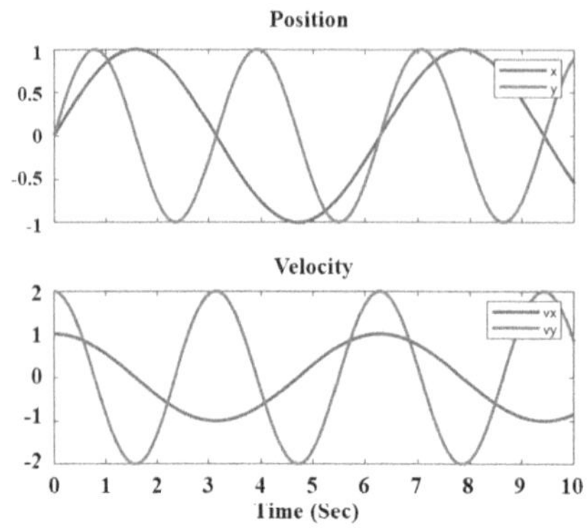

(a) Visualization using MATLAB

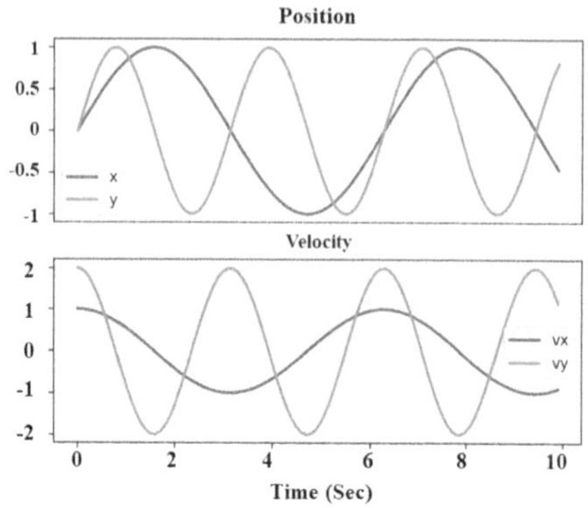

(b) Visualization using Python

Fig. 10.1 Velocity, position and their first derivative of the particle moving in the form of an infinity symbol

Table 10.3 Statics comparison of velocity and position of a particle

Parameter	MATLAB	Python
Mean	0.1792	0.1865
Standard deviation	0.6680	0.6641

From Table 10.3, it can be analyzed that MATLAB and Python can generate nearly identical waveforms. In measurement, a signal is always recorded with unwanted signals called noise. Behavior analysis of noise is necessary to suppress it from the original signal and retrieve the original information.

MATLAB provides a powerful tool to analyze and remove the noise from the original signal. The behavior of noise can be analyzed with Python using the modules available. In this article, white noise is added to the generated positional signal, as shown in Fig. 10.2 and statics of noisy signal are listed in the table to compare MATLAB and Python.

Uniformly distributed pseudorandom numbers add noise to the signal with the same dimension as the original signal. Table 10.4 summarizes the difference in static parameters of both signals. Mean, standard deviation, and Median varies by an absolute difference of 0.0161, 0.0309 and 0.013, respectively.

The signal-to-noise ratio (SNR) represents the power of the desired signal to that of undesired noise. A high SNR value indicates a low noise level in the original signal, which is desirable for accurate analysis. Analyzing simulated signals with different SNR values is crucial for developing effective filtering techniques in real-time applications. In MATLAB, the "*awgn*" function allows the direct addition of white Gaussian noise to a signal using a specified SNR value. However, in Python, no built-in function or method is available. Hence, this article introduces a user-defined function in Python that adds white Gaussian noise to the signal with a specified

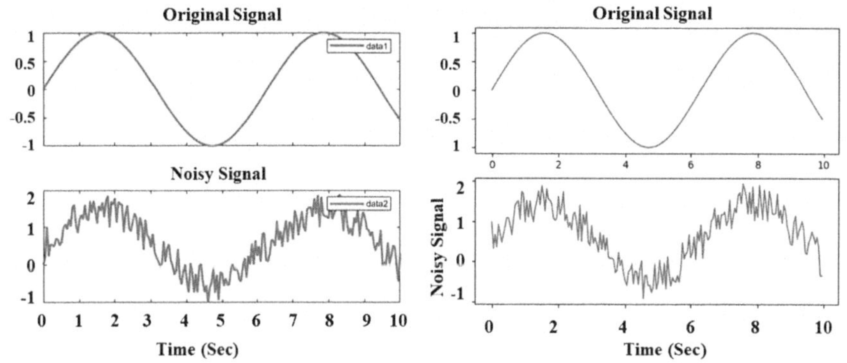

(a) Noise added using MATLAB (b) Noise added using Python

Fig. 10.2 Uniformly distributed pseudorandom noise added to the signal

Table 10.4 Difference in static parameters of noisy signal generated by distributed pseudorandom numbers

Parameter	MATLAB	Python	Absolute difference
Mean	0.6694	0.6855	0.0161
Standard deviation	0.7133	0.7442	0.0309
Median	0.7529	0.7399	0.0130

SNR value. The outcome of this function is compared with the results obtained using MATLAB. By comparing the results between the two programming languages, researchers can assess the effectiveness and accuracy of the user-defined Python function for adding white Gaussian noise with a specific SNR value, ensuring reliable signal analysis in real-life applications. A sinusoidal signal is simulated with an amplitude of 10, and a noisy signal with an SNR value of 5dB is generated. For Python, *the awgn* function has been created based on the mathematical equations for the same sinusoidal signal, as shown in Fig. 10.3. Python's *awgn* function (*AWGN_function.py*) is listed in the appendix.

As Python's function is created using mathematical equations, the noise added to the signal is not the same. The mean of the signal varies by an absolute difference of 1.19, which is slightly higher than a noisy signal simulated using distributed pseudorandom numbers. Table 10.5 lists the absolute difference in mean, standard deviation and median.

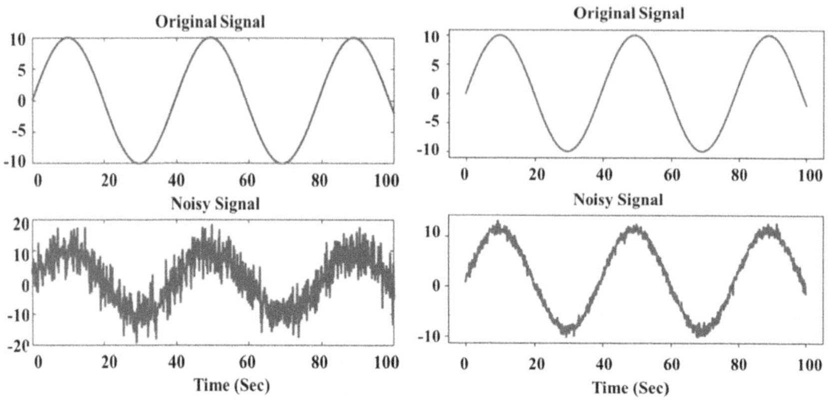

(a) Noise added by awgn function of MATLAB

(b) Noise added by awgn function implemented on Python

Fig. 10.3 White Gaussian noise added to the simulated signal with SNR value

Table 10.5 Difference in static parameters of the noisy signal generated by the Gaussian method

Parameter	MATLAB	Python	Absolute difference
Mean	1.3000	2.4990	1.1990
Standard deviation	7.9720	6.9790	0.9930
Median	2.0970	3.5890	1.4920

10.4 Signal Processing

As previously mentioned, recorded data from DAQ devices often contains unwanted signals that must be eliminated. Filtering is a commonly used technique in signal processing to remove such unwanted signals and can also be employed to smooth the signals if needed. Different types of filtering techniques are available to perform the desired task. The choice of filtering technique depends on the application and the kind of noise present in the original signal. If a strain response of any structure is being recorded, the signal may have high-frequency unwanted signals due to environmental vibrations. A low-pass filter (LPF) is a perfect choice to remove high-frequency components from the original signal. Once the filter type is decided, a filter design technique is a critical choice, and we can go for an averaging or windowing method [19].

In this article, the authors have focused on implementing digital filters using MATLAB and Python for filtering applications associated with SHM. As mentioned earlier, signals always get recorded with high-frequency unwanted signals, and to remove high-frequency components, LPF is used. An infinite impulse response (IIR) low pass filter is typical for filtering applications where a linear phase is not required.

10.4.1 IIR Filter Implementation

The impulse response of the IIR filter is infinite in duration, which can be used as a frequency-selective filter. All parameters must be specified in the frequency domain to implement the frequency-selective filters. If linear phase characteristics are not required, the IIR filter is the best choice as they have lower side lobes in the stopband, and implementation has fewer parameters and requires less memory. The desirable characteristics of the practical frequency-selective filters are the passband ripple, stopband ripple, transition band, and bandwidth. All the input parameters are decided based on desirable characteristics. The filters should be designed in such a way that they can meet the requirements of the causality. There are several methods to develop digital filters having infinite impulse response, like the design of IIR filters from analog filters, design by approximation of derivatives, etc. [4]. A Butterworth filter is most commonly used for the filtering application as it has a flat frequency response in the passband. Consider a typical case of a Butterworth filter with a magnitude of −3 dB at the cutoff frequency, and the transfer function of such filter is defined in Eq. 10.3:

$$\left| H(j\Omega) = \frac{1\omega_p}{1+\Omega^{2n}} \right| \tag{10.3}$$

where, Ω (Normalized Frequency) = $\frac{\omega}{\omega_p}$, and n is the order of the filter. As n increases, the magnitude response becomes flatter in the transition region, as shown in Fig. 10.4, but as the order of the filter increases, phase linearity decreases.

MATLAB's signal processing toolbox offers a function to design a digital low-pass IIR Butterworth filter. Similarly, the Scipy module provides methods to implement MATLAB-style filters in Python [20]. A digital Butterworth IIR low pass filter is

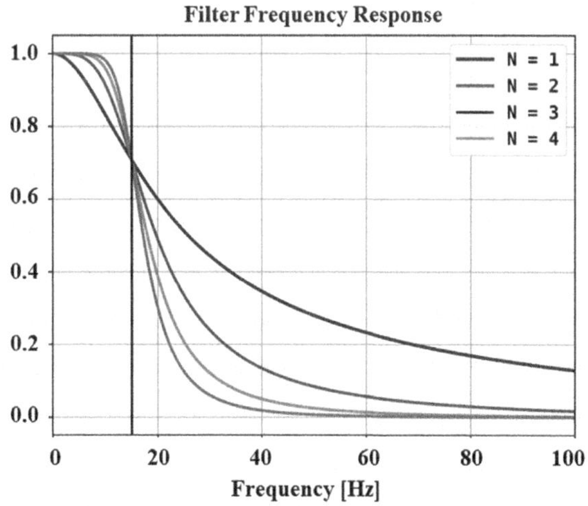

Fig. 10.4 Frequency response of the LPF with a different order, N

analyzed to remove the high-frequency noise from recorded signals. First, a sinewave of 5Hz is simulated, and then the noise is added to the simulated signal. The same noisy signal is used to analyze the behavior of the filter design in both MATLAB and Python. A user-defined function (lowpassButterworthFilter.py, listed in the appendix) is created in Python with input parameters such as order, cutoff frequency, and sampling rate. This function is also used to plot the magnitude response of the Butterworth filter with different filter orders. MATLAB script (*butterworth_LPF_IIR.m*) uses the same waveform imported from Python. A fourth-order Butterworth filter with a sampling frequency of ten times the original signal frequencies is implemented. As shown in Fig. 10.5, there is a phase change between the original and filtered signal, but we can recover the original signal with the fourth-order filter, and the same can be done using MATLAB. The maximum and minimum values of the filtered signal are not the same for both packages. Maximum values deviate by 0.451 and minimum by 0.203. Table 10.6 lists the errors with the original signal, and MATLAB is close to the original signal in terms of standard deviation.

10.4.2 Frequency Response and Fast Fourier Transformation (FFT)

The cutoff frequency is an important parameter to configure the frequency-selective filters, which decides the point or value of the frequency that needs to be eliminated or considered. The FFT algorithm is a wise tool for determining the signal's frequency components and helps perform Discrete Fourier Transformation (DFT). Once frequency components are known, the cutoff frequency of the filters can be decided. MATLAB and Python both have a built-in function to calculate the FFT. Figure 10.6 shows the frequency response of the sinusoidal having 2 Hz frequency and 20 Hz using MATLAB and Python, respectively. MATLAB and Python codes used to simulate sinusoidal waves and perform FFT are listed in the appendix (*fftFun.m and fftfun.py*). Both packages use the same algorithm to calculate an FFT, and the results are almost identical.

10.5 Signal Analysis

Signal analysis helps in determining the features of the signal. Finding the features of the signal is necessary, like finding out the threshold or determining the similarities between two signals of structure, etc. In SHM, cross-correlation is often performed to calculate the similarity of two signals as a function of the displacement of one relative to the other.

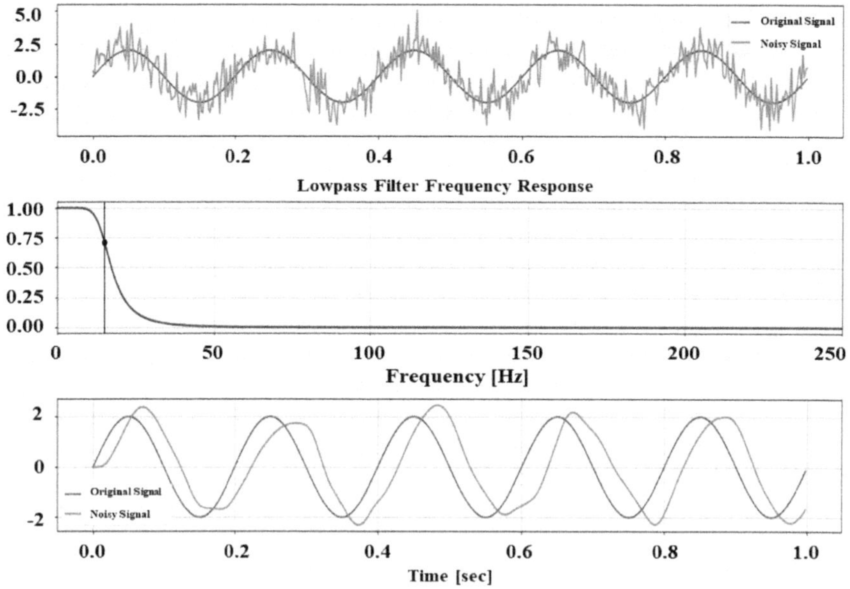

(a) Python Implementation

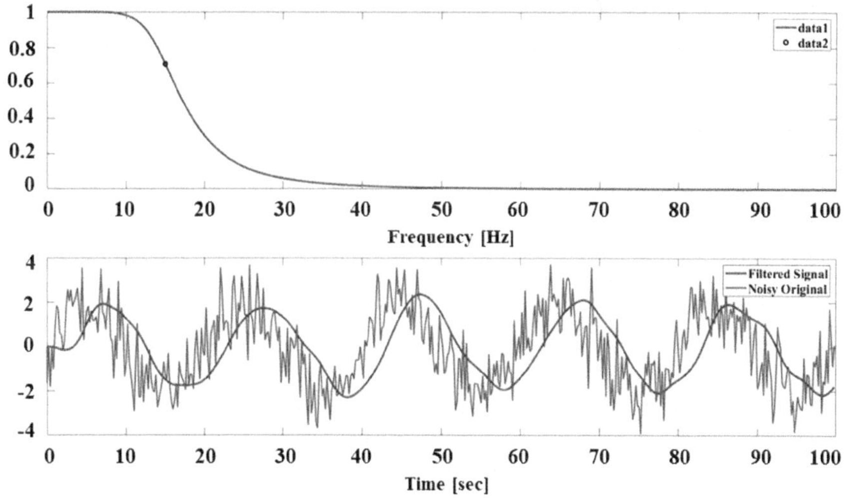

(b) MATLAB Implementation

Fig. 10.5 Filtering of the noisy signal using a fourth-order Butterworth filter

Table 10.6 Error in statics of original and noisy signal

Parameter	Original signal	Filtered signal		Error	
		MATLAB	Python	MATLAB	Python
Mean	~0.0	−0.0290	−0.0090	NA	NA
Standard deviation	1.4170	1.3900	1.3320	5.76%	5.99%
Median	~0.0	−0.0357	0.0319	NA	NA
Maximum value	2.0	2.3600	1.9090	18%	4.55%
Minimum value	−2.0	−2.2800	−2.4824	14%	24%

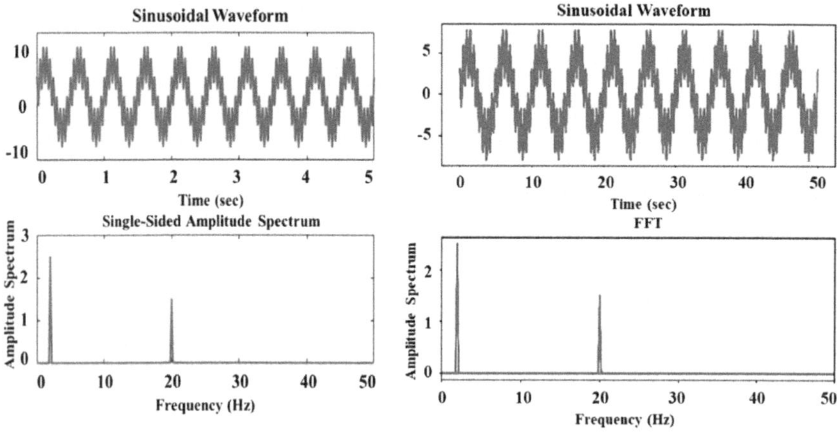

(a) FFT of the signal using MATLAB (b) FFT of the signal using Python

Fig. 10.6 Frequency response of the sinusoidal signal having 2 and 20 Hz frequency components

10.5.1 Correlation Coefficient

A cross-correlation of a signal with itself is called auto-correlation. It is valuable to detect the periodicity of signals. The term correlation coefficient is a measure of the correlation between two signals or variables. The correlation coefficient (Eq. 10.4) measures the linear dependency of signals with each other, and for N, observation is defined as [21]:

$$\rho(x, y) = \frac{1}{N-1}\sum_{i=1}^{N}\left(\frac{x_i - \mu_x}{\sigma_x}\right)\left(\frac{y_i - \mu_y}{\sigma_y}\right) \tag{10.4}$$

where μ_x and μ_y are the mean of the signal x and y, and σ_x and σ_y are the standard deviation of signal x and y, respectively. The correlation coefficient matrix (R) of signals can be given as:

$$R = \begin{pmatrix} \rho(x, x) & \rho(x, y) \\ \rho(y, x) & \rho(y, y) \end{pmatrix} \tag{10.5}$$

The cross-correlation coefficient of the signal itself will always be 1, as they are directly correlated. MATLAB and Python are handy tools available to visualize the correlation matrix. In this article, a correlation matrix of four sensors has been created. In this section, the authors have only focused on the mathematical and programming part, and in the next section, real-life SHM application of the correlation matrix has been discussed. The code to compute and visualize the correlation matrix for MATLAB is listed in the appendix.

10.5.2 Interpolation

Interpolation is a technique to convert the sampled signal into higher sampling rate signals. Interpolation is a combination of digital LPF and decimation. Two signals recorded with different sampling rates can be combined for comparison using interpolation. Lost signals during accruing data can also be incorporated using interpolation. Linear interpolation is the simplest form of interpolation, where linear connections are used to obtain relations with adjacent points. Another form of interpolation is spline interpolation, a piecewise polynomial function that can be used for interpolation and data smoothing. In SHM, interpolation is very useful for proper frequency responses and comparing signals recorded with different frequencies.

10.6 Experimental Analysis

Non-destructive vibration-based global structural health monitoring and damage detection methods are model-based and signal-based. Various sensors like accelerometers and strain gauges are used to capture the vibrations of the structures. Before implementing the damage detection algorithms, post-processing of the sensor response is needed. This study considers a simple supported steel beam with a length of 3000 mm, width of 65 mm, and thickness of 8.5 mm for experiments. The structural steel has a density of 7.85 × 103 kg/m^3, and its Young's modulus is 2 × 105 MPa with 0.3 of Poison's ratio. Fiber Bragg Grating (FBG) sensors are used to measure the strain response of the beam. Four sensors were mounted on the beam to record

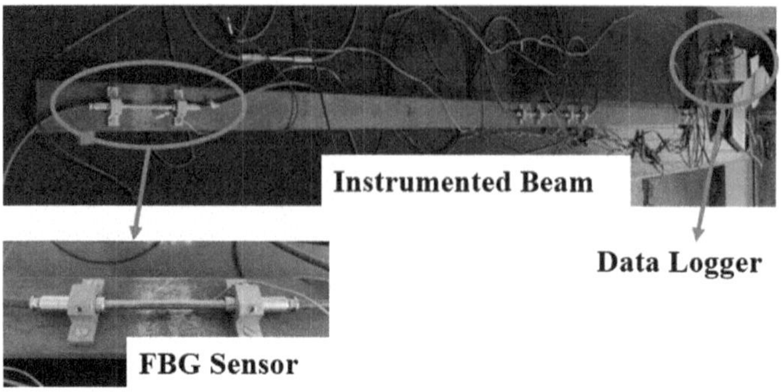

Fig. 10.7 Beam instrumented with FBG strain sensors for damage identification

strain response under natural and excited conditions. The beam has been excited by an impact hammer, and strain data is recorded for 1 min with a sampling rate of 1000 Hz. HBK Make data logger is used to record the responses of the sensors, which is capable of saving data files in CSV format. Once data is saved, signal processing and analysis have been done using MATLAB and Python to verify the discussed techniques (Fig. 10.7).

FBG sensors are widely recognized for their exceptional sensitivity, accuracy, and signal-to-noise ratio (SNR). The strain response was captured using short gauge strain sensors with a length of 10cm, manufactured by Micron Optics. The response is recorded for all four sensors, and the impulse hammer is used for excitation. Figure 10.8 shows the response of FBG sensors for a one-minute duration of the strain generated in the beam. After recording data using the DAQ system, post-processing becomes necessary to eliminate noise and extract information. The approach involves converting the recorded time series data into the frequency domain, and the frequency modes of the structures can be identified. In this article, the FFT and cross-correlation coefficient calculations have been performed for all sensors. MATLAB and Python packages have been employed to load the experimental data files and determine the frequency response. The complete Python code for these operations is in the appendix (*fileFFTPython.py*). Figure 10.9 showcases the FFT of the strain recorded by the second sensor, visualizing the frequency components present in the signal.

The first and second modes of the frequency are found to be 9.75 and 15.40 Hz, respectively. To design the low pass IIR filter, the cutoff frequency is decided based on the frequency modes of the structures and the filter is implemented as discussed in the above paragraph. In structural health monitoring, damage in the structure can be identified by different methodologies, such as changes in model frequencies and curvature mode shapes or by detecting the dissimilarities (Cross-correlation coefficient) between two sensors attached. The cross-correlation coefficient measures dissimilarities between two data sets and may be used for damage detection of the structures

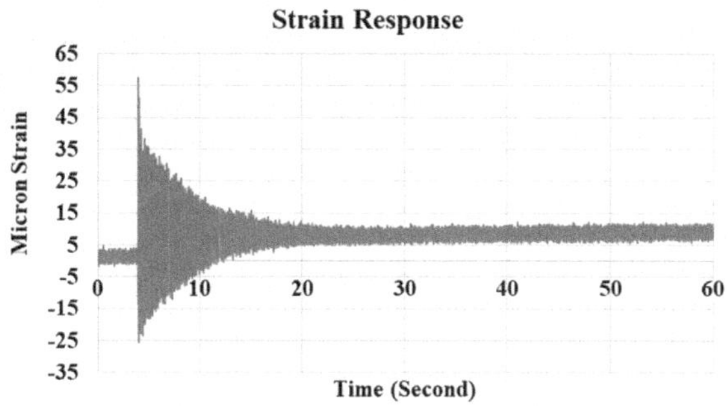

Fig. 10.8 Strain response of FBG

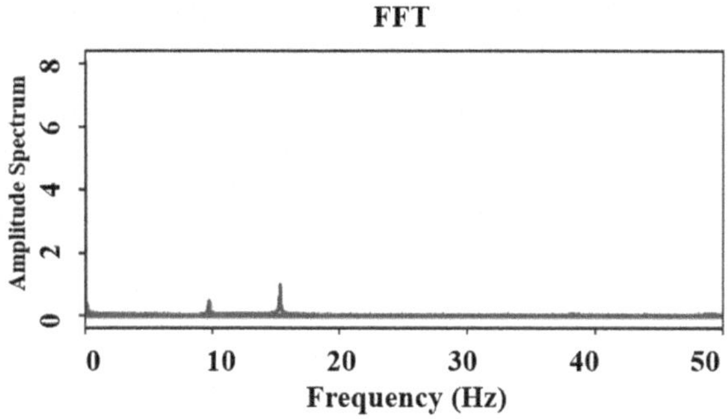

Fig. 10.9 Frequency response (FFT)

[22, 23]. Here, the calculation of the cross-correlation coefficient is presented using MATLAB and Python for SHM applications. Cross-correlation coefficients among all the sensors installed on the beam have been calculated to compare the outcomes of MATLAB and Python packages. For four sensors, a 4×4 matrix is created as the cross-correlation coefficient of one sensor is calculated with respect to the other three sensors. The value of the Cross-correlation coefficient is always one when it is calculated for the same sensor. MATLAB and Python codes to calculate the cross-correlation coefficient are attached in the *crossCoMatlab.m* and *crossCoPython.py* appendix, respectively.

Tables 10.7 and 10.8 illustrate the cross-correlation coefficients among all four sensors. Both software packages indicate minimal discrepancies in these coefficient values. Therefore, either can be preferred for SHM applications, as the maximum absolute difference (Table 10.9) between coefficients generated from Python and MATLAB is approximately 0.0001, which is negligible. Damage can be identified by comparing the cross-correlation coefficient matrices of undamaged and damaged structures. The developed code will streamline the research process by reducing the time for file handling, signal processing (such as filtering signals for noise), and signal

Table 10.7 Cross-correlation coefficients matrix of sensors mounted on a beam using Python

Sensors number	Sensor 1	Sensor 2	Sensor 3	Sensor 4
Sensor 1	1	−0.0799	−0.0278	−0.6431
Sensor 2	−0.0799	1	0.4600	−0.4601
Sensor 3	−0.0278	0.4600	1	−0.4767
Sensor 4	−0.6431	−0.4601	−0.4767	1

Table 10.8 Cross-correlation coefficients matrix of sensors mounted on a beam using MATLAB

Sensors number	Sensor 1	Sensor 2	Sensor 3	Sensor 4
Sensor 1	1	−0.08	−0.0279	−0.6431
Sensor 2	−0.08	1	0.4600	−0.4601
Sensor 3	−0.0279	0.4600	1	−0.4768
Sensor 4	−0.6431	−0.4601	−0.4768	1

Table 10.9 Absolute difference between cross-correlation coefficients values calculated from Python and MATLAB

Sensors number	Sensor 1	Sensor 2	Sensor 3	Sensor 4
Sensor 1	0.0	0.0001	0.0	0.0
Sensor 2	0.0001	0.0	0.0	0.0
Sensor 3	0.0001	0.0	0.0	0.001
Sensor 4	0.0	0.0	0.0001	0.0

analysis (computing cross-correlation coefficients). The techniques discussed, and the provided codes may assist researchers in accelerating their analysis and aid in selecting the appropriate tool for their analysis needs.

10.7 Conclusion

The use of software in structural health monitoring is discussed. The implementation and analysis of signal processing techniques using MATLAB and Python for SHM have been done. The following conclusions can be obtained from the study:

- When Uniformly distributed pseudorandom numbers are used to add noise to the signal, the mean of the noisy signal varies by an absolute difference of 1.199 between MATLAB and Python.

- In filtering applications, MATLAB and Python show deviations in mean, standard deviation, maximum, and minimum values apart from phase change. MATLAB exhibits a lower standard deviation of 5.76% in the recovered signal from the noisy signal compared to Python, which is 5.99%.
- Both MATLAB and Python can be used for frequency domain analysis as the results are identical for real-life applications.
- Cross-correlation coefficients in both packages are identical up to two decimal points, with a maximum absolute difference of 0.0001, which is considered negligible.

Implementing code for SHM applications can provide better performance and save much time for the researchers. This study can help decide on software packages to perform analysis needs in SHM and other applications. Developed codes can help in damage identifications, FFT is used to calculate curvature damage factor, and damage identification algorithms can be implemented using a cross-correlation coefficient matrix. This comparative analysis provides concrete evidence for users to make informed decisions based on accuracy and performance, ensuring optimal outcomes for their research and analysis.

Acknowledgements This work was supported by the Council of Scientific & Industrial Research (CSIR), India and done under project GAP-0212. The authors would like to thank Ritika and Sonali for their help in supporting all the tests.

Competing Interests The authors have no competing interests to declare that are relevant to the content of this article.

Submission Declaration and Verification The preprint DOI for the paper is https://doi.org/https://doi.org/10.21203/rs.3.rs-1856394/v1. The work described in this paper has not been published previously in any journal.

Appendix

All the MATLAB and Python codes are listed here, name of MATLAB code end with.*m* and name of Python's code end with *.py*.

I. ***txtToXls.m***

```
clc; clear all; close all;
fileId = fopen("text_file.txt",'r'); %open text file in matlab
txtData = fscanf(fileId, '%d %f', [2, inf])
fclose(fileId); %close file
data = txtData';
writematrix(data,'convertedfile.xlsx') % file will be saved to current directory
```

II. ***AWGN_function.py***

```
import numpy as np

def awgn(SNRinDB, originalSignal):
    signalMean = np.mean(originalSignal)
    signalPower = 10 * np.log10(abs(signalMean))  # signalPower
    print('signalMean: ', signalMean, ', signalPower: ', signalPower)
    # Noise calculation
    noisePower = signalPower - SNRinDB
    print('noisePower: ', noisePower)
    # DB to scaller
    noiseScaler = 10 ** (noisePower / 10)
    print('noise Power Scaler', noiseScaler)
    noiseStd = np.sqrt(noiseScaler)

    # Generate Noise sample
```

VI. ***txtToXls.py***

```
import pandas as pd
txtDataFrame = pd.read_csv("text_file.txt", sep="\t")
excelFile = pd.ExcelWriter('convertedfile.xlsx')
txtDataFrame.to_excel(excelFile, sheet_name='Sheet1', index=False)
excelFile.save()
```

VII. ***lowpassButterworthFilter.py***

```
import scipy.signal as signal
def lowpassButterworthFilter(order, cutofFreq, Fs):
    """
    Butterworth filter and frequency response of it
    :param order: filter order N
    :param cutofFreq: cutoff Freq of filter in Hz
    :param Fs: Sampling Frequency in Hz
    :return:
    """
    NnyqFreq = Fs*0.5
    normalCutOfFreq = cutofFreq/NnyqFreq
    b, a = signal.butter(N= order, Wn= normalCutOfFreq, btype= 'lowpass', analog= False)
```

```
    NoiseMean = signalMean
    noiseSignal =
np.random.normal(NoiseMean, noiseStd,
len(originalSignal))
    return noiseSignal
```

III. ***butterworth_LPF_IIR.m***

```
clc; close all;
f = 5;
N = 4;
Fs = f*100;
time = 0:1/Fs:2
cutOffFre = 15; %Hz
NyqFreq = Fs*0.5
Wn = cutOffFre/NyqFreq
[b, a] = butter(N, Wn,"low");
[h, w] = freqz(b, a,[], Fs);
%NoisySinwave = NoisySinwave{:,:}
y = filter(b, a, NoisySinwave)
%NoisySinwavetime =
NoisySinwavetime{:,:}
subplot(211)
plot(w,abs(h), LineWidth= 2)
xlabel("Frequency (Hz)")
hold
plot(cutOffFre, 0.5*sqrt(2), 'ko',
LineWidth= 2)
legend
grid
xlim([0 100])
subplot(212)
plot(NoisySinwavetime, y, LineWidth= 2)
hold
```

```
    w, h = signal.freqz(b, a, worN=8000)
    y = signal.lfilter(b, a, noisySignal)
    return w, h, y
```

VIII. ***fftFun.m***

```
function [FFT,frequency] =
fftFun(signal,timeaxis, fs)
n = length(timeaxis);
frequency = (0:n/2-1)*fs/n; %fs = sampling
frequency
FFTfull1 = fft(signal,n);
FFT_Full = FFTfull1((1:n/2));
FFT = abs(FFT_Full);
figure;
subplot 211
plot(timeaxis,signal,'linewidth',1.2)
title('Strain Response')
xlabel('time(s)')
subplot 212
plot(frequency, FFT,'linewidth',1.2);
xlim([0 120])
title('FFT')
xlabel('Frequency (Hz)');
end
```

IX. ***fileFFTPython.py***

```
def loadFile(fileName):
    file = pd.read_excel(fileName)
    time = (file.Time)
    data = (file.Strain)
    return time, data

fileName = 'experimentFile.xlsx'
```

```
plot(NoisySinwavetime, NoisySinwave,
LineWidth= 2)
xlabel("Time (Sec)")
legend(["Filtered Signal" , "Noisy
Original"])
```

iv. ***fftfun.py***

```
def fftFun(signal, timeaxis):
    signal = np.array(signal)
    timeaxis = np.array(timeaxis)
    samplingRate = 1 / round(timeaxis[1], 3)
    duration = timeaxis[len(timeaxis)-1]
    print('SR', samplingRate)
    N = round(samplingRate * duration)
    yf = fft.fft(signal)
    xf = fft.fftfreq(N+1, 1 / samplingRate)
    print(len(yf), len(xf))
    plt.subplot(2, 1, 1)
    plt.plot(timeaxis, signal, linewidth=2)
    plt.xlabel('Time (Sec)')
    plt.title('Strain Response')
    plt.subplot(2, 1, 2)
    plt.plot(xf, np.abs(yf / len(signal)),
linewidth=2)
    plt.xlim(0, 50)
    plt.xlabel('Frequency (Hz)')
    plt.ylabel('Amplitude Spectrum')
    plt.title('FFT')
    plt.subplots_adjust(hspace=0.5)
    plt.savefig('FFT_Python_fileFFT.tif',
dpi=1200)
    plt.show()
time, data = loadFile(fileName)
fftFun(signal = data, timeaxis = time)
```

x. **crossCoPython.py**

```
import numpy as np
import pandas as pd
fileName = 'experimentFile1.xlsx'
file = pd.read_excel(fileName)
time = file.Time
data1 = file.Strain1
data2 = file.Strain2
data3 = file.Strain3
data4 = file.Strain4
data1 = np.array(data1)
data2 = np.array(data2)
data3 = np.array(data3)
data4 = np.array(data4)
array = [data1, data2, data3, data4]
co = np.corrcoef(array)
print(co)
```

```
fftOfSignal = np.abs(yf / len(signal));
return xf, fftOfSignal
```

v. ***crossCoMatlab.m***

```
clc;
clear all;
file = xlsread('experimentFile1.xlsx')
time = file(:,1);
data1 = file(:, 2);
data2 = file(:, 3);
data3 = file(:, 4);
data4 = file(:, 5);
array = [data1, data2, data3, data4]
co = corrcoef(array)
```

References

1. Baral S, Adhikari S, Negi P, Bhalla S (2022) Development and evaluation of reusable piezo sensors for health. J Civ Struct Health Monit 12(3):647–657. https://doi.org/10.1007/s13349-022-00567-3
2. Caponero MA, Dell'Erba D, Kropp C (2019) Use of fibre optic sensors for structural monitoring of temporary emergency reinforcements of the church S. Maria delle Grazie in Accumoli. J Civ Struct Health Monit 9:353–360. https://doi.org/10.1007/s13349-019-00335-w
3. Oppenheim AV (1999) Discrete-time signal processing. Pearson Education India
4. Proakis JG, Manolakis DG (1996) Digital signal processing. Prentice-Hall
5. Naeem Ahmed S (2015) 8—Signal processing. In: Physics and engineering of radiation detection, 2nd Edn. Elsevier, pp 477–540
6. Nebeker N (1998) Fifty years of signal processing: the IEEE signal processing society and its technologies. IEEE Signal Process Soc
7. Hoffmann M (2007) Digital signal processing maths. Cern Accel Sch Proceed Sigtuna, Sweden
8. Paul WF, Lynn (1998) A introductory digital signal processing with computer applications. John Wiley & Sons
9. Tseng EJ (2001) Toward nanocomputers. Science 1293–1294
10. Oppenheim AV (1978) Applications of digital signal processing. Englewood Cliffs
11. Wikibooks, (2018) Digital signal processing/software tool [Online]. Available: https://en.wikibooks.org/wiki/Digital_Signal_Processing/Software_tool
12. Mathworks (2022) What Is MATLAB? [Online]. Available: https://in.mathworks.com/discovery/what-is-matlab.html
13. W3school, Python Introduction, [Online]. Available: https://www.w3schools.com/Python/Python_intro.asp
14. Patel C, Panigrahi SK, Chourasia A, Roy TB, Bagchi A, Tirca L (2020) Wavelet transformation approach for damage identification of steel structure model. In: Recent trends in wave

mechanics and vibrations. Springer, pp 245–257. https://doi.org/10.1007/978-981-15-0287-3_18
15. Frerking M (2013) Digital signal processing in communications systems. Springer
16. Mathworks (2022) Supported file formats for import and export, [Online]. Available: https://in.mathworks.com/help/matlab/import_export/supported-file-formats-for-import-and-export.html
17. Rao RN (2022) Core Pyhton programming, 3rd edn, Dreamtech Press
18. Cbtnuggets, 14 File Types You Can Import Into pandas, [Online]. Available: https://www.cbtnuggets.com/blog/technology/programming/14-file-types-you-can-import-into-pandas
19. Haslwanter T (2021) Hands-on signal analysis with Python: an introduction. Springer Nature
20. The SciPy community (2022) Signal processing (scipy.signal). Available: https://docs.scipy.org/doc/scipy/reference/signal.html
21. Haslwanter T (2016) An introduction to statistics with python. Stat Comput
22. Anwar A, Adarsh S (2022) Structural health monitoring using multifractal cross-correlation analysis. Recent Adv Civ Eng 172:1087–1097. https://doi.org/10.1007/978-981-16-4396-5_90
23. Hamidian P, Soofi Y, Bitaraf M (2022) A comparative machine learning approach for entropy-based damage detection using output-only correlation signal. J Civ Struct Heal Monit. https://doi.org/10.1007/s13349-022-00587-z

Chapter 11
A Review of Structural Health Monitoring and Damage Detection Techniques in Frames and Bridges

Saeed Hamidi, Atefeh Soleymani, and Maria Rashidi

Abstract Structures can experience localized damages for various reasons, and if the location of these damages remains unknown, they can be exacerbated due to natural disasters such as earthquakes or artificial factors like improper excavations, leading to their destruction. Therefore, monitoring the health of structures and their components is considered one of the most important research topics in civil engineering, mechanical, and aerospace disciplines. Numerous research studies have been published so far in the field of structural health monitoring. Nowadays, with the use of structural health monitoring methods, it is possible to accurately identify structural damages in the early stages and prevent financial and life-threatening damages from occurring. Structural health monitoring is essential to ensure the safe performance of a structure throughout its operational lifespan. Due to the typically uncertain location and severity of damage in structures, significant efforts have been made to achieve an accurate, reliable, and cost-effective method for damage detection. The detection of damage in structures, especially in recent years, has received significant attention. A comprehensive review of published research studies has been conducted in this chapter and an overall description was provided of various methods and compared them in terms of accuracy and correctness.

Keywords Structural health monitoring · Damage detection · Damage localization · Damage severity identification · Frames · Bridges

---

S. Hamidi
Department of Civil Engineering, University of Birjand, Birjand, Iran
e-mail: saeedhamidin79@birjand.ac.ir

A. Soleymani
Department of Civil Engineering, Shahid Bahonar University of Kerman, Kerman, Iran
e-mail: Atefeh_Soleymani@eng.uk.ac.ir

M. Rashidi (✉)
Centre for Infrastructure Engineering, Western Sydney University, Sydney, NSW, Australia
e-mail: m.rashidi@westernsydney.edu.au

H. Jahangir et al. (eds.), *Damage Detection and Structural Health Monitoring of Concrete and Masonry Structures*, Springer Tracts in Civil Engineering,
https://doi.org/10.1007/978-981-97-8975-7_11

11.1 Introduction

Structural health monitoring and damage identification in structures have always been the focus of researchers. Inspection and proactive testing are common approaches in structural health monitoring for detecting the progression of fatigue-induced damages leading to structural failure. Accurately identifying damaged elements in a structure and reinforcing or replacing them can significantly increase the useful life of the structure. Offshore platforms are predominantly used for oil and gas extraction and are exposed to water and vulnerabilities.

Rahami et al. [1] using the recorded acceleration response signals from a sensor installed on the jetty platform's deck, identified the presence of its damages. Darvishan [2] investigated damage detection in cable-stayed bridges using signal processing and machine learning techniques at a low cost. They have utilized only one sensor for structural vibration monitoring, which has resulted in cost reduction. They have subsequently introduced a new damage index based on cross-wavelet conversion. Samaei et al. [3] utilized the modal strain energy method, which relies on dynamic response, for detecting damage in a prototype wind turbine structure's foundation. This method can detect low-intensity injuries in specific locations. However, it may incorrectly identify damage in other intact components. Amiri et al. [4] have examined the moment frame connections by utilizing a combination of the modal analysis method and Hilbert transforms to investigate the change in curvature. In this method, natural frequencies and mode shapes of the structure are estimated, and damage is identified by comparing their curvature. In the same research work, Amiri et al. [5] examined the failure in a shear frame by employing signal processing techniques, mixed window transforms, and phase modulation. It is an efficient tool for detecting damage in both the presence and absence of signal disturbances. Ahmadi et al. [6] have utilized neural networks and square time–frequency functions for damage identification in concrete slab girders of bridges. Kazemian et al. [7] used methods based on minimizing the difference between measured responses and responses obtained from analytical models. They have examined the detection of damages in the trusses using static responses with disturbances. They concluded that by using the vibration mode shapes and dynamic responses of the structure, one can determine the extent of the damage. Daneshvar et al. [8] employed information derived from dynamic responses to identify damage in frames. Kavoosi et al. [9] used the Contourlet transform, as one of the modern signal processing methods, for damage detection in prestressed concrete beams. With a focus on time series and the history of vibration responses, it is possible to identify the locations of structural damages with varying geometric shapes. Mokhtari et al. [10] have utilized vibration responses in the time domain to detect damage in prestressed concrete beams. Jahangir et al. [11] employed wavelet analysis and modal data to detect damage in reinforced concrete beams. Recently multi-objective optimization methods have attracted the attention of researchers. This method results in improved performance in damage localization and quality evaluation. Jahangiri et al. [12] employed multi-objective optimization algorithms, including MOPSO and MOEA/D, to detect damages. Darvishan [13] utilized blind source separation and multifractal analysis degraded of detrended fluctuations for damage detection. Shirzadi et al. [14] focused on damage identification in structures using an adaptive controller by utilizing control force as a damage-sensitive characteristic. Jahangir et al. [15] focused on gradual damage detection using modal vibration data. After an earthquake, access to the structure is not possible. Some methods only detect damage using the output responses. Beheshti et al. [16] employed signal processing and artificial neural networks for damage identification.

11.2 Damage Detections in Frames

In civil engineering, a frame is a structural element that is designed to withstand and transfer loads and forces. Frames are composed of direct structural components such as columns and beams that are interconnected as a system. These structural elements can inherently possess high structural resistance and can absorb and distribute dynamic and seismic forces. To identify damage in frames, researchers and engineers employ various methods and techniques. These methods include experimental and field studies, structural analysis, signal processing, and modal analyses. One of the methods used to identify damage in frames is the use of structural analysis and numerical modeling. In this approach, engineering software and analytical techniques are employed to examine the behavior of the structure under normal and damaged conditions. By comparing experimental data with analytical results, damage in frames can be identified. In cases where access to the structure is challenging, non-destructive methods are employed for damage identification. These methods include the use of imaging techniques such as radar imaging, thermography, ultrasonic waves, and other similar techniques. By analyzing the information obtained from images and signals, damage in frames can be detected. In addition to the aforementioned methods, the use of sensors and damage indicators is also employed for damage identification in frames. Various sensors such as vibration sensors, strain sensors, pressure sensors, temperature sensors, and humidity sensors can help in recording changes in the structure. By analyzing the data obtained from sensors and utilizing damage indicators such as frequency shift, and relative deformation, damage in frames can be detected.

11.2.1 Hilbert Transform

The Hilbert transform is a novel method for decomposing complex signals into a limited number of intrinsic mode functions. By decomposing the signal into a set of spectral processing, the Hilbert transform is capable of determining the time–frequency locations in the signal. This transform can be employed in the analysis and processing of non-stationary signals, such as vibration signals and sounds [4]. The

Hilbert transform involves empirical mode decomposition (EMD) and Hilbert spectral analysis (HSA). It is a novel method for analyzing non-stationary and non-linear data. Determining the frequency and magnitude range is a method for describing the non-stationary state. For this reason, spectral analysis of a portion of the signal is considered part of the Hilbert transform. The Hilbert transform is defined below, where P represents the singular integral value:

$$y(t) = \frac{1}{\pi} P \int_{-\infty}^{+\infty} \frac{x(\tau)}{t-\tau} d\tau \tag{11.1}$$

The marginal spectrum represents the amplitude range within the entire data range and indicates the contribution of each frequency. It is obtained from Eq. (11.2), where the [0, T] is defined within the specified range of data.

$$h(\omega) = \int_{0}^{T} H(\omega, t) dt \tag{11.2}$$

For modal decomposition, it is assumed that at each point in time, modes exhibit oscillations with different frequencies and all data reside between upper and lower envelopes. A component is recognized as an intrinsic mode function when it satisfies the following conditions:

- In all signals, the number of extrema and the number of crossings of the curve with the horizontal axis differ by no more than one unit.
- At each point, the mean envelope defined by local maxima and the mean envelope defined by local minima is both zero.

There are different criteria for terminating the separation process. One of them is presented below:

$$SD_k = \frac{\sum_{t=0}^{T} |h_k - 1(t) - h_k(t)|^2}{\sum_{t=0}^{T} h_{k-1}^2(t)} \tag{11.3}$$

The average upper and lower envelopes are referred to as m_1. The difference between the envelopes and the data is the first significant deviation, denoted by h_l. After repeating the k iterations, h_1 transforms into an intrinsic mode function.

$$h_{1(k-1)} - m_{1k} = h_{1k} \tag{11.4}$$

By using the number of degrees of freedom and the number of modal decompositions, we calculate the modal shapes of the structural response using the following equations. Where n is degrees of freedom, r is the reference degrees of freedom and m is the number of modal decompositions:

$$\begin{aligned}
\phi_{11} &= \frac{IMF_{11}.IMF_{1r}}{IMF_{2r}^2}, \ldots, \phi_{1n} = \frac{IMF_{1n}.IMF_{1r}}{IMF_{1r}^2} \\
\phi_{21} &= \frac{IMF_{21}.IMF_{2r}}{IMF_{2r}^2}, \ldots, \phi_{2n} = \frac{IMF_{2n}.IMF_{2r}}{IMF_{2r}^2} \\
\phi_{n1} &= \frac{IMF_{n1}.IMF_{nr}}{IMF_{nr}^2}, \ldots, \phi_{2n} = \frac{IMF_{nn}.IMF_{nr}}{IMF_{nr}^2}
\end{aligned} \tag{11.5}$$

11.2.2 Signal Transformation

Signal transformation is a process in which the input signal is converted into a different format or a different domain [5]. The purpose of signal transformation is to extract useful information from the signal, analyze its properties, or prepare it for further processing. The continuous wavelet transforms (CWT) is a mathematical technique used for signal analysis and processing. It allows for the decomposition of a continuous-time signal into a set of wavelet functions, which are small, localized waveforms that can capture both frequency and temporal information of the signal. The CWT provides a time–frequency representation of the signal, allowing for the detection of transient events, oscillations, and other localized features in the signal. The CWT is a powerful tool for analyzing signals with non-stationary or time-varying characteristics. The relationship of CWT is given below:

$$cwt(\tau, a) = \int_{-\infty}^{+\infty} x(t)\omega(\tau - t, a)dt \tag{11.6}$$

In Eq. (11.6), the τ is the transfer parameter and a is the scale parameter. In the generalized signal transformation method, the input signal is converted into a different format or domain, in such a way that the necessary information in the original signal is preserved, and new information is also extracted, as presented below:

$$s(\tau, f, p) = \int_{-\infty}^{+\infty} x(t)\omega_{gen}(\tau - t, f, p)\exp(-2\pi ift)dt \tag{11.7}$$

In Eq. (11.7), p is the generalized window shape control parameters set. The mixed Gaussian window consists of two parts: a real part that is a Gaussian function and an imaginary part that has a phase function. By using this window, the input signal is transformed into a different format or a different domain. The relationship given below is provided:

$$s_{CW}(\tau, f, \sigma) = \int_{-\infty}^{+\infty} h(t)\{\omega_{CG}(\tau - t, f, \sigma) \times \exp(-2\pi ift)\}dt \quad (11.8)$$

In Eq. (11.8), the ω_{CG} is a mixed Gaussian window that can be obtained from the following equation:

$$\omega_{CG}(\tau - t, f, \sigma) = \begin{cases} P \exp\left[-\frac{f^2(\tau-t)^2}{2}\right]\exp\{-2\pi if(\tau - t\} & , t < \tau + \sigma/|f| \\ +2\pi isign(f)\sigma \log\left[\sigma + |f|(\tau - t)\right] & \\ 0 & , t \geq \tau + \sigma/|f| \end{cases} \quad (11.9)$$

In Eq. (11.9), the pre-coefficient P can be obtained from the following equation:

$$P^{-1} = \int_{-\sigma/|f|}^{\infty} \exp\{-\frac{f^2\tau^2}{2} - 2\pi if\tau + 2\pi i\, sign(f) \times \sigma \log\left[\sigma + |f|\tau|\right]d\tau \quad (11.10)$$

where, the P is the phase modulation degree controller.

11.2.3 Vibrational Modes

Each vibration mode is associated with a specific frequency at which the structure naturally vibrates [8]. Each vibration mode has a distinct vibrational pattern that is determined by the shape and motion of the structure. Vibration modes are used in the analysis of dynamic behavior and the response of structures to vibrational energy. They provide valuable information about the behavior and performance of structures under different forces and vibrations. The energy index can indicate changes in modal strains due to damage and is obtained from the following equation:

$$EI_k = \sum_{i=1}^{M} \left| \frac{U_k^d - U_k^0}{\max\left(U_k^0\right)} \right| \quad (11.11)$$

In Eq. (11.11), the M is the vector length, U^d is the measured energy from the damaged structure and U^0 is the measured energy from the undamaged structure.

11.2.4 Structural Controllers

In structures with active control, the effects of damage on structural behavior are reduced due to the application of control forces [14]. This reduces the effectiveness of classical methods that rely on the structural response for damage detection. The task of control algorithms is to change the input values of a system in such a way that the output level becomes equal to the desired value. In the design of adaptive controllers, in addition to the main control loop related to the control of the structure, there is another loop that is responsible for adjusting the parameters based on the conditions. For control design, a closed-loop system is considered with adjustable parameters for the controller θ. The control force at each moment is obtained from the following equation [14]:

$$u_i = m_i u_c(\theta_i - 1) \tag{11.12}$$

The use of the Lyapunov stability theory is another design method. In this approach, the difference between the output of the differential equation and the output of the calculated model is evaluated. The Lyapunov potential equation is given as follows:

$$V(e, \theta) = \frac{1}{2}\left(ye^T Pe + \left(\theta - \theta_0\right)^T (\theta - \theta_0)\right) \tag{11.13}$$

In Eq. (11.13), the P is a matrix that should be positive definite.

11.3 Damage Detection in Bridges

The identification of damage in bridges has been a focus of attention for engineers and researchers. These structures are considered vital components of transportation and economic production infrastructure. By detecting and identifying damages, it is possible to implement corrective and repair measures to improve the safety and service life of bridge structural elements.

11.3.1 Energy Entropy

The energy entropy method operates based on the concept of local absorbed energy, which is a measure of diversity and order in the distribution of energy. By analyzing and examining the changes in energy entropy in structures, damages, and abnormal variations can be detected. As damages increase in a structure, the diversity and order in the energy distribution decrease, which serves as an indicator of the occurrence

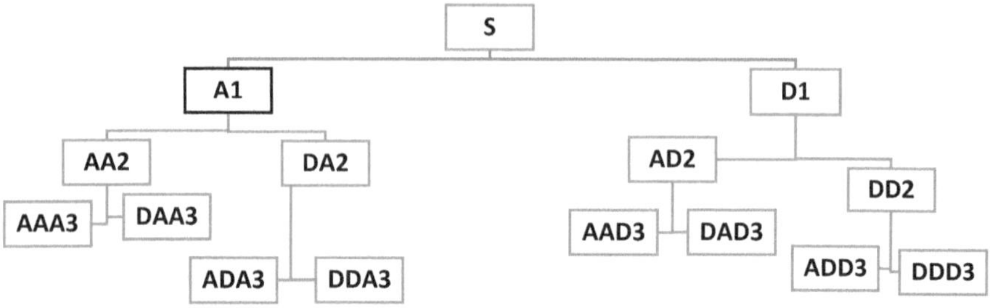

Fig. 11.1 The sequentially signal decomposition

of damage. This method can be used as an important tool in detecting and monitoring damage in structures. To this end, damage-sensitive features are determined by utilizing the concept of energy entropy and performing sensitivity analysis on the components obtained from wavelet packet decomposition. This transformation is a wavelet transform that allows more filters to pass through the signal compared to a discrete wavelet transform. In the wavelet packet transform, a family of wavelet functions and their associated scaling functions are used to decompose the original signal into sub-bands. As presented in Fig. 11.1, the signal is sequentially decomposed in both high-frequency and low-frequency parts, leading to further decomposition into subsequent wavelet packets.

Entropy is a measure of the amount of information or randomness in a signal. A higher entropy value indicates a greater amount of stored information. The damage-sensitive entropy index is defined below:

$$EDSI_n = \left| \frac{E^u_{j,n} - E^d_{j,n}}{E^u_{j,n}} \right| \times 100 \tag{11.14}$$

In Eq. (11.14), the $EDSI_n$ is the damage sensitivity index of component (n) at level (j) of the wavelet packet decomposition tree, $E^u_{j,n}$ is the entropy of energy in the healthy state, and $E^d_{j,n}$ is the entropy of energy in the damaged state. The definition of this index is based on the principle that the occurrence of damage in a structure leads to significant changes in the entropy values of several components of the wavelet packet energy entropy.

11.3.2 *Signal Processing*

The empirical mode decomposition (EMD) method, which was previously explained, will now be supplemented with the presentation of other methods. One of the challenges of the EMD method is the phenomenon of mode mixing, where multiple oscillations appear within a single mode. Another issue is the sensitivity of this method to noise. The combination of time-varying filters and the EMD (TVF-EMD) method utilizes a time-varying filter to overcome the limitations associated with modes. The signal in this space is obtained as below:

$$g_m^n(t) = \sum_{k=-\infty}^{\infty} c(k)\beta^n\left(\frac{t}{m} - k\right) \tag{11.15}$$

In Eq. (11.15), the c(k) represents the spline coefficients.

Another approach is the empirical wavelet transform (EWT) method. To apply the EWT method, a set of bandpass filters that divide the Fourier spectrum into (n) partitions, should be utilized. In this way, the decomposed modes have compact supports, and their spectra are centered around a specific frequency. The empirical modes are obtained from the following equations:

$$\begin{aligned} f_0(t) &= W_f^{\varepsilon}(0, t) * \phi_1(t) \\ f_k(t) &= W_f^{\varepsilon}(0, t) * \phi_k(t) \end{aligned} \tag{11.16}$$

The required parameters in Eq. (11.16) are calculated from the following equations:

$$\begin{aligned}
\underset{n}{\widehat{\phi}(\omega)} &= \begin{cases} 1 & \textit{if } |\omega| \le \omega_n - \tau_n \\ \cos\left[\frac{\pi}{2}\beta\left(\frac{1}{2\tau_n}(|\omega| - \omega_n + \tau_n)\right)\right] & \textit{if } \omega_n - \tau_n \le |\omega| \le \omega_n + \tau_n \\ 0 & \textit{otherwise} \end{cases} \\
&\begin{cases} 1 & \textit{if } \omega_n + \tau_n \le |\omega| \le \omega_{n+1} - \tau_{n+1} \\ \cos\left[\frac{\pi}{2}\beta\left(\frac{1}{2\tau_{n+1}}(|\omega| - \omega_{n+1} + \tau_{n+1})\right)\right] & \textit{if if } \omega_{n+1} - \tau_{n+1} \le |\omega| \le \omega_{n+1} + \tau_{n+1} \\ \sin\left[\frac{\pi}{2}\beta\left(\frac{1}{2\tau_n}(|\omega| - \omega_n + \tau_n)\right)\right] & \textit{if } \omega_n - \tau_n \le |\omega| \le \omega_n + \tau_n \\ 0 & \textit{otherwise} \end{cases} \\
\widehat{\phi_n}(\omega) &= \begin{cases} 0 & \textit{if } x \le 0 \\ \textit{and } \beta(x) + \beta(1-x) = 1 & \\ 1 & \textit{if } x \ge 1 \end{cases} \\
W_f^{\varepsilon}(n, t) &= \langle f, \psi_n \rangle = \int f(\tau)\overline{\psi_n}(t - \tau)\,d\tau = \left(\hat{f}(\omega)\widehat{\psi_n}(\omega)\right)^{\vee} \\
W_f^{\varepsilon}(0, t) &= \langle f, \phi_n \rangle = \int f(\tau)\overline{\phi_1}(t - \tau)\,d\tau = \left(\hat{f}(\omega)\widehat{\phi_n}(\omega)\right)^{\vee}
\end{aligned} \tag{11.17}$$

The natural excitation technique, also known as NExT, utilizes auto-correlation or cross-correlation to extract the free vibration decay functions of a system under ambient excitation. The correlation function is considered the response of the structure's free vibrations, which is obtained from the following equation:

$$R_{w_f}, w_{fi}(t) = \frac{1}{N-k}\sum_{i=1}^{l-k} w_f(l)w_{fi}(l-k) \tag{11.18}$$

In Eq. (11.18), N is the total number of samples and k is the time delay.

11.3.3 Modal Strain Energy

When a structural element undergoes the action of a force, stress is generated in it. Additionally, this force causes the structural element to deform, and the positions of its different points change relative to the initial state. Modal strain energy is a state in which no external force is applied to the structure, and the structure is in a state of free vibration. By performing dynamic analysis and solving the provided equations, the modal strain energy of each component can be obtained. Damage in a structure typically leads to a reduction in its stiffness and does not have an impact on the mass matrix of the structure. The damage index for each element is defined below:

$$\beta_j = \frac{\sum_{i=1}^{NM}\left(\phi_i^{*T} C_{jo}\phi_i^* + \sum_{i=1}^{NE}\phi_i^{*T} C_{ko}\phi_i^*\right)}{\sum_{i=1}^{NM}\left(\phi_i^T C_{jo}\phi_i + \sum_{i=1}^{NE}\phi_i^T C_{ko}\phi_i\right)} \tag{11.19}$$

In Eq. (11.19), the ϕ_i is the mode shape vector of the structure and C is the stiffness matrix of the structure.

11.3.4 Interference Distribution

The interface distribution method consists of three stages [6]:

1. Vibration data is collected in both the healthy and damaged states.
2. Using the reduced interference distribution (RID) method, time–frequency planes are extracted.
3. In the decision-making stage, the status of the structure is determined using a neural network.

The interface distribution method can process all types of signals, including stationary, non-stationary, linear, and nonlinear signals. This method for analyzing the analytical signal $x(t)$ is defined by the following equations:

$$RID(t,\omega) = \int_{-\infty}^{+\infty} h(\tau)R_x(t,\tau)e^{-i\omega\tau}d\tau \tag{11.20}$$

In Eq. (11.20), the $h(\tau)$ is the time smoothing window and R_x is defined as:

$$R_x(t,\tau) = \int_{\frac{-\tau}{2}}^{\frac{\tau}{2}} \frac{g(v)}{|\tau|}\left(1 + Cos\frac{2\pi v}{\tau}\right) \times x\left(t + v + \frac{\tau}{2}\right)x^*\left(t + v - \frac{\tau}{2}\right)dv \quad (11.21)$$

In Eq. (11.21), the $g(v)$ is the frequency smoothing window. The Hanning window is used to estimate spectra in time–frequency functions. The function of this window is defined as below:

$$Hann(v) = \frac{1}{2} + \frac{1}{2}Cos\left(\frac{2\pi v}{\tau}\right) \quad (11.22)$$

11.4 Damage Detection in Concrete Structures

Damage identification in concrete structures refers to the process of detecting and determining the presence and location of damages or defects resulting from various factors in concrete structural elements such as beams and columns. To identify damage in concrete structural elements, various methods, and techniques are used, including visual inspection, non-destructive testing (such as ultrasonic testing and electrical resistivity testing), vibration analysis, and the use of advanced technologies like smart sensors and neural networks.

11.4.1 Contourlet Transform

This transformation has broad applications in the field of image processing. These methods are based on the Fourier transform. Wavelets are the first and most important method after the Fourier transform and have been proposed to overcome the low accuracy of the Fourier transform. For image representation in high dimensions, a flexible and multiscale, orientation-sensitive method called Contourlet is used. This transformation uses a Laplacian pyramid and an orientation filter bank in the decomposition stage [9]. Initially, the original signal is decomposed by a Laplacian pyramid into an approximation signal, called the low-pass signal, and the details of the signal, called the high-pass signal. Then, by applying an orientation filter bank on the high-pass signal, it provides directional information. The displacements of each block are calculated using the following averages:

$$CF = \sqrt{\frac{1}{M}\sum_{n=1}^{N}\sum_{m=1}^{M}(F(m,n) - F(m - 1,n))^2}$$

$$RF = \sqrt{\frac{1}{M}\sum_{n=1}^{N}\sum_{m=1}^{M}(F(m,n) - F(m,n-1))^2}$$

$$SF = \sqrt{CF^2} + \sqrt{RF^2} \tag{11.23}$$

Then, to determine the fused signal of the merged image, the Contourlet transformation will be applied as below:

$$\begin{cases} F_i = D_i & SF_i^A > SF_i^B + TH \\ F_i = D_{i+8} & SF_i^A > SF_i^B - TH \\ F_i = (D_i + D_{i+8})/2 & Otherwise \end{cases} \tag{11.24}$$

11.4.2 Vibrational Responses

The vibrational responses method aims to identify damages by focusing on time series and the history of vibrational responses. The input of this method consists of acceleration histories of degrees of freedom in two damaged and undamaged states. The extent of damage is determined based on the area under the curve of the acceleration history. This method considers two cases of tensile cracking and concrete softening under compression. The structural relationship of the damage parameters for tensile and compressive damages is expressed below:

$$\sigma_t = (1 - d_t)E_0\left(\epsilon_t - \epsilon_t^{pl}\right)$$

$$\sigma_c = (1 - d_c)E_0\left(\epsilon_c - \epsilon_c^{pl}\right) \tag{11.25}$$

In Eq. (11.25), the E is the ductility coefficient, ε_t is tensile strain, ε_c is compressive strain, d_t is tensile damage index and d_c is the compressive damage index. Upon entering the region of strain softening, the values of damage indices are obtained from the following relationships:

$$d_t = 1 - \frac{\sigma_t}{f_t}$$

$$d_c = 1 - \frac{\sigma_c}{f_c^{`}} \tag{11.26}$$

11.5 Damage Detection Techniques and Results

In the following, we will discuss the models presented in each method and the results obtained from them.

11.5.1 Damage Detection Techniques in Frames

Gudari Amiri et al. [4] utilized the Hilbert method to identify damages in a seven-story two-dimensional frame and a numerical model of a laboratory-based experimental specimen. The proposed algorithm is capable of identifying damages in structures based on the measurement of free vibration acceleration responses of the floors. They concluded that using the Hilbert method, the structural frequency and corresponding mode shape can be extracted. Based on the observations, it was found that the calculated frequencies and mode shapes using the proposed method show a good agreement with the results obtained from finite element software in the undamaged state. Due to the nature of the proposed relationship in determining the curvature of mode shapes, this method has limitations in detecting damages in the connections of beams to columns in the end-story levels. In another similar research work, Gudari Amiri et al. [5] used the signal transmission method to detect damages in two shear wall frames of four and eight stories that were subjected to seismic loading. The main earthquakes were used as the input excitation in the identification of damages caused by the main earthquake. The desired signal on each floor was extracted in undamaged and damaged states. After calculating the damage index for each signal, the amount of damage on each floor was identified. The proposed method accurately estimates the damage in each floor and each mode with an acceptable approximation. In some floors without damage, the damage value is negative, which considering the small and negative values, can be disregarded. The results showed that the proposed method is a powerful tool for processing the relative displacement signals of floors and detecting stiffness changes in the floors due to the presence of damage.

Daneshvar et al. [8] applied vibrational modes to detect damage scenarios in a reinforced concrete frame. They considered vibrational mode shapes as inputs for the proposed damage index. In the numerical finite element models, progressive damages are created by applying loading in multiple stages to the frame. In this method, a reinforced concrete frame was used. The concrete materials in this sample are considered non-linear and capable of reducing stiffness and resistance as damages occur. These damages gradually occur by applying concentrated loads at the mid-span of the frame and are based on the strains induced in the concrete. After each loading stage, modal analysis is performed on the specimen, and the mode shapes of vibration are obtained. Their proposed damage index did not demonstrate the ability to detect damages caused in the first loading stage. However, damages created in the second and third stages were well detected.

Shirzadi and Bitaraf [14] applied the structural control method to damage detection in the numerical model a three-story structure in the form of a shear wall frame with active controllers on all three floors was modeled, and the control forces were obtained. Then, using the proposed coupled model, damage-sensitive features were extracted from the control forces and classified using the second-order differential analysis method. This method correctly classifies all healthy states as healthy and categorizes all damaged states into the damaged categories. It should be noted that the severity and location of damages have not been accurately identified. In this method, the controllers were considered ideal, and the effects of time delay and limited applied force were not considered.

11.5.2 Damage Detection Techniques in Bridges

Rahami et al. [1] used the entropy energy method to identify damages in a two-dimensional model with mass and spring elements, having one degree of freedom in each direction, considered along the horizontal axis. To determine the mass and stiffness, the time period and kinetic energy in each vibration mode were matched with their corresponding values in the actual shear model. In order to monitor the health of the studied platform, the structure is first excited using white Gaussian noise disturbance. In this stage, by employing wavelet packet analysis, the response taken from the decomposed structure is analyzed, and wavelet packet coefficients are calculated at different levels. After reconstructing the signals, the entropy energy, according to the provided equation, is calculated, and the entropy-based damage-sensitive index is determined for each component. The obtained results indicate the favorable performance of this method in damage identification for various patterns. The sensitivity of the desired index shows an ascending trend with increasing damage severity, indicating that the proposed damage index will increase as the damage intensity rises. The proposed method's capabilities include not needing to determine mass and stiffness matrices and only requiring one sensor on the platform deck to record the structural response.

In another research work, Darvishan [2] utilized signal processing techniques to evaluate the health condition of the Yong Bridge, which is one of the first cable-stayed bridges located in China. In this method, only one sensor was used, so the selection of an appropriate location for sensor placement is crucial. The best location is on the points that have the maximum amplitude of vibration. The novel signal processing method, named empirical wavelet transform (EWT) combined with empirical mode decomposition (EMD), was applied in the study. The proposed method exhibits less dispersion in extracting frequencies. The results indicate that using Support Vector Machines (SVM), it is possible to accurately separate damage indices in healthy and damaged states with high accuracy, despite significant oscillations.

Samai et al. [3] applied the modal strain energy method on a three-legged wind turbine platform model with horizontal and flexural restraints to detect its damages. This platform is one of the most common types. After modeling the platform, the eigenvectors and eigenvalues, which represent the mode shapes and natural frequencies of the structure, respectively, were extracted. Then, the natural frequencies are sorted in ascending order, with the smallest frequency representing the first natural frequency of the structure, and its corresponding mode shape is the first mode shape of the structure. The results of the model evaluation indicate that the modal strain energy effectively identifies the location of damages for the single-damage state. In cases where there are multiple damages, it is possible that besides the damaged elements, neighboring undamaged elements may exhibit suspicious behavior, and their damage index may indicate a slightly higher value than one unit. The results showed that the proposed method can detect both small and large damages.

Ahmadi et al. [6] presented the interference distribution method to identify the possible damages of the numerical model based on the Shahid Madani Bridge in Tabriz City. A finite element model is considered with ten identical elements and eleven nodes. The nodes represent the locations of sensors. After processing and computing time–frequency plots, the dynamic characteristics of the system are extracted. Based on the utilized algorithm, before the occurrence of damage, a driving force is applied to the bridge deck, and the response of the bridge at the mid-span of the Shah Teyr is recorded. In this method, only based on the measured response from the bridge structure, its dynamic characteristics are extracted, and the damage is determined. By reducing the number of sensors, the costs of health monitoring software and hardware are significantly reduced. The obtained results demonstrate the satisfactory capability of the proposed method and the accuracy of diagnosing the health condition and location of the damage.

11.5.3 Damage Detection Techniques in Concrete Structures

Kavousi et al. [9] applied the Contourlet transform as a novel signal-processing technique to detect the damage scenario in a prestressed concrete slab model. By applying impacts to each degree of freedom and measuring the response using accelerometers, natural frequencies, and mode shapes of the specimens have been calculated using frequency response functions. For damage detection using the Contourlet transform, the first mode shape of the numerical vibration pattern was obtained for both undamaged and damaged conditions for each type of artificially induced damage. In order to increase sensitivity, the modal deflections or strains were calculated using curve fitting with the spline function. They created three damage scenarios. Then, they considered the difference in modal strains of the first mode of vibration as an input to the Contourlet transform. They used the outputs to identify the location of the damage. The sensitivity of this transformation is higher around the corners and the vicinity of the corners. Moreover, unlike many other methods, it has the ability to detect damage in points close to the corners. Mokhtari et al. [10] used a similar model of the prestressed concrete slab and signal processing of the vibrational responses in the time domain to detect damage scenarios. They have used the plastic damage model for concrete modeling from the ABAQUS software library. Edges of the underside surface of the slab are placed on hinge supports to ensure zero displacements along the coordinate axis. The input to the proposed damage index is the acceleration histories of the degrees of freedom of the pre-stressed concrete slab specimen. With the input force and the output accelerations from the structure, the mechanical properties of the structure can be inferred. The obtained results demonstrate that the proposed method has the capability to identify the location of structural damage in both the middle and corner regions of the pre-stressed concrete slab, and it can be used for damage detection purposes.

11.6 Conclusions

This chapter presented a comprehensive review of damage detection and structural health monitoring techniques utilized to identify the health conditions of various types of structures, including frames, bridges, and the overall concrete structural elements. The reviewed methods included the Hilbert transform, signal transformation, vibrational modes, structural controllers, energy Entropy, signal processing, modal strain energy, interference distribution, Contourlet transform, and vibrational responses. The advantages and disadvantages of each damage detection and structural health monitoring method as well as its applications in identifying damages in frames, bridges, and concrete structural elements were discussed and reviewed.

References

1. Rahami H, Amini Tehrani H, Akhavat M, Ghodrati Amiri G (2016) Damage detection in offshore fixed platforms using concepts of energy entropy in wavelet packet transform. Amirkabir J Civ Environ Eng 48(3):241–248. https://doi.org/10.22060/ceej.2016.597. http://ceej.aut.ac.ir/article_597.html
2. Darvishan E (2020) Low-cost damage detection of cable-stayed bridges using signal processing and machine learning. Amirkabir J Civ Eng 52(1):257–274. https://doi.org/10.22060/ceej.2018.14648.5711
3. Samai SR, Hassanabad MQ, Zahraei AK (2024) Damage identification in offshore wind turbine three-legged platform using modal strain energy method. Struct Anal Earthq 18(3), 1400:51–62. https://doi.org/10.30495/civil.2021.686956
4. Amiri GG, Talebzadeh M, Talebi M, Tabrizian Z (2015) Damage assessment in flexural frame joints using Hilbert-Huang transform. Sharif Imran Eng (Spec Civ Eng Sharif) 32–2(1/1):11–3. SID. https://sid.ir/paper/128225/fa
5. Gholamreza Amiri G, Akhavat M, Amini Tehrani H (2015) Identification of structural damage in shear frames based on signal processing using generalized S transform with mixed window. Sharif Imran Eng (Spec Civ Eng Sharif) 32–2(1/2):19–28. SID. https://sid.ir/paper/128181/fa
6. Ahmadi HR, Allahyari AM, Allahyari HM (2022) Identifying damages in girders of bridges using square time-frequency distribution and neural network. Amirkabir J Civ Eng 54(9):3287–3312. https://doi.org/10.22060/ceej.2022.19812.7262

7. Kazemian M, Jahangir H, Ghalehnovi M (2015) Damage detection of trusses using static response in the presence of excitation. In: The 10th international civil engineering conference, Faculty of Civil Engineering, Tabriz
8. Daneshvar Khorram M, Jahangir H, Ghari Ghuran A (2015) Damage identification in frames using data obtained from dynamic responses. In: The 10th international civil engineering conference, Faculty of Civil Engineering, Tabriz
9. Kavousi M, Khatibinia M, Jahangir H (2018) Using the contourlet transform for damage identification in prestressed concrete beams. In: International conference on civil engineering, architecture, and urban development management in Iran, Tehran, University of Tehran
10. Mokhtari M, Jahangir H, Khatibinia M (2018) Identification of damage in prestressed concrete beams using vibration responses in the time domain. In: International conference on civil engineering, architecture, and urban development management in Iran, Tehran, University of Tehran
11. Jahangir H, Esfahani M, Hasani H (2021) Wavelet-based damage localization and severity estimation of experimental RC beams subjected to gradual static bending tests 34:3055–3069. ISSN 2352-0124. https://doi.org/10.1016/j.istruc.2021.09.059
12. Jahangiri M, Ahmadi Nadushan B (2016) Identifying damage in structures using MOPSO and MOEA/D multi-objective evolutionary optimization algorithms. Civ Eng (Faculty of Engineering) 30(1(series 19)):63–78. SID. https://sid.ir/paper/362101/fa
13. Darvishan E (2021) Damage detection of structures using blind source separation and multifractal detrend fluctuation analysis. Amirkabir J Civ Eng 53(4):1367–1382. https://doi.org/10.22060/ceej.2020.16919.6392
14. Shirzadi H, Bitaraf M (2023) Damage detection of model reference adaptively-controlled structures using control force as a damage sensitive feature. Amirkabir J Civ Eng 55(2):301–318. https://doi.org/10.22060/ceej.2022.20786.7526
15. Jahangir H, Daneshvar M, Esfahani M (2014) Identification of progressive damage in structures using modal vibration data. In: The 4th international conference on acoustics and vibrations, held at the Iran University of Science and Technology, Tehran, Iran
16. Beheshti Aval SB, Ahmadian V, Darvishan E (2018) Damage detection of structures using signal processing and artificial neural networks. J Model Eng 16(52):253–265. https://doi.org/10.22075/jme.2018.2939

Chapter 12
Application of Image Processing Techniques for Analyzing Strength of Concrete Structures—A Novel Approach

Navdeep Mor, Pawan Kumar, Madhu, Charity Aliyinza, and Jamiu Ajibola Busari

Abstract Generally, a myriad of reasons leads to the falling apart of concrete structures over time creating unsafe situations in the environment. The circumvention of this problem requires recurring and constant evaluation involving the utilization of simple and efficient tools for the surveillance of structural health allowing for the early detection of threatening changes that have occurred over time in these structures that may pose as unsafe to the environment using not only empirical data but also data collected in real time. Surveillance of structural health is dynamic for the efficient functioning and complete safety of significant civil structures throughout their stipulated time as well as thereafter. Visual inspection is traditionally way being used for the evaluation of structural wellness, performing a thorough inspection and evaluation, allows for the extraction of enough details that can be linked in an array of strategies when assessing structure integrity. In this study, an attempt has been made to present the applications of image processing techniques for analyzing the strength of concrete structures. The study explains in detail about the Complex,

N. Mor (✉)
Department of Civil Engineering, Guru Jambheshwar University of Science and Technology, Hisar, Haryana, India
e-mail: navdeepmor@gjust.org

P. Kumar
Department of Computer Science and Engineering, Guru Jambheshwar University of Science and Technology, Hisar, Haryana, India
e-mail: pkyadav@gjust.org

Madhu
School of Chemistry, University of Hyderabad, Hyderabad, India

C. Aliyinza
Civil Engineering Department, Pan African University Institute of Basic Sciences Technology and Innovation, Jomo Kenyatta University of Agriculture and Technology, Juja, Kenya

J. A. Busari
Department of Civil Engineering, University of Ilorin, Ilorin, Nigeria

H. Jahangir et al. (eds.), *Damage Detection and Structural Health Monitoring of Concrete and Masonry Structures*, Springer Tracts in Civil Engineering,
https://doi.org/10.1007/978-981-97-8975-7_12

laborious and costly problems in civil engineering primarily related to the strength of concrete structures that could be solved through employment of various image processing methods. Towards the end, it has been concluded that the image processing technique is capable enough to solve complex problems related to the structural health monitoring and accordingly the prevention measures could be taken by various stakeholders associated with this field to improve the utility of the structures.

Keywords Image processing · Structural health monitoring · Strength analysis of structures · Prediction of accuracy · Object recognition · Construction safety

12.1 Introduction

The use of image processing techniques cuts across all fields of science, however, it is frequently employed in the field of civil engineering for dealing with issues that are tedious, costly, and complex. Image processing is the transformation of an image to a digital form that allows for the execution of certain actions enabling the retrieval of relevant information. All structures including but not limited to residential ones, bridges, and tunnels, weaken over time due to a vast range of reasons. One such is fatigue damage caused by recurring traffic loads, environmental influences as well as natural disasters like earthquakes [1]. This, on the one hand, necessitates the need to have constant, periodic evaluations of these structures to ascertain their health and strength statuses.

On the other hand, if the structure under scrutiny has a large surface area like a bridge, then, its thorough monitoring becomes quite challenging. To solve this challenge, continuous monitoring of the structure rather than routine check-ups to get information on its dynamics over time and prove its safety in the environment using real time data is pertinent. Throughout the years, to cut expenses costs, techniques adopted for the non-discrete monitoring of structure and reflexive detection of vulnerability have been established. These methods frequently rely on the non-stopping use of sensors to retrieve data that are considered to reveal a structure's condition of health [2].

Identifying appropriate technology applications for assessing the health and safety of structures is a global concern. Current structural inspection standards, for example, require an inspector to make a trip to the site of a target structure and evaluate on there. Using a robotic system that not only inspects structures but does so many times and yet untiringly and accompanying detect damage that are autonomous in nature, would be a less laborious and more economical replacement for the current monitoring techniques. The use of optical instruments is one of such less laborious and economical methods [1]. Historically, visual inspection techniques have been used by competent engineers to manually identify and categorize structural problems both during and after construction [3]. However, to accurately monitor structural performance, particularly in terms of its long-term performance, more advanced, proficient and unbiased structural examinations are a necessity.

12.1.1 Structural Health Monitoring

Structural health monitoring (SHM) is the combined use of sensors, transmission of data, smart components, processing and ability to undertake computation, throughout structures entailing the use of a damage detection strategy. SHM is a merged framework of linked sensing and actuation, data interrogation, and statistical analysis that aids in the assessment of structural health [4]. The practices involved are critical to the operational security of penitent civil structures not only during the span of life specified but even beyond such time [1]. This process involves monitoring a structure over time using regularly spaced measurements, gathering inputs from these quantifications, and then analyzing these inputs to assess the structural system's current state of health. In recent times, keen observation of a structure's health has been an increasingly significant area of study in the structural engineering domain. Current structural engineering practice encompasses SHM, assessment of structural behavior, health prognosis, and structural sensing [5].

The primary goals of SHM are to monitor the conduct of a structure, assess the behavior of a structure while exposed to more than one service load, undertake detection of impairment, and reveal structure condition [6]. SHM deals with either local health monitoring or global health monitoring. A great deal of health monitoring methods are based on identifying fluctuations in the shape of structural mode, modifications occurring to the initial shape of the structure, or natural frequencies. Fluctuations in natural frequencies and structural mode shape modifications happen due to damage to structural elements [7]. The typical mode of an oscillating system is a motion pattern in which every component moves in a sinusoidal fashion at an equal frequency and with a steady phase relation. The natural frequency of a system is the rate at which it oscillates in its typical state. Previous studies in the monitoring of health discovered that the loss of a single structural component in a structure can cause alterations in even the basic natural frequency from up to thirty percent. It was concluded then by researchers that if the structure was determined statically, then any loss of structural component would inevitably yield an unsteady structure.

12.1.2 Image Processing

Image processing is a technique for tweaking an image to create a better image or draw pertinent details from it. In this type of signal processing, a photograph is used as the input, and the output could be another image, characteristics, or attributes associated with the input image. Image processing implies analysis to improve graphic information for better meaning.

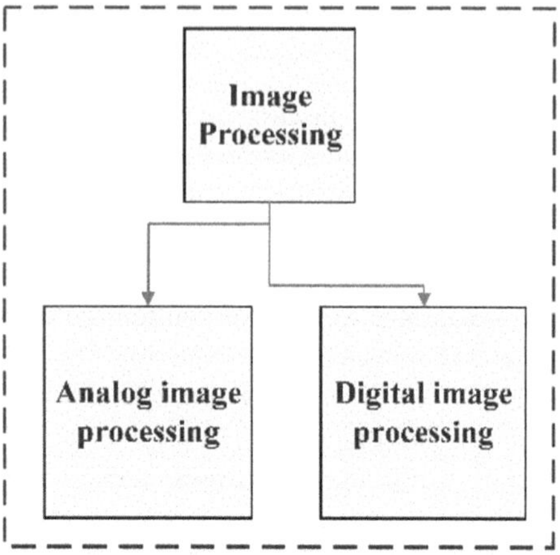

Fig. 12.1 Types of image processing

Basically, image processing comprises the following three steps [8]:

i. Importing an image using image capturing software;
ii. Image analysis and editing;
iii. outcome that can be used to modify the final report or picture through the analysis of the image.

The method of image processing is categorized into two types; analog and digital image processing as presented in Fig. 12.1. This chapter mostly focuses on digital image processing techniques.

12.1.2.1 Analogue Image Processing

It is a processing technique in which images can be altered by varying electrical signals. Analog image processing is done on analogue signals, encompassing two-dimensional analogue signals. Image analysts use an array of interpretational fundamentals when using these visual techniques.

12.1.2.2 Digital Image Processing

Digital image processing is a technique that is used in the manipulation and analysis of digital images through an algorithm. In addition, this technique is used to improve and encode images as well as allowing users to acquire pertinent information and save it in various formats. The advancement of computer network technology, mathematics, and the continuous need for digital image processing technology from a wide range of societal sectors have all led to the advancement of digital image processing technology, thereby creating new avenues for its progression.

12.1.2.3 Digital Image Processing Benefits

- Improves image edges by making them appear sharper. The second image appears cleaner and more appealing upon closer inspection. Sharpening, which is an important part of printing, is usually done to make a picture look appealing on the printed page.
- Removing noise from an image that could interfere with data transmission. The effects of different electrical components on data traveling through them may be undesirable.
- Removes image motion blur. Motion blur can take place when the speed of the camera's shutter is too slow in comparison to the speed of the object.
- It is helpful when obtaining image edges. It can be necessary for the quantification of structural parts. An edge detection algorithm can be used as the first step in edge enhancement.
- Helps in removing details from an image. It can be used for measuring and counting purposes.

12.1.2.4 Analog Image Processing Versus Digital Image Processing

Analog and digital image processing has been broadly described in Table 12.1.

12.1.3 Methods of Processing Image

Several methods of processing images employ pictures as input and even also as output. Other methods will take pictures as input but the resultant output would be the attributes of the picture. It is not mandatory to employ all these methods

Table 12.1 Analog image processing versus digital image processing

Analog image processing	Digital image processing
It is limited to handling two-dimensional signals when applied to analog signals	It is used with digital signals for picture analysis and manipulation
Because of the time-varying inputs, the pictures produced by analog image processing change	It enhances the image's digital clarity and has a precise intensity distribution
Processing analog images is slower and more expensive	The digital method of processing images is a more efficient and cost-effective way of maintaining and accessing pictures
It is commonly non-discrete and not disintegrated into small pieces	It employs the approach of segmenting image approach to detect discontinuity caused by a broken connecting route
It is limited to optical principles generated and cannot be processed as desired	It can be applied to any operation, be it linear or non-linear

at one time when it comes to image processing [9]. The methods selected use the information on the application in question. The following make up the main details of various methods used for the processing of images.

12.1.3.1 Acquisition of Image

The importance of the acquisition of images in picture processing cannot be downplayed. Acquiring the image is essentially the process of collecting images from multiple sources. Even with the availability of various image enhancement techniques, if the pictures are not properly collected, the various methods used for processing pictures may not be as effective as they should otherwise be. The most common approach for acquiring images is real-time acquisition. This method generates a pool of files that are automatically processed.

12.1.3.2 Pre-processing of Image

The main aim of pre-processing a picture is to enhance the image data through the enhancement of some features while repressing others. Recordings of image data may be acquired with some errors in pixel shape and settings with regards to the brightness. These errors are fixed during image pre-processing through the use of suitable mathematical models, which can be either statistical or definite. Pre-processing of image also consists of basic operations such as noise reduction, picture sharpening and smoothing, enhancement of contrast, and advanced operations such as segmentation of image.

12.1.4 Image Enhancement

Due to the limits of imaging subsystems and lighting circumstances, some of the captured photographs from conventional digital cameras and satellite shots may be overly dark or too bright. One of the most straightforward methods to get around this problem is image enhancement. It improves some hidden features or highlights certain characteristics that are noteworthy for succeeding analysis of the image. This technique can either be methods of spatial domain or frequency domain. The technique of spatial domain modifies or aggregates the pixels that make up the image, whereas the technique of frequency domain enhances the picture linearly by setting the invariant operator.

12.1.5 Image Restoration

Image restoration is the process of improving the appearance of a corrupted or degraded image by removing noise or blur. Prior information on the disturbance or noise causing the picture to degrade is required for the restoration of that image. Image restoration is carried out in a spatial or frequency domain. The spatial domain contains the process of filtering, which is executed directly on the picture's operating pixel whereas that of the frequency domain is executed by a systematic mapping of the spatial domain right into the frequency domain using the transformation of Fourier. After filtering, the picture is remapped also into a spatial domain using Fourier transform but in this case, the inverse, to get the picture that has been restored. Some of the most significant findings where image restoration has been done to detect image cracks are presented in Table 12.2.

12.1.5.1 Morphological Processing

It is a group of linear procedures used to separate picture elements that help in representing and describing shapes. Dilation, erosion and their combinations make up the fundamental morphological activities. Structures of a specific size and shape that are provided by the structuring element may be removed using erosion, whilst holes of a specific size and shape can be filled using dilation.

Image processing should be used in conjunction with corrosion and expansion. The target image is first corroded followed by its expansion in an open operation. The target image is expanded after being corroded in a closed operation. The linear boundary of the fracture is obtained by morphologically processing the fracture image. Some of the recent significant research carried out for crack detection in structures using the Morphological Processing technique has been presented in Table 12.3.

Table 12.2 Significant research on cracks detection in concrete structures using image restoration

Name of author and year	Technique used	Dataset used	Remarks/conclusion
Khan et al. (2023) [10]	Image restoration	Critical review	Deep learning based image restoration technique showed accurate results in detection of cracks
Vivekananthan et al. (2023) [11]	Image processing method	Various dataset	Results related to cracks in structure were accurate using this technique
Reghukumar and Anarbasi (2020) [12]	Image processing and deep learning	40,000 images	Vision-based crack detection methodology showed almost exact information
Hoang (2018) [13]	Shape analysis algorithms	Various dataset	Precise results in the prediction of cracks

Table 12.3 Significant research on cracks detection in structure using morphological processing

Name of author and year	Technique used	Dataset used	Remarks/conclusion
Jin et al. (2023) [14]	Morphological processing	Critical review	Mean average precision showed almost real results with high accuracy of results in the detection of cracks
Licun et al. (2022) [15]	Deep learning-assisted image processing method	Various dataset	Bounding box crack detection methodology showed almost exact information
Suliman et al. (2021) [16]	Shape analysis algorithms	Various dataset	The morphological processing algorithm showed better results in the prediction of cracks
Shivaprasad et al. (2015) [17]	Image processing morphological method	Various dataset	Surface cracks were precisely detected using the proposed edge detection technique

12.1.5.2 Image Segmentation

This process involves the division of a digital image into segments for its simplification as well as its transformation from one form to another which is more meaningful and understandable. The process may employ single methods like thresholding, statistical classification, detection of region and edge, or an amalgamation of these methods. Image segmentation can be region-based or edge-based. The segmentation relies on the grouping of regions based on their anatomical or functional responsibilities, unlike previous algorithms that relied on a similarity of pattern levels in terms of intensity within a cluster of surrounding pixels. Due to the reliance on methods that are based on the edge of discontinuities seen in the image data amid discrete areas, the segmentation methodology seeks to precisely determine the border separating these segments. Some of the recent significant research carried out for crack detection in structures using Image Segmentation technique has been presented in Table 12.4.

12.1.6 Object Recognition

It is a method of labeling an object in a video or digital picture through the use of its descriptors. Recognition of an object is the process of teaching a computer to recognize a specific object from different viewpoints, lighting conditions, and backgrounds. Scene clutter, photometric effects, shape changes, and object viewpoints can all have an impact on how an object looks.

Table 12.4 Significant research on the detection of cracks in structure using image segmentation

Name of author and year	Technique used	Dataset used	Remarks/conclusion
Mostafa et al. (2023) [18]	Morphological processing	Various structures	crack inspection done using Image Segmentation and neural networks showed best results in the prediction of cracks
Taehee et al. (2023) [19]	Deep learning-AI model	1235 images	Proposed model was efficient in detecting cracks in concrete structures
Jordan et al. (2022) [2]	Digital image correlation	Various dataset	Image Segmentation helped in understanding different properties of structures
Li et al. (2015) [20]	Digital image segmentation	Various dataset	Bayesian segmentation helped in better Understanding of structural properties

12.1.7 Strength Analysis of Structures

Structural analysis is a study that relates to the external loads that are exerted on the structure and analysis of the deformation and internal response of the different parts of the structure. The aim for structural strength analysis is to determine how the structure will respond to internal or exterior static loads with boundary conditions. Strength analyses can be conducted to a complete structure, a local part that defines the overall structure, or a specific area of a structure that has to be looked at. In terms of boundary conditions, analyses for the complete structure are unquestionably the most trustworthy. However, due to limits on computer resources, modeling time, and analysis step length, analyzing a component of the structure that characterizes the entire structure or a necessary local element of the structure is the most common approach used.

The analysis must be orthodox in the sense that the expected failure load must be lower than the maximum load that the structure can carry. This is pertinent due to variations in applied loads and the properties of materials within normal boundaries, as well as unpredictability in the analytical assumptions. The idea of an overall safety factor (SF) is made known for the accounting of many unpredictability factors, and there is an increment of the limit loads in correspondence to the SF (Ultimate Load = SF x Limit Load). It is critical to have clear in mind all the possibilities of failure modes and inculcate them in the design [14]. The most common modes of failure for civil engineering structures are:

(a) Absolute failure and disintegration caused by stresses exceeding the material's maximum capacity
(b) Unfavorable yielding induced by strains greater than the material's strength of yield, which alters structural integrity or performance
(c) Instability (buckling) caused by an amalgamation of deformations, loads, and part geometry like the structure collapses before the limit of material strength is met
(d) Fatigue of structural elements or the structure as a whole caused by crack in the material
(e) Material failure modes influenced by time which include stress rupture, creep, thermal fatigue, and stress corrosion.
(f) Creep of structural members that cause strains and deformations due to either elevated temperatures or constant stress.

The primary objective of structural health monitoring (SHM) technology was to ensure the operational safety of engineering structures. This was achieved by implementing multiple kinds of sensors, monitoring a variety of physical quantities, assessing the structural condition and performance, as well as offering guidance for regular examination and maintenance. Cutting-edge sensing technologies from various fields, including mechanics, optics, chemistry, electricity, electromagnetism, and thermology, contribute significantly to the accurate collection of tremendous quantities of original data reflecting really structural and environmental variables in a large-scale SHM system [17]. Monitoring and sensing using Machine vision technology has come to prominence as a cutting-edge research topic, attracting growing interest from the civil engineering community, as a result of exceptional advancements in computer science and optics technology. Significant research done by various researchers using image processing has been presented in Table 12.5.

12.1.8 Purpose of Image Processing Techniques for SHM

The role of image processing techniques to not only evaluate but also monitor the strength of a structure is presented in Fig. 12.2.

(a) **To identify and discover**: this technique is used to identifying and discover defects and anomalies of the structure this can be reinforcement corrosion, and cracks among others [15]. Developed a vision-based approach to find concrete fractures without computing the fault attributes, utilizing a deep architecture of convolutional neural networks. The results showed that the proposed method portrayed better performance and could indeed be used to find concrete cracks in realistic situations.
(b) **To measure and monitor**: A proper approach to monitoring structure enables the long-term assessment of structural performance, maintenance planning, and occupant safety. Long-term monitoring of high-rise buildings connected by link

Table 12.5 Significant research on SHM using image processing

Name of author and year	Technique used	Dataset used	Remarks/conclusion
Guo and Vladimir (2012) [21]	Ultrasound-stimulated IR processing	Single building	2D and 3D image simulation showed accurate results
Hamrat et al. (2016) [22]	Digital image correlation	3 type of concrete	Results related to cracks in structure were accurate
Heideklang and Pariasa (2015) [23]	various signal processing methods	Concrete structures	Near-surface defects could be detected very precisely
Sokratis et al. (2015) [24]	Digital Image processing	Single concrete building	Precise results in prediction of cracks
Mohammad and Marsi (2012) [25]	Robotic system of 2D and 3D images	Concrete buildings	Vision-based crack detection methodology was significant
Shahid (2010) [26]	ANN and image processing	10 year concrete block	Highly accurate results using image processing technique
Xueqin et al. (2014) [27]	Fast discrete curvelet transform analysis	Multiple tiles	Crack defects were significantly observed on magnetic tile using contours of the textures
Zakrya et al. (2023) [28]	SHM using image processing	Concrete structures	Image processing algorithms revealed almost accurate results
Jianghua et al. (2022) [29]	Reviewed existing techniques	Review	Concluded that image processing techniques is precise technique

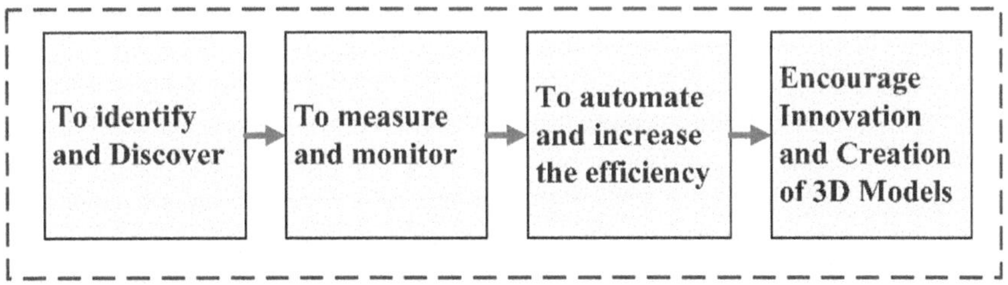

Fig. 12.2 Purpose of image processing techniques

bridges [16]. Barros in the year 2021 used 3D digital image correlation to evaluate the deflection caused by the static load on the pedestrian bridge. They used cameras installed by windows on different floors of an adjacent building [30].

(c) **To automate while efficiency**: is also increased; automation of structural health monitoring is gaining popularity in the civil engineering world. The vision-based strategy, on the other hand, has significant demerits. Deep learning is being used to solve most in several departments to solve problems. Deep learning refers to machine learning approaches that use hidden layers in artificial neural networks to increase performance [31, 32] employing a convolutional neural network approach for the quasi-real-time image and video processing to detect and locate forms of damage.

(d) **Encourage innovation and the creation of 3D models**: structural health monitoring systems are crucial for identifying aberrant behavior and removing or reducing its impacts. For the non-contact distant measurement of structural reactions, camera and computer vision-based sensors have shown promise according to research [33]. Crack identification on the pixel level is possible using a number of techniques. Structural flaws may be quickly and precisely detected using CNN-based frameworks. An image-based rendering technique has been created for the creation of three-dimensional models of the structures in order to assess damage and carry out non-contact, optical-based measures for catastrophe prevention.

12.1.9 Picture Processing Techniques Used for Strength Analysis of Structures

Structures have exceptional maintenance, conservation, and restoration requirements. This calls for a mandatory detailed diagnosis and documentation of the conservation, restoration, and rehabilitation works. Finding the right technological methods to evaluate the health and safety of structures is a major problem everywhere in the globe. A robotic system might take the role of the present conventional monitoring methods, examining structures more frequently and performing autonomous damage diagnosis more quickly and affordably. The growth of imaging technology in equal to growth of computer technology has improved the use of image production techniques as evaluation and observation methods to test structural output.

Image production procedures are conveniently employed in civil engineering to evaluate hard-to-assess, time-taking, and costly engineering issues. Researchers in civil engineering have always made used of these procedures in the study of structural engineering, concrete technology, and building materials [34]. This procedure is to make a thorough inspection to announce information necessary to make different connections to examine structural honesty, which gives numerous subsets to improve picture assessment applications.

12.2 Structural Repair and Maintenance

Any built building will eventually need maintenance to prolong its useful life. The focus of this period of time is on building upkeep. Checking, creating reports, and, if necessary, determining what measures to take are some of the infrastructure problems for building and sustaining the work. As a result, more personnel need to be involved in the maintenance phase. The vital task is determining the present status of the structure and essential maintenance for significant civil constructions like bridges, high-rise buildings, highways, etc. The requirement for maintenance may result from the need to: check for structural cracks; monitor and analyze pavement cracks; or conduct real-time structural monitoring, particularly for large-scale civil engineering projects.

The number of engineering constructions has expanded due to industrial progress. As a result of the uncomplicated and correctness of image-based structural health monitoring techniques, the common dividend among researchers in growing more easy techniques is developing all day. Present research on bridges, dams, and edifice structures has also revealed a precise examination of the service lifespan and late condition that is required to increase the structure's durability. Image-based evaluation is more efficient than traditional examination in this situation [35].

A genuine and long-term remote monitoring system was developed by Fukuda and Vasileva for the reactions of large civil constructions [36, 37]. Employed the target panel's structure to transmit the network as a result of movement recorded by an attached camera. The program for digital image processing extracted these signals in dynamic reactions to calculate dislocation in real-time when employing a gate array that is field-programmable, for high yield. A total station assisted with images for the computation of absolute 3D coordinates and structural natural frequencies examinations, as well as a dynamic deformation monitoring system for 3D motions was built in the year 2021 [38]. However, sensors like accelerometers are employed to infer the vibration behavior of the building. The technique involves using IATS and fixed, labelled targets. Real-time video broadcasts were enabled via the total station telescopic camera. The total station's electronic distance measurement (EDM) provides accuracy during the conversion of image loading to length sizes. It was able to obtain accuracy far from 0.05 mm for static applications and more than 0.2 mm for applications that are dynamic.

12.3 Assessment of Structural Cracks

Crack location, pattern, and identification are the first steps in any structural examination. Cycle load, fatigue stress, and tensile stress are three factors that can cause fractures in buildings, particularly those made of concrete. Engineering constructions' fractures lower local stiffness and result in material discontinuity. Extra care should be taken on cracks in the concrete surface because surface cracks are very important indicators of structural dilapidation [39]. However, the present assessment methods rely on people, which increases execution costs and inaccuracies. Detecting Crack through processing of image is a procedure of detecting surface crack, which particularly employs algorithms that are based on image loading to analyze cracks against the surface of engineering structures surfaces. The loading of image can adequately examine to identify crack properties including crack width, length, and area [40].

A laser scanning with image processing techniques was developed by Valenca to inspect bridge flaws [41]. They employed MCrack-TLS, which combines the capabilities of terrestrial laser scanning (TLS) with image processing for crack discovery (MCrack). According to the location and kind of structure, photos are acquired using various tools and advance laser scanning tools in ortho-rectifying the photos for loading. The author introduced a model based on methods of loading images for automated crack analysis and detection. In this current model, the picture thresholded by the Otsu approach is preprocessed using a method of gray intensity adjustment known as Min–Max Gray Level Discrimination (M2GLD). The usefulness of the gray intensity remodeling technique is to enhance the outcomes of displacement identification in terms of accuracy. M2GLD integrated with the Otsu technique, then more shape differentiating algorithms, may effectively detect fracture in digital pictures, according to experimental results.

Vijayan and Geethalakshmi [42] (2018) opined a simplified architecture image processing-based fracture identification. Creation of the initial data set or collection of images is the first stage of their suggested architecture. After that, preprocessing techniques including filtering and smoothing were applied to the database's photos. Image processing and fracture identification were performed in a single step. Here, processing strategies including statistical methods, thresholding techniques, and Otsu thresholding are applied. Finally, convolutional neural network (CNN) or Fuzzy-based algorithms are purposely useful in analyzing the identified images. The summary of all steps has been presented in Fig. 12.3.

Ren and others [43] in the year 2020 recommended a fully convolutional neural network that has been optimized for fracture detection in tunnels. The network's moniker was Crack Seg Net. Encoder, decoder route, and 3×3 convolutions were all used in the neural network's modular construction, which was then followed by a 2×2 max pooling layer. A new, bigger dataset of 919 photos was created by adding to the original dataset's 409 photographs. For training purposes, simplification of the dataset into a training set to validation set at 4 ratio 1. Prior to being employed in the detection procedure, the original RGB photos were transformed into binary and grayscale versions. Before being introduced into the network, they were lastly treated to noise reduction. The photos with just cracks were returned by the crack segmentation method.

Sankarasrinivasan [44] in the year 2015 proposed a novel approach which combines hat reformation and overall appearance, saturation, and value thresholding techniques for crack detection. UAVs have an image-capturing system, and the recorded information is manually transformed. Although there is practicability in this procedure, accurately collecting and interpreting the collected data requires a lot

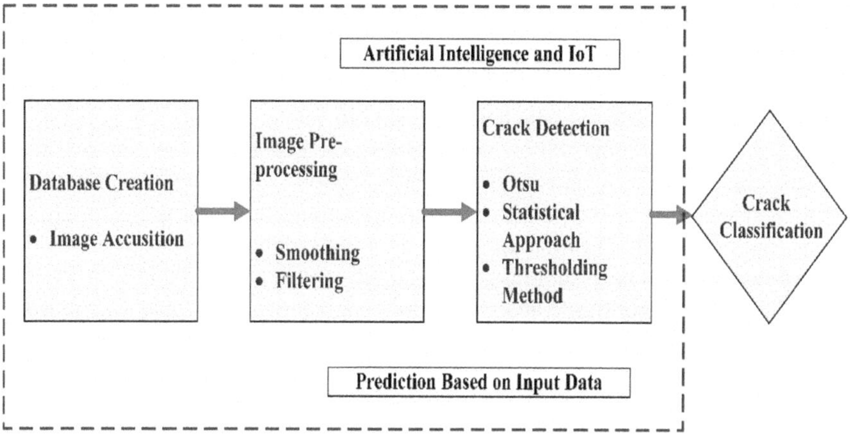

Fig. 12.3 Architecture of fracture identification through image loading [43]

of work. Meanwhile, the evaluation devices are installed on UAVs, real-time information transmission and ground station inspection are made possible. With the process, interfacing in real-time with systems of image loading can enhance the loading of many pictures which can effectively accelerate the structural outcome process.

12.3.1 Other Image Loading Procedures Being Used to Acquire the Strength of Structures

Image Loading Technology and Manual Neural Networks; Concrete is the most important structure material that researchers have studied for a long. The influence of its features on its mechanical performance and the generation of its mechanical performance theories are done regularly to check the authenticity of the results [45]. Propound the use of image loading with artificial neural networks to access the mechanical characteristics of concrete such as concrete durability, elasticity module, and cracks. The observation was made in the data got from the research experiment that the concrete's elasticity module developed as its strength in compression went up. The concrete's deformation value decreased as its durability in compression grew. In this study, estimation of the crack value of each sample was done directly when the failure occurred. Infrared Building Analysis Using Image Segmentation and Retina Net; due to its capacity to identify a variety of structural flaws, thermography is being utilized in building inspection more frequently since doing so permits the formulation of improvement strategies for effective energy use. Thermography is an appropriate tool for energy audits in the building sectors to enhance indoor comfort and construction resistance. Infrared (IR) inspections can identify early-stage insulation flaws like heat losses, air leakage, and moisture that are not visible and are challenging to locate [46]. Object identification and image subsystem allow to observe objects that are useful for the enhancement of the automatic estimation of thermal building character. IR imaging makes it possible to identify system that happen on object bases which are opposite invisible, typically by noticing the contrast between the impacted elements. Unexpected or undesirable thermal reactions may be indicated by unusual color representations. Different anomalies can be localized depending on the process's visual output and the color scales used [47]. In a nutshell, the abstract of the applications of algorithms and tools in the field of image processing has been summarized in the subsequent tables. Table 12.6 presents the application of algorithmic analysis in civil engineering and Table 12.7 presents the applications of tools for image processing in the domain of civil engineering.

Table 12.6 Application of algorithmic analysis in civil engineering

Researchers	Application	Algorithm	Comment
Laefer (2020) and Kim et al. (2006) [34]	Pavement crack classification via intelligent approach	Machine learning (ML) and processing of image algorithms	Too used for image processing: steerable filter, nonlocal means, projective integral, and image thresholding ML tool: artificial bee colony used to optimize a support vector machine
Yu et al. (2020) [48]	Algorithm for crack detection in the skeleton of the crack	Skeleton extraction algorithm proposed by Zhang-Suen	Gamma picture improvement, automated thresholding, and filtering by OTSU approach
Zhao and Niu (2012) [49]	Evaluation of damage via tunnel crack	A fuzzy set is employed for the segmentation of the image threshold segmentation	Cracks are used as input for the picture analysis
Coster and Chermant (2001) [50]	Image analysis of materials in civil engineering	Probabilistic models: polyphaser and tessellation	picture capture, binary image processing, segmentation, and feature extraction
Fukuda et al. (2010) [39]	Using displacement data, to develop a monitoring system for large civil constructions	Image processing target panel identification algorithm	Geometric analysis of the target panel using the panel's white spot placement
McCormick and Lord (2012) [51]	Deformation of structure measurement	processing two photos under various loading conditions	cameras with high resolution to evaluate displacement

Table 12.7 Applications of tools for image processing in the domain of civil engineering

Researchers	Tools and Methods	Area of application	Essence of study
Leu and Chang (2005) [52]	Slice dice, Borland C++	Borland C++ created a subsystem for the processing of images, while Slice dicer incorporated not just 3D visualization but also 3D reconstruction into the general system	The technique may efficiently direct aspirants in depicting features that are geologically unearthed while tunnel excavation is ongoing
Fukuda et al. (2010) [53]	A notebook computer and digital camcorder coupled alongside DirectShow technology (a product of Microsoft corporation) via the programming language C++	Modules for processing images were set up through the use of DirectShow software. The module aids in the detection of geometric patterns and this is done by the use of trigonometric transforms	The system is able to measure responses that are dynamic in real time and is suitable for the monitoring of civil engineering structures
McCormick and Lord (2012) [54]	ARAMIS software	To further focus the deformation measurement in subpixel interpolation, the Newton Raphson method and 2D spline fitting were used	Civil structure full-field deformation measurements using DIC are efficient and can be evaluated at multiple sensor locations
Liu et al. (2014) [55]	Unmanned aerial vehicle (UAV), ground controlling station (GCS)	Aerial perspective data is collected by UAV, and simulation is carried out using GCS software	Potential uses for UAV in mapping, geometric analysis, and civil engineering
Bidgoli et al. (2020) [56]	MATLAB (version: R2014)	Programming using MATLAB employed in calculating the proportion of pixels existing for every surface fracture	Damage research in the midst of aggregate and boinder was carried out using a registration image-processing technique

12.4 Conclusion

The idea of SHM has evolved and is regarded as one of the most important technologies for ensuring the health monitoring of various structures. A significant occurrence can result in perpetual losses, so it is critical to be well informed in advance. Lifeline infrastructure, such as hospitals, bridges, tunnels, and other buildings, should be monitored because their failure results in the most severe damages and this is where structural health monitoring comes in. Monitoring structural health is essential for all structures to keep track of their performance and lifespan. It provides a way to overcome these flaws with minimal damage and loss. Image-based monitoring for structural monitoring has previously been demonstrated. Since measurements can be taken with no contact made on the structure, it is useful for tough measuring assignments where is seen no solid basis for employing displacement gauges. The only limitation to a camera sensor measuring an extremely large number of points is sensor resolution. This allows for the tracking of intricate deformation patterns. Many camera sensors can detect three-dimensional deformation. The study strongly recommends the use of image processing technique for structural health monitoring of concrete structures as present study explored all the major contribution of image processing technique that could be used to detect the cracks and other relevant features of concrete structure and using the analytical observation out of it, various preventive measures can be taken to ensure the better utility of concrete structures. As most of the aspects of the technique have been explored, this study will help the people working in the field of structural health monitoring from a different point of view and for the betterment of society. The study recommends the use of image processing techniques for getting the most desirable information out of concrete structures. It will reduce human efforts as well as human error and improve the efficiency of monitoring the health of any concrete structure.

References

1. Raksha M (2010) A review on–structural health monitoring and image processing. Int J Sci Res Eng Dev
2. Curt J et al (2022) An algorithm for structural health monitoring by digital image correlation: proof of concept and case study. Opt Lasers Eng 151:106842
3. Kumar A, Mor N (2021) An approach-driven: Use of artificial intelligence and its applications in civil engineering. In: Artificial intelligence and IoT: smart convergence for eco-friendly topography, pp 201–221
4. Nandal M, Mor N, Sood H (2020) An overview of use of artificial neural network in sustainable transport system. In: Computational methods and data engineering: proceedings of ICMDE, vol 1, pp 83–91
5. Jahanshahi MR, Masri SF (2012) Adaptive vision-based crack detection using 3D scene reconstruction for condition assessment of structures. Autom Constr 22:567–576
6. Roshan P et al (2015) Review paper on structural health monitoring: its benefit and scope in India. J Civ Eng Environ Technol 2(2):109–112
7. Mor N, Sood H, Goyal T (2020) A statistical model for prediction of road accidents in the State of Punjab. J Interdisc Math 23(1):229–236
8. Kumar A, Mor N (2020) Prediction of accuracy of high-strength concrete using data mining technique: a review. In: Proceedings of international conference on IoT inclusive life (ICIIL 2019), NITTTR Chandigarh, India. Springer Singapore
9. Joseph S (2018) Image processing techniques and its applications: an overview. Int J Adv Res Innov Ideas Educ (IJARIIE) 4(3):2168–2174
10. Khan MAM et al (2023) Image processing techniques for concrete crack detection: a scientometrics literature review. Remote Sens 15(9):2400

11. Vivekananthan V et al (2023) Concrete bridge crack detection by image processing technique by using the improved OTSU method. Mater Today Proc 74:1002–1007
12. Reghukumar A, Anbarasi LJ (2020) Crack detection in concrete structures using image processing and deep learning. In: International conference on advances in electrical and computer technologies. Springer Nature, Singapore
13. Hoang ND (2018) Detection of surface crack in building structures using image processing technique with an improved Otsu method for image thresholding. Adv Civ Eng
14. Jin Q et al (2023) A deep learning and morphological method for concrete cracks detection. J Circ Syst Comput:2350271
15. Yu L et al (2022) Intelligent crack detection and quantification in the concrete bridge: A deep learning-assisted image processing approach. Adv Civ Eng
16. Munawar HS et al (2021) Image-based crack detection methods: a review. Infrastructures 6(8):115
17. Shivaprasad K, Vishwanath MK, Narasimha K (2015) Morphology based surface crack detection. J Adv Res Sci 1(1):15–20
18. Iraniparast M et al (2023) Surface concrete cracks detection and segmentation using transfer learning and multi-resolution image processing. Structures 54. Elsevier
19. Lee T et al (2023) Improvement of concrete crack segmentation performance using stacking ensemble learning. Appl Sci 13(4):2367
20. Li G et al (2015) Bayesian based damage segmentation for structural health monitoring. In: Smart materials and structures (submitted 2014)
21. Guo X, Vavilov V (2013) Crack detection in aluminum parts by using ultrasound-excited infrared thermography. Infrared Phys Technol 61:149–156
22. Hamrat M et al (2016) Flexural cracking behavior of normal strength, high strength and high strength fiber concrete beams, using Digital Image Correlation technique. Constr Build Mater 106:678–692
23. Heideklang R, Shokouhi P (2015) Multi-sensor image fusion at signal level for improved near-surface crack detection. NDT & E Int 71:16–22
24. Iliopoulos S et al (2015) Detection and evaluation of cracks in the concrete buffer of the Belgian nuclear waste container using combined NDT techniques. Constr Build Mater 78:369–378
25. Sangwan A, Mor N. A critical investigation for improving the ductile behaviour of concrete structures using different materials
26. Kabir S (2010) Imaging-based detection of AAR induced map-crack damage in concrete structure. NDT & E Int 43(6):461–469
27. Li X, Jiang H, Yin G (2014) Detection of surface crack defects on ferrite magnetic tile. NDT and E Int 62:6–13
28. Azouz Z, Asli BHS, Khan M (2023) Evolution of crack analysis in structures using image processing technique: a review. Electronics 12(18):3862
29. Deng J et al (2022) Review on computer vision-based crack detection and quantification methodologies for civil structures. Constr Build Mater 356:129238
30. Punia A, Mor N (2021) For a better tomorrow: smart and sustainable cities using artificial intelligence. In: Handbook of green engineering technologies for sustainable smart cities. CRC Press, pp 173–189
31. Barros F et al (2022) Displacement monitoring of a pedestrian bridge using 3D digital image correlation. Procedia Struct Integrity 37:880–887
32. Cha Y-J, Choi W, Büyüköztürk O (2017) Deep learning-based crack damage detection using convolutional neural networks. Comput Aided Civ Infrastruct Eng 32(5):361–378
33. Roussel M et al (2014) Long-term monitoring of high-rise buildings connected by link bridges. J Civ Struct Health Monit 4:57–67
34. Laefer DF (2019) Harnessing remote sensing for civil engineering: then, now, and tomorrow. In: Applications of geomatics in civil engineering: select proceedings of ICGCE 2018. Springer, Singapore
35. Payawal JMG, Kim DK (2023) Image-based structural health monitoring: a systematic review. Appl Sci 13(2):968

36. Chen ZQ, Hutchinson TC (2007) Urban damage estimation using statistical processing of satellite images. J Comput Civ Eng 21(3):187–199
37. Coster M, Chermant J-L (2001) Image analysis and mathematical morphology for civil engineering materials. Cement Concr Compos 23(2–3):133–151
38. Garber D, Shahrokhinasab E (2019) Performance comparison of in-service, full-depth precast concrete deck panels to cast-in-place decks No. ABC-UTC-2013-C3-FIU03-Final. Accelerated Bridge Construction University Transportation Center (ABC-UTC)
39. Fukuda Y, Feng MQ, Shinozuka M (2010) Cost-effective vision-based system for monitoring dynamic response of civil engineering structures. Struct Control Health Monit 17(8):918–936
40. Vasileva AV, Vasilev AS, Konyakhin IA (2018) Vision-based system for long-term remote monitoring of large civil engineering structures: design, testing, evaluation. Meas Sci Technol 29(11):115003
41. Başyiğit C, Çomak B, Kilinçarslan S (2012) A new approach to compressive strength assessment of concrete: image processing technique. In: AIP conference proceedings, vol 1476(1). American Institute of Physics
42. Zawad MRS et al (2021) A comparative review of image processing based crack detection techniques on civil engineering structures. J Soft Comput Civ Eng 5(3):58–74
43. Kim H et al (2017) Comparative analysis of image binarization methods for crack identification in concrete structures. Cem Concr Res 99:53–61
44. Valença J et al (2017) Assessment of cracks on concrete bridges using image processing supported by laser scanning survey. Constr Build Mater 146:668–678
45. Geethalakshmi SN (2018) A survey on crack detection using image processing techniques and deep learning algorithms. Int J Pure Appl Math 118(8):215–220
46. Ren Y et al (2020) Image-based concrete crack detection in tunnels using deep fully convolutional networks. Constr Build Mater 234:117367
47. Sankarasrinivasan S et al (2015) Health monitoring of civil structures with integrated UAV and image processing system. Procedia Comput Sci 54:508–515
48. Yu X et al (2020) Crack detection algorithm of complex bridge based on image process. In: CICTP 2020, pp 1341–1353
49. Zhao L, Niu L (2012) Study on key techniques of image processing and automatic recognition of tunnel cracks. In: ICLEM 2012: logistics for sustained economic development—technology and management for efficiency, pp 427–433
50. Gaich A, Pischinger G (2016) 3D images for digital geological mapping: focussing on conventional tunnelling. Geomech Tunn 9(1):45–51
51. Walter I, Tanasković M, Stanković M (2022) IR building analysis with extraction of elements using image segmentation and RetinaNet. Buildings 13(1):109
52. Dogan G, Arslan MK, Ceylan M (2017) Concrete compressive strength detection using image processing based new test method. Measurement 109 (2017): 137–148
53. Hedayat A, Ashur S (2015) Digital image correlation and its application in an undergraduate civil engineering materials laboratory. US: Fort Wayne
54. Anding K, Garten D, Linß E (2013) Application of intelligent image processing in the construction material industry. Acta Imeko 2(1):61–73
55. Leu S-S, Chang S-L (2005) Digital image processing based approach for tunnel excavation faces. Autom Constr 14(6):750–765
56. Arbabpour Bidgoli M et al (2020) Introducing adhesion–cohesion index to evaluate moisture susceptibility of asphalt mixtures using a registration image-processing method. J Mater Civ Eng 32(12):04020376

Part IV
Advanced Technologies and Applications

Chapter 13
Structural Health Monitoring: Glass Recycle Material

Rakesh Choudhary and P. V. Ramana

Abstract Structural Health Monitoring (SHM) is crucial for ensuring the safety and reliability of infrastructure by continuously assessing structural integrity and detecting potential damage. This chapter explores the application of SHM to concrete specimens reinforced with glass powder, a sustainable building material. Through comprehensive testing and analysis, the study aims to evaluate the effect of glass powder on concrete properties and assess its suitability for practical use. The research contributes to environmentally friendly construction practices and infrastructure development, aligning with principles of sustainability and responsible engineering. The study also discusses challenges associated with SHM, highlighting the need for accurate sensor placement, effective data management, and advanced data analysis techniques. Additionally, the chapter introduces Digital Image Correlation (DIC) as a cutting-edge technique within SHM for measuring and analysing full-field displacements and strains. DIC offers non-contact, high-precision measurements suitable for various materials and loading conditions. While DIC provides significant insights into material characteristics and structural performance, challenges such as calibration complexity and sensitivity to environmental factors must be addressed. Overall, the integration of SHM and DIC presents promising opportunities for enhancing infrastructure safety, sustainability, and resilience.

Keywords SHM · DIC · Acoustic emission · Compressive strength · Split tensile strength · Flexural strength

---

R. Choudhary · P. V. Ramana (✉)
Department of Civil Engineering, MNIT Jaipur, Jaipur, India
e-mail: pvramana.ce@mnit.ac.in

R. Choudhary
e-mail: 2022rce9534@mnit.ac.in

H. Jahangir et al. (eds.), *Damage Detection and Structural Health Monitoring of Concrete and Masonry Structures*, Springer Tracts in Civil Engineering,
https://doi.org/10.1007/978-981-97-8975-7_13

13.1 Introduction

Structural Health Monitoring (SHM) involves the continuous or periodic monitoring of systems using various sensing technologies to assess their health and detect any potential damage, flaws, or changes in behaviour over time [1]. The primary objective of SHM is to enable early identification of structural faults and ensure timely maintenance and repairs, reducing the risk of catastrophic failures and enhancing overall structural safety and reliability. Structures exposed to various environmental and operational conditions are subject to deterioration, wear, and potential damage [2]. To address this, SHM employs a range of sensors that monitor parameters such as strain, displacement, temperature, acceleration, and vibration. These sensors are thoughtfully positioned at critical spots within the structure to collect data on its performance and behaviour. The collected data is then transmitted to a data acquisition system, which is processed, analysed, and saved for further evaluation [3]. In the data processing and analysis stage, advanced algorithms and techniques, including machine learning and statistical methodologies, are applied to identify anomalies or deviations from the structure's typical behaviour [4]. Engineers and maintenance staff can use this information to plan and carry out necessary repairs or maintenance tasks promptly, thus prolonging the structure's lifespan and optimizing its performance. The application of SHM is not limited to civil engineering alone. It finds applications in various industries, including aerospace, energy, transportation, and manufacturing, where structures' integrity and safety are paramount [5]. For example, in the aerospace industry, SHM monitors aircraft components for signs of fatigue, corrosion, or damage, ensuring safe and reliable flight operations. There has been growing interest in sustainable construction practices that minimize environmental impact [6]. As a result, incorporating recycled materials in building materials has gained significant attention. Glass, in the form of waste glass powder, has emerged as a promising recycled material with potential applications in construction [7]. Concrete reinforced with glass powder has gained recognition due to its mechanical properties and numerous economic benefits. Glass powder provides structural restraint, increases toughness, and potentially eliminates or reduces the conventional reinforcement required in concrete structures [8]. This promotes waste recycling and reduces the carbon footprint associated with traditional construction materials. Integrating SHM techniques with glass powder-reinforced concrete offers new possibilities for sustainable infrastructure management. By subjecting glass powder concrete specimens to various loading conditions and environmental factors, engineers can gain valuable insights into the behaviour of these materials. These insights can lead to improved design practices, a better understanding of material behaviour under stress, and the optimization of construction methods for enhanced structural performance and durability. In this study, we explore the application of SHM to concrete specimens reinforced with varying percentages of glass powder [9]. By conducting comprehensive testing and analysis, we aim to determine the glass powder effect on concrete properties. Through an in-depth examination of the performance of glass powder concrete in different loading conditions, we aim to assess its suitability as a practical and sustainable building material [10]. The significance of this research lies in its potential to contribute to more environmentally friendly construction practices and infrastructure development [11]. Ultimately, this research endeavours to ensure safer and more resilient infrastructure for the future, aligning with the principles of sustainability and responsible engineering [12].

13.1.1 Structural Health Monitoring

Structural Health Monitoring (SHM) entails continuous or periodic monitoring of structures using diverse sensing technologies to assess their health and detect any potential damage or changes in behaviour over time [13]. The primary objective of SHM is to identify structural issues early, ensuring timely maintenance and repairs and reducing the risk of catastrophic failures [14]. By strategically placing various sensors, SHM captures critical data on the structure's physical characteristics, such as strain, displacement, temperature, acceleration, and vibration. This data is then collected and analysed using advanced algorithms and techniques, including machine learning and statistical methodologies [15]. The analysis aims to detect anomalies or deviations from the structure's expected behaviour, which may indicate underlying damage or weaknesses [16]. By continuously monitoring the structure's condition, SHM empowers engineers and maintenance personnel to make informed decisions and plan proactive maintenance strategies [17]. This approach minimizes the risk of unexpected structural failures and extends the structure's life, enhancing its safety and performance. In essence, SHM acts as a vigilant guardian, constantly overseeing the health and integrity of structures, providing valuable insights that facilitate timely interventions, and ensuring the longevity and reliability of critical infrastructures [18].

A non-destructive testing method known as acoustic emission (AE) continuously observes and evaluates structural integrity in materials and components, as shown in Fig. 13.1. It depends on recognizing and assessing acoustic signals or waves produced when a material is deformed or damaged [19]. AE is frequently used in various sectors, including manufacturing, power generation, aerospace, and civil engineering.

13.1.1.1 Elements of SHM

Typical key elements of structural health monitoring include:

- The structure's physical characteristics are measured using various sensors that monitor strain, displacement, temperature, acceleration, and vibration [20]. These sensors are thoughtfully positioned at crucial spots to collect information about the activity of the structure.

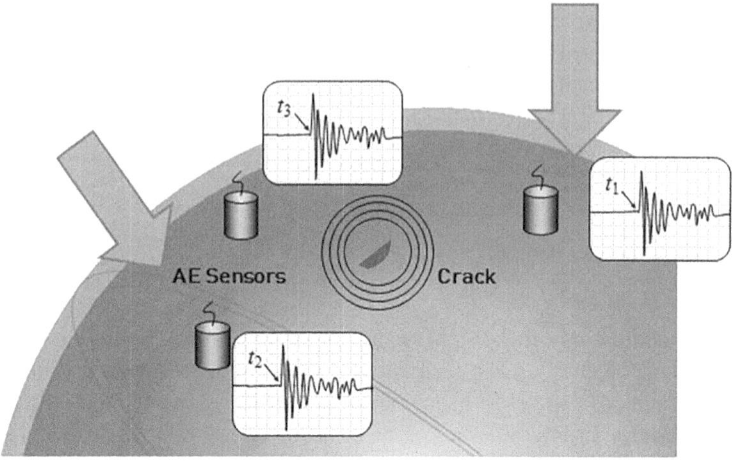

Fig. 13.1 Acoustic emission technique of SHM

- Data acquisition system: A data acquisition system is connected to the sensors and gathers, analyzes, and saves the sensor data.
- Data processing and analysis: The data is analyzed for anomalies or departures from usual behavior. The data may be interpreted, and possible damage detected using cutting-edge algorithms and techniques like machine learning and statistical methodologies [21].
- Damage detection and localization: By analyzing the data, SHM systems may determine the precise position and degree of damage within a structure [22]. Engineers and maintenance staff can use this information to plan and carry out necessary repairs or maintenance tasks.
- Making decisions and planning maintenance: SHM data offers insightful information about structural health, enabling planners to allocate resources wisely and choose effective maintenance practices [23].

13.1.1.2 Advantages of SHM

SHM provides several benefits, including:

- Early Damage Detection: SHM allows for the early identification of structural damage or deterioration, enabling timely intervention to prevent accidents and unforeseen failures.
- Condition Assessment: SHM provides a continuous or periodic assessment of a structure's condition, offering real-time data on its health and integrity [24].
- Proactive Maintenance: By detecting issues early, SHM enables proactive maintenance, reducing the need for costly and disruptive emergency repairs.
- Cost Savings: Implementing SHM can lead to cost savings in the long run by optimizing maintenance efforts and extending the life of structures.

- Enhanced Safety and Reliability: SHM enhances the safety and dependability of critical infrastructure such as bridges, buildings, and aircraft parts, minimizing the risk of structural failures [25].
- Structural Performance Optimization: By monitoring the structure's behavior under different loading conditions, SHM allows for optimizing structural design and materials.
- Risk Mitigation: SHM helps identify potential risks and vulnerabilities in structures, allowing engineers to implement risk mitigation strategies.
- Remote Monitoring: Many SHM systems can be monitored remotely, reducing the need for physical inspections and enabling efficient data collection [26].
- Improved Asset Management: SHM facilitates better asset management by providing detailed information on the structural health of multiple assets, enabling prioritization of maintenance and resource allocation [27].
- Research and Innovation: SHM data contributes to research and innovation in structural engineering, developing new technologies and methodologies for safer and more efficient structures.
- Enhanced Structural Resilience: By continuously monitoring structures, SHM helps identify and address issues that may compromise their resilience to extreme events, such as earthquakes or harsh environmental conditions [28].

13.1.1.3 Disadvantages of SHM

The following is a list of some of the most significant disadvantages associated with structural health monitoring:

- Positioning and Coverage of Sensors: Inaccurate sensor placement may lead to incomplete data or missed damaged areas, affecting monitoring system accuracy.
- Calibration and Accuracy: Regular calibration of sensors is essential for reliable data; however, environmental factors can influence sensor performance [29].
- Data Analysis Complexity: Managing and interpreting SHM data can be challenging, especially with large datasets, leading to longer processing times.
- Initial Implementation Cost: Setting up a comprehensive SHM system can involve significant initial investment and infrastructure requirements [30].
- Technical Expertise: SHM systems may require specialized installation, operation, and data analysis knowledge.
- Data Privacy and Security: SHM collects and transmits sensitive data, requiring robust security measures.
- Sensor Sensitivity to Environmental Factors: Environmental conditions can affect sensor readings and accuracy [31].
- False Positives or Negatives: Automated analysis techniques may produce inaccurate results, requiring manual verification.
- System Maintenance: Regular maintenance is essential to ensure SHM systems remain operational and reliable.
- Limited Application Scope: SHM may not be suitable for all structures or have limitations in certain environmental conditions [32].

Structural health monitoring is frequently used in many industries, such as civil engineering, aircraft, transportation, and energy, where the safety and dependability of structures are crucial [33]. The sophistication, reliability, and cost-effectiveness of SHM systems are improving as technology develops, making them essential tools for infrastructure management [34].

13.1.2 Challenges

The challenges are due to the complexity of large-scale structures and the need for accurate and reliable assessment. Some key challenges include:

- Sensor Placement and Density: Determining the optimal locations for sensors is crucial to capture accurate data on structural behaviour [35]. Achieving sufficient sensor density across a large structure can be challenging, as different parts may experience varying loads and stress patterns.
- Data Acquisition and Management: Collecting data from numerous sensors distributed across the structure requires effective data acquisition systems [36]. Managing and storing the massive amount of data generated can be complex and require robust data management strategies.
- Sensor Calibration and Reliability: Sensors must be accurately calibrated to ensure their measurements are reliable and consistent over time. Environmental factors, such as temperature changes and exposure to harsh conditions, can impact sensor performance and longevity.
- Data Interpretation and Analysis: Interpreting the collected data to extract meaningful information about structural health can be challenging [37]. Advanced signal processing and data analysis techniques are required to distinguish between normal variations and potential damage indicators.
- Damage Detection and Localization: Identifying damage's presence, extent, and location within a structure is complex. Different types of damage, such as cracks, corrosion, or material degradation, can manifest differently in sensor data, necessitating sophisticated algorithms for detection and localization.
- Data Fusion and Integration: Many structures have multiple sensors (e.g., accelerometers, strain gauges, displacement sensors) to monitor structural health [38]. Integrating data from various sensors and sources to provide a comprehensive assessment requires data fusion techniques.
- Wireless Sensor Networks: Deploying wired sensors in large civil structures can be cumbersome [39]. Using wireless sensor networks introduces challenges regarding power management, communication reliability, and signal interference.
- Environmental Effects: Structures are subjected to varying environmental conditions such as temperature, humidity, and wind loads. These factors can impact sensor readings and complicate distinguishing between damage-related changes and ecological effects.

- Validation and Calibration: Ensuring the accuracy of SHM systems requires a thorough validation process. This includes comparing sensor data to physical measurements and conducting controlled experiments to validate the performance of damage detection algorithms.
- Integration with Structural Design and Maintenance: Integrating SHM into civil structures' design, construction, and maintenance processes requires collaboration between structural engineers, sensor experts, and data analysts. Implementing effective decision-making processes based on SHM data can be challenging [40].

13.1.3 Digital Image Correlation

SHM has made remarkable strides in ensuring the safety and reliability of various structures, ranging from civil engineering marvels like bridges and buildings to critical components in the aerospace and transportation industries [41]. SHM employs diverse sensing technologies to monitor systems continuously, enabling early detection of potential damage or deviations from normal behaviour [42]. This proactive approach allows for timely interventions, reducing the risk of catastrophic failures and extending the life of structures. One of the cutting-edge techniques within SHM is DIC. This method has emerged as a powerful tool for measuring and analysing full-field displacements and strains on the surfaces of objects. With its ability to capture complex deformation patterns and behaviours, DIC has gained substantial attention in engineering, materials science, biomechanics, and beyond [43]. At its core, the DIC process involves analysing successive images of an object's surface before and after deformation. By comparing these images, DIC algorithms extract displacement and strain information at sub-pixel levels, offering exceptional measurement precision [44]. This high level of accuracy, coupled with the technique's non-contact nature, makes DIC particularly well-suited for scenarios involving fragile materials or intricate strain distributions.

13.1.3.1 Stages of DIC Process

The following are the stages, as shown in Fig. 13.2, that can be deconstructed into the process.

- Acquiring an Image: Take pictures of the object before and after the deformation. It is essential that the photos be of high quality and possess a resolution that is high enough to catch even the finest of details.
- Image preprocessing: It involves correcting for differences in illumination, noise, and any other artifacts that can impact the accuracy of the correlation [45]. This is done to get the images ready for analysis.
- Image correlation: It employs correlation algorithms to compare the reference images (taken before the deformation) and the deformed images. These algorithms

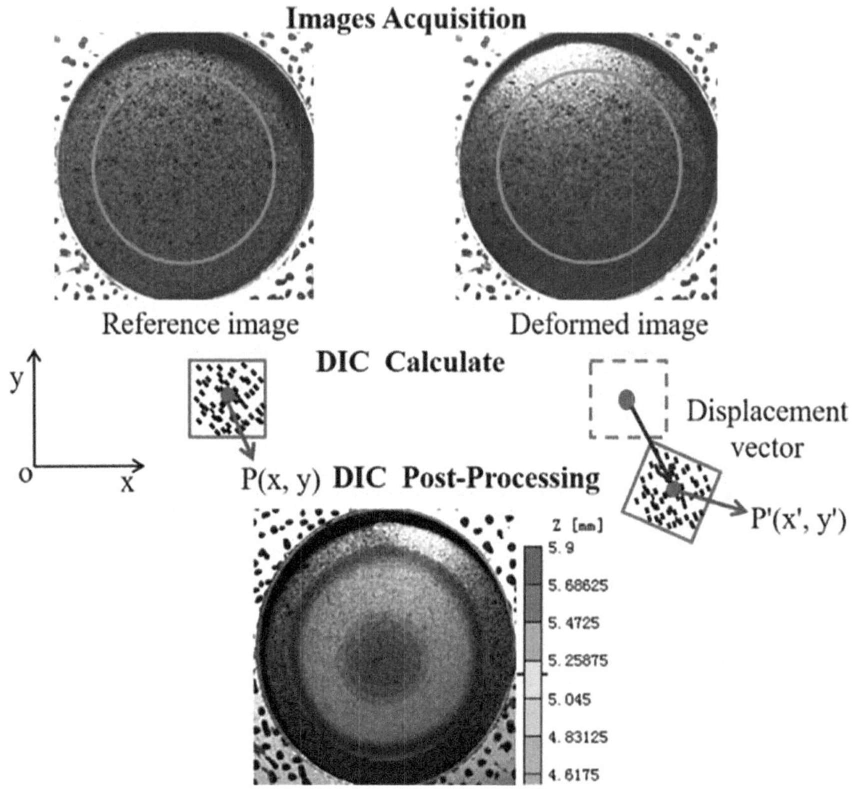

Fig. 13.2 DIC process

look for patterns within the photos to ascertain the displacements and strains on the object's surface.

- Post-processing: It consists of analyzing the data received from the correlation process to derive information regarding displacement and strain [46]. To accomplish this, the results may need to be filtered, strain maps may need to be generated, primary strains may need to be calculated, and various other essential values may need to be derived.

13.1.3.2 Advantages of DIC

The following are some of the benefits that come from using digital image correlation:

- Non-Contact Measurement: DIC is a method of measurement that does not involve making physical contact with the object being measured; as a result, it is appropriate for use with fragile or soft materials.

- Full-field measurement: DIC offers a thorough knowledge of the object's behavior by providing information on displacement and strain across the whole surface of the object.
- Great degrees of precision: DIC can achieve significant displacement and strain measurement accuracy when adequately calibrated and using advanced algorithms.
- Material Diversity: DIC applies to various materials, including metals, composites, polymers, and biological tissues. This versatility allows researchers and engineers to study various materials under different loading conditions.

13.1.3.3 Disadvantages of DIC

The following is a list of some of DIC's limitations:

- The accuracy of DIC measurements can be affected by factors such as environmental conditions, surface roughness, and image quality [47]. DIC measurements are sensitive to noise.
- Image correlation can be a computationally costly operation, mainly when working with massive datasets or photos with a high resolution, especially if both factors are present.
- The intricacy of the calibration process lies in the fact that precise outcomes depend on correct calibration, which can be a time-consuming process that necessitates specialized knowledge.
- Sensitivity to Environmental Factors: DIC measurements can be influenced by environmental conditions, such as lighting variations, surface roughness, and temperature changes, affecting measurement accuracy.
- Computational Complexity: Image correlation can be computationally intensive, especially when dealing with large datasets or high-resolution images, leading to longer processing times.
- Calibration Complexity: Accurate DIC results depend on proper calibration, which can be time-consuming and specialized, requiring expertise to obtain precise outcomes.
- Sensitivity to Noise: DIC measurements are sensitive to noise or artifacts present in the images, which may impact the accuracy of the results.
- Limited Measurement Depth: DIC is primarily a surface-based measurement technique and may not capture deformations within the object's interior.

Researchers and engineers can acquire significant insights into the characteristics of materials, structural integrity, and performance under various loading circumstances by using a digital image correlation technique.

13.1.4 Glass Powder

Concrete reinforced with glass powder (GP) is a building material quickly gaining acceptance in manufacturing concrete [48]. Recent years have seen many thorough investigations to observe the mechanical implementation of GP. As a result, the mechanical properties of glass powder concrete (GPC) are essential for a more extensive application of the substance [49]. The mechanical properties of specimens can be observed using one or two basic procedures. Experimental analyses were conducted on test samples. The behaviour of residual compression is very similar in their journals. GP is widely used to produce consumer items, including water bottles. It is lightweight, adaptable, robust, moisture and chemical-resistant, and a long-lasting material that can tolerate high temperatures, making it perfect for various applications [50]. GP is an explicit, durable, lightweight material that is simple to produce and mould into different shapes and forms. It has outstanding mechanical qualities such as strong tensile strength and wear resistance. GP is also noted for its transparency and low gas permeability, making it an excellent material for packaging and storage. GP is commonly utilized to manufacture glass bottles for drinks and other liquids. Because they are lightweight, shatter-resistant, and have high barrier qualities, these bottles are an excellent option for packing. Glass bottles are also recyclable. They may be used to make various items. GP is widely utilized to manufacture numerous products, including bottles, packaging, and materials. It offers several advantages, including hardness, flexibility, and transparency. However, the extensive usage of GP has several negative environmental and public health consequences. One of the most severe issues with GP is its disposal. GP is not biodegradable and cannot be degraded by natural processes [51]. Glass powder significantly reduces plastic and drying shrinkage cracking. They successfully stop the concrete from losing crushing strength. GP, in contrast, reduced the concrete's toughness despite having a minor impact [52]. This study examines how glass powder-reinforced concrete responds to temperature changes at 27 °C (room temperature). This study uses glass powder as support to examine specific statistical characteristics. On a weight basis, six concrete combinations were created with glass powder limits of 0, 18, 19, 20, 21, and 22%. A w/c ratio of 0.4 is given. The results indicate that GP presence can increase concrete properties by replacing fine aggregate up to 20%. The crushing strength (CS), split tensile strength (STS) and flexural strength (FS) of glass powder concrete are increased as compared to conventional concrete. When glass powder is included in a weight of more than 20%, the mechanical strength of concrete decreases.

13.1.5 Introduction to the Software (MATLAB®)

While commercial software solutions exist for DIC, they often come with limitations in customization, accessibility, and cost. In response to these challenges, this chapter presents a MATLAB® coding procedure for implementing the digital image correlation technique. This approach empowers researchers and engineers with the flexibility to adapt the methodology to their specific experimental setups, thereby addressing the unique strain measurement requirements encountered in various domains. The integration of DIC into the broader field of SHM holds immense promise. By incorporating full-field displacement and strain data, SHM practitioners can better understand how structures respond to external forces, environmental conditions, and potential damage. This newfound insight enhances the accuracy of structural health assessments, enabling more informed decision-making and maintenance strategies. This chapter will delve into the intricacies of the MATLAB® coding procedure for digital image correlation. By understanding the steps, algorithms, and considerations involved, readers will be equipped with a versatile toolset that harmonizes accurate strain analysis with computational control and experimental adaptability. Ultimately, the fusion of SHM and DIC can revolutionize how we perceive, analyse, and ensure the integrity of structures in diverse industries. In structural analysis and material testing, the necessity often arises to measure strain in scenarios where conventional contact-based techniques are inadequate, such as samples of varying sizes, compliances, temperatures, or intricate strain distributions. Addressing this challenge, the digital image correlation (DIC) technique emerges as a versatile solution capable of tackling diverse strain measurement requirements encountered in structural mechanics and material science. The fundamental principle of digital image correlation involves extracting strain information from successive images, enabling sub-pixel level resolution in strain measurement. This methodology, explored in detail in the subsequent sections, offers a holistic approach to deciphering complex strain fields intrinsic to various structures and materials. Despite the availability of codes in domains like image registration, artificial intelligence, and robotics, their direct applicability to strain measurement still needs to be improved. Commercial software solutions do exist, presenting advantages such as assurance of functionality, meticulous design, well-crafted user interfaces, and often accelerated processing speeds. However, these benefits come at a cost, both monetarily in terms of high pricing and operationally, as they often require specific hardware configurations. Additionally, the flexibility to tailor such commercial solutions to specific testing setups is often constrained due to the lack of programmable interfaces and the risk of data inaccessibility stemming from license issues or technology obsolescence.

In response to these challenges, this MATLAB® coding procedure aims to bridge the gap by presenting a technical pathway for implementing the digital image correlation technique. The provided code offers a bespoke approach to strain measurement, offering control and adaptability that caters directly to the needs of the strain measurement community. By adopting this approach, practitioners gain the flexibility to fine-tune the methodology to fit diverse experimental setups, thus circumventing the limitations of off-the-shelf software solutions. In the following sections, delve into the intricacies of the MATLAB® coding procedure for digital image correlation, elucidating the steps, algorithms, and considerations that empower researchers and engineers to undertake precise and tailored strain measurements. This endeavour

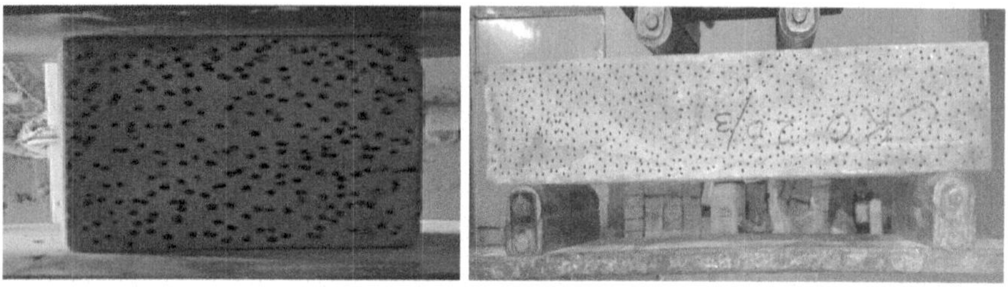

(a) Concrete cube (b) Concrete beam

Fig. 13.3 Concrete cube and beam

aims to equip the community with a potent toolset that harmonizes accurate strain analysis with experimental adaptability and computational control.

13.2 Methodology

The methodology adopted for this study involved a comprehensive experimental program to assess waste GP impact on concrete. The evaluation encompassed both fresh and hardened concrete specimens, with a particular focus on mechanical strength parameters. The experimental approach included the following key steps: experiment program, properties of aggregates, proportion of materials, mix design, and digital image correlation analysis.

13.2.1 Experiment Program

The experimental program revolved around testing concrete specimens in their fresh and hardened states. The evaluation of hardened concrete was conducted through the testing of 100 mm-sized cubes to determine the CS. Additionally, STS tests were performed on 300 mm × 150 mm concrete cylinders, while FS tests were carried out on beams sized 500 mm × 100 mm × 100 mm, as shown in Fig. 13.3.

13.2.2 Properties of Aggregates

The properties of aggregates, specifically fine and coarse aggregates, played a crucial role in shaping the concrete mixture. Fine aggregates exhibited a specific gravity of 2.887, a fineness modulus of 3.051, and a moisture content of 2.843% as shown in Table 13.1. The bulk density of fine aggregates was 1.935 kg/liter, and their water absorption capacity was 3.8%. Coarse aggregates displayed a specific gravity of 2.87,

Table 13.1 Fine aggregate

Properties	Fine aggregate (FA)
Specific gravity	2.887
Fineness modulus	3.051
Moisture content	2.843%
Bulk density	1.935 kg/liter
Water absorption	3.8%
Material finer than 75 μ	2.7%

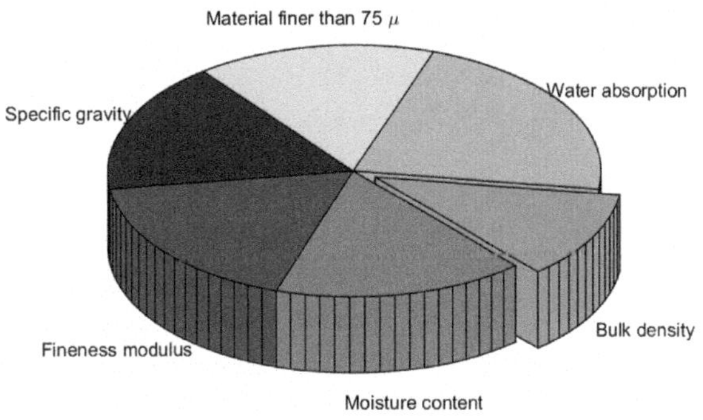

a flakiness and elongation index of 14.69% and 34.83%, respectively, and a loose bulk density of 1.39 kg/liter. The water absorption capacity of coarse aggregates was measured at 1.38%, while impact and crushing values were determined as 16.19% and 21.71%, respectively, as shown in Table 13.2.

13.2.3 Proportion of Materials

The proportion of materials in the concrete mixture was a critical factor in shaping the properties of the resulting concrete. Different mix designs were formulated, each incorporating varying percentages of glass powder waste. The mix designs included GP0, GP18, GP19, GP20, GP21, and GP22, with the corresponding proportions of cement, fine aggregate, coarse aggregate, glass powder, and admixture clearly defined in Table 13.3.

13.2.4 Mix Design

The mix design was established for concrete grade (M25) and water-cement ratio (0.4). The coarse aggregates were selected from 10 to 20 mm. The compaction factor was set at 0.9. Specific gravity values for sand, cement, and coarse aggregate were considered 2.67, 3.12, and 2.57, respectively. The sand fell into the Zone II classification; water absorption for coarse aggregates was 0.5% and fine aggregates 1%, respectively.

Table 13.2 Coarse aggregate

Properties	Coarse aggregate (CA)
Specific gravity	2.87
Flakiness and elongation index	14.69% and 34.83%
Loose bulk density	1.39 kg/liter
Water absorption	1.38%
Impact value	16.19%
Crushing value	21.71%

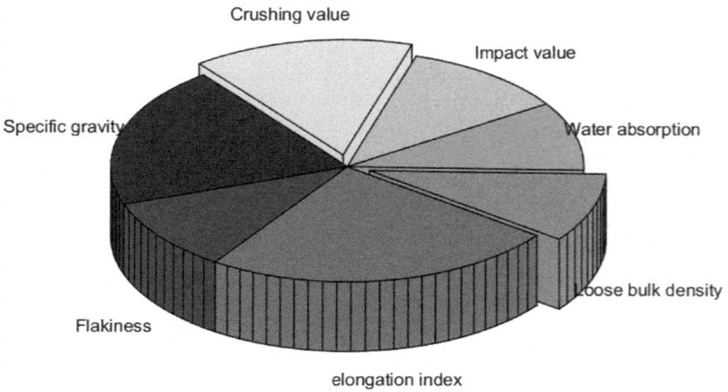

Table 13.3 Proportions of materials (kg/m^3)

Mix	Cement	FA	CA	Glass powder (%)	Admixture (%)
GP0	1	2.11	2.92	0	1.5
GP18	1	1.73	2.92	18	1.5
GP19	1	1.71	2.92	19	1.6
GP20	1	1.69	2.92	20	1.6
GP21	1	1.67	2.92	21	1.7
GP22	1	1.65	2.92	22	1.7

13.2.5 DIC

The DIC technique was employed to analyse concrete specimens' deformation and displacement patterns. This technique involved a systematic process, as outlined below.

Step 1: Generating filename list

The filename list for the images to be analyzed was generated using the 'filelist_generator.m' script. This script allowed for manual input and automated sequence generation of image filenames. The filenames were structured using a specific naming convention for efficient organization.

Step 2: Grid generation for correlation

The 'grid_generator.m' script facilitated the precise placement of markers for image correlation. Users could choose between various grid types, including rectangular and circular grids. The grid positions were stored in text-based files for reference.

Step 3: Correlation automation

The core function, 'automate_image.m,' processed all markers and images. The script correlated successive images, generating displacement data. Users could choose to continue or cancel the correlation process at each step. The time required for processing all images was displayed and could vary based on marker count and image complexity.

Step 4: Displacement analysis

The 'displacement.m' function enabled post-processing and analysis of the collected displacement data. Users could review displacement fields, calculate strains, and manage marker data quality. The function allowed for 3D displacement visualization and analysis, with options to rotate orientations for accurate strain measurements.

To facilitate a streamlined and efficient image analysis process, this guide is a concise resource for swiftly conducting analysis using DIC. Before the study, meticulous attention should be paid to the image format and file naming conventions. Notably, the recommended image format is denoted as x.tif, which can be effectively compressed using the pack bits compression technique. It is crucial to note that formats like JPEG or MPEG video compression should be avoided, as they sacrifice sub-pixel resolution for space efficiency during image processing.

Step 1: Generating filename list

The initial step involves generating a comprehensive list of filenames using the 'filelist_generator.m' script. At this juncture, two distinct approaches can be employed:

Option 1: Manual input of image numbers

Opting for manual input necessitates the input of image numbers that require processing. This can be executed directly in the command line interface. In contrast, if the preference is for an automated generation of image sequences, the following alternative is recommended:

Option 2: Automated image list generation

The user designates the first image within the sequence in this approach. The 'filelist_generator.m' script automatically detects and collates images with numerically increasing names preceding the fixed point in the filename. For instance, if the initial image is labelled 'IMAGE00001.Tif,' the script will efficiently compile all relevant images within the sequence. Opting for the approaches above prompts a window allowing manual input or showcasing automatically generated image sequences. This interface significantly streamlines the filename list generation process, enhancing user convenience and accuracy. A structured foundation is laid for subsequent Digital Image Correlation analysis by adhering to this initial step, fostering a robust and productive workflow that aligns with the precision and demands of technical image analysis methodologies. The steps involved in entire process is shown in Figs. 13.4, 13.5, 13.6, 13.7, 13.8, 13.9, 13.10, 13.11, 13.12, 13.13, 13.14, 13.15, 13.16, 13.17, 13.18, 13.19, 13.20, 13.21, 13.22, 13.23, 13.24 and 13.25.

This process begins with inputting the first and last image numbers into the 'filelist_generator.m' script. The numerical value at the end of each filename is required for this input. Subsequently, upon entering these numerical values, a dialog

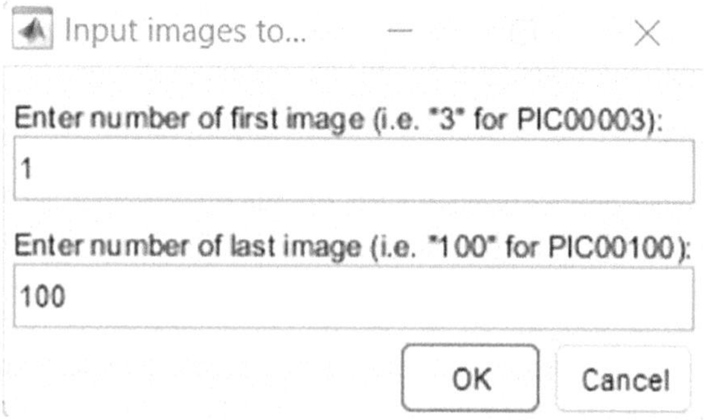

Fig. 13.4 Create an image list

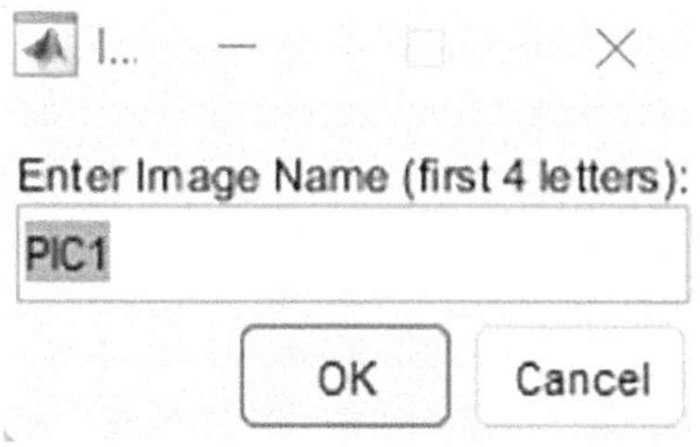

Fig. 13.5 Input image name

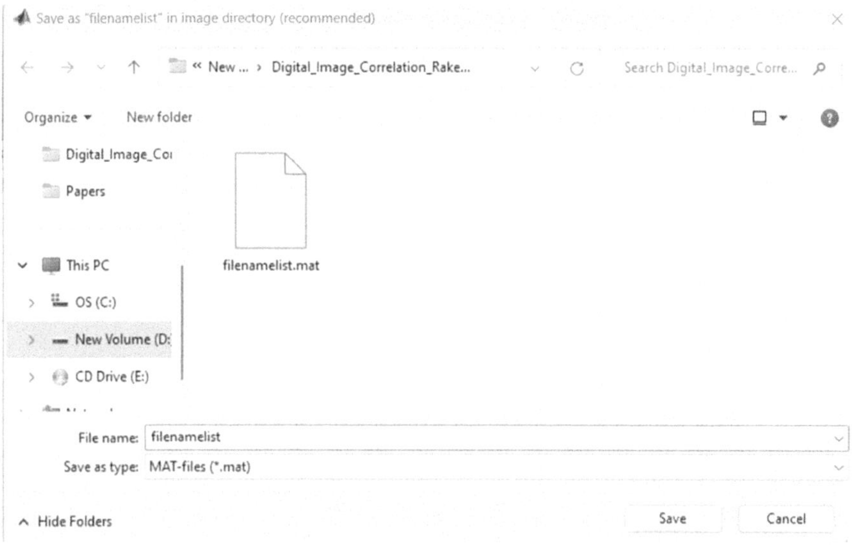

Fig. 13.6 Save the file to analyse with images in a folder

prompts the user for the first four letters of the filenames to be included in the sequence.

In this stage, the user inputs the filenames' initial four letters through the 'filelist_generator.m' script. This critical input aids in the precise compilation of the image sequence.

The ensuing step involves preserving the generated filename list. Opting for the filename 'filenamelist.mat' is recommended, as the subsequent scripts will actively seek this specific filename list within the designated folder containing the images intended for analysis. By meticulously following these successive stages, the 'filelist_generator.m' script efficiently generates a systematic and ordered filename list, seamlessly aligning the images for subsequent Digital Image Correlation analysis. This methodical approach ensures accuracy, precision, and ease in handling the crucial image sequencing necessary for comprehensive analysis.

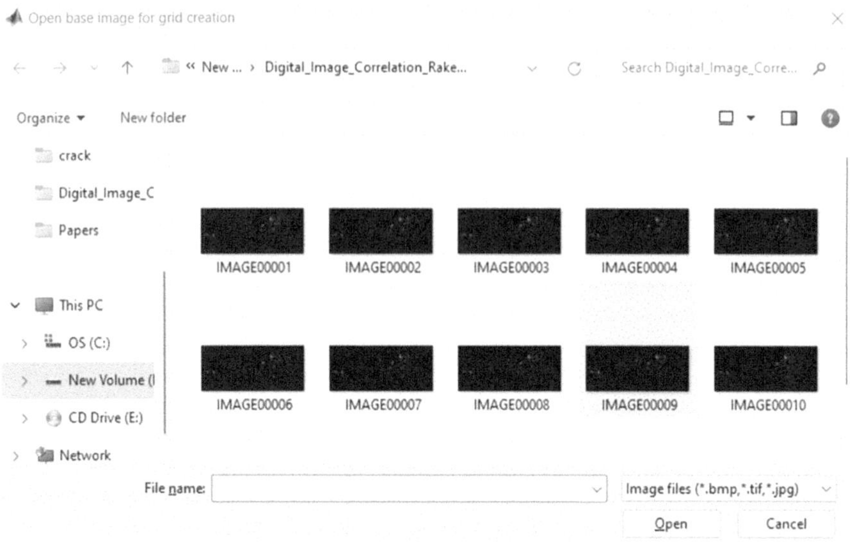

Fig. 13.7 Open base image

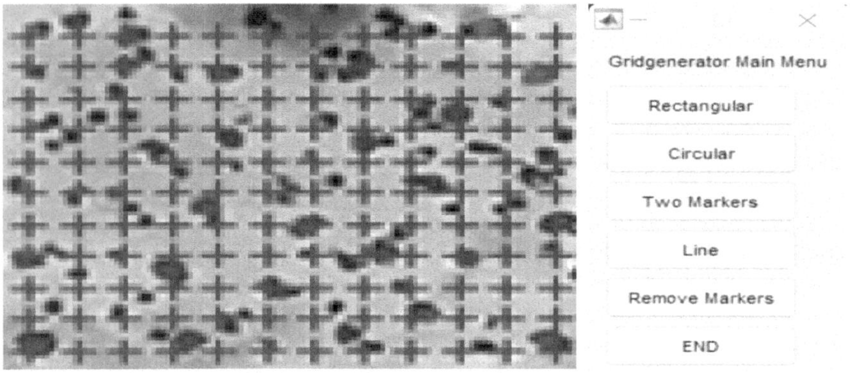

Fig. 13.8 Grid generator menu

Step 2: Grid generation for correlation

An indispensable step in pursuing precise image correlation is meticulously generating a grid for marker positioning. This process, facilitated by the 'grid_generator.m' script, allows users to tailor their marker positions for optimal accuracy. Notably, the pixel coordinates of the markers are stored in a text-based format, with 'grid_x.dat' capturing the x-positions and 'grid_y.dat' recording the y-positions.

The initial step involves selecting the first (base) image to serve as the reference for grid generation.

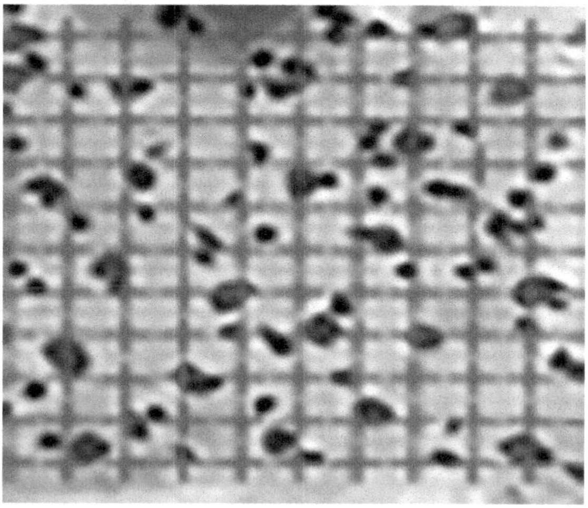

Fig. 13.9 Grid positioning

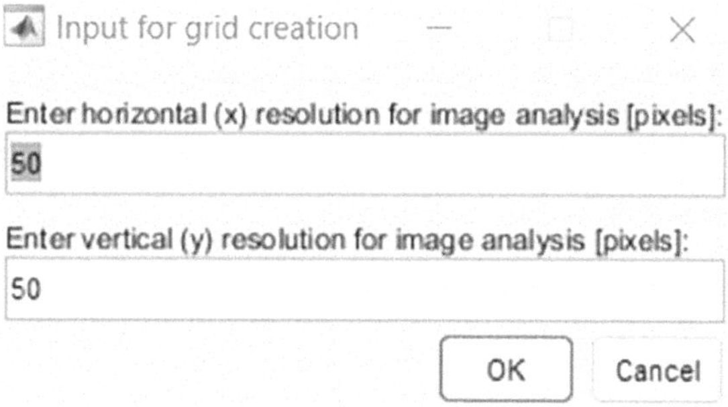

Fig. 13.10 Grid creation input

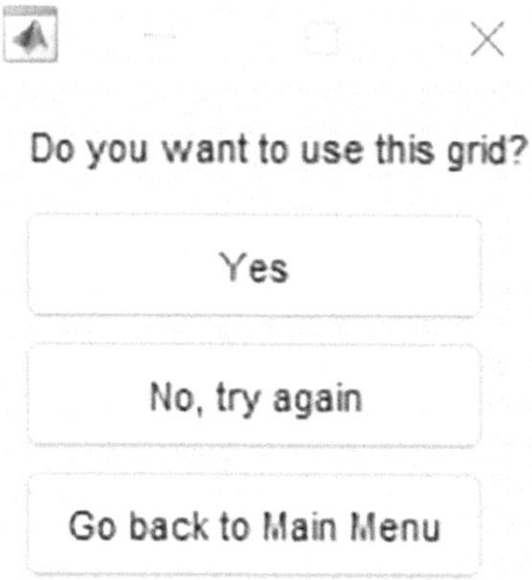

Fig. 13.11 Grid selection

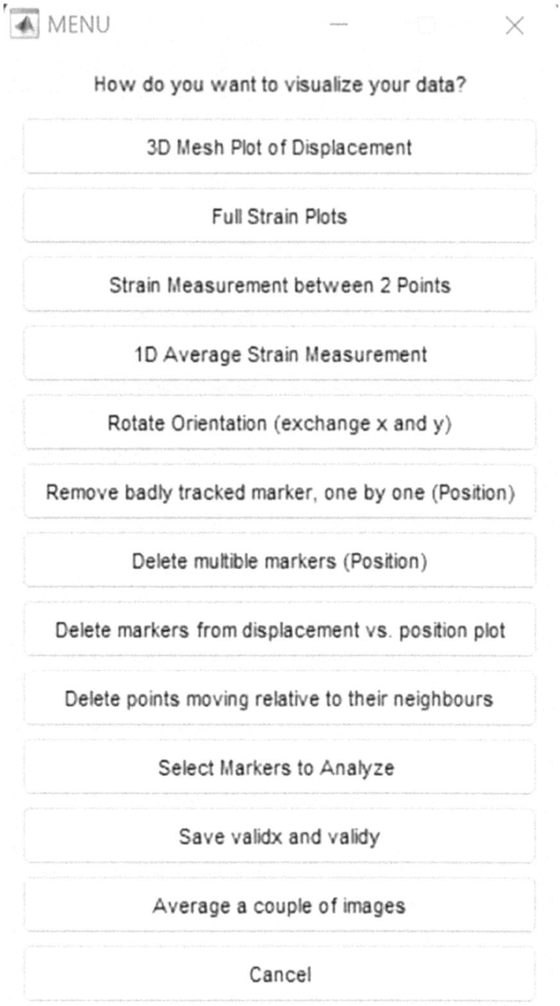

Fig. 13.12 Data visualization menu

After base image selection, a dialog emerges, offering a choice of grid types. These encompass rectangle or circle shape grids. Opting for a rectangular grid alters the cursor into a horizontal and vertical line, aiding precise placement. This method entails clicking on the two diagonal corners to delineate the outer boundaries of the grid-containing box.

This interface transitions to display orthographic lines, enabling the designated grid's exact location. This user-friendly feature ensures meticulous alignment, fostering the establishment of a reliable reference framework.

Upon defining the grid's outer dimensions, a dialog emerges, prompting input of the distance of the raster point in x and y both directions. This critical parameter determines the grid's granularity and precision.

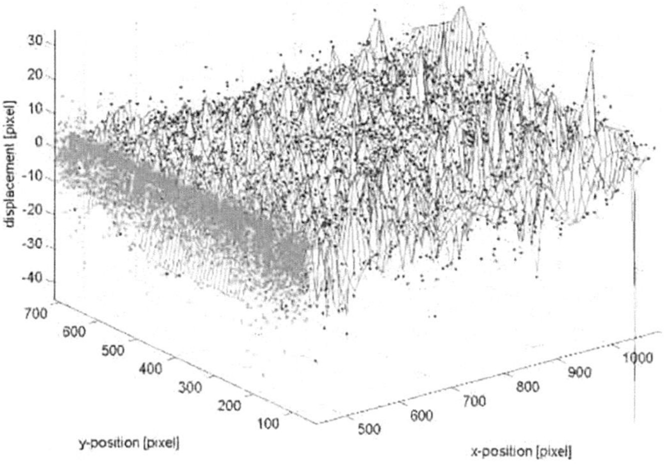

Fig. 13.13 3D displacement graph

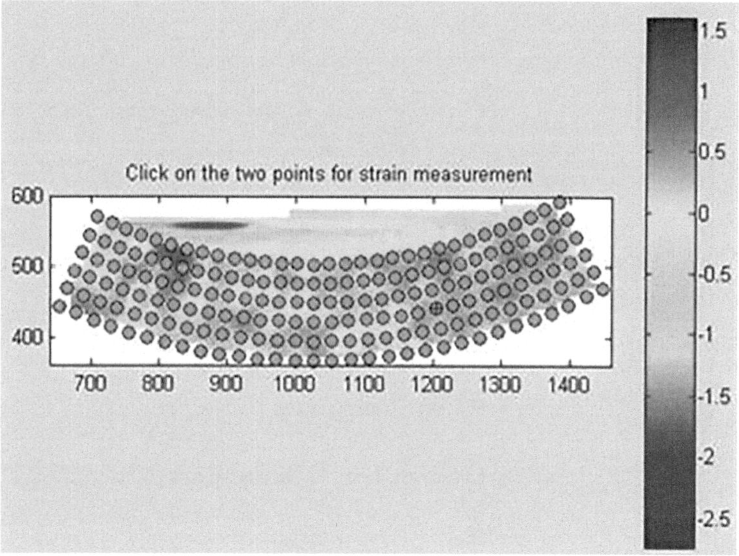

Fig. 13.14 Point selection for strain measurement

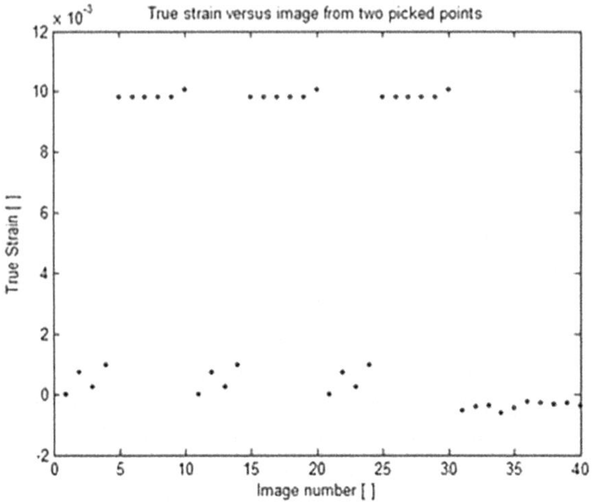

Fig. 13.15 Strain versus image number

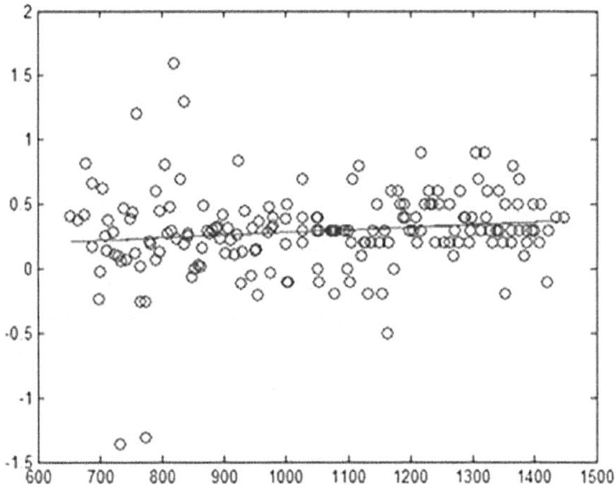

Fig. 13.16 Displacement versus position

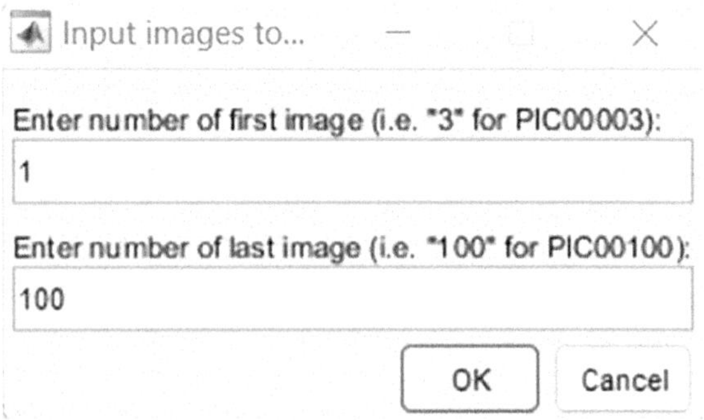

Fig. 13.17 Create an image list

Fig. 13.18 Input image name

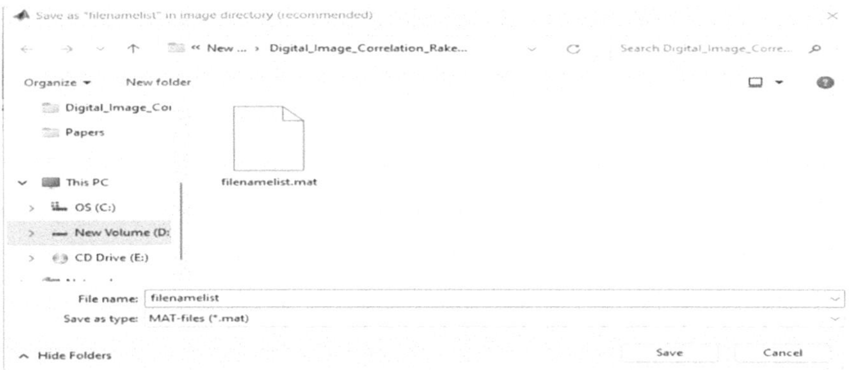

Fig. 13.19 Save the file to analyse with images in a folder

Fig. 13.20 Open base image

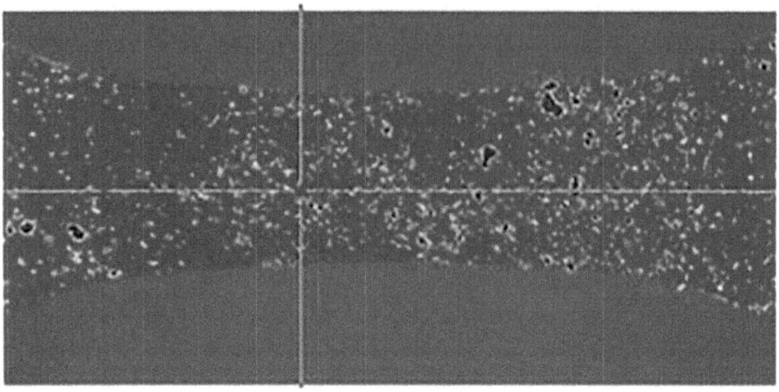

Fig. 13.21 Area selection for peaks

Fig. 13.22 Manual peak selects

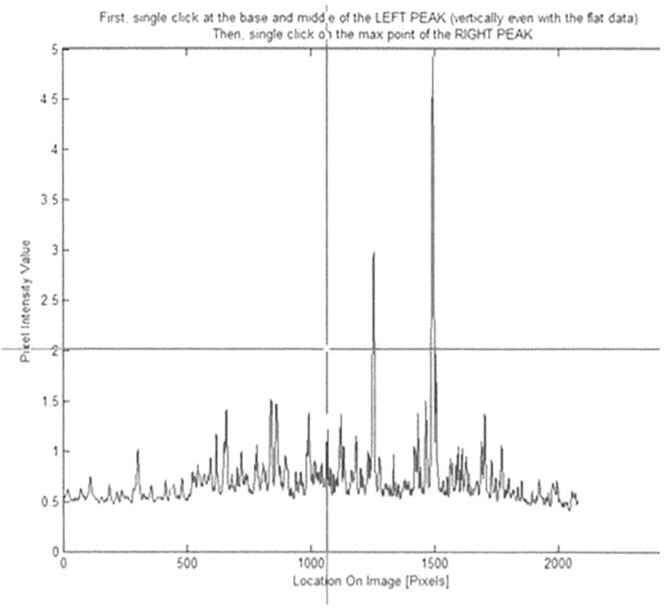

Fig. 13.23 Grayscale line profile

Fig. 13.24 Open base image for correlation

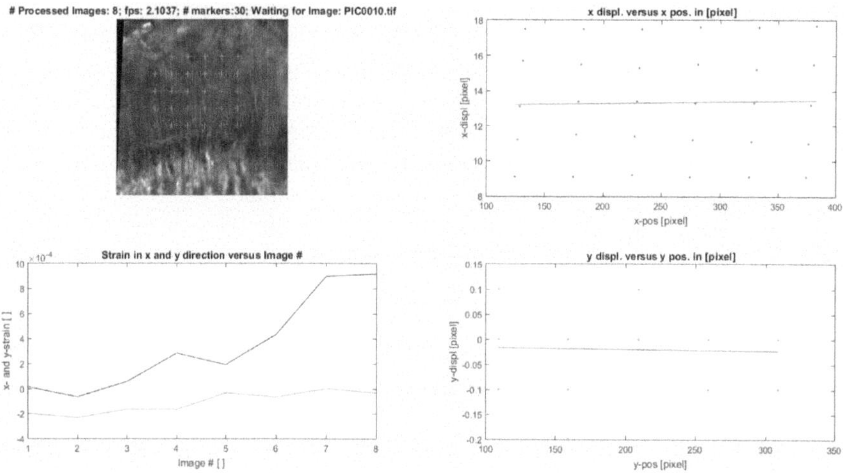

Fig. 13.25 Subplots graph

The script generates and superimposes the selected grid onto the base image. Subsequently, a final dialog surfaces, offering the option to adopt the grid and save the corresponding files. Users can also opt to revisit the process, try again, or choose a different grid type. Executing these systematic stages via the 'grid_generator.m' script ensures an accurately established grid framework that underpins subsequent Digital Image Correlation operations with meticulous precision and confidence.

Step 3: Correlation automation

The updated raster coordinates will be shown as red crosses when pictures one and two correlate. The green crosses are atop the image. You can keep or stop this correlation when the dialog box asks you to do so. It is possible to anticipate how long it will take to process all of the photos by knowing how quickly raster points process data. This procedure could take a few seconds or several days, depending on how many images and markers are recorded. A competent computer should take 200 s to process 100 photos and 200 marks. For a better solution, you can always run processes with higher resolutions overnight (for example, 6000 markers in 1000 photographs).

'CORRSIZE,' modified in 'cpcorr.m,' will limit the key, so keep that in mind. If the 15 pixels are used as recommended, a marker distance of 30 pixels will result in complete coverage of the strain field. Interpolation will result from choosing closer marker distances since two nearby markers share pixels.

Step 4: Displacement analysis

Post-processing, which comes last, is the most exciting and satisfying phase since it allows for analyzing the gathered displacement data. With the help of the displacement.m function, we can examine the displacement field, determine the strain, or remove markers that weren't well correlated or tracked.

Enter "displacement" into the command box to get started. or '[validx and validy] = displacement'; if you wish to save the altered files from the workspace. A box will appear asking for the file containing the x-displacement of all markers in all photos. This is followed by a dialog asking for the file, which provides the y data.

Enter 'displacement'; in command to start the analysis. The necessity to save the updated files from the workspace can also be combined with 'displacement'; as an alternative. The following dialog will ask for the file named "validx.dat," which provides x-displacement information for all markers across images. This will be followed by a similar dialogue for a file named "validy.dat," which contains y-displacement information. The user is then presented with various analysis options via 'displacement.m', which incorporates both files. In a typical analysis, some markers that performed poorly must be removed. This may occur for various reasons, including marker movement throughout the test, alterations in the lighting, or problems with the correlation technique because the sample surface did not offer enough features.

For a comprehensive analysis, it's vital to eliminate markers exhibiting poor tracking performance, often attributed to marker movement, changing lighting conditions, or inadequate surface characteristics. The process involves meticulous marker curation through trial and error, achieved via the following methods:

Remove Badly Tracked Markers, One by One (Position): Clicking on markers out of alignment prompts a color-coded plot update, aiding in marker selection and exclusion.

Delete Multiple Markers (Position): Selecting a rectangular region facilitates the removal of multiple markers.

The analysis can commence after refining the dataset and possibly saving it. Two approaches are available.

Strain Measurement between 2 Points: This method involves selecting two points, with the function determining the closest points and plotting the strain against image numbers. Results can be saved as an ASCII file. *1D Average Strain Measurement:* A linear fit of x-displacement versus x-position is performed for each image.

The ‘Select Markers to Analyse’ button in the ‘displacement.m’ menu proves valuable for precise sample analysis. This option provides a tailored approach to strain measurement within specific regions, offering the flexibility to choose different marker sets. Subsequently, the ‘1D Average Strain Measurement’ can compute strain.

Engaging with these advanced analysis features requires diligent practice and meticulous data assessment. Mastering the techniques and ensuring marker accuracy, the ‘displacement.m’ tool empowers users to unlock profound insights into displacement, strain, and structural behaviour.

13.2.6 Digital Image Tracking

This succinct guide aims to streamline the process of image analysis. Before initiating the study, validate the image format and file naming. The recommended structure is ‘x.tif’, which can be compressed using pack bits compression for optimal results. Avoid using compression formats like JPEG or MPEG video as they can compromise sub-pixel resolution.

13.2.6.1 Generating Filename List

Execute ‘filelist_generator’; in the MATLAB® command line. This prompts a dialog to appear, where input the first and last image numbers.

Subsequently, prompted to provide the first four letters of the filename.

The next step is to store the file in a folder containing the photographs to be processed. Save the created file in the same folder as the photographs.

13.2.6.2 Grid Generation for Correlation

For automated peak labelling, execute ‘[validx, validy] = peak_labelling’; to scan and process images. This function identifies peaks exceeding a set grey value, enabling peak fitting. Select a base image for peak identification and define an area for peak detection.

The image will be scanned, the backdrop should be removed because it should be black, and maxima with values more than a particular value of grey will be found.

The function will plan power, perform peak identification, and fit the peaks. Relevant files ('fitxy.dat,' 'validx.dat,' 'validy.dat') will be saved.

'imagekpeak.m'.

Execute 'imagekpeak'; and select an image for peak tracking. Define boxes around the peaks to track.

One must define a package for each peak by clicking on each peak's lower left and upper right. A blue circle will highlight the center of each box. Choose a comprehensive package for the curve fitting to get enough data points. After defining the boxes, the function will do the peaks and save the data as in 'peak_labelling.m'.

13.2.6.3 Displacement Analysis

- 'stress_strainmatch.m': Matches stress and strain files.
- 'Markerplotting.m': Plots markers on analyzed images.
- 'resume_automate_image.m': Resumes stopped automate_image jobs.
- 'jobskript.m': Automates job creation and execution.
- 'RTCorrCode.m': Real-time image correlation and strain calculation.

RTCorrCode aids in real-time image correlation. It dynamically generates a filenamelist, tracks images, and calculates strain. Users can define stop conditions and process images on-the-fly.

Subplots in RTCorrCode display the image, displacement versus marker position, and x- and y-strain versus image number. This provides real-time insight into displacement and strain behavior.

13.3 Problem Statements

Numerous studies have been done on using waste materials, such as glass powder, in creating concrete. When utilized as an alternative to fine aggregate, researchers have looked into how these materials affect the characteristics of concrete. Researchers have demonstrated that waste components can change how hardened and durable typical Portland cement-based concretes are. However, more analysis is required to determine how well these waste materials work. The primary goal is to reduce the amount of glass powder in our environment. There currently needs to be more procedures for decomposing glass powder. Sand can also result in water contamination, landslides, and flooding in nearby places, which is another issue. The cost of obtaining sand has also increased due to the need for more contractors and builders. Researchers are looking into alternative building materials due to the dearth of sand and the environmental problems linked with its mining. The project aims to reduce the amount of glass powder in the environment by using it as a functional building material. Our society would benefit if waste material could be used in buildings. To

Table 13.4 Concrete samples with varying percentages of glass powder

Glass powder (GP) %	GP particle size (μ)
0	595
18	595
19	595
20	595
21	595
22	595

tackle this scenario, GP is used with concrete, partially replacing sand. Compression test (CS), split tensile test (STS) and flexural test (FS) can determine the best GP percentage replacing sand in concrete. Glass powder % and particle size are shown in Table 13.4. These problem statements and research can further contribute to understanding the structural behavior of glass powder-modified concrete and its potential applications in the construction industry.

13.3.1 Problem Statement for Compressive Strength Analysis

While the effects of varying percentages of glass powder on compressive strength have been extensively evaluated, there is a need to comprehensively understand the impact of glass powder content on concrete durability and stability. Specifically, there needs to be more understanding of the long-term behavior of concrete with 18 to 22% glass powder replacement. This research investigates how these higher levels of glass powder incorporation affect concrete's compressive strength and structural integrity over extended periods, shedding light on the viability of using such concrete in applications with higher structural demands.

13.3.2 Problem Statement for Split Tensile Strength Analysis

While the study has provided insights into the split tensile strength of concrete containing varying glass powder proportions, there still needs to be more detailed exploration regarding the strain distribution and crack propagation mechanisms in these specimens. This research investigates the internal stresses and failure patterns within concrete examples with different glass powder content to address this gap, utilizing advanced digital image correlation techniques. By analyzing the micro-level responses of these specimens under tensile loading, this study aims to provide a deeper understanding of the fracture behavior and crack development in glass powder-modified concrete.

13.3.3 Problem Statement for Flexure Strength Analysis

The flexural strength analysis has offered valuable insights into the effects of glass powder replacement on concrete beams. However, there needs to be more investigation into the post-failure behavior and residual strength of these beams after experiencing loading beyond their elastic limit. This research addresses this gap by studying the flexural behavior of glass powder-modified concrete beams subjected to progressively increasing loads until failure. By analyzing these beams' load–deflection response and failure modes, this study provides a comprehensive understanding of the structural performance and potential applications of glass powder-enhanced concrete in flexural applications.

13.3.4 Problem Statement for Overall Structural Performance Evaluation

Although the individual aspects of compressive, split tensile, and flexural strengths have been examined, there needs to be a more holistic analysis regarding the overall structural performance of glass powder-modified concrete. This research aims to integrate the findings from the individual strength analyses to evaluate the comprehensive structural performance of concrete specimens with varying glass powder content.

Figure 13.26 shows concrete cube specimens with different waste Glass powder percentages, i.e., 0%, 18%, 19%, 20%, 21% and 22% with replacement of fine aggregate.

(a) *Cube, beam and cylinder with 0% glass powder*

Replaced 0% of sand by weight with glass powder waste in this case.

(b) *Cube, beam and cylinder with 18% glass powder*

Replaced 18% of sand by weight with glass powder waste in this case. The particle size of the glass powder waste was 30 meshes, approximately equal to 595 μ.

(c) *Cube, beam and cylinder 19% glass powder*

Replaced 19% of sand by weight with glass powder waste in this case. The particle size of the glass powder waste was 30 meshes, approximately equal to 595 μ.

(d) *Cube, beam and cylinder 20% glass powder*

Replaced 20% of sand by weight with glass powder waste in this case. The particle size of the glass powder waste was 30 meshes, approximately equal to 595 μ.

(e) *Cube, beam and cylinder 21% glass powder*

Replaced 21% of sand by weight with glass powder waste in this case. The particle size of the glass powder waste was 30 meshes, approximately equal to 595 μ.

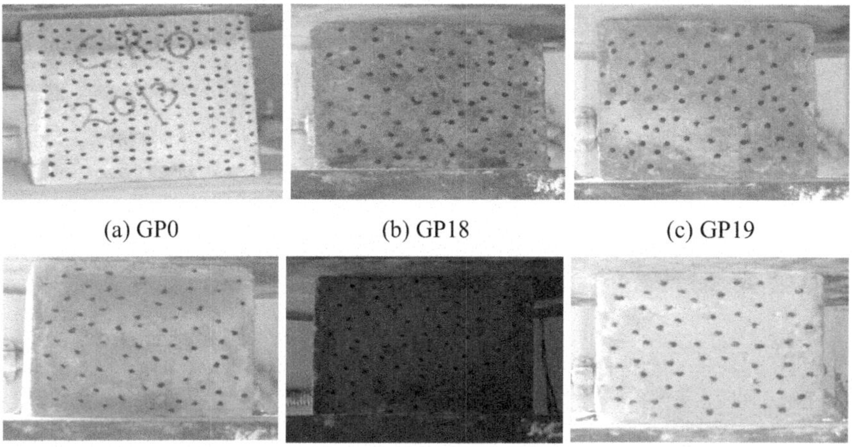

(a) GP0 (b) GP18 (c) GP19

(d) GP20 (e) GP21 (f) GP22

Fig. 13.26 Specimens with different % of glass powder

(f) *Cube, beam and cylinder 22% glass powder*

Replaced 22% of sand by weight with glass powder waste in this case. The particle size of the glass powder waste was 30 meshes, approximately equal to 595 μ.

13.4 Results and Discussion

13.4.1 Compressive Strength (CS)

To execute the CS test, one should cast three cubes and perform the test after 28 days of casting.

13.4.1.1 Cube with 0% Glass Powder

Cubes develop cracks as a result of this progressively rising stress over time. Images of damages appear in Fig. 13.27. Displacement graphs at different loads are shown in Fig. 13.28. The concrete cube's 2D surface displacements and strain have been measured using the DIC procedure. Selected photographs were analysed using the mathematical tool.

A square grid pattern has been considered for correlation. Using DIC-acquired strain, cube stress was calculated. The process's reduction factor, i.e., 5, relies on the grid size employed. From Fig. 13.29, the maximum strain acquired is 8×10^{-3}.

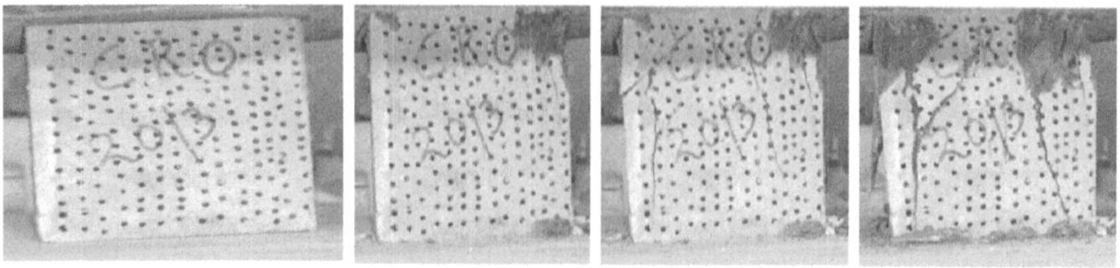

(a) (b) (c) (d)

Fig. 13.27 Specimen surface showing cracks at different loads with 0% glass powder

Table 13.5 indicates that the predicted value (DIC), i.e., 40 MPa, is approximately equal to the experimental value, i.e., 34.3 MPa.

13.4.1.2 Cube with 18% Glass Powder

One should perform a test on cubes (GP18) with 18% glass powder as a replacement for sand. And one should find the average compressive strength value of cubes, i.e., 35.2 MPa, according to the IS code. If one should compare the result with GP0, one should find little change in strength because the quantity and size of glass powder are minimal, which has a negligible effect on stability.

Images of cracks appear in Fig. 13.30. Displacement graphs at different loads are shown in Fig. 13.31. From Fig. 13.32, the maximum strain acquired is 4×10^{-3}. Using DIC-acquired strain, cube stress was calculated. The process's reduction factor, i.e., 5, relies on the grid size employed. Table 13.5 indicates that the predicted value (DIC), i.e., 35.8 MPa, is approximately equal to the experimental value, i.e., 35.2 MPa.

13.4.1.3 Cube with 19% Glass Powder

One should perform a test on cubes (GP19) with 19% glass powder as a replacement for sand. And one should find the average value of compressive strength of cubes, i.e., 36 MPa according to IS code. If one should compare the result with GP0, one should find little change in strength because the quantity and size of glass powder are minimal, which has a negligible effect on stability.

Images of cracks appear in Fig. 13.33. Displacement graphs at different loads are shown in Fig. 13.34. From Fig. 13.35, the maximum strain acquired is 7.2×10^{-3}. Using DIC-acquired strain, cube stress was calculated. The process's reduction factor, i.e., 5, relies on the grid size employed. Table 13.5 directs that the predicted value (DIC), i.e., 36 MPa, is approximately equal to the experimental value, i.e., 36 MPa.

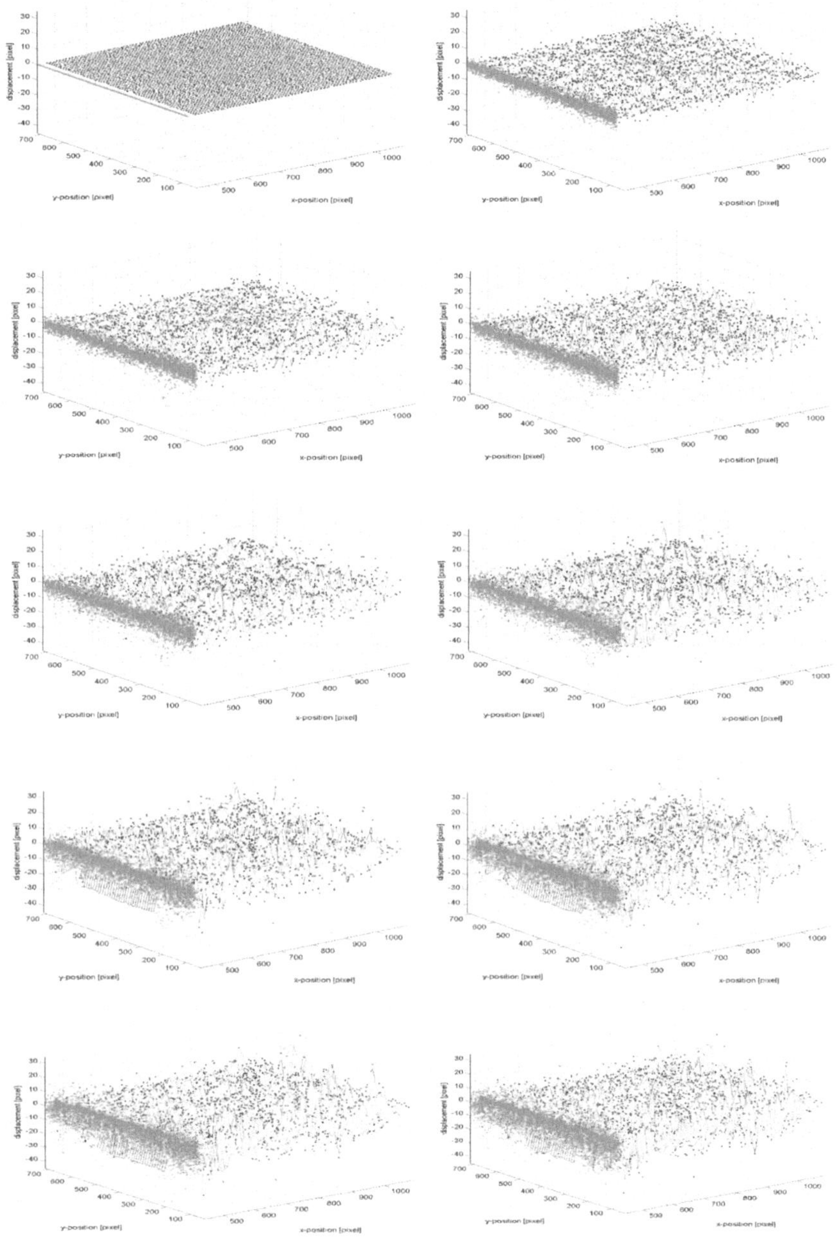

Fig. 13.28 Displacement graphs at different loads with 0% glass powder

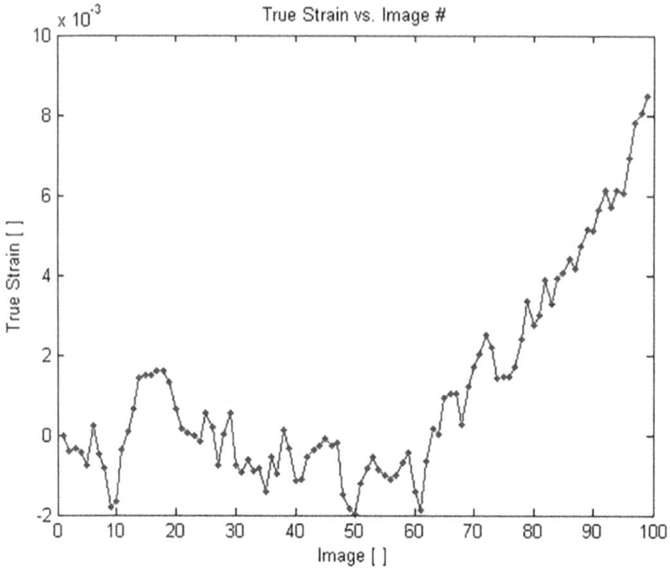

Fig. 13.29 True strain versus images graph with 0% glass powder

Table 13.5 Compressive Stress comparison for cubes

S. no	Glass powder (%)	Stress (N/mm^2)	
		Experimental	DIC predicted
1	GP0	34.3	40
2	GP18	35.2	35.8
3	GP19	36	36
4	GP20	37.5	39.5
5	GP21	36.2	34.5
6	GP22	33.9	34

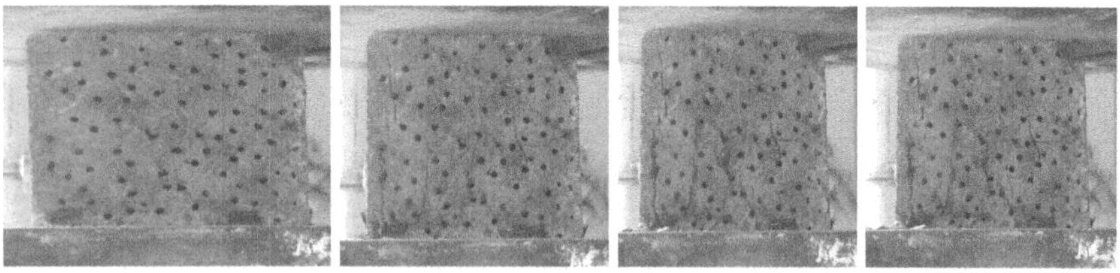

(a) (b) (c) (d)

Fig. 13.30 Specimen surface showing cracks at different loads with 18% glass powder

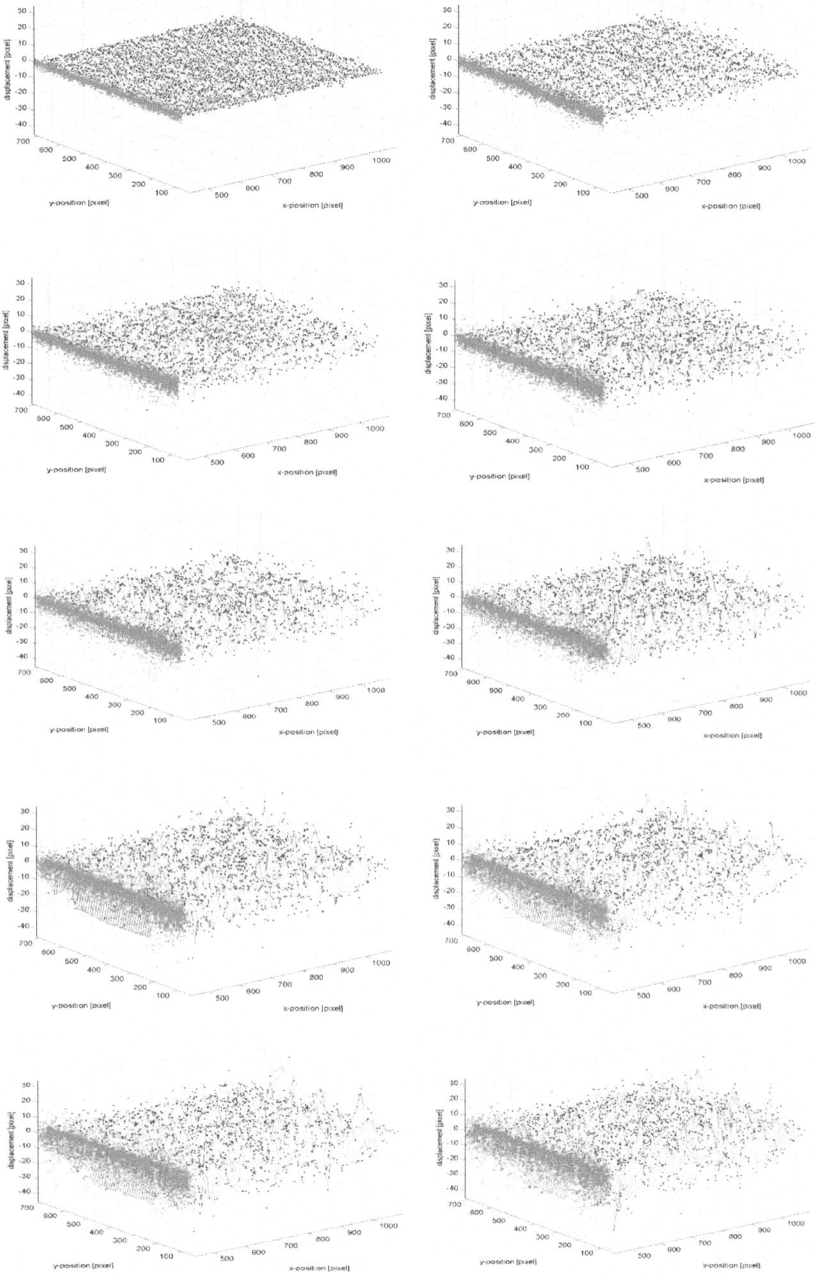

Fig. 13.31 Displacement graphs at different loads with 18% glass powder

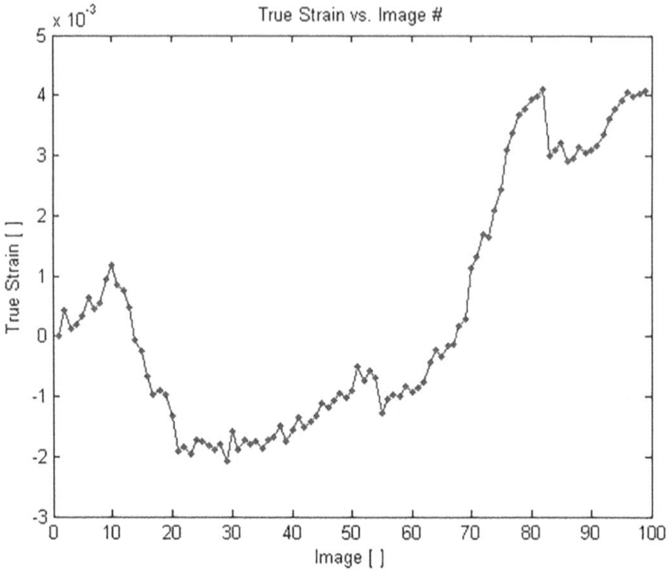

Fig. 13.32 True strain versus images graph with 18% glass powder

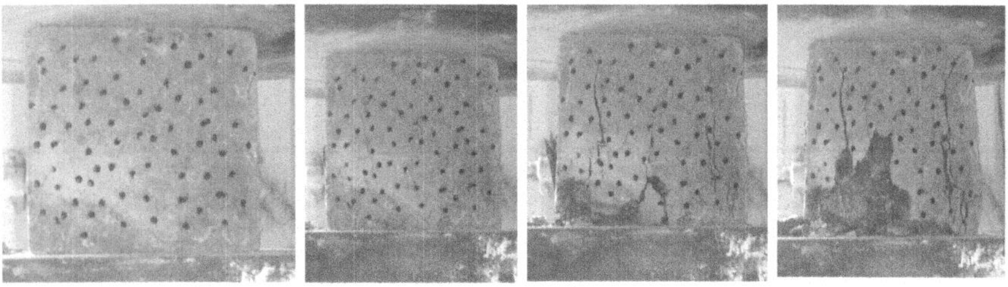

(a) (b) (c) (d)

Fig. 13.33 Specimen surface showing cracks at different loads with 19% glass powder

13.4.1.4 Cube with 20% Glass Powder

One should perform a test on cubes (GP20) with 20% glass powder as a replacement for sand. And one should find the average value of compressive strength of cubes, i.e., 37.5 MPa according to IS code. If one should compare the result with GP0, one should find little change in strength because the quantity and size of glass powder are minimal, which has a negligible effect on stability.

Images of cracks appear in Fig. 13.36. Displacement graphs at different loads are shown in Fig. 13.37. From Fig. 13.38, the maximum strain acquired is 7.9 × 10^{-3}. Using DIC-acquired strain, cube stress was calculated. The process's reduction factor, i.e., 5, relies on the grid size employed. Table 13.5 indicates that the predicted

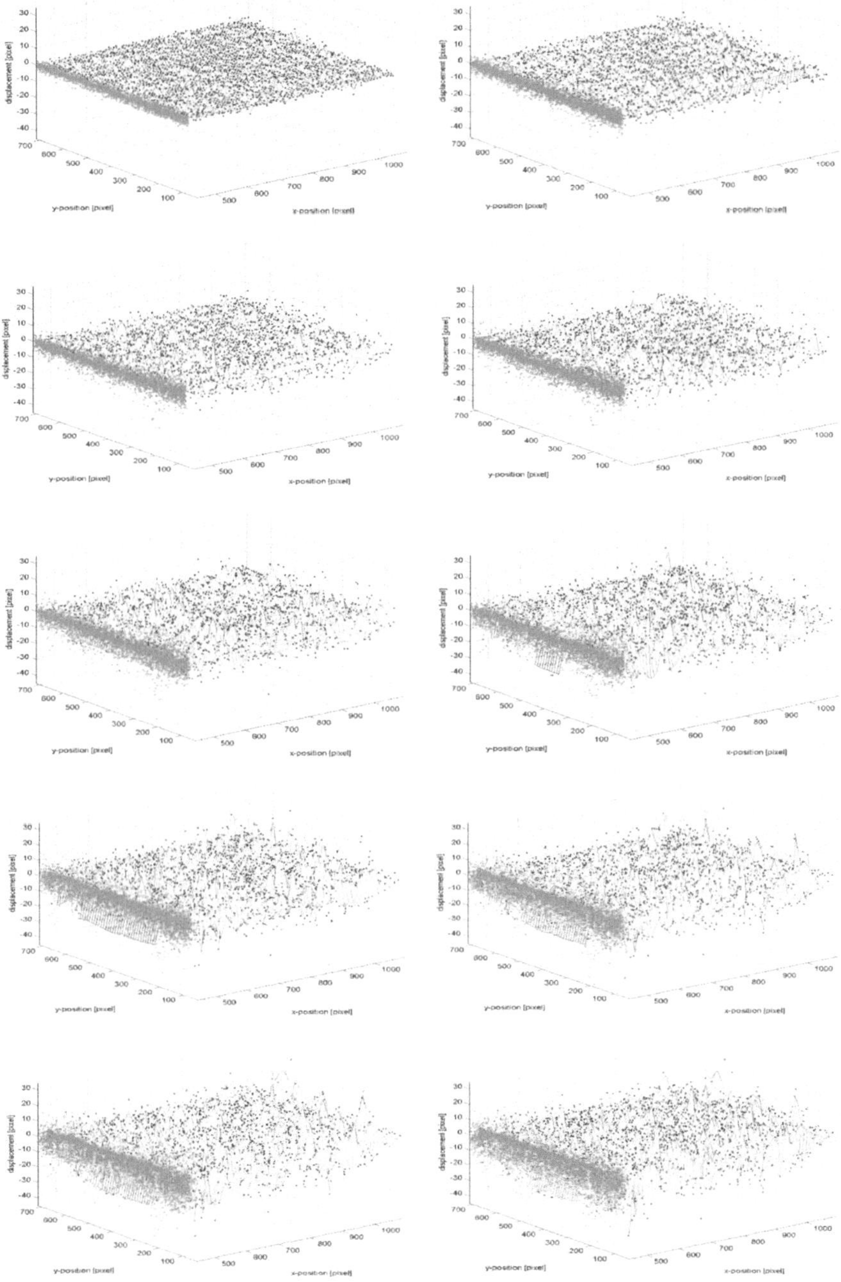

Fig. 13.34 Displacement graphs at different loads with 19% glass powder

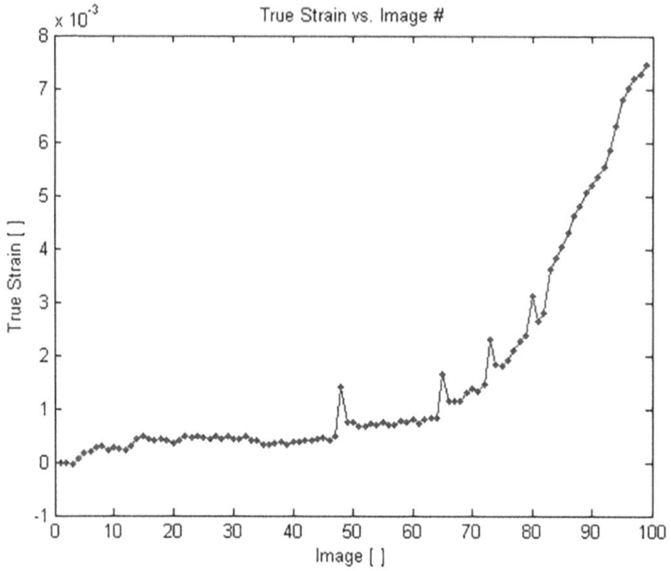

Fig. 13.35 True strain versus images graph with 19% glass powder

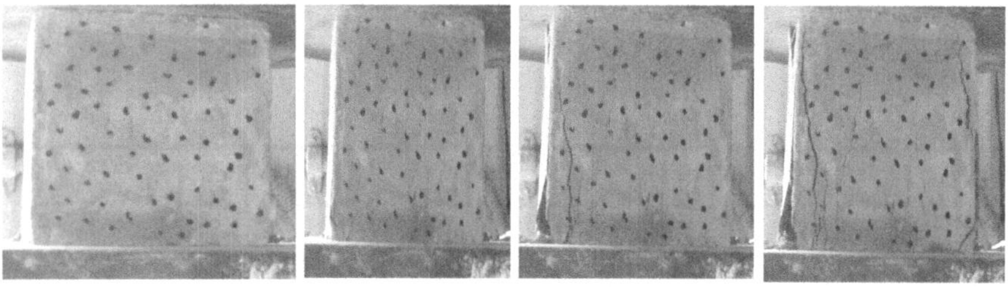

(a) (b) (c) (d)

Fig. 13.36 Specimen surface showing cracks at different loads with 20% glass powder

value (DIC), i.e., 39.5 MPa, is approximately equal to the experimental value, i.e., 37.5 MPa.

13.4.1.5 Cube with 21% Glass Powder

One should perform a test on cubes (GP21) with 21% glass powder as a replacement for sand. And one should find the average value of compressive strength of cubes, i.e., 36.2 MPa according to IS code. If one should compare the result with GP0, one should find little change in strength because the quantity and size of glass powder are minimal, which has a negligible effect on stability.

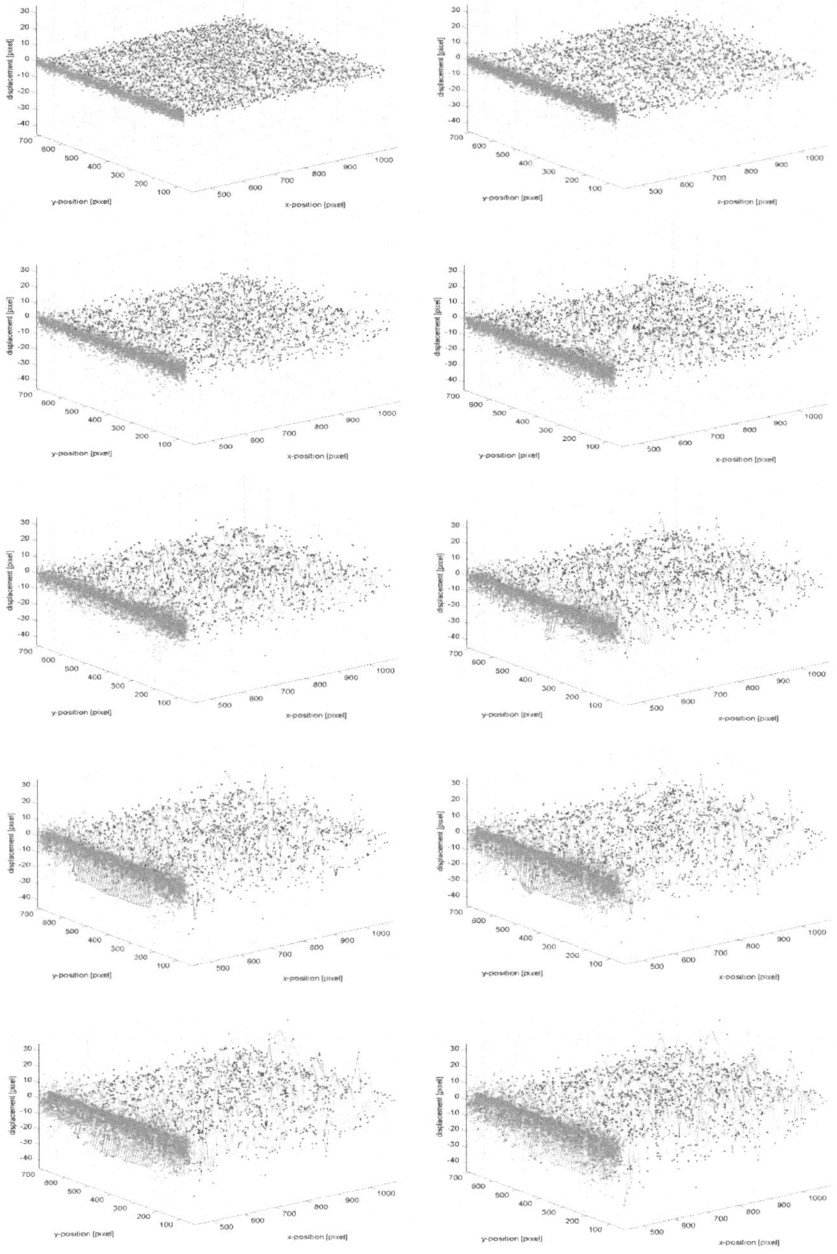

Fig. 13.37 Displacement graphs at different loads with 20% glass powder

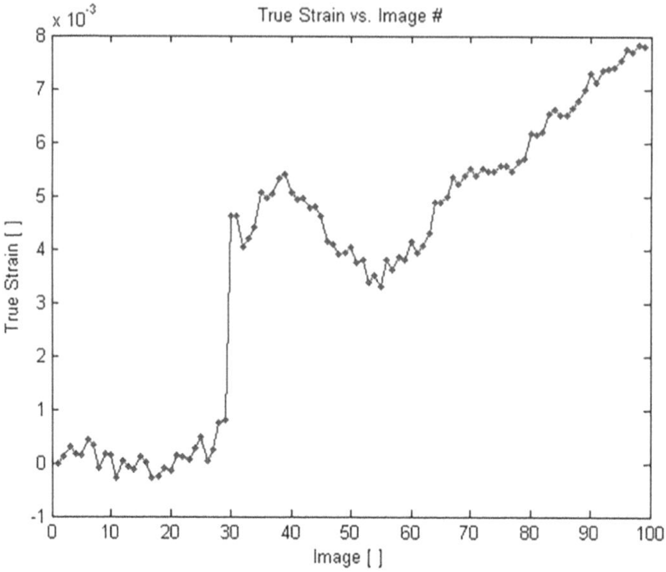

Fig. 13.38 True strain versus images graph with 20% glass powder

Images of cracks appear in Fig. 13.39. Displacement graphs at different loads are shown in Fig. 13.40. From Fig. 13.41, the maximum strain acquired is 6.9×10^{-3}. Using DIC-acquired strain, cube stress was calculated. The process's reduction factor, i.e., 5, relies on the grid size employed. Table 13.5 directs that the predicted value (DIC), i.e., 34.5 MPa, is approximately equal to the experimental value, i.e., 36.2 MPa.

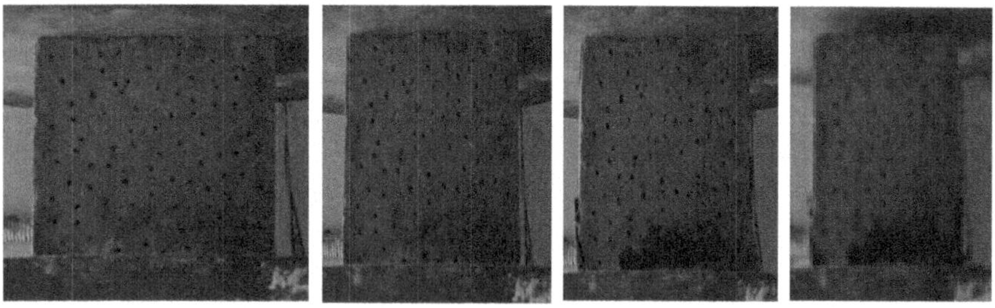

(a) (b) (c) (d)

Fig. 13.39 Specimen surface showing cracks at different loads with 21% glass powder

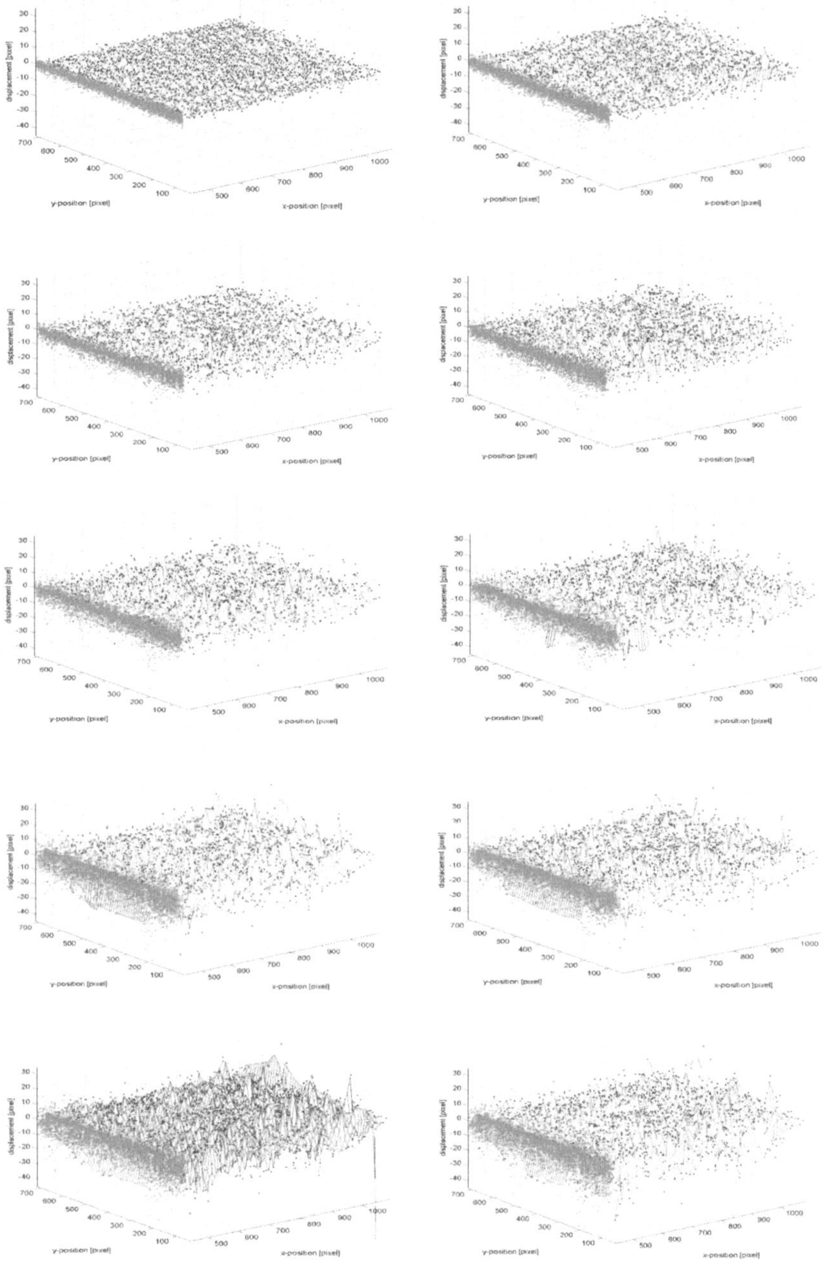

Fig. 13.40 Displacement graphs at different loads with 21% glass powder

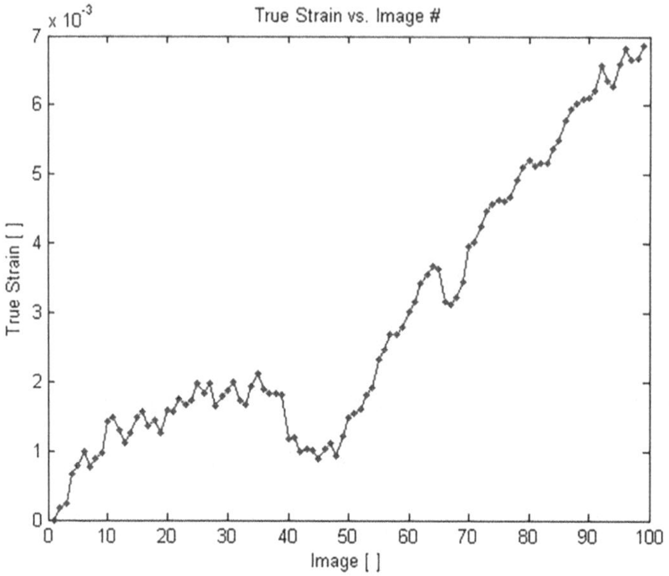

Fig. 13.41 True strain versus images graph with 21% glass powder

13.4.1.6 Cube with 22% Glass Powder

One should perform a test on cubes (GP22) with 22% glass powder as a replacement for sand. And one should find the average value of compressive strength of cubes, i.e., 33.9 MPa according to IS code. If one should compare the result with GP0, one should find little change in strength because the quantity and size of glass powder are minimal, which has a negligible effect on stability.

Images of cracks appear in Fig. 13.42. Displacement graphs at different loads are shown in Fig. 13.43. From Fig. 13.44, the maximum strain acquired is 4.8×10^{-3}. Using DIC-acquired strain, cube stress was calculated. The process's reduction factor, i.e., 5, relies on the grid size employed. Table 13.6 indicates that the predicted value (DIC), i.e., 34 MPa, is approximately equal to the experimental value, i.e., 33.9 MPa (Fig. 13.45).

13.4.2 Split Tensile Strength (STS)

To execute the STS test, one should cast three cylinders of size 300 mm × 150 mm and perform the test after 28 days of casting.

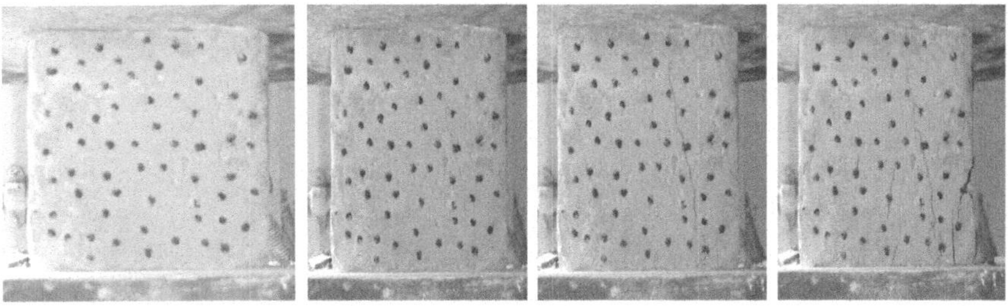

(a) (b) (c) (d)

Fig. 13.42 Specimen surface showing cracks at different loads with 22% glass powder

13.4.2.1 Cylinder with 0% Glass Powder

The trial started from the control mix (GP0), i.e., the glass powder percentage is 0. Found the average value of split tensile strength of cylinders, i.e., 2.7 MPa according to IS code. DIC test was not performed for cylinders.

13.4.2.2 Cylinder with 18% Glass Powder

One should perform a test on cylinders (GP18) with 18% glass powder as a replacement for sand. And one should find the average split tensile strength value of cylinders, i.e., 2.8 MPa, according to the IS code. If one should compare the result with GP0, one should find little change in strength because the quantity and size of glass powder are minimal, which has a negligible effect on stability. DIC test was not performed for cylinders.

13.4.2.3 Cylinder with 19% Glass Powder

One should perform a test on cylinders (GP19) with 19% glass powder as a replacement for sand. And one should find the average split tensile strength value of cylinders, i.e., 2.9 MPa, according to the IS code. If one should compare the result with GP0, one should find little change in strength because the quantity and size of glass powder are minimal, which has a negligible effect on stability. DIC test was not performed for cylinders.

13.4.2.4 Cylinder with 20% Glass Powder

One should perform a test on cylinders (GP20) with 20% glass powder as a replacement for sand. And one should find the average split tensile strength value of cylinders, i.e., 3.1 MPa, according to the IS code. If one should compare the result with

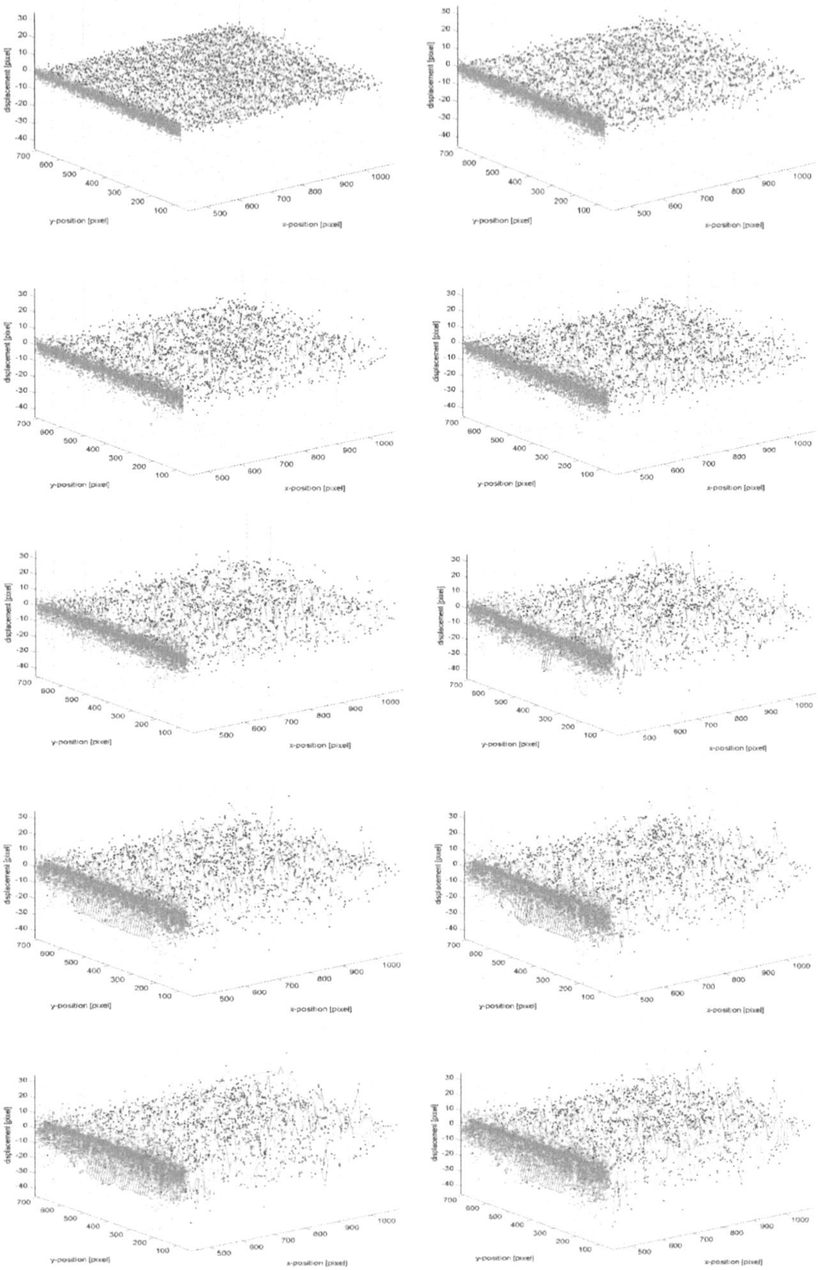

Fig. 13.43 Displacement graphs at different loads with 22% glass powder

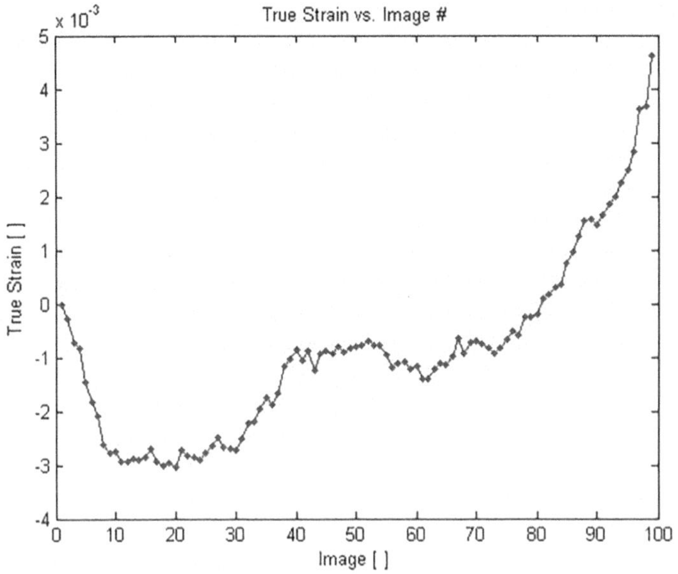

Fig. 13.44 True strain versus images graph with 22% glass powder

Table 13.6 Split tensile strength for cylinders

S. no	Glass powder (%)	Experimental stress (N/mm^2)
1	GP0	2.7
2	GP18	2.8
3	GP19	2.9
4	GP20	3.1
5	GP21	2.9
6	GP22	2.6

GP0, one should find little change in strength because the quantity and size of glass powder are minimal, which has a negligible effect on stability. DIC test was not performed for cylinders.

13.4.2.5 Cylinder with 21% Glass Powder

One should perform a test on cylinders (GP21) with 21% glass powder as a replacement for sand. And one should find the average split tensile strength value of cylinders, i.e., 2.9 MPa, according to the IS code. If one should compare the result with GP0, one should find little change in strength because the quantity and size of glass powder are minimal, which has a negligible effect on stability. DIC test was not performed for cylinders.

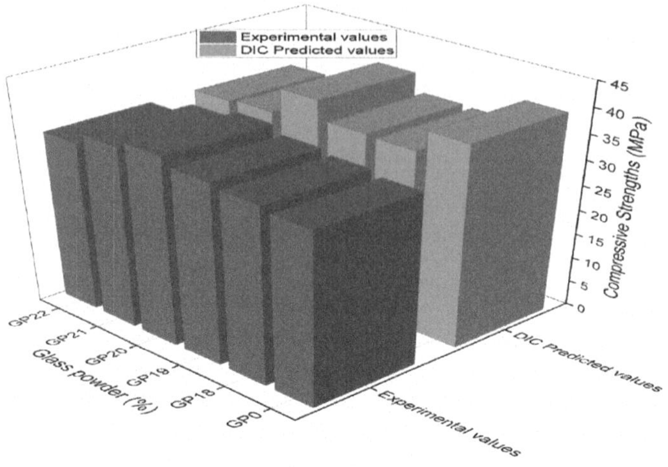

Fig. 13.45 Compressive strength comparison chart

13.4.2.6 Cylinder with 22% Glass Powder

One should perform a test on cylinders (GP22) with 22% glass powder as a replacement for sand. And one should find the average split tensile strength value of cylinders, i.e., 2.6 MPa, according to the IS code. If one should compare the result with GP0, one should find little change in strength because the quantity and size of glass powder are minimal, which has a negligible effect on stability. DIC test was not performed for cylinders (Fig. 13.46).

13.4.3 Flexure Strength (FS)

To execute the FS test, one should cast three beams and perform the test after 28 days of casting.

13.4.3.1 Beam with 0% Glass Powder

The trial started from the control mix (GP0), i.e., the glass powder percentage is 0. Found the average value of flexural strength of beams, i.e., 4.7 MPa according to IS code. DIC test was not performed for beams.

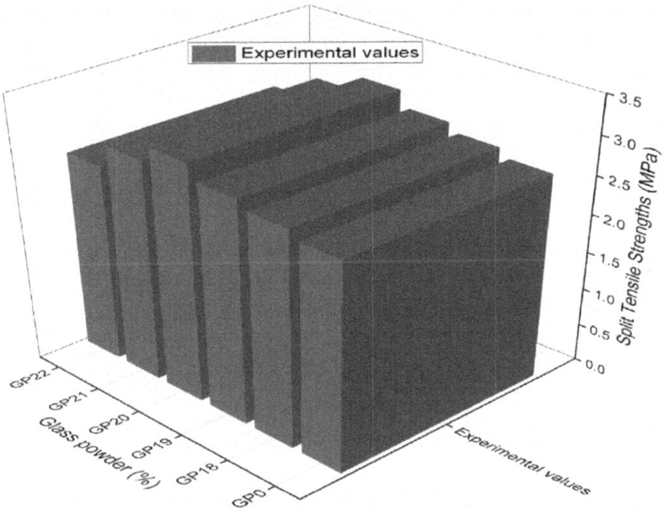

Fig. 13.46 Split tensile strength chart

13.4.3.2 Beam with 18% Glass Powder

One should perform a test on beams (GP18) with 18% glass powder as a replacement for sand. And one should find the average flexural strength value of beams, i.e., 5.3 MPa, according to the IS code. If one should compare the result with GP0, one should find little change in strength because the quantity and size of glass powder are minimal, which has a negligible effect on stability. DIC test was not performed for beams.

13.4.3.3 Beam with 19% Glass Powder

One should perform a test on beams (GP19) with 19% glass powder as a replacement for sand. And one should find the average flexural strength value of beams, i.e., 5.8 MPa, according to the IS code. If one should compare the result with GP0, one should find little change in strength because the quantity and size of glass powder are minimal, which has a negligible effect on stability. DIC test was not performed for beams.

13.4.3.4 Beam with 20% Glass Powder

One should perform a test on beams (GP20) with 20% glass powder as a replacement for sand. And one should find the average flexural strength value of beams, i.e., 5.9 MPa, according to the IS code. If one should compare the result with GP0, one should find little change in strength because the quantity and size of glass powder are minimal, which has a negligible effect on stability. DIC test was not performed for beams.

13.4.3.5 Beam with 21% Glass Powder

One should perform a test on beams (GP21) with 21% glass powder as a replacement for sand. And one should find the average flexural strength value of beams, i.e., 5.8 MPa, according to the IS code. If one should compare the result with GP0, one should find little change in strength because the quantity and size of glass powder are minimal, which has a negligible effect on stability. DIC test was not performed for beams.

13.4.3.6 Beam with 22% Glass Powder

One should perform a test on beams (GP22) with 22% glass powder as a replacement for sand. And one should find the average flexural strength value of beams, i.e., 4.5 MPa, according to the IS code. If one should compare the result with GP0, one should find little change in strength because the quantity and size of glass powder are minimal, which has a negligible effect on stability. DIC test was not performed for beams (Fig. 13.47 and Table 13.7).

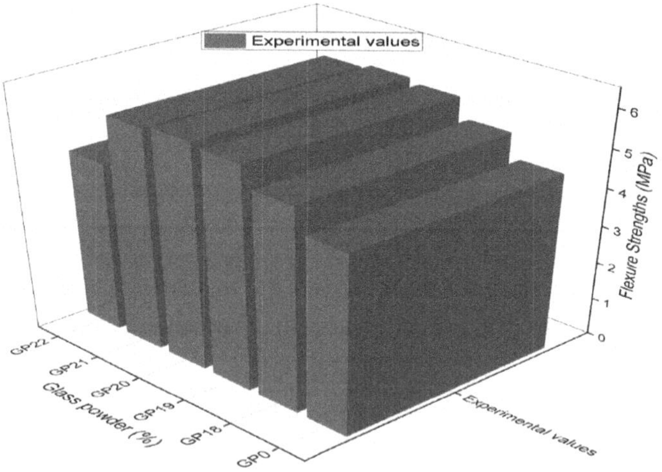

Fig. 13.47 Flexure strength chart

Table 13.7 Flexure strength of beams

S. no.	Glass powder (%)	Experimental stress (N/mm^2)
1	GP0	4.7
2	GP18	5.3
3	GP19	5.8
4	GP20	5.9
5	GP21	5.8
6	GP22	4.5

Discussion:

- Table 13.8 and figure show CS, STS, and FS for 3 samples with 0% glass powder.
- The concrete cubes without glass powder (GP) replacement exhibit excellent CS, STS, and FS.
- The control group (0% GP) is a reference for comparison with other cases.

Discussion:

- Table 13.9 and figure show CS, STS, and FS for 3 samples with 18% glass powder.
- With 18% GP replacement, the concrete's CS, STS, and FS remain good.
- There is a slight increase in strengths compared to the control group, which is within an acceptable range.
- The addition of GP at 18% demonstrates its feasibility as a partial replacement for sand.

Discussion:

- Table 13.10 and figure show CS, STS, and FS for 3 samples with 19% glass powder.
- At 19% GP replacement, the concrete's CS, STS, and FS remain satisfactory.
- There is a gradual increase in strengths compared to the control group and 18% GP replacement cases.
- The concrete with 19% GP is suitable for specific applications, but further optimization may be needed.

Discussion:

- Table 13.11 and figure show CS, STS, and FS for 3 samples with 20% glass powder.
- At 20% GP replacement, the concrete's CS, STS, and FS are excellent.
- The concrete with 20% GP exhibits a noticeable strength gain compared to the previous cases.
- While still usable, caution should be exercised in applications with higher structural demands.

Table 13.8 Samples with 0% glass powder

Samples	Strength (MPa)			Remarks
	CS	STS	FS	
GP0S1	34.2	2.7	4.6	Excellent
GP0S2	34.6	2.8	4.7	Excellent
GP0S3	34.1	2.6	4.6	Excellent

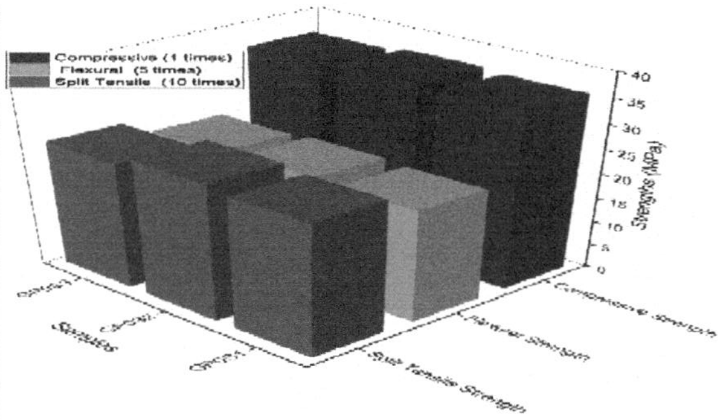

Table 13.9 Samples with 18% glass powder

Samples	Strength (MPa)			Remarks	
	CS	STS	FS		
GP18S1	35.3	2.9	5.3	Good	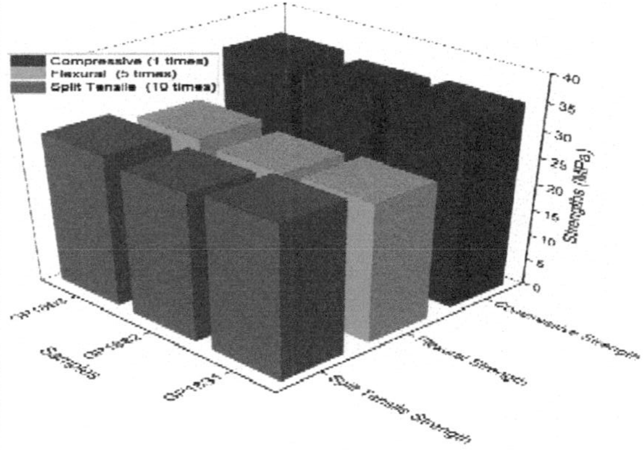
GP18S2	34.8	2.8	5.2	Good	
GP18S3	35.5	2.9	5.4	Good	

Table 13.10 Samples with 19% glass powder

Samples	Strength (MPa)			Remarks
	CS	STS	FS	
GP19S1	35.7	2.7	5.9	Satisfactory
GP19S2	35.9	2.8	5.9	Satisfactory
GP19S3	36.4	2.9	6.1	Satisfactory

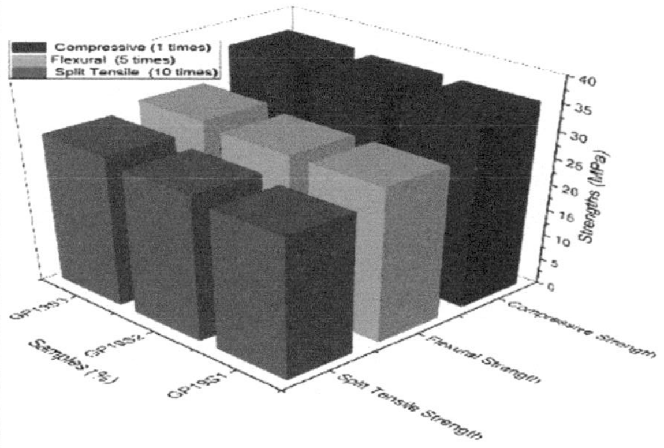

Table 13.11 Samples with 20% glass powder

Samples	Strength (MPa)			Remarks	
	CS	STS	FS		
GP20S1	37.9	3.2	6.7	Excellent	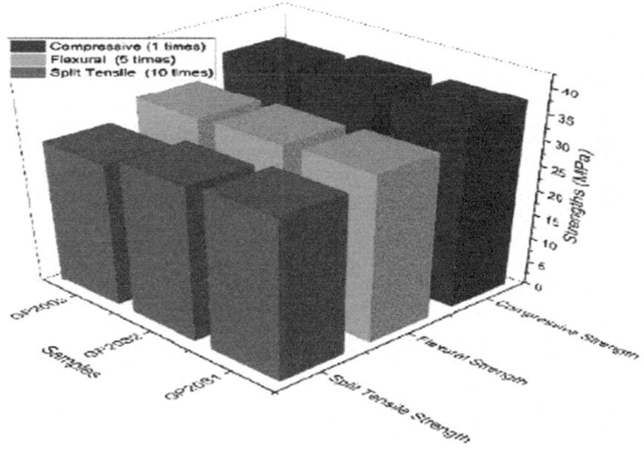
GP20S2	37.8	3.1	6.6	Excellent	
GP20S3	36.8	2.9	6.5	Excellent	

Discussion:

- Table 13.12 and figure show CS, STS, and FS for 3 samples with 21% glass powder.
- At 21% GP replacement, the concrete's CS, STS, and FS are marginal.
- The concrete with 21% GP demonstrates significant reductions in strengths compared to the previous case of 20% replacement.
- Its use may be restricted to non-structural or less demanding applications.

Discussion:

- Table 13.13 and figure show CS, STS, and FS for 3 samples with 22% glass powder.
- At 22% GP replacement, the concrete's CS, STS, and FS are not ideal.
- Concrete with 22% GP shows a substantial decline in strengths compared to all previous cases.
- Its use should be carefully evaluated, considering non-structural elements.

Table 13.14 and figure shows the average CS, STS, and FS of samples with varying percentage of glass powder as a replacement for sand (Fig. 13.48).

13.4.4 Microstructural Analysis

13.4.4.1 Scanning Electron Microscope (SEM)

Microstructural examination using Scanning Electron Microscopy (SEM) offers a visual insight into the material's composition and structure. As depicted in Fig. 13.49, SEM micrographs of the control mix (GP0) and the mixture containing 20% glass powder (GP20) were obtained. The SEM micrograph of the control mix reveals the typical cementitious matrix with well-distributed aggregates. In contrast, the micrograph of the GP20 mix showcases an interplay between the glass particles and the cementitious phases. The presence of glass particles is evident, which suggests the integration of glass powder within the concrete matrix.

The Scanning electron microscope micrograph in Fig. 13.49 provides a detailed view of the microstructure of concrete samples, highlighting the impact of waste glass powder (WGP) on the composition and arrangement of the materials. The two images, labelled (a) and (b), represent two different concrete mixes: (a) the control mix with 0% waste glass content and (b) the GP20 mix with 20% waste glass content.

(a) ***Control mix—0% waste glass***

In the SEM micrograph labelled as (a), which corresponds to the control mix without any waste glass powder (GP0), specific characteristics of the concrete microstructure are evident. The matrix of the concrete is composed of cementitious particles, aggregates, and pores. Pores appear as void spaces within the matrix, representing areas where the concrete is less dense. These pores can significantly impact the material's

Table 13.12 Samples with 21% glass powder

Samples	Strength (MPa)			Remarks	
	CS	STS	FS		
GP21S1	36.1	2.9	6.1	Marginal	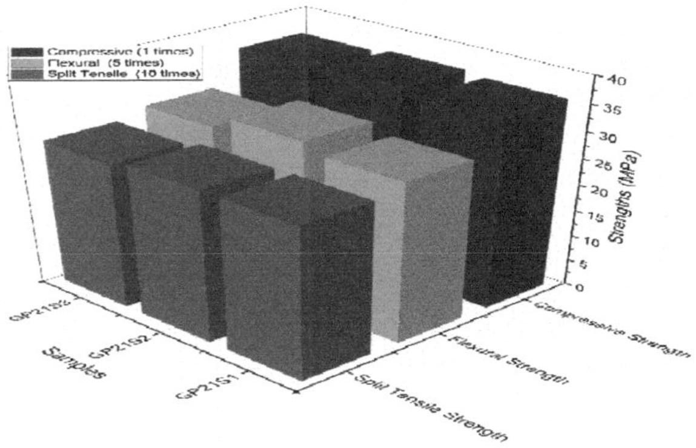
GP21S2	36.6	2.9	6.4	Marginal	
GP21S3	35.9	2.8	5.9	Marginal	

Table 13.13 Samples with 22% glass powder

Samples	Strength (MPa)			Remarks
	CS	STS	FS	
GP22S1	34.2	2.7	4.7	Not Ideal
GP22S2	34.1	2.6	4.6	Not Ideal
GP22S3	33.4	2.5	4.5	Not Ideal

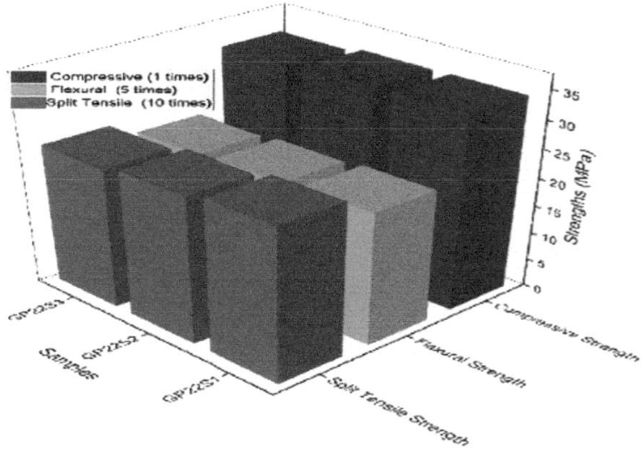

Table 13.14 Average stresses

GP (%)	Strength (MPa)		
	CS	STS	FS
GP0	34.3	2.7	4.7
GP18	35.2	2.8	5.3
GP19	36	2.9	5.8
GP20	37.5	3.1	5.9
GP21	36.2	2.9	5.8
GP22	33.9	2.6	4.5

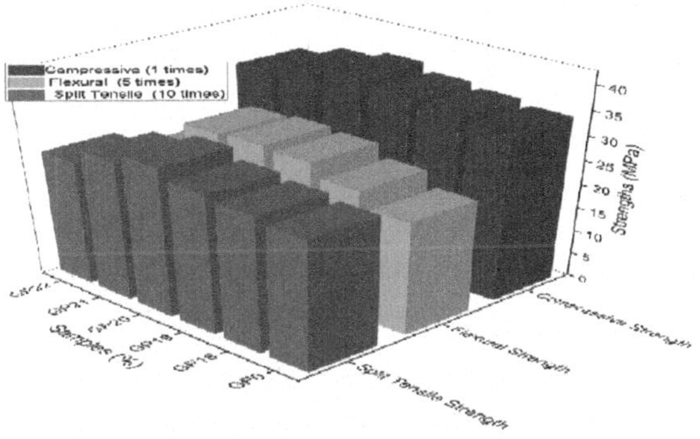

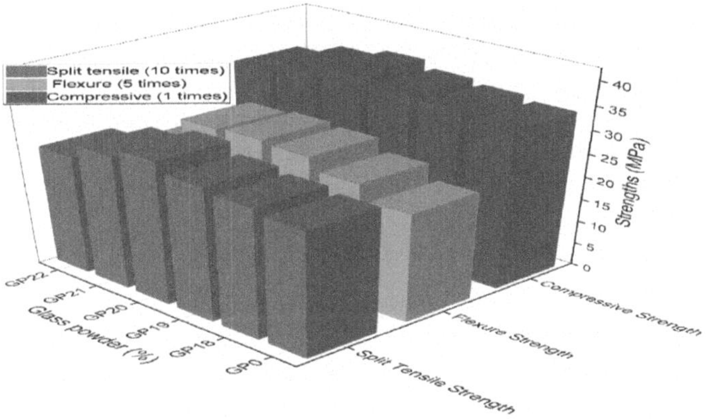

Fig. 13.48 Average stresses chart

Fig. 13.49 SEM micrograph of concrete **a** control mix and **b** GP20 mix

overall strength and durability by providing pathways for water and other potentially harmful agents to infiltrate the structure. The absence of waste glass particles means that the microstructure lacks any additional filler material that could contribute to reducing these pores.

(b) ***GP20 mix—20% waste glass***

The SEM micrograph labelled as (b) represents the GP20 mix, where waste glass powder has replaced 20% of the sand content. This addition of waste glass introduces a notable change in the microstructure. Tiny particles of waste glass are dispersed within the matrix, effectively filling some of the void spaces that would otherwise exist. As a result, the microstructure appears to be more compact and less porous than the control mix. The waste glass particles themselves are visible as distinct particles within the matrix, contributing to the overall densification of the material. Their presence appears to have facilitated a better packing arrangement of the cementitious particles and aggregates, thus reducing pore size and interconnected porosity. The cementitious particles, aggregates, and overall matrix structure remain consistent in both micrographs. However, introducing waste glass particles in the GP20 mix (image labelled as b) influences the microstructure by modifying the distribution and arrangement of voids and particles. The SEM micrographs vividly capture the impact of waste glass powder on concrete microstructure, with the GP20 mix showcasing the potential benefits of waste glass as a filler material for improving the material's porosity and possibly enhancing its mechanical properties.

13.4.4.2 Raman Spectrum

The Raman spectrum provides insight into materials' molecular vibrations and crystal lattice characteristics. Figure 13.50 illustrates the Raman spectrum graph for the control mix (GP0). The spectrum exhibits characteristic peaks associated with conventional concrete components. However, in future studies, it would be valuable to examine the Raman spectra of concrete mixes containing different percentages of glass powder to identify any distinct signatures related to glass integration.

Figure 13.50 illustrates the Raman spectroscopy graph for the control mix (GP0), which consists of 0% waste glass. Raman spectroscopy is a technique used to study molecular vibrations and provides information about a material's molecular structure and composition based on how it scatters light. The x-axis of the graph represents the Raman shift. It is typically measured in reciprocal centimeters (cm^{-1}) and indicates the material's vibrational frequencies of molecular bonds. The y-axis shows the intensity of the Raman scattering, which measures the amount of light scattered at different Raman shifts. The Raman spectroscopy graph for the control mix (GP0) provides insights into the molecular structure of the concrete mixture without any waste glass addition. It identifies the vibrational modes and chemical bonds present in the materials that constitute the concrete. The detailed interpretation of the graph's peaks and vibrational modes would require specific information about the Raman shift values and their corresponding molecular vibrations. Raman spectroscopy is

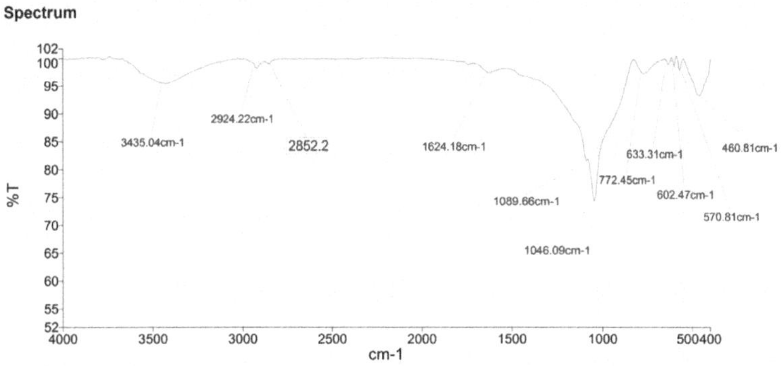

Fig. 13.50 Raman spectrum graph for control mix (GP0)

precise to the material being studied, and the accurate interpretation of the peaks often requires understanding the molecular structures of the compounds involved. Figure 13.50 presents the Raman spectroscopy graph for the control mix (GP0), offering valuable information about the molecular vibrations and composition of the concrete mixture without waste glass. This information can aid researchers in understanding the interaction and behaviour of the components in the concrete mix at the molecular level.

13.4.4.3 Interpreting the Graph

Intensity Peaks: The graph displays a series of peaks, each corresponding to specific vibrational modes in the material. These peaks are the result of molecular vibrations within the sample.

Baseline: The baseline of the graph represents the background signal. The intensity peaks rise above this baseline, indicating specific vibrational modes.

Peak's Position and Shape: Each peak is associated with specific molecular vibrations, corresponding to the material's arrangement of atoms and bonds. The position (Raman shift) and shape of the peaks provide information about the types of chemical bonds present and the molecular environments of the atoms.

Fingerprint Region: Different materials have unique Raman spectra, often called their "fingerprint" region. This region contains peaks that are specific to the material's molecular composition.

Raman Modes: In the context of your description, the Raman modes may refer to specific vibrational modes associated with the glass and mineral components present in the concrete mix.

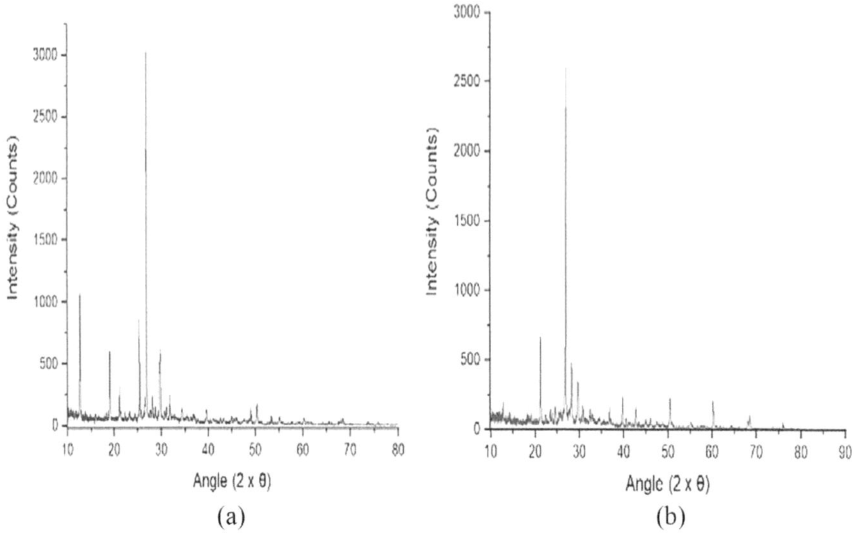

Fig. 13.51 XRD of waste glass powder **a** control cube and **b** GP20 cube

13.4.4.4 X-ray Powder Diffraction (XRD)

X-ray Powder Diffraction (XRD) provides information about the crystalline phases present within the concrete. Figure 13.51 displays XRD images of the control cube (GP0) and the cube with 20% glass powder (GP20). The XRD pattern of the control cube corresponds to the well-known peaks of typical cement phases. However, in the GP20 cube, additional peaks corresponding to glass-related phases are discernible. This indicates the successful inclusion of glass powder and its interaction with the existing cementitious compounds.

Figure 13.51 showcases the X-ray Diffraction (XRD) patterns of waste glass powder, illustrating the qualitative composition of the materials used in the study. The XRD technique is employed to identify the crystalline phases present in a given material, providing insights into its mineralogical composition.

(a) ***Control cube (0% waste glass)***

In the XRD pattern labelled as (a), representing the control cube without any waste glass addition, the diffraction peaks correspond to the mineralogical phases in the brick clay used in the study. The XRD graph's peaks indicate the crystal lattice structures of these phases. The major crystalline phases identified in the control cube are calcite, clinochlore, muscovite, and quartz. These peaks represent the characteristic scattering of X-rays by the crystalline lattice planes of these minerals, allowing for their qualitative identification.

(b) ***GP20 cube (20% waste glass)***

The XRD pattern marked as (b) pertains to the GP20 cube, wherein waste glass particles have replaced 20% of the sand in the brick clay. In this XRD graph, the presence of crystalline peaks related to minerals is minimal, indicating glass's amorphous nature. Unlike the control cube, no distinct diffraction peaks correspond to mineral phases like calcite, clinochlore, muscovite, or quartz. This absence of crystalline peaks suggests that the waste glass particles used in the GP20 mix are primarily amorphous, lacking the well-ordered lattice structures that give rise to diffraction patterns in crystalline materials. Figure 13.49 provides an XRD-based comparison between the mineralogical composition of the control cube (0% waste glass) and the GP20 cube (20% waste glass). The control cube's XRD pattern displays peaks representing the crystalline phases of the brick clay's minerals, while the GP20 cube's pattern lacks such heights due to the amorphous nature of the waste glass particles. This comparison aids in understanding the material differences introduced by the inclusion of waste glass in the concrete mix, highlighting the transformation from crystalline to amorphous phases.

13.5 Conclusions

Integrating microstructural analysis techniques, such as Scanning Electron Microscopy (SEM), X-ray Powder Diffraction (XRD), and Raman Spectroscopy, enriches understanding of the complex interactions between glass powder and the conventional concrete matrix. These techniques offer a multidimensional perspective on incorporating glass powder, shedding light on the material's structural modifications and potential phase changes. The microstructural analysis validates the physical integration of glass powder into the concrete matrix. Observing glass particles within the SEM micrographs and additional peaks in the XRD pattern corresponding to glass phases underlines the successful blending of glass powder with the cementitious matrix. Further microstructural analysis investigations could explore the distribution of glass particles within the matrix, potential chemical interactions between glass and cementitious phases, and the impact of glass powder on forming hydration products. These insights would contribute to a holistic understanding of the composite material's behaviour and could guide optimization efforts for achieving desirable mechanical and microstructural properties. The comprehensive analysis of the experimental data has led to the following key conclusions:

- GP Content Effect on Mechanical Properties: The glass powder generally enhances concrete properties compared to conventional concrete. This enhancement was particularly pronounced for concrete mixes containing up to 20% GP content.
- Improvement in Mechanical Properties: Concrete's crushing strength exhibited a notable increase with higher GP content, with mixes containing 20% GP showing a substantial 9.33% improvement over regular concrete. Similarly, concrete's

split tensile and flexural strength demonstrated remarkable improvements as GP content increased to 20%, with increases of 14.81% and 25.53%, respectively.

- Optimal GP Content: The study highlighted an optimal GP content of 20% for achieving the most favourable mechanical properties. Beyond this threshold, there was a decline in mechanical strength, suggesting that an excessive GP content might compromise concrete performance.
- Potential for Sustainable Concrete: The findings of this research signify the potential of incorporating waste glass powder into concrete to produce environmentally friendly and structurally-efficient building materials.

13.5.1 Future Scope

The study has unveiled several avenues for further exploration, including investigating different types and particle sizes of glass powder, variations in cement forms and water-cement ratio, and supplementary materials or additives to enhance concrete properties. The durability characteristics of concrete with glass powder also warrant further research, encompassing its resistance to environmental factors, chemical attacks, and long-term aging effects.

This research provides valuable insights into using glass powder as a sustainable additive in concrete production. The optimization of GP content to achieve desired mechanical properties aligns with the growing emphasis on waste recycling and eco-conscious construction practices. By addressing the outlined future research directions, the construction industry can progress toward more resilient and eco-friendly building methods, ushering in a new era of sustainable infrastructure development. As structural health monitoring advances, applying Digital Image Correlation for analysing concrete behaviours opens doors to a deeper understanding of material response under varying conditions. This research marks a significant step forward in advancing glass powder use in concrete and using advanced monitoring techniques to ensure the structural integrity of modern construction materials.

References

1. Novák J, Kohoutková A (2017) Fiber-reinforced concrete is exposed to elevated temperatures. IOP Conf Ser Mater Sci Eng 246(1):012045. IOP Publishing
2. Peng GF, Yang WW, Zhao J, Liu YF, Bian SH, Zhao LH (2006) Explosive spalling and residual mechanical properties of fiber-toughened high-performance concrete subjected to high temperatures. Cem Concr Res 36(4):723–727
3. Ramana PV, Surendranath A (2021) Recycled substance evaluation and effect on fire resistance structures. Mater Today Proc
4. Orouji M, Zahrai SM, Najaf E (2021) An environmental approach is the effect of glass powder & polypropylene fibers on concrete's compressive and flexural strengths, toughness, and ductility. Structures 33:4616–4628. Elsevier

5. Surendranath A, Ramana PV (2021) Mathematical approach on recycled material strength performance via statistical mode. Mater Today Proc 38(Part 5):2943–2952
6. ASTM Standard Specification for chemical admixtures for concrete. ASTM C494/C494M (2019)
7. Singh V, Ramana PV (2013) The magnitude of linear problems. In: International conference on emerging trends in engineering and applied sciences (ICETEAS), published in Int J Adv Eng Comput Technol. ISSN: 2249-4928
8. Agnihotri A, Ramana PV (2021) GGBS: fly-Ash evaluation and mechanical properties within high strength concrete. Mater Today Proc 38(Part 5):2943–2952
9. Chen M, Sun Z, Tu W, Yan X, Zhang M (2021) The behavior of recycled tire polymer fiber reinforced concrete at elevated temperatures. Cem Concr Compos 124:104257
10. Uğur AE, Ünal A (2022) Assessing the structural behavior of reinforced concrete beams produced with macro synthetic fiber reinforced self-compacting concrete. Structures 38:1226–1243. Elsevier
11. Meena A, Ramana PV (2021) Impact of blast loading over reinforced concrete without infill structure. Mater Today Proc
12. Bamonte P, Gambarova PG (2012) A study on the mechanical properties of self-compacting concrete at high temperature and after cooling. Mater Struct 45(9):1375–1387
13. Bangi MR, Horiguchi T (2011) Pore pressure development in hybrid fiber-reinforced high-strength concrete at elevated temperatures. Cem Concr Res 41(11):1150–1156
14. Singh V, Ramana PV (2021) Statistical concert of strength effect on fiber concrete. Mater Today Proc 38(Part 5):2943–2952
15. IS 1489 (Part 1) (2015) Portland-Pozzolana cement specification
16. IS 10262 (2019) Concrete mix proportioning guidelines
17. IS 383 (2016) Coarse and fine aggregate for concrete specification
18. Ramana PV (2020) Functioning of bi-material interface intended for polypropylene fiber concrete. Mater Today Proc
19. Jameran A, Ibrahim IS, Yazan SHS, Rahim SNA (2015) Mechanical properties of steel-polypropylene fiber reinforced concrete under elevated temperature. Proc Eng 125:818–824
20. Murahari K, Rao R (2013) Effects of polypropylene fibers on the strength properties of fly ash-based concrete. Int J Eng Sci Invent 2(5):13–19
21. ACI-318 (2019) Building code requirements for structural concrete and commentary
22. Eurocode 2 (2004) Design of concrete structures—Part 1–1: general rules and rules for buildings. EN 1992-1-1
23. Rani SY, Nusari MS, Bin Non J, Poddar S, Bhaumik A (2022) The durability of geopolymer concrete with adding polypropylene fiber. Mater Today Proc 56:2846–2851
24. Dheyaaldin MH, Mosaberpanah MA, Alzeebaree R (2022) Shrinkage behavior and mechanical properties of alkali-activated mortar incorporating nanomaterials and polypropylene fibers. Ceram Int
25. Agnihotri A, Ramana PV (2021) Mathematical model for recycled polyethylene terephthalate material mechanical strengths. Mater Today Proc 38(Part 5):2943–2952. https://doi.org/10.1016/j.matpr.2020.09.260
26. Karimipour A (2020) Effect of untreated coal waste as a fine and coarse aggregate replacement on steel and polypropylene reinforced concrete's properties. Mech Mater 150:103592
27. Ramana PV, Agnihotri A (2021) Modified adomian decomposition method for uni-directional fracture problems. Sadhana, unique addition, pp 1–9
28. Bureau of Indian Standards (1959) Methods of tests for strength of concrete. IS:516
29. Surendranath A, Prasad BKR, Ramana PV (2021) Interpretation of bi-material interface through the mechanical and microstructural possessions. Mater Today Proc
30. Bureau of Indian Standards (2000) Plain and reinforced concrete code of practice. IS:456
31. Ramana PV, Meena A (2021) High-rise structural stalling, and drift effect owe to lateral loading. Mater Today Proc 38(Part 5):2943–2952
32. KG A (2020) Mechanical, ballistic impact, and water absorption behavior of luffa/graphene reinforced epoxy composites. Polym Compos 41(11):4716–4726

33. Ashok KG, Kani K (2021) Experimental studies on interlaminar shear strength and dynamic mechanical analysis of luffa fiber epoxy composites with nano PbO addition. J Ind Text:15280837211052317
34. Ashok K, Ajith D, Bibin C, Sheeja R, Nishanth R (2021) Influence of nanofiller lignite fly ash on the tribo-mechanical performance of Sansevieria roxburghiana fiber reinforced epoxy composites. J Nat Fibers:1–15
35. Bayraktar OY, Eshtewı SST, Benli A, Kaplan G, Toklu K, Gunek F (2021) The impact of RCA and fly ash on the mechanical and durability properties of polypropylene fiber-reinforced concrete exposed to freeze-thaw cycles and $MgSO_4$ with ANN modeling. Constr Build Mater 313:125508
36. El Ouni MH, Shah SHA, Ali A, Muhammad S, Mahmood MS, Ali B, Raza A et al (2022) Mechanical performance, water and chloride permeability of hybrid steel-polypropylene fiber-reinforced recycled aggregate concrete. Case Stud Constr Mater 16:e00831
37. Prasad BKR, Ramana PV (2014) Modified adomian decomposition method for Van der Pol equations. Int J Nonlinear Mech 65:121–132
38. Gencel O, Kazmi SMS, Munir MJ, Kaplan G, Bayraktar OY, Yarar DO, Ahmad MR et al (2021) Influence of bottom ash and polypropylene fibers on the physicomechanical, durability and thermal performance of foam concrete: an experimental investigation. Constr Build Mater 306:124887
39. Wang C, Wu H, Li C (2022) Hysteresis and damping properties of steel and polypropylene fiber reinforced recycled aggregate concrete under uniaxial low-cycle loadings. Constr Build Mater 319:126191
40. Yan P, Chen B, Afgan S, Haque MA, Wu M, Han J (2021) Experimental research on ductility enhancement of ultra-high performance concrete incorporation with basalt fiber, polypropylene fiber, and glass fiber. Constr Build Mater 279:122489
41. Meena A, Ramana PV (2021) Mathematical model for recycled polyethylene terephthalate material mechanical strengths. Mater Today Proc 38(Part 5):2943–2952
42. Karimipour A, Ghalehnovi M (2021) Comparison of the effect of the steel and polypropylene fibers on the flexural behavior of recycled aggregate concrete beams. Structures, Elsevier 29:129–146
43. Ali B, Qureshi LA, Kurda R (2020) Environmental and economic benefits of steel, glass, and polypropylene fiber-reinforced composite cement application in jointed plain concrete pavement. Compos Commun 22:100437
44. Bentegri I, Boukendakdji O, Kadri EH, Ngo TT, Soualhi H (2020) Rheological and tribological behaviors of polypropylene fiber reinforced concrete. Constr Build Mater 261:119962
45. Ramana PV, Meena A (2021) Assessment of endurance and microstructural properties effect on polypropylene concrete. Mater Today Proc 38(Part 5):2943–2952. https://doi.org/10.1016/j.matpr.2020.09.260
46. Yuan Z, Jia Y (2021) Mechanical properties and microstructure of glass fiber and polypropylene fiber reinforced concrete: an experimental study. Constr Build Mater 266:121048
47. Karimipour A, Ghalehnovi M, De Brito J, Attari M (2020) The effect of polypropylene fibers on the compressive strength, impact, and heat resistance of self-compacting concrete. Structures, Elsevier 25:72–87
48. Singh V, Ramana PV (2014) The emerging solution for partial differential problems, SEC2014, IIT Delhi, India (SEC14BAT). ISBN: 978-81-322-2189-0
49. Abdul-Rahman M, Al-Attar AA, Hamada HM, Tayeh B (2020) Microstructure and structural analysis of polypropylene fiber reinforced reactive powder concrete beams exposed to elevated temperature. J Build Eng 29:101167
50. Amin M, Tayeh BA (2022) Investigating fiber-reinforced lightweight concrete's mechanical and microstructure properties under elevated temperatures. Case Stud Constr Mater 13:e00459
51. Krishna A, Kaliyaperumal SRM (2021) Effect of elevated temperature on strength and ductility of axially loaded hybrid fiber reinforced concrete columns. Structures, Elsevier 34:3548–3556
52. Rasoul ZMA, Radhi MS, Alsaad AJ, Muhannad H (2020) The high-temperature performance of reinforced concrete beams containing waste polypropylene fibers. Case Stud Therm Eng 21:100705

Chapter 14
A Comprehensive Review of Utilizing Smart Bricks in Structural Health Monitoring and Damage Detection of Masonry Structures

S. M. Priok Rashid, Atefeh Soleymani, and Mohammadreza Mofidi

Abstract Most of damage detection and structural health monitoring (SHM) methods need sensors which are attached to the structure surfaces or gather required data via remote sensing process. The principal objective of this article is to examine novel and recent concepts regarding smart bricks which can be classified as a robust structural health monitoring method in masonry structures. Moreover, recent advances in smart sensor technology are discussed, along with various field applications for monitoring strain in a full/small-scale masonry building prototype subjected to controlled damage. In recent years, innovative piezoresistive smart sensors that bear resemblance to traditional clay bricks in appearance, and are known as "smart bricks", have been developed to monitor diffuse strain in masonry structures, thereby enabling the deployment of monitoring systems to be significantly facilitated for this type of building. The article further elaborates on the manufacturing process of smart bricks, their integration into masonry load-bearing structures, and the numerous experimental outcomes. In order to maximize such improvements to the structure, SHM functions may offer significant details on the structural stability of brick structures to use. These systems are designed at regaining or enhancing their structural integrity for timely preservation and ensuring the inhabitants' safety. Through laboratory experimentation utilizing a full or small-scale wall sample subjected to diverse loading scenarios and controlled impairment, it has been evidenced that smart bricks possess the potential for multifarious applications, including the capacity for identifying fractures within masonry walls. In essence, the findings demonstrate that a

S. M. Priok Rashid
Department of Civil Engineering, Faculty of Engineering and Built Environment, Universiti Kebangsaan Malaysia, UKM Bangi 43600, Malaysia
e-mail: p131825@siswa.ukm.edu.my

A. Soleymani (✉) · M. Mofidi
Department of Civil Engineering, Shahid Bahonar University of Kerman, Kerman, Iran
e-mail: Atefeh_soleymani@eng.uk.ac.ir

M. Mofidi
e-mail: Mofidi@eng.uk.ac.ir

H. Jahangir et al. (eds.), *Damage Detection and Structural Health Monitoring of Concrete and Masonry Structures*, Springer Tracts in Civil Engineering,
https://doi.org/10.1007/978-981-97-8975-7_14

small number of strategically positioned smart bricks can effectively monitor the strain magnitude within a wall and generate data that can be utilized for identifying the presence of cracks.

Keywords Smart bricks · Damage detection · Structural health monitoring · Acoustic emissions · Manufacturing process · Electrical experiments · Seismic behavior · Strain-sensing behavior

14.1 Introduction

The implementation of effective monitoring systems is crucial for structural evaluation and identification of damages in order to actively conserve historical masonry structures. Such technologies strive to optimize maintenance and restoration efforts while operating within the confines of established resources allocated for historical preservation [1]. Customized Structural Health Monitoring (SHM) systems have proven to be advantageous in preserving the integrity of masonry structures over their operational lifespan. Such systems aid in the timely identification of structural alterations in monitored edifices, and thus play a key role in detecting the onset of damages or precarious conditions, and in devising suitable retrofitting and maintenance measures [2–4].

The static or dynamic response characteristics of masonry structures are the basis of their structural health monitoring [5]. Vibration-based monitoring, a long-term method that requires only a few sensors, is commonly employed for monitoring historical structures. This method identifies variations in modal features, such as natural frequencies, to detect anomalies in the overall structural response caused by local enhancements in structural stiffness from the development of crack patterns [6–8]. Static parameter-based techniques, on the other hand, are used to identify the local damages and cracks propagation and growth through the use of specialized sensors and measurement instruments such as linear variable differential transformers (LVDTs), resistive strain gauges (RSGs) and inclinometers [9]. These methodologies address a limitation that dynamic monitoring systems face by analyzing low-severity local damages to impact modal properties. However, the installation of sensors directly on the vulnerable area of the building or across a fissure, as well as their maintenance and inspection throughout the useful masonry structures lifespan, may present challenges and incur high costs.

Additionally, methods that use contactless sensors can be used to locate cracks, such as infrared thermography14 and ground-penetrating radar for rapid periodic inspections [10]. Digital image correlation (DIC), also known as digital image processing or, more generally, digital image processing, is another contactless technology for monitoring strain in masonry buildings [11]. The utilization of DIC has been employed in the assessment of the efficacy of composite reinforcements [12] and may be employed to supervise the evolution of cracks and deteriorations in masonry structures [13]. In order to recognize and determine damage in masonry edifices, acoustic emissions [14] and examination methods that depend on modifications in the speed of elastic waves [15] have also been proposed. Additionally, terrestrial laser scanning [16–18] is a highly efficient technique for accurately mapping surface crack patterns and reconstructing the shape of masonry buildings.

Rapid detection of the behavior is preferred since critical occurrences like earthquakes can endanger the integrity of the structures and result in serious cracks, damages, and collapses. Because of its brittle failure mechanism and non-homogeneous behavior, masonry buildings appear to be most vulnerable to seismic stresses among all construction typologies [19, 20].

Unfortunately, there aren't many large-scale examples of continuous SHM systems being used on brick structures right now. This is mainly because off-the-self sensing methods have its limitations. Traditional sensors are sensitive to ambient variables when used in situ, which over time may significantly decrease their effectiveness. Moreover, there are constraints associated with the affordability of the sensing systems, the difficulty of installing the sensors on the structure, and the need for extensive wiring to connect to the data collection system. When analyzing historical buildings, the aforementioned limitations are particularly significant as it is vital to safeguard their visual and structural attributes.

SHM systems may not be able detect local changes or crack develops since they only employ commercial sensors placed at a small number of sites. These sensors also have concerns with durability, installation costs, and maintenance since they are externally connected and have physical and mechanical characteristics that are significantly unlike from those of the structure they are monitoring. Recent advances in materials science have made a variety of multifunctional building materials with improved physical, chemical, electrical, and mechanical capabilities available. Piezoresistive fillers in particular have the potential to improve the self-sensing properties of the matrix in which they are published in [21–23]. These transformed materials are less expensive, more widely and simply applicable, and have durability comparable to the original matrices.

Carbon-based nanofillers, especially carbon nanotubes and nanofibers, have been employed extensively to enhance the electrical properties and piezoresistive abilities of cement-based materials [24, 25]. These materials have been utilized to fabricate sensors that, when damaged, can produce a noticeable change in the electrical output. Contemporary self-sensing technologies facilitate the identification of initial and progressing localized damage by offering real-time information on a building's overall health status [26–28]. Automated Structural Health Monitoring (SHM) systems can leverage this advanced technology for concrete construction applications.

Downey et al. [29] made the initial suggestion for "smart bricks". They also demonstrated the piezoresistive behavior and gave an example of how they might be used to determine the condition of a small-scale wall that is being subjected to eccentric compression loads. These were created by the authors to perform diffuse strain monitoring in masonry activities, making it possible to install monitoring systems in all types of structures that fall under this category.

This research endeavors to achieve a broader comprehension of the smart bricks strain monitoring properties. The main aim is to develop an enhanced electromechanical model that establishes a correlation between the electrical output of the bricks and their strain condition. This model will incorporate both the internal piezoresistivity and the sensing that takes place at the electrode contact interface. The study posits that by utilizing global and concentrated damage indicators that are established on the electrical outputs of smart bricks, it is possible to detect and pinpoint damage in the aftermath of each seismic event, including its initial phases. The research recognizes that a more comprehensive characterization of the behavior of smart bricks is necessary, as the elementary linear strain gauge equation is insufficient in describing the strain-sensing conduct of smart bricks under relatively low compressive conditions.

14.2 Structural Health Monitoring

The goals of SHM have been used by numerous authors to describe or explain the idea. The following paragraphs provide more instances using the concept first presented in the seminal research paper [30]:

- Saisi et al. [5] defined SHM as the ongoing monitoring of sensors attached to a structure in order to determine characteristics that reflect the current structural condition and health.
- According to Bassoli et al. [31], 'Characterizing existing civil structures by SHM allows for the identification of their structural components and damage detection purposes.'
- According to Guidorzi et al. [32], 'SHM seeks to continuously diagnose the "state" of the various parts and of the entire ensemble of the components that make up the structure as a whole.'

The core idea of SHM is to characterize a structural system by monitoring how it responds to loads. Model-based approaches can be utilized by SHM, however, these models comprise of various indeterminacies and reductions, such as in materials, geometry, and boundary conditions, which may result in disparities between the predicted outcomes of the model and the actual structural behavior. SHM uses structural identification methods to verify the models and predict how the structure will respond in different situations. Ongoing SHM can detect damage or determine how well the methods are working [33].

Foti [34] demonstrated damage detection was initially done through visual inspection or methods that required accessing easily visible areas. Techniques like X-rays, magnetism, acoustics, and thermal measurements could measure damage. In recent years, new sensing technologies have enabled analyzing factors that provide more insight into structural performance and inform decision making. This is because there is a need to know the overall damage state or response to specific scenarios. Although growing in number due to affordable monitoring systems with new sensors, microprocessors, and wireless communication, SHM is enabling long-term monitoring of structures. Many new structures have real-time monitoring systems during construction and operation to ensure safety and reliability, especially for large infrastructure like bridges [35], dams [36], tunnels [37], buildings [38], etc. Modern sensor technologies are often combined with effective identification techniques under the control of a processor that can be accessed online in real time.

In summary, the key ideas are: (1) SHM characterizes structures by monitoring their response to loads; (2) Models contain uncertainties, so SHM uses methods to verify them; (3) Damage detection was first visual but now uses sensors; (4) SHM provides insight into structural performance and damage; (5) SHM enables long-term monitoring, especially for infrastructure; and (6) SHM combines sensors, identification techniques, and real-time processing.

Smart bricks have been created using self-sensing concrete technology to keep attention on masonry structures. By utilizing steel fillers with high-temperature resistance, comparable to self-sensing concrete, smart piezoresistive bricks are demonstrated in Fig. 14.1 to be strain sensors. The sensor can detect damage in both new and old structures, which might have significant effects on the preservation of historic masonry structures [29, 39]. The detection of fire alarms and the detection of moisture in hazardous liquids under changing temperatures can both be facilitated by using cement-based self-sensing composites [40].

Figure 14.1 shows a possible SHM monitoring system using smart bricks in a masonry structure. Smart bricks would be placed in key areas, such as facades or locations with high loads like connections, lintels, or openings. The smart bricks' ability to sense changes would be used by measuring variations in their electrical resistance. Each smart brick could use either a two-electrode or four-electrode setup to take these resistance measurements, following the principles of standard resistivity testing.

In summary, the main concepts are: (1) Smart bricks would be installed in important areas of a masonry structure. (2) The smart bricks can sense changes, which would be measured by monitoring changes in their electrical resistance. (3) Each smart brick could use a two-electrode or four-electrode configuration to measure

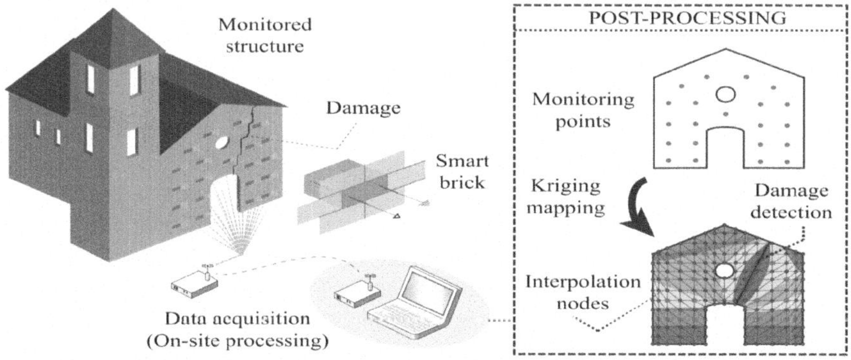

Fig. 14.1 Diagram of the smart brick-based structural health monitoring system [40]

resistance, similar to conventional resistivity testing approaches. (4) This SHM system using smart bricks could monitor key locations in the structure that experience high loads. (5) The smart bricks take advantage of their ability to sense changes by measuring variations in their own electrical resistance.

In summary, the concept of continuous SHM pertains to the examination of data data that has been obtained through the use of sensors at a sampling rate determined by external loads impacting the building. This data is then used to validate or compare with a model, while utilizing statistical analysis to evaluate the system's condition, deterioration progress, or the effectiveness of remedial measures.

14.3 Acoustic Emission

The utilization of acoustic emission (AE) monitoring is widely prevalent in the domain of materials research and site monitoring, primarily to detect damage. AE monitoring exhibits remarkable potential in identifying masonry damage and determining cracking during laboratory and field testing. In particular, unreinforced masonry is frequently exposed to compressive loads, and AE monitoring has been employed to evaluate walls that have undergone uniaxial creep [41, 42], fatigue [43], and cyclic testing [44]. Verstrynge et al. [45] studied the use of piezoelectric sensors and optical fibers for AE sensing in a deep masonry beam under bending, comparing the results with various strain monitoring methods. AE sensing has also been used to assess the durability and effectiveness of traditional masonry [46] maintenance techniques, including grout injection and externally bonded strengthening [47] with polymer- or mortar-based composites.

14.4 Potential of Smart Brick Technology

Smart bricks are a novel class of composite materials that possess exceptional capabilities in terms of smart sensing, including the capacity to identify cracks and self-monitor the inner strain state of a component or system, at an economical price point for concrete and masonry—two of the most common structural materials. The potential of these inventive composites lies in their ability to transform civil constructions into self-sensing systems, thereby enabling the cost-effective large-scale application of SHM solutions to a variety of structures, including newly constructed and historical buildings, as well as important infrastructures. The implementation of smart brick technology provides a feasible approach for continuous analysis of structural issues. This approach offers the added advantage of minimizing the costs involved in deploying traditional sensing equipment along with its associated operation and maintenance expenses. Additionally, the sensing mechanisms developed are highly resilient, overcoming the durability issues typically found in conventional sensing solutions. The implementation of smart bricks for obtaining real-time diagnostic and prognostic information can enable proprietors and administrators to more efficiently supervise maintenance and renovation plans, implement preventive conservation measures, and improve overall safety protocols.

14.5 Smart Bricks

14.5.1 Concept

Smart bricks, which are made of nanocomposite clay (as seen in Fig. 14.2), serve as strain sensors in structural masonry. These bricks are created by substituting standard clay bricks with conductive or semi-conductive elements, similar to smart concrete. The inclusion of such elements should be resistant to high temperatures of baking process that bricks undergo during the production process. One option is to use titanium dioxide (titania), while the other is to use micro- or macro-fibers which constructed of specific high temperature resistant metals without occurring any oxidation [48].

According to Meoni et al. [20], a smart brick is a unique, long-lasting sensor device for masonry construction SHM. The clay brick shown in Fig. 14.3a has been improved to include electrically conductive nanofillers or microfillers during production, resulting in a strain monitoring instrument which has similar mechanical properties to conventional bricks. This enhanced brick can detect changes in electrical resistance when deformed, thereby augmenting its electrical features and piezoresistive potential. Real-time analysis of data generated by intelligent bricks is a crucial factor in the assessment of the condition of monitored infrastructure,

(a) (b)

Fig. 14.2 The process of: **a** manufacturing smart bricks including conductive inclusions; **b** testing smart bricks after being in high temperature conditions [48]

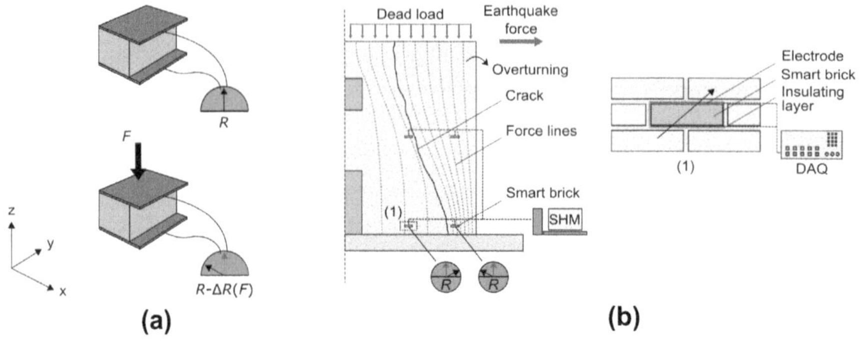

Fig. 14.3 The technology used in smart bricks: **a** simple demonstration of piezoresistive behavior; **b** a representation of how intelligent bricks may be used in a masonry wall [20]

and is particularly important in identifying any safety hazards that may arise after a seismic event. Smart bricks possess the ability to perceive local damage and monitor its progression over time. These sensors can be strategically situated within masonry buildings to detect prolonged alterations in the conditions of strain influence by dead loads produced after seismic events, by comparing genuine strain data with those pertaining to the undamaged structure, as depicted in Fig. 14.3b. Theoretically, Smart bricks can be installed internally and cover the entire width of the masonry wall, rendering them more appropriate for identifying changes in load patterns that cause internal stress redistributions than off-the-shelf sensors like strain gauges that are attached externally. It is noteworthy to mention that the utilization of smart brick technology extends beyond seismic events and can be expanded to include all scenarios in which structural damage leads to dead loads' redistribution, such as persistent changes in deformation conditions. Hence, smart bricks can be employed for monitoring settlements or damages caused by dynamic loads which are prevalent issues in the context of masonry constructions.

Downey et al. undertook an experimental investigation wherein a smart brick variant, possessing electrical conductivity and piezoresistive properties, was utilized. The smart brick exhibited measurable changes in electrical resistance in response to external loads. During brick manufacturing, adding an appropriate conductive filler can enhance the properties of a typical clay brick. This research delved into the exploration of titania-doped clay bricks as promising fillers. Titania was preferred owing to its electrical resistivity that lies between 0.1 and 10 Ωcm [44], whereas the electrical resistivity of ordinary commercial clay bricks approximates 1000 kΩm [45]. In addition, titania showcases remarkable efficiency even at elevated temperatures of approximately 1000 °C, as demonstrated by thermo-gravimetric analysis [46]. This study suggests the integration of select intelligent bricks at key points throughout a building for the purpose of gauging the degree of strain experienced by the brick, and furthermore, observing regional changes in strain that may arise as a result of structural adjustments or probable failure mechanisms, such as those that may arise after an earthquake. The technology of smart bricks is deemed to be superior to the currently utilized monitoring systems, owing to its capability to offer extremely precise readings (as the brick can detect stresses on the brick) and long-lasting durability, as it demonstrates the stability of the structural material.

Smart bricks have an architectural design similar to regular bricks in appearance, but with the added advantage of being able to conceal the electrodes by placing them on the internal surface of the bricks. Figure 14.4 depicts different viable uses for intelligent bricks, which comprise situating them in a continuous wall for the purpose of monitoring alterations in the load path (Fig. 14.4a), situating them at significant positions to scrutinize variations in concentrated loads, like those which are under lintels or at connections shown if Fig. 14.4b, and situating them in an arch to examine the placement of the pressure line and the potential initiation mechanism of collapse which shown in Fig. 14.4c.

14.5.2 Manufacturing Process

Meoni et al. (2020) created a set of twenty smart bricks through the utilization of newly extracted clay from a quarry located in Italy. The construction blocks were fabricated into a geometric figure with a rectangular foundation of 5 cm on all sides and a vertical altitude of 7 cm. The bricks were strengthened by steel fibers amount of 0.25%, specifically the R.STAT/S model. D'Alessandro et al. [49] indicate that the microfiller being discussed consists of an exclusive alloy (AISI 316 L) that exhibits low values of electrical resistivity around 60–80 Ω/cm and remarkable thermal resilience. The fibers utilized in this study were composed of stainless steel and possessed a diameter measuring 12 μm while extending 5 mm in length.

A smart brick's production process is shown in Fig. 14.5. Fresh clay and steel fibers were combined through mechanical means for a duration of 10 min using an electrical mixer, as depicted in Fig. 14.5a, b. Subsequently, the doped clay was placed into lubricated cubic models and covered with sand to facilitate the extrusion of the brick, as illustrated in Fig. 14.5c, d. Following this, the brick underwent a drying process at 90 °C for 6 h, as shown in Fig. 14.5e, followed by being burnt for 6 h at 900 °C, an event depicted in Fig. 14.5f. To act as external conductive electrodes, two planar copper electrodes were situated on opposing ends of the sample. Insulating tape was employed on either side of the sensor which is connected by the mortar to restrict the spread of current flow to the brick samples and preclude any outflows, as indicated in Fig. 14.5g.

According to Downey et al. [29], Wet clay from the initial stage of processing by an Italian brick manufacturer was used for producing the bricks. Titania particles, at 5% of the wet clay's weight, are added to the clay to create nanocomposite clay bricks (Fig. 14.6a). The filler material was subjected to mechanical circulation for a duration of 15 min through employment of a mixing machine of 1000 watts (refer to Fig. 14.6b, c). After smoothing, the prismatic steel molds were filled with the composite material afterwards that had been coated with oil and sanded (Fig. 14.6d). The specimens utilized in the study were square prisms that possessed a base side

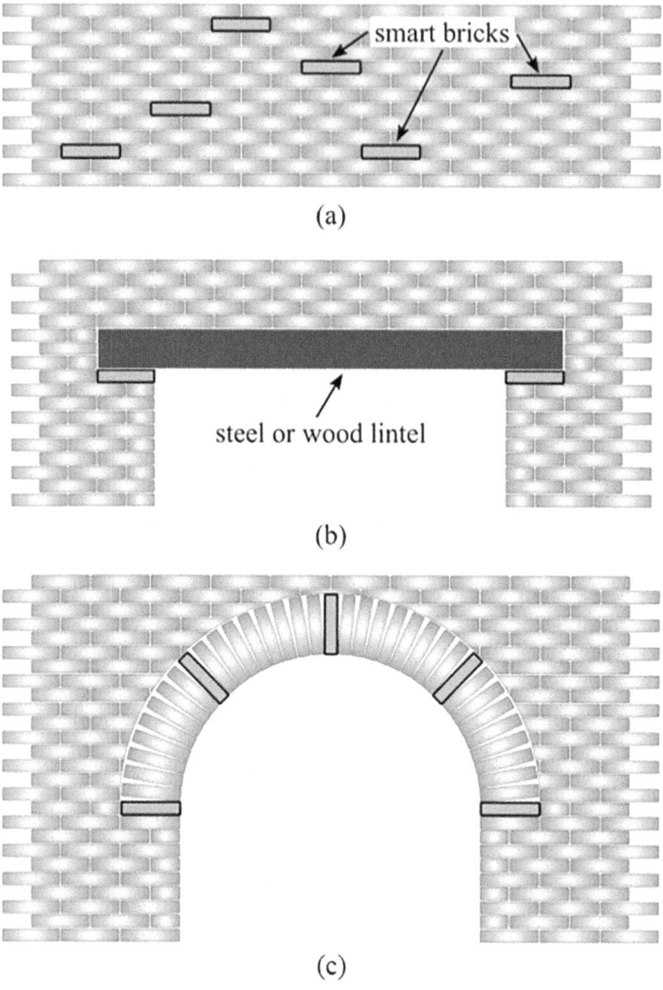

Fig. 14.4 Smart brick placement options: **a** distributed in wall surface; **b** steel and wooden lintels (concentrated loading points); and **c** at strategic positions inside masonry arches [29]

measuring 50 mm and a length of 70 mm. Four symmetrically inserted, 2.2 mm-diameter Kanthal steel wire electrodes with high-temperature resistance were placed at intervals of 10, 20, and 10 mm along the central axis of the samples (Fig. 14.6e). Two thermal increment stages were used to dry the clay pieces in an oven: 150 min at 50 °C and 120 min at 90 °C (Fig. 14.6f). The dried specimens were burnt at 900 °C for 12 h after being given time cooling (Fig. 14.6g). After baking, the bricks' color turned to red-brown (Fig. 14.6h). Ultimately the examination of the inner structure of nanocomposite and plain bricks was conducted by means of a scanning electron microscope (SEM). Figure 14.6 depicts the results of the analysis of material, which was completed after the burning process. Figure 14.6a displays

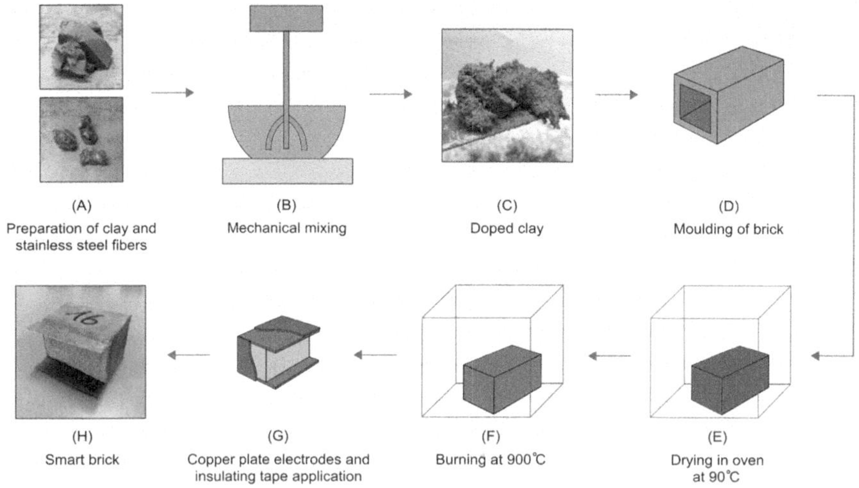

Fig. 14.5 The process of manufacturing smart brick [21]

the clean brick, Fig. 14.6b displays titania-doped bricks. The smart brick's SEM image evinces the evident existence and superb dissemination of the titania dioxide particles (diminutive white spheres).

D'Alessandro et al. [49] developed new clay bricks that contain varying values of conductive micro inclusions composed of stainless-steel fibers. These inclusions ranged from 0 to 1%, with intervals of 0.25%, 1.5%, and 2%, and were added to the fresh clay while considering its overall weight. The fibers, which were constructed

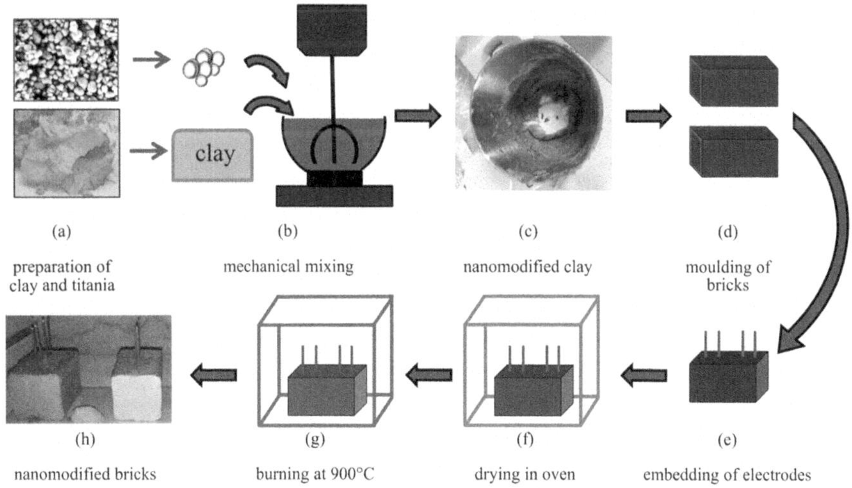

Fig. 14.6 The process for creating smart clay bricks that incorporate nanocomposites [29]

by a special alloy (AISI 316L) and measured 12 μ in diameter and 5 mm in length, were the model of R.STAT/S. Even when exposed to high temperatures, like those attained when baking clay at a temperature of around 1000, the alloy maintained low electrical resistivity values of the order of 60–80Ω/cm. The fibers' performance at high temperatures was examined through thermogravimetric experiments. The brick preparation temperature of 900 °C was attained by heating small amounts of fibers in sealed containers (Fig. 14.7) in an oven. At upward trend of maximum temperatures ranging from 300 to 900 °C, four further thermal cycles were performed (Fig. 14.7a). Figure 14.7b shows that there was no weight loss in the fibers. Using a multimeter equipped with two clamps, the measurement of the electrical resistance of the fibers was conducted before and after any heat cycles. No appreciable reduction is observed (Fig. 14.7c). Stainless steel fibers were validated as candidates for incorporation into conductive clay bricks by thermogravimetric experiments.

The smart bricks utilized in this experiment had the same geometrical dimensions as the traditional clay bricks used to build the masonry wall examples, which were 250 × 55 × 120 mm^3 [50]. The electrically conductive fillers, stainless steel microfibers, were amalgamated with fresh clay using ratio of 0.25% by mass. The main steps in the preparation process for smart bricks are depicted in Fig. 14.8. Clay and steel microfibers that have been distributed were subjected to mechanical mixing in order to produce a homogeneous composite material with an optimal dispersion of the inclusions, as depicted in Fig. 14.8a. The resulting mixture was then placed in sand-dusted wooden molds to produce fresh smart bricks (Fig. 14.8b). The materials in a commercial oven underwent a process of initial drying at 90 °C for six hours, followed

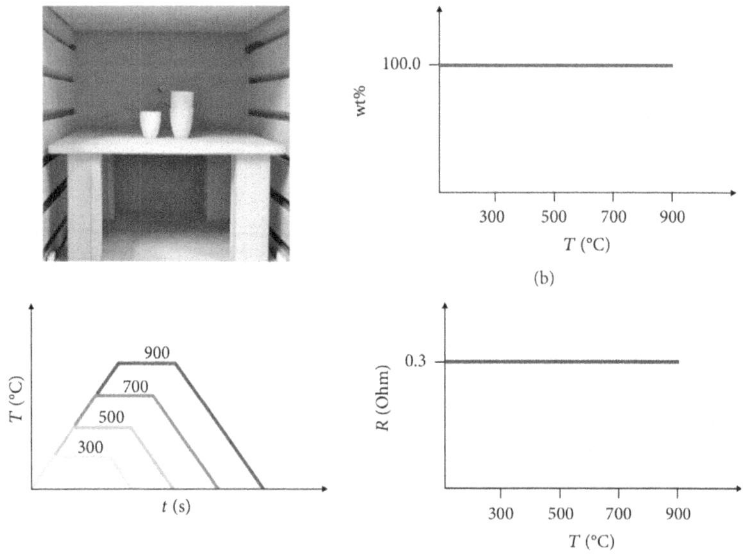

Fig. 14.7 Thermogravimetric experiment on stainless steel fibers: **a** imposed thermal cycles waveforms; **b** the weight remained after thermal cycles [49]

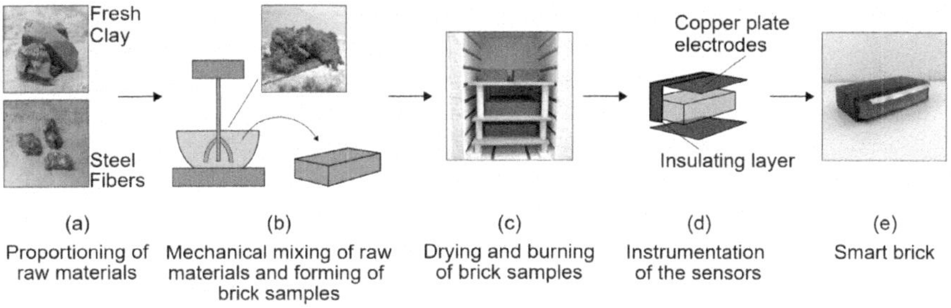

Fig. 14.8 Production process of a smart brick sample [50]

by burning at 900 °C for another six hours, as illustrated in Fig. 14.8c. Burned bricks were initially connected to two copper plate electrodes at their top and lower sides once they had cooled down (Fig. 14.8d), followed by a tape of insulation to protect the sensors from outside the electrical fields. The final smart-brick example is shown in Fig. 14.8e.

Smart bricks with prismatic nominal shape of $70 \times 50 \times 50$ mm^3 were produced by considering the manufacturing procedure reported in Fig. 14.9 [51]. Freshly prepared clay was subjected to a mechanical mixing process whereby conductive stainless steel micro-fibers, specifically model R.STAT/S, were incorporated at different ratios ranging from 0 to 1% by weight of the clay. This process is depicted in Fig. 14.9a. Samples were initially created through the process of pouring the composite material (Fig. 14.9b) into prismatic molds that had been treated with oil and dusted with sand. These samples were subsequently subjected to a thermal cycle that involved drying at a temperature of 90 °C and being burned up to 900 °C (Fig. 14.9c). Afterwards, each specimen was instrumented by applying two external copper plate electrodes on its opposite sides, covering them with an insulating layer (Fig. 14.9d).

Smart bricks were investigated by Meoni et al. [52] and are constructed of a composite material based on clay that exhibits improved piezoresistive performance and optimized strain-sensing capabilities [49]. A specific quantity of the chosen electrically conductive filler, stainless steel microfibers, model R.STAT/S, was mixed mechanically with fresh clay to create the desired customized composite (Fig. 14.10a). Following this, wet wooden molds were utilized to create brick samples that were dusted with sand (as depicted in Fig. 14.10b). These samples were then subjected to firing at 900 °C for an extra six hours after undergoing a drying process in a muffle at 90 °C for six hours (as illustrated in Fig. 14.10c). Initially, copper plate electrodes are affixed to smart bricks prior to cooling, thereby coating them with an insulating layer to impede the current flow towards other structural elements of masonry within which they are situated (as depicted in Fig. 14.10d). Throughout this research, thirteen smart bricks with standard dimensions of 25 cm × 5.5 cm × 12 cm were made utilizing steel fibers at a ratio of 0.25% to the weight of fresh clay [53].

To manufacture the smart bricks that were studied in this research [54], the bricks were created following the manufacturing process illustrated in Fig. 14.11. Using

Fig. 14.9 Smart brick production procedure [51]

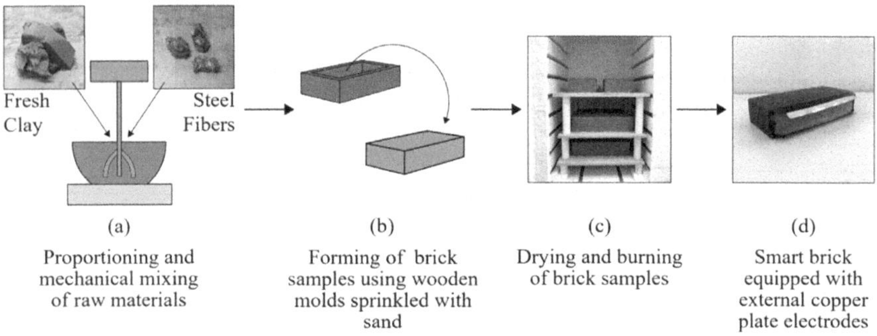

Fig. 14.10 The process of manufacturing smart bricks with standard dimensions [52]

a filler percentage of 0.50% relative to the weight of the new clay, electrically conductive stainless steel microfibers, model R.STAT/S were mixed with fresh clay (Fig. 14.11a, b). To create brick samples, the resultant composite material was put into prismatic hardwood molds measuring 7.0 × 5.0 × 5.0 cm^3, which had been thoroughly wetted and dusted with sand (Fig. 14.11c). The latter specimens underwent an additional six-hour heating process at a temperature of 900 °C subsequent to an initial six-hour drying period in an oven set at 90 °C. Finally, two external copper plate electrodes were added to the smart bricks to conduct electrical measurements.

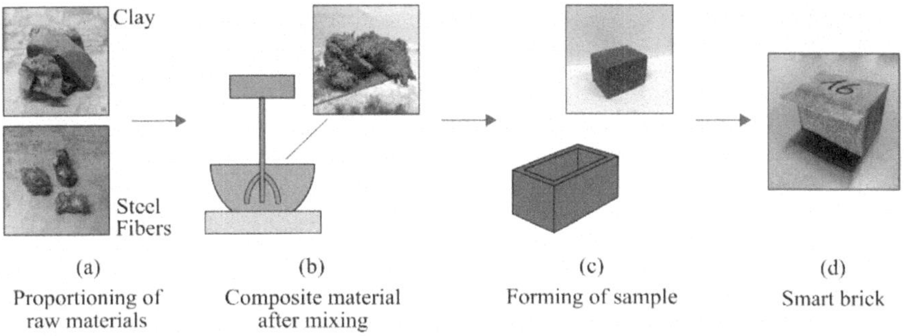

Fig. 14.11 The depiction of the process of manufacturing smart bricks [54]

These electrodes were positioned on their opposing horizontal sides with dry contact. Each sample had an insulating coating applied to prevent current flow when it was placed within the masonry structures (Fig. 14.11d).

14.5.3 Electrical Measurements

Electrical evaluations were performed with the intention of entirely mitigating the polarization phenomenon, which is commonly observed in dielectric substances such as clay-based materials [20]. By employing a voltage square wave to facilitate the charging and the material discharging in every cycle, the temporal drift of the electrical output generated by this phenomenon is mitigated. Additionally, the implementation of this method induces an electric field that works against the one being utilized. In the experimental testing, every brick received a voltage square wave of ± 10 V (20 V peak-to-peak), 50% duty cycle, and frequency of 1 Hz from a RIGOL DG1022 function generator. A chassis with the model number NI PXIe-1073, with the model number NI PXI-4071 of digital multimeter, and a sampling frequency of 10 Hz represented the data gathering system.

The authors utilized a biphasic direct current (DC) technique, which was specifically designed for monitoring carbon-doped cement composites with multiple channels, to prevent any polarization effects on resistance measurements [29]. However, they applied a modified measurement strategy to enable a single-channel acquisition utilizing just two external electrodes, known as a two-probe measurement. To counteract the material's polarization effect, they employed a voltage square wave with a duty cycle of 50%, which charges and discharges the sample. Through the positive phase of the biphasic signal, they obtained a current sample (i) for analysis. By having knowledge of the applied voltage, which is represented by V = 1/2Vpp, it is possible to compute the resistance (R) of the smart brick.

$$R = \frac{V}{i} \tag{14.1}$$

Electrical tests were carried out to determine how the electrical conductivity of the smart bricks with varying filler contents varied and how the design of the electrodes affected the data [49]. Both embedded and exterior electrode designs were looked into for this aim. The experiments also sought to confirm the measurement reproducibility. A digital multimeter and an NI PXIe-1073 chassis made up the data acquisition system. In order to create charge and discharge cycles on the specimens and minimize polarization, a voltage square wave possessing a $\pm$ 10 V (20 V peak-to-peak) amplitude, 50% duty cycle, and 1 Hz frequency was administered utilizing a Rigol DG1022 function generator. During the biphasic signal's constant phase, the current strength was gauged to be approximately 80% in the positive area. The square wave voltage applied was configured to have the identical frequency as the sampling rate expressed in samples per second (S/s). According to Ohm's law (Eq. 14.1), the electrical resistance was calculated.

Electrical measurements were conducted by utilizing the RIGOL DG1022 function generator and a biphasic DC measuring methodology. The waveform input comprised of a square voltage wave with an amplitude of approximately $\pm$10 V (20 V peak-to-peak), a 50% duty cycle, and a frequency of 1 Hz. The resultant current signal was recorded at a sampling rate of 10 Hz with the aid of the NI PXIe-1073 DAQ and NI PXI-4071 digital multimeter [29]. Using Ohm's law and the current intensity measurement recorded at 80% of the positive steady component of the collected output signal, the electrical resistance was computed.

14.5.4 Strain Sensing Principle of Smart Bricks

The sensitivity of standard-sized smart bricks strain when sustain in-plane shear load in masonry walls was studied by the researchers. The ability of smart bricks to record the strain is due to the combination of the doped clay matrices piezo-resistivity and the contact resistance that develops at their electrodes [53]. Despite this, the primary driver of the sensing mechanism in these innovative sensors is the piezoresistive behavior, which is further enhanced by incorporating particular conductive electrically fillers into the matrices of clay through a customized of the smart bricks' mix design. Smart bricks can be utilized in masonry structures for the purpose of evaluating strain changes via their electrical outputs, specifically electrical resistance, by utilizing the piezoresistive principle. When integrated into coarse sensor networks, smart bricks are capable of monitoring permanent strain changes caused by seismic activity or differential settlements, by comparison of the current strain measurements to those of the undamaged building [20]. The strain outputs of these bricks can also be analyzed to recreate strain fields in both masonry structures and structural components, provided that these advanced sensors are widely distributed within load withstanding systems. In this regard, the identifying of damages can be executed by detecting irregularities within the reconstructed strain field maps [39, 55].

The current study investigates the use of smart bricks in a wall sample subjected to eccentric compression load. These experiments were conducted at UniPG LabDyn,

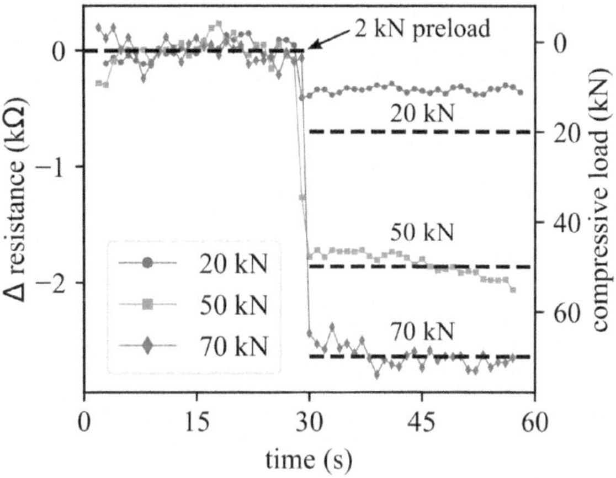

Fig. 14.12 The smart brick 2 response placed in the surface of wall sample, subjected to a concentrated load (20, 50 and 70 kN)

and the findings are presented in this section [29]. The findings indicate that smart bricks have the potential to operate as a self-sensing structural component incorporated in the surface of the wall. The graphical representation presented in Fig. 14.12 illustrates the variation in brick 2's resistance data for three different loading scenarios (20, 50, and 70 kN). The outcomes are presented as a change in resistance to account for a slight variation in the sensors' nominal resistance due to tests being performed on different days. The findings suggest that the variation in resistance is proportionate to the load in cases of greater loading magnitude (50 and 70 kN). Nonetheless, in the instance of the smallest loading magnitude (20 kN), the resistance changes not correlate with the process of loading. This inconsistency is ascribed to the heterogeneous load routes within the wall, specifically when the wall is subjected to comparably low loading conditions.

Smart bricks are a novel type of clay bricks with electrically conductive and piezoresistive properties. These bricks are capable of exhibiting measurable changes in their electrical properties when subjected to deformation under compressive loads. Specifically, the smart bricks' electrical resistance decreases when there is an increase in compressive strain due to applied loads, thus giving them strain-sensing capabilities. Moreover, the piezoresistive and electrical properties of smart bricks are further enhanced by the incorporation of appropriate conductive nano- or micro-fillers into the clay matrix. This results in a composite material with superior piezoresistive performance compared to a standard clay brick, while keep identical mechanical and physical properties (Fig. 14.13a) [53].

Smart bricks utilize the same clay matrix and production process as regular bricks, enabling them to be incorporated into masonry structures for monitoring purposes. Through their strain-sensing capabilities, smart bricks can detect permanent strain changes caused by alterations in load paths, which may result from the formation of crack patterns. By analyzing the electrical outputs of smart bricks in comparison to a reference condition or an undamaged structure, it is possible to detect areas of considerable strain concentrations which may signify the beginning of damage. This

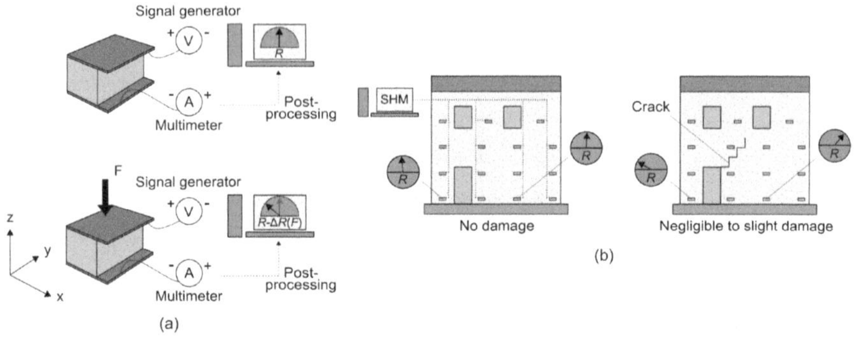

Fig. 14.13 The technology of smart bricks: **a** Smart brick in constant electrical resistance; **b** no damage or slight damages conditions [53]

real-time analysis is crucial for the early detection and localization of damages, thus playing a significant role in preventing premature brittle fracture mechanisms. The technology behind smart bricks is particularly effective in situations where structural issues in masonry constructions lead to a static load paths' redistribution, resulting in permanent changes in strain conditions.

Smart bricks utilize their improved piezoresistive features to generate quantifiable changes in their electrical characteristics, including electrical resistance, which correspond to changes in their strain state. As the compressive strain state increases, the electrical resistance decreases thus exhibiting piezoresistive behavior. The incorporation of smart bricks into masonry structures can facilitate the detection of changes in loads resulting from structural damage. A recent proposal and validation of an electromechanical model has demonstrated its suitability in establishing a correlation between alterations in the electrical properties of smart bricks and their compressive strain state. This model is comprised of a linear combination of the optimized piezoresistivity of smart bricks and the contact resistance that is developed at the external electrodes, which effectively governs their strain-sensing behavior.

14.6 Summary of Smart Bricks

This study includes a novel sensing technique that has recently been proposed by numerous authors with various applications of smart bricks. Overall, research has revealed that smart bricks have the potential to be a new smart sensing strategy for masonry structures' SHM, allowing for the early detecting and localizing of earthquake-induced damage. As long as they cause strain redistribution, the suggested monitoring technology may theoretically predict both local and global failure modes, with local damage modes being somewhat harder to detect because smart bricks must be placed in the essential location. Table 14.1 summarizes the recent studies of smart bricks.

Table 14.1 Summary of recent smart bricks research works

References	Year	Name	Investigated parameters
[20]	2019	Meoni et al.	• Characterization of the sensors • Masonry building specimen • Shaking table tests • Numerical model
	Observation	To generate a progressive damage on the unreinforced masonry structure specimen during the shaking table tests, the ground motion intensity was gradually increased over the course of three seismic phases. The authors' recent proposal of a novel sensor technology for the seismic monitoring of masonry buildings is given in the study as the smart bricks' first full-scale use	
[29]	2017	Downe et al.	• Characterization tests • Utilizing a smart brick in a wall for strain monitoring • The electrical current distribution in a wall • Eccentric compression experiments on wall specimen
	Observation	Experimental findings indicate that by cross-analyzing the results of diverse smart bricks, it becomes feasible to identify changes in strain subsequent to the development of compression cracks in circumstances that are nearing the ultimate limit state. Smart bricks possess the capability to gauge variations in strain within the masonry as a result of alterations in the load's eccentricity. After an earthquake, these traits may be utilized to identify changes in loading routes, which may indicate the development of structural damage or the start of a concentrated failure. The authors suggested that smart bricks might be an effective and long-lasting sensor technology for masonry structures, delivering essential data for structural prediction while retaining architectural and structural integrity	
[50]	2023	Meoni et al.	• Basic sensing principle • Experimental application – Masonry wall specimens – Diagonal compression tests • Numerical simulations
	Observation	To understand the experimental findings, numerical simulations were conducted. At their peak, numerical models produced strain levels that were almost equal to 10^{-4} and followed similar patterns to those of smart bricks. Overall, it is shown that smart bricks are good in damage detection and strain monitoring	
[51]	2019	D'Alessandro et al.	• Experimental tests – Single smart bricks with different percentages of fibers – Small-scale walls with bricks of different percentages of fibers – Middle-scale wall with bricks doped with 0.25% of steel microfibers • Numerical simulations

(continued)

Table 14.1 (continued)

References	Year	Name	Investigated parameters
	Observation	Compressive experiments on brick with varied filler quantity demonstrated their sensitivity to externally applied strains and stresses, allowing determination of the ideal composition. The sensitivity of the novel sensors to changes in the load route and the emergence of fractures was then examined by incorporating smart bricks into small- and medium-scale walls and applying them under various loading situations. The ideal smart brick arrangement for detecting the strain field and load route's changes of the brickwork was examined using numerical simulations. The capabilities of smart bricks for detecting damage and reconstructing strain fields were assessed through a comparison of numerical and experimental data	
[52]	2022	Meoni et al.	• Experimental investigation – Smart bricks calibration – Full-scale masonry building prototype
	Observation	The authors presented a brief overview of the smart bricks' preparation process and self-sensing behavior. The issue was addressed by presenting the findings of the electromechanical tests that were conducted and the calibration of a set of smart bricks. The building prototype's masonry load-bearing structure deployment of smart bricks as well as the initial findings from that testing were also covered. The calibration results allow the series resistor model to be described in terms of the strain-sensing capabilities of the manufactured set of smart bricks. The initial results of the field application revealed how changing environmental conditions impact the outputs of smart bricks	
[53]	2020	Meoni et al.	• Strain sensing principle • Analytical model and experimental methodology – The suggested electromechanical model – Study the smart bricks' percolation threshold – Strain-sensing performance characterization of the smart brick – Small-scale wall under eccentric compression loads
	Observation	This paper proposes a novel theoretical model that characterizes the strain-sensing response of smart bricks, a unique class of piezoresistive smart sensors, for monitoring the structural health of masonry constructions. Electromechanical measurements of the individual smart bricks were used to experimentally validate the proposed concept. An example application is also provided to demonstrate how the model can be used in situations where smart bricks are included in masonry walls	

(continued)

Table 14.1 (continued)

References	Year	Name	Investigated parameters
[54]	2021	Meoni et al.	• Experimental investigation – A single smart brick's axial compression tests – Small-scale masonry wall subjected to eccentric axial compression load
	Observation	This study described the efficiency of smart bricks for the assessment of strain under compression loads, and has been the subject of experimental and numerical studies. In order to evaluate the innovative sensors' capacity to detect strain, increasing axial compression loads were first applied to individual samples, and then smart bricks were used to track the strain state of a small-scale masonry wall sample under eccentric axial compression pressures. The new sensors were constructed using stainless steel microfibers with a composition of 0.50% by weight of fresh clay. Several typical strain gauges were installed on the exterior surfaces of each new sensor to compare the strain data. The results reveal that smart bricks may be used to estimate strain subject to compression loads, demonstrating the potential of unique sensors as a strong novel sensing strategy suitable for masonry structure's SHM	

14.7 Conclusions

There has been a novel method for strain monitoring in masonry constructions suggested. The concept relies on the innovative concept of smart bricks. The smart bricks are unique clay bricks which are piezoresistive and may provide quantifiable electrical output when a mechanical load is applied. They offer high-fidelity strain monitoring at critical sites inside a masonry building by recognition of the loading changes patterns and correlating the outputs of various smart bricks, for example, before and after a seismic event. This information may then be utilized to evaluate stress as well as discover cracks and damage. The strain-sensing theory and manufacturing procedure of revolutionary clay-based sensors were presented to provide key ideas fundamental to smart brick technology.

Results show that the innovative theoretical model is useful for estimating strain and detecting damage in a conventional structural situation, as well as for properly simulating the behavior of bricks when subjected to compressive stresses. As more pressure was placed on the tested wall specimens, nearly all of the smart bricks produced a rising compressive strain. Moreover, the innovative sensors provided modifications in the pattern of their strain outputs that could be straightforwardly associated with the commencement of fractures in wall samples. The smart bricks are suitable for identifying the earthquake-induced damage and also to localize damage because they may effectively demonstrate small permanent changes in structural circumstances caused by gradual damage.

The present research project's experimental and numerical findings indicate that standard-sized smart bricks possess the potential to function as monitoring devices for strain and early fracture detection in structural elements of masonry building that are exposed to in-plane shear loads. The potential of smart brick technology lies in its monitoring capabilities, which render it a highly viable alternative for SHM of masonry projects. In summary, based on state-of-the-art research initiatives and data, the smart brick technology has evolved to a degree where the utilization of this method is applicable for conducting exploratory SHM practices on full-scale masonry buildings that are exposed to genuine loads and environmental conditions.

References

1. Masciotta M-G, Ramos L, Lourenço PB (2017) The importance of structural monitoring as a diagnosis and control tool in the restoration process of heritage structures: a case study in Portugal. J Cult Heritage 27:36–47
2. Masciotta MG, Roque JC, Ramos LF, Lourenço PB (2016) A multidisciplinary approach to assess the health state of heritage structures: the case study of the church of Monastery of Jerónimos in Lisbon. Constr Build Mater 116:169–187
3. Gentile C, Saisi A (2007) Ambient vibration testing of historic masonry towers for structural identification and damage assessment. Constr Build Mater 21(6):1311–1321
4. Carpinteri A, Lacidogna G (2006) Damage monitoring of an historical masonry building by the acoustic emission technique. Mater Struct 39(2):161–167
5. Saisi A, Gentile C, Ruccolo A (2018) Continuous monitoring of a challenging heritage tower in Monza, Italy. J Civil Struct Health Monit 8(1):77–90
6. Ramos L, Mevel L, Lourenço PB, De Roeck G (2008) Dynamic monitoring of historical masonry structures for damage identification. In: Proceedings of the 26th international modal analysis conference; Orlando, Fl, USA
7. Saisi A, Gentile C, Guidobaldi M, Xu M (2014) Dynamic monitoring and seismic response of a historic masonry tower. Key Eng Mater 628:55–60
8. Krstevska L, Tashkov L, Naumovski N et al (2010) In-situ experimental testing of four historical buildings damaged during the 2009 L'aquila earthquake. In: Cost action C26: urban habitat constructions under catastrophic events—proceedings of the final conference, vol 26. Napoli, Italy, pp 427–432
9. Vermeltfoort A, Martens D, Van Zijl G (2007) Brick–mortar interface effects on masonry under compression. Can J Civ Eng 34(11):1475–1485
10. Bavusi M, Soldovieri F, Piscitelli S, Loperte A, Vallianatos F, Soupios P (2010) Ground-penetrating radar and microwave tomography to evaluate the crack and joint geometry in historical buildings: some examples from Chania, Crete, Greece. Near Surf Geophys 8(5):377–387
11. Moropoulou A, Labropoulos KC, Delegou ET, Karoglou M, Bakolas A (2013) Non-destructive techniques as a tool for the protection of built cultural heritage. Constr Build Mater 48:1222–1239
12. Ghiassi B, Xavier J, Oliveira DV, Lourenço PB (2013) Application of digital image correlation in investigating the bond between FRP and masonry. Compos Struct 106:340–349
13. Ghorbani R, Matta F, Sutton MA (2014) Full-field deformation measurement and crack mapping on confined masonry walls using digital image correlation. Exp Mech 55:227–243
14. Han Q, Xu J, Carpinteri A, Lacidogna G (2014) Localization of acoustic emission sources in structural health monitoring of masonry bridge. Struct Control Health Monit 22:314–329
15. Serra M, Festa G, Vassallo M, Zollo A, Quattrone A, Ceravolo R (2016) Damage detection in elastic properties of masonry bridges using coda wave interferometry. Struct Control Health Monit 24:e1976
16. Schueremans L, Genechten BV (2009) The use of 3d-laser scanning in assessing the safety of masonry vaults—a case study on the church of saint-jacobs. Opt Lasers Eng 47:329–35

17. Riveiro B, Morer P, Arias P, de Arteaga I (2011) Terrestrial laser scanning and limit analysis of masonry arch bridges. Constr Build Mater 25:1726–1735
18. Laflamme S, Turkan Y, Tan L (2015) Bridge structural condition assessment using 3d imaging. In: Proceedings of the 2015 conference on autonomous and robotic construction of infrastructure
19. Formisano A, Vaiano G, Fabbrocino F, Milani G (2018) Seismic vulnerability of italian masonry churches: the case of the nativity of blessed virgin mary in stellata of bondeno. J Build Eng. 20:179–200
20. Meoni A, D'Alessandro A, Cavalagli N, Gioffré M, Ubertini F (2019) Shaking table tests on a masonry building monitored using smart bricks: damage detection and localization. Earthq Eng Struc 48(8):910–928
21. Coppola L, Buoso A, Corazza F (2011) Electrical properties of carbon nanotubes cement composites for monitoring stress conditions in concrete structures. Appl Mech Mater 82:118–123
22. Han B, Ding S, Yu X (2015) Intrinsic self-sensing concrete and structures: a review. Measurements 59:110–128
23. Rainieri C, Song Y, Fabbrocino G, Markand JS, Shanov V (2013) CNT-cement based composites: fabrication, self-sensing properties and prospective applications to structural health monitoring. In: Proceedings SPIE, fourth international conference on smart materials and nanotechnology in engineering, vol 8793, p 10
24. Azhari F, Banthia N (2012) Cement-based sensors with carbon fibers and carbon nanotubes for piezoresistive sensing. Cem Concr Compos 34(7):866–873
25. Galao O, Baeza FJ, Zornoza E, Garcés P (2014) Strain and damage sensing properties on multifunctional cement composites with CNF admixture. Cem Concr Compos 46:90–98
26. D'Alessandro A, Ubertini F, García-Macías E et al (2017) Static and dynamic strain monitoring of reinforced concrete components through embedded carbon nanotube cement-based sensors. Shock Vib 2017:11
27. Konsta-Gdoutos MS, Aza CA (2014) Self sensing carbon nanotube (CNT) and nanofiber (CNF) cementitious composites for real time damage assessment in smart structures. Cem Concr Compos 53:162–169
28. Saafi M (2009) Wireless and embedded carbon nanotube networks for damage detection in concrete structures. Nanotechnology 20(39):395502
29. Downey A, D'Alessandro A, Laflamme S, Ubertini F (2017) Smart bricks for strain sensing and crack detection in masonry structures. Smart Mater Struct 27:015009
30. Sohn H, Farrar CR, Hemez FM, Shunk DD, Stinemates DW, Nadler BR, Czarnecki JJ (2004) A review of structural health monitoring literature: 1996–2001. Los Alamos National Laboratory Report, LA-13976-MS
31. Bassoli E, Vincenzi L, Bovo M, Mazzotti C (2015) Dynamic identification of an ancient masonry bell tower using a MEMS-based acquisition system. In: Conference: EESMS 2015 environmental, energy and structural monitoring systems (Trento, Italy). https://doi.org/10.1109/EESMS.2015.7175882
32. Guidorzi R, Diversi R, Vincenzi L, Mazzotti C, Simioli V (2014) Structural monitoring of a tower by means of MEMS-based sensing and enhanced autoregressive models. Eur J Control 20:4–13. https://doi.org/10.1016/j.ejcon.2013.06.004
33. Ramos LF, Marques L, Lourenço PB, De Roeck G, Campos-Costa A, Roque J (2010) Monitoring historical masonry structures with operational modal analysis: two case studies. Mech Syst Signal Process 24(5):1291–1305. https://doi.org/10.1016/j.ymssp.2010.01.011
34. Foti D (2013) Dynamic identification techniques to numerically detect the structural damage. Open Constr Build Technol J 7:43–50
35. Magalhães F, Cunha Á, Caetano E (2012) Vibration based structural health monitoring of an arch bridge: from automated OMA to damage detection. Mech Syst Sig Process 28:212–228
36. Singh R, Choudhury S, Singh B (2015) Wireless disaster monitoring and management system for dams. Procedia Comput Sci 48:381–386. https://doi.org/10.1016/j.procs.2015.04.197

37. Hoult NA, Soga K (2014) Sensing solutions for assessing and monitoring tunnels. Sensor technologies for civil infrastructures. In: Applications in structural health monitoring, vol 2. Woodhead Publishing Series in Electronic and Optical Materials, pp 309–346. https://doi.org/10.1533/9781782422433.2.309
38. Ni YQ, Xia Y, Liao WY, Ko JM (2009) Technology innovation in developing the structural health monitoring system for Guangzhou new TV tower. Struct Control Health Monit 16(1):73–98
39. García-Macías E, Ubertini F (2019) Earthquake-induced damage detection and localization in masonry structures using smart bricks and Kriging strain reconstruction: a numerical study. Earthq Eng Struct Dyn 48(5):548–569
40. Demircilioğlu E, Teomete E, Schlangen E, Baeza FJ (2019) Temperature and moisture effects on electrical resistance and strain sensitivity of smart concrete. Constr Build Mater 224:420–427
41. Verstrynge E, Schueremans L, Van Gemert D, Wevers M (2009) Monitoring and predicting masonry's creep failure with the acoustic emission technique. Ndt & E Int 42(6):518–523
42. Verstrynge E, Adriaens R, Elsen J, Van Balen K (2014) Multi-scale analysis on the influence of moisture on the mechanical behavior of ferruginous sandstone. Constr Build Mater 54:78–90
43. Andreev K, Shetty N, de Smedt M, Yin Y, Verstrynge E (2019) Correlation of damage after first cycle with overall fatigue resistance o refractory castable concrete. Constr Build Mater 206:531–539
44. Shetty N, Livitsanos G, Van Roy N, Aggelis DG, Van Hemelrijck D, Wevers M et al (2019) Quantification of progressive structural integrity loss in masonry with acoustic emission-based damage classification. Constr Build Mater 194:192–204
45. Verstrynge E, De Wilder K, Drougkas A, Voet E, Van Balen K, Wevers M (2018) Crack monitoring in historical masonry with distributed strain and acoustic emission sensing techniques. Constr Build Mater 162:898–907
46. Anzani A, Binda L, Carpinteri A, Lacidogna G, Manuello A (2008) Evaluation of the repair on multiple leaf stone masonry by acoustic emission. Mater Struct 41(6):1169–1189
47. Verstrynge E, Wevers M, Ghiassi B, Lourenço PB (2016) Debonding damage analysis in composite-masonry strengthening systems with polymer- and mortar-based matrix by means of the acoustic emission technique. Smart Mater Struct 25(1):015009
48. Ubertini F, D'Alessandro A, Downey A, García-Macías E, Laflamme S, Castro-Triguero R (2017) Recent advances on SHM of reinforced concrete and masonry structures enabled by self-sensing structural materials. In: Proceedings, vol 2(3). MDPI, p 119
49. D'Alessandro A, Meoni A, Ubertini F (2018) Stainless steel microfibers for strain-sensing smart clay bricks. J Sensors 2018(7431823):8. https://doi.org/10.1155/2018/7431823
50. Meoni A, D'Alessandro A, Saviano F, Lignola GP, Parisi F, Ubertini F (2023) Strain monitoring and crack detection in masonry walls under in-plane shear loading using smart bricks: first results from experimental tests and numerical simulations. Sensors 23(4):2211
51. D'Alessandro A, Meoni A, Ubertini F (2019) Recent results on the use of smart bricks for earthquake-induced damage detection in masonry structures, pp 104–111
52. Meoni A, D'Alessandro A, Virgulto G, Buratti N, Ubertini F (2022) Smart bricks for monitoring strain in full-scale masonry structures: recent advances and first field application. In: European workshop on structural health monitoring: EWSHM 2022, vol 1. Springer International Publishing, Cham, pp 835–843
53. Meoni A, D'Alessandro A, Ubertini F (2020) Characterization of the strain-sensing behavior of smart bricks: a new theoretical model and its application for monitoring of masonry structural elements. Constr Build Mater 250:118907
54. Meoni A, D'Alessandro A, Ubertini F (2021) Recent advances and open issues on the use of smart bricks for seismic monitoring of masonry buildings: experimental tests and numerical simulations. In: European workshop on structural health monitoring: special collection of 2020 papers, vol 1. Springer International Publishing, pp 851–860
55. Meoni A, D'Alessandro A, Kruse R, De Lorenzis L, Ubertini F (2021) Strain field reconstruction and damage identification in masonry walls under in-plane loading using dense sensor networks of smart bricks: experiments and simulations. Eng Struct 239:112199

Chapter 15
Bridge Health Monitoring: A Review of Utilizing the Internet of Things, Digital Twin, and Advanced Technologies

Hamed Hasani and Francesco Freddi

Abstract Structural monitoring of bridges is commonly carried out using conventional methods. However, with the advancements in the internet of things (IoT) and digital twin (DT) technologies, this process has been revolutionized. The implementation of IoT enables enhanced sustainability by monitoring various parameters through interconnected devices and sensors. Simultaneously, DT technology creates a synchronized virtual representation of the physical bridges, improving safety, maintenance, and decision-making. This review paper explores foundational concepts of IoT and DT in bridge monitoring, discussing the synergy between DT and IoT for heightened precision in damage detection. It also covers innovative methodologies such as photogrammetry, terrestrial laser scanning, ground penetrating radar, building information modeling, and unmanned aerial vehicles/drones. These methodologies produce detailed 3D models, improving precision, practicality, and efficacy in the monitoring process. The paper serves as a guide for researchers and practitioners, illustrating a progression from foundational concepts to the state-of-the-art technologies. This envisions a future where bridges are monitored with unprecedented accuracy to ensure safer and more resilient infrastructure on a global scale.

Keywords Bridge structural health monitoring (BSHM) · Internet of things (IoT) · Digital twin (DT) · Light detection and ranging (LiDAR) · Unmanned arial vehicle (UAV) · Ground penetrating radar (GPR) · Terrestrial laser scanning (TLS) · Building information modeling (BIM) · Smart infrastructures

H. Hasani (✉) · F. Freddi
Department of Engineering and Architecture, University of Parma, Parma, Italy
e-mail: hamed.hasani@unipr.it

F. Freddi
e-mail: francesco.freddi@unipr.it

H. Jahangir et al. (eds.), *Damage Detection and Structural Health Monitoring of Concrete and Masonry Structures*, Springer Tracts in Civil Engineering,
https://doi.org/10.1007/978-981-97-8975-7_15

15.1 Introduction

There have been advancements in the field of bridge construction over the past decades. Considerable economic and social progress has been made in countries due to the construction of these large-scale structures. However, with their proliferation, bridge engineering has encountered inherent safety concerns, including structural damages, cracks, corrosion of materials, degradation of performance, functional failure, and other forms of deterioration caused by load conditions, natural disasters, and environmental effects [1, 2]. In addition to disrupting bridge operation, such damages also decrease the bridge's structural safety and pose a significant security risk [3].

In this regard, visual inspections have traditionally been used to detect surface defects and evaluate bridge structural health. However, the effectiveness of this method heavily relies on the experience and expertise of inspectors. Additionally, its labor-intensive and time-consuming nature presents further drawbacks [4]. Furthermore, delayed action or neglect of maintenance can incur substantial future costs, particularly for structures of special significance [5]. As a solution to these limitations, structural health monitoring (SHM) has become one of the prominent ways of monitoring civil engineering constructions [3, 6, 7], which is based on three crucial steps. Structural monitoring with sensor data, extracting the features, and an analytical phase that uses the information from the previous steps to evaluate the structural condition [8, 9]. This technique has led to the installation of real-time bridge structural health monitoring (BSHM) systems in recent years. The process of bridge SHM involves acquiring data, assessing the technical condition, and employing algorithms for analysis to ascertain the safety, durability, and suitability of bridges [7].

The recent progress in BSHM, including technologies such as the internet of things (IoT) and digital twin (DT), along with other advanced tools for constructing a 3D model of bridges, has revolutionized the real-time monitoring methods. Using a network of interconnected devices and sensors to collect and transmit data on the internet, IoT is employed in the monitoring process [10, 11]. This method is utilized in civil engineering to capture and detect various parameters such as acceleration, displacement, strain, humidity, temperature, and more. The primary objective of incorporating IoT in BSHM is to enhance service life, thereby contributing to the development of sustainable infrastructure on a global scale [12]. IoT-based monitoring sensors are supported by machine learning (ML) technologies and do not require human intervention [13].

Furthermore, as another concept in bridge monitoring, DT technology, a virtual replica of the physical bridge asset synchronized with its real-world structure, has been introduced in recent years [14]. During its sensing stage, IoT is used to capture the behavior of the structures through sensors and transmit the data to the DT, which then simulates and mirrors the real-world bridge behavior. Therefore, it is possible to simulate and enhance the comprehension of the structural response of a physical bridge by integrating the real-time sensor data (IoT) with its virtual model (DT). Recent advancements in technology, along with enhancements in various methodologies such as light detection and ranging (LiDAR) [15], unmanned aerial vehicles (UAVs/drones) [16–18], terrestrial laser scanning (TLS) [19–27], ground-penetrating radar (GPR) [23], and photogrammetry [28], have significantly improved the data capturing process. This development has resulted in the implementation of an ongoing monitoring system for bridges, facilitating the generation of 3D models for these structures.

The flowchart in Fig. 15.1 illustrates the constituent sections of the paper. This review article presents an overview of recent developments in BSHM, emphasizing the benefits of utilizing IoT and DT technologies, and exploring their applications, and potential in enhancing bridge safety, maintenance, and decision-making.

By integrating these innovative technologies into civil engineering, the path is paved for more resilient and sustainable infrastructure development. Moreover, the abilities of techniques in BSHM for making the 3D model of the bridge are mentioned

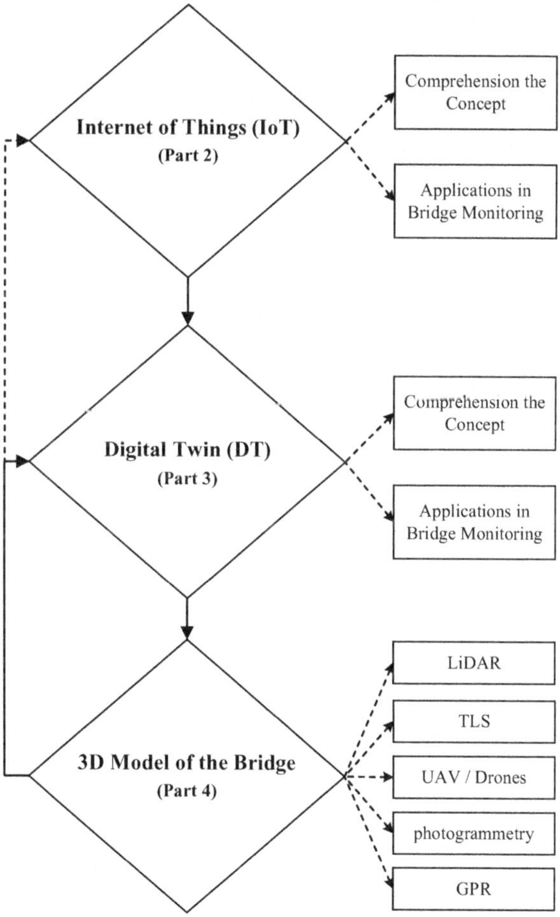

Fig. 15.1 Overview of the workflow and sections in the paper

through examples. Starting with IoT, this review then explores DT and covers techniques such as LiDAR, TLS, UAV, photogrammetry, and GPR. Following a description of each method's concept, the paper presents examples of their applications for each.

15.2 Internet of Things (IoT)

Professor Kevin Ashton first introduced the idea of the IoT back in 1991 at the Massachusetts Institute of Technology (MIT). At MIT, the "Automatic Identification Center" defined it as the transmission of information on all objects using radio frequency. To enable intelligent identification and management, sensing devices were connected to the internet. The IoT has undergone significant evolution over time, encompassing various devices today, such as infrared sensors, radio frequency identification (RFID) devices, laser scanners, and global positioning systems. Combined with network facilities and the internet, these devices are capable of perception, computation, and execution. Leveraging this network, information can be transmitted, coordinated, and processed more easily. Through this technology, objects can be automatically identified, located, tracked, and monitored in real-time. The IoT connects all objects through the power of the internet, bridging the physical and digital worlds, enabling seamless networking of "things and things" [29].

There are four distinct layers in IoT, namely perception, network, processing, and application [29], which are illustrated by Fig. 15.2. The perception layer serves as the foundation for the IoT. Sensors and devices make up this layer, which collects data and recognizes objects. The network layer facilitates data transmission between the perception layer and the processing layer. This is achievable via using a variety of communication networks. An important function of the processing layer is to process vast amounts of data. This is accomplished using IoT management centers, cloud platforms, and expert systems. In the IoT framework, the application layer is the final stage. The application layer is responsible for deploying smart applications across various domains.

IoT offers distinct advantages over traditional internet. Firstly, it incorporates a wide range of sensing technologies. The IoT deploys various sensors that serve as sources of information. Different sensors capture information in different formats and content. The data obtained from these sensors is real-time, as it collects environmental data periodically and updates continuously. The second reason is its classification as an internet-based ubiquitous network. The internet continues to serve as the foundation and core of the IoT. Wired and wireless networks are integrated to deliver real-time and accurate information about objects. Sensor data must be transmitted through the network for the IoT to function. The large quantity of information generated leads to the formation of massive amounts of data. To ensure accuracy and timeliness, the data transmission process must adapt to heterogeneous networks and protocols. Additionally, IoT offers intelligent object control as well as sensor

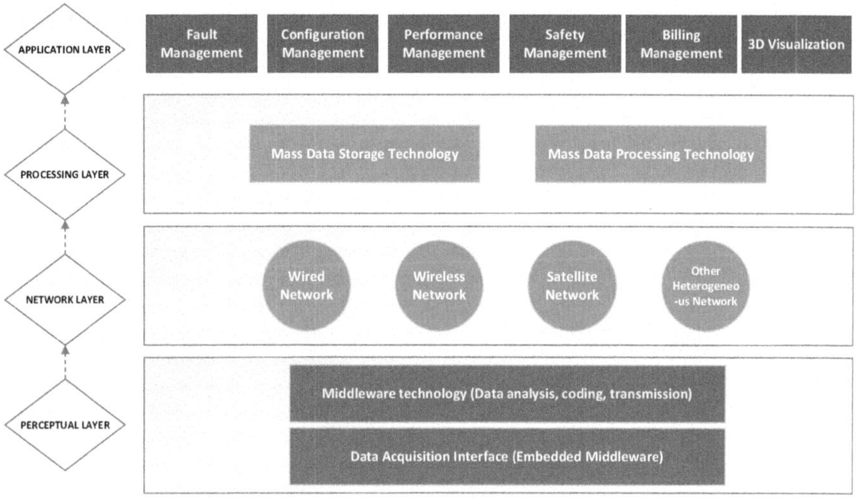

Fig. 15.2 The IoT structure [29]

connectivity. Combining sensors, intelligent processing, and cloud computing allows for the expansion of IoT applications.

By creating a network for identifying bridge structure damage, a significant breakthrough is made over the traditional methods. BSHM is easier through the reduction of management blind spots. Using IoT, a seamless convergence is made to connect the informational gaps that used to exist between the bridge structure and management. Integrating these technologies leads to improved safety and monitoring capabilities for digital SHM [29]. Figure 15.3 illustrates the framework for identifying structural damage to bridges.

To enhance safety management for construction using IoT technology, three key points must be addressed [29]:

1. The first point involves collecting comprehensive information about objects, including their position, displacement, and overall condition, through intelligent terminal technology. The capability to capture and analyze data related to facilities, equipment, the environment, structures, or components enables a thorough assessment of the bridge's safety status.
2. Secondly, the health of the bridge structure needs to be monitored in real-time and on a proactive basis. The network layer facilitates easy and reliable data transfer and information sharing between IoT objects. This capability allows for timely and accurate insights into the security status of the construction site, enabling the implementation of effective safety measures to prevent potential hazards.
3. IoT applications depend on the efficiency of data processing to deal with huge amounts of collected data. Data and information acquired by IoT are processed and analyzed efficiently using intelligent computing technology. Monitoring and processing platforms receive data collected by the objects via the network, which

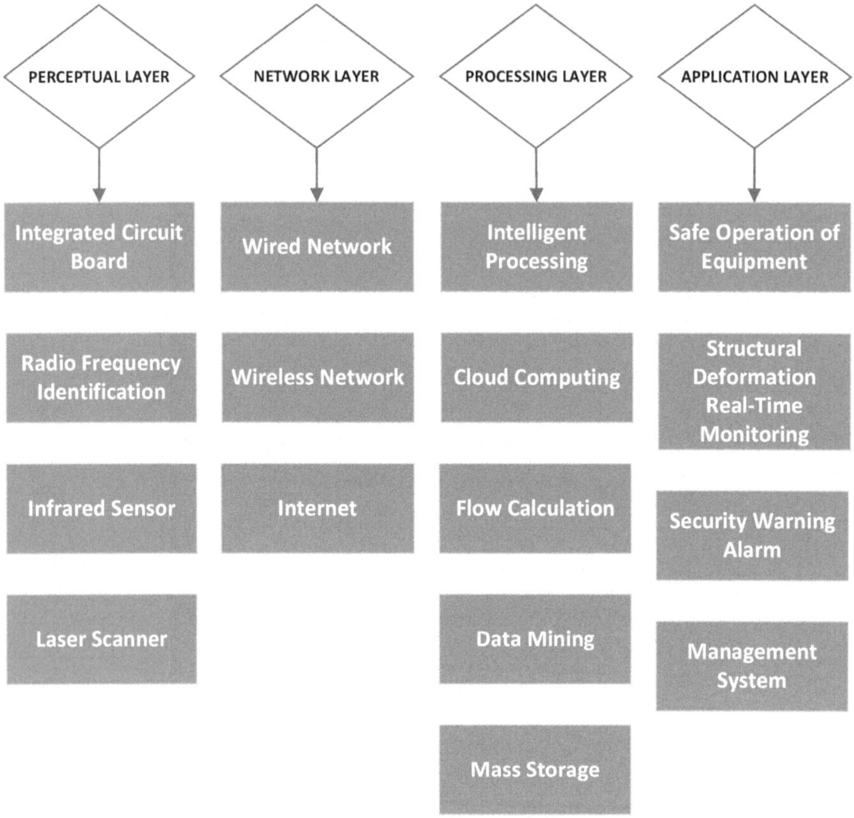

Fig. 15.3 IoT structure for bridge damage identification [29]

is equipped with algorithms, databases, and fuzzy rules to provide insightful analyses of the field's status swiftly.

Table 15.1 outlines diverse applications of IoT in BSHM. Utilizing IoT, researchers have investigated a range of approaches, technologies, and methodologies to enhance bridge safety and collect real-time data. Certain papers employed specific technologies, such as ZigBee, NB-IoT, and Wi-Fi modules, and others explore the incorporation of IoT with AI, 3D printers, and drone technology.

The collective evidence from numerous studies underscores the effectiveness of IoT applications in bridge monitoring. The utilization of low-power wireless sensors, ZigBee technology, and NB-IoT has proven instrumental in real-time data transmission, allowing for the monitoring of acceleration, water levels, air quality, and bridge displacement. Integration with cloud computing platforms, advanced sensing schemes such as FBG optical sensing, and the incorporation of AI and machine learning further enhance the capabilities of bridge monitoring systems. Additionally, IoT facilitates the early detection of structural issues, anomaly detection, and

Table 15.1 Applications of IoT in bridge monitoring

References	Description	Results
Tong et al. [30]	• Low-power wireless acceleration sensor that is connected to the cloud platform	• Validated in a field test on the Chijing bridge in Shanghai • Facilitated the transmission of data • A complete, practical, available, and low-cost system
Lee et al. [31]	• Using ZigBee technology • Monitoring water levels, air quality, pipelines and other safety conditions	• Enhanced safety and maintenance • Mobile telecommunication devices can be used
Hou et al. [32]	• Using NB-IoT (Narrowband Internet of Things) technology for bridge displacement monitoring	• Collecting accurate displacement data and transmit it to a server
Yi et al. [33]	• How to extend the operational lifespan of a network efficiently while ensuring the attainment of the desired coverage • A novel sensor scheduling strategy for partial confident information coverage (CIC), combining a reinforcement learning model named learning automata (LA)	• Sensors were able to learn an optimal scheduling strategy adaptively
Sun et al. [34]	• Railway bridge monitoring platform to enable communication and connectivity between sensor nodes	• Safety test conducted by China Railway Corp.; the proposed system achieved its desired purpose
Naraharisetty et al. [35]	• IoT devices and the AWS cloud computing platform synergistically	• Using AWS to enable real-time SHM with accelerometers, demonstrating live data collection, analysis, and visualization through a web interface
Thedy et al. [36]	• An innovative method for monitoring bridge health, employing a displacement reconstruction algorithm	• Damaged and undamaged bridges could be identified • Distinguish between stiffened and unstiffened bridges in the second experiment
Zhao et al. [37]	• Introducing an innovative strategy utilizing BIM and IoT tools to augment intelligent management throughout the lifecycle of bridges	• System capable of assessing bridge quality, reducing construction risks, and speeding up construction
Farhangdoust et al. [38]	• A reliable power supply for an internet of things platform that uses wireless sensor networks on bridge support cables to effectively collect and send data	• The bimorph piezoelectric energy harvester (BPEH) harnessed ambient vibrations

(continued)

Table 15.1 (continued)

References	Description	Results
Lazo et al. [39]	• Deploy a system for real-time communication among devices tasked with monitoring the structural health of a bridge	• Integrated with a real bridge management system, the prototype integrated existing infrastructure
Mohapatra et al. [40]	• Proposing an FBG (fibre bragg grating) optical sensing scheme for real-time strain distribution monitoring and load localization	• Models of suspension bridges were used to validate the proposed machine learning algorithms
Moallemi et al. [41]	• Introducing a full-stack deployment of an efficient and scalable anomaly detection pipeline for real-time SHM systems	• The approach significantly reduces network traffic, demonstrating successful deployment on a highway bridge in Italy with minimal impact on anomaly detection accuracy
Zhang et al. [40]	• Detecting cracks in bridge structures and monitor their structural health	• Collected images of bridge cracks underwent preprocessing, leading to the creation and training of a convolutional neural network (CNN) classification model
Zinno et al. [42]	• A number of state-of-the-art technologies were discussed, including AI, drone technology, and 3D printers	• Described how AI and other technologies may be incorporated into future SHM systems
Pawar et al. [43]	• Using IoT into construction to enable continuous feedback and monitoring	• Compression sensors were proposed to be placed to monitor the impact of loading on the bridge
Fernando et al. [44]	• To detect signs of degradation and potential failure in bridge bearings, proposing a system design using MEMS accelerometers based on inclination sensing	• On an under-construction bridge in Kingston, Ontario, Canada, the prototype system was implemented
Li et al. [45]	• Analyzing the application to optimize hardware components, improve data collection capabilities, and improve highway bridge monitoring performance	• The design of highway bridge health monitoring data acquisition system based on Internet of things is completed in this study
Aono et al. [46]	• A quasi-self-powered sensor system for continuous and long-term monitoring of bridge	• Mackinac Bridge is studied as a case study in northern Michigan
Raj et al. [47]	• A remote sensor system has been proposed and tested on forty-five sensors of three types which are widely used in suspension scaffolds	• The proposed remote sensor arrangement-based extension wellbeing checking framework was designed

(continued)

Table 15.1 (continued)

References	Description	Results
Xu et al. [48]	• A fault monitoring device was incorporated in the study to enable the acquisition, transmission, and analysis of data about expansion joint faults in highway bridges	• High detection accuracy and high fault monitoring accuracy were demonstrated
Tahir et al. [49]	• Using wireless sensor networks (WSNs) for monitoring to make a smart environment • Whenever temperatures, light, or humidity levels change, the system sends alerts on cell phone	• Demonstrated in field tests that the WSN used in the bridge monitoring system has excellent package delivery and high-power efficiency
Lamonaca et al. [50]	• Reviewing various types of SHM systems used to identify, detect, and characterize degradation and damage	• The study demonstrated that IoT-based SHM systems reduce structural failure risks
Paul et al. [51]	• A portable and robust SHM solution is proposed • Flex sensors are mounted on structures such as concrete beams, metal structures, and bridge joints	• IoT connectivity and flex sensors allow it to detect deformation and damage, alerting inhabitants as soon as possible • Sending data to a smartphone app app through Bluetooth Low Energy (BLE) and cloud storage
Malik et al. [52]	• Using an IoT-based approach for early detection of damages and weaknesses, and real-time monitoring in the bridge were enabled	• Demonstrated high accuracy for detecting structural health in an indoor pedestrian steel bridge • The data was gathered from various sensors and sent to the cloud platform
Ding et al. [53]	• A framework for integrating BIM and IoT into smart steel bridge construction	• The proposed approach promised improved quality, cost management, and development of steel bridges in the context of Industry 4.0
Aono et al. [54]	• Monitoring the health of large civil infrastructure, such as multi-span bridges, the paper uses specifically self-powered RFID sensor networks • "Sense now, retrieve later" is a paradigm for continuously detecting and logging essential structural health statistics	• IoT coverage can be extended to monitor different segments of large civil infrastructure • Through self-powering, essential structural statistics can be continuously monitored and logged

continuous, long-term monitoring through self-powered sensor systems. The overall impact of IoT on bridge monitoring extends beyond traditional parameters, encompassing intelligent management, resilience, and adaptability throughout the lifecycle of bridges.

15.3 Digital Twin (DT)

Several technological advancements have ushered in the industry 4.0, including IoT, big data, cyber-physical systems (CPS), and AI. Consequently, these are used to increase economic growth, safety, productivity, and overall performance across various industries. The DT's concept has emerged as a prominent technology trend to meet this demand [14, 55]. Digital twins are virtual replicas of physical objects, whether they are processes, systems, assets, or other entities. This digital equivalent accurately captures the dynamics of the physical structure during its whole life cycle, maintaining its operations, properties, and status [56]. Researchers and practitioners are focusing on digital twin technology because it has the potential for transforming industries' operations and innovation [14]. Leveraging predefined control outcomes derived from simulations, DT enables optimal management. Data, artificial intelligence, and a digital model of the actual structure are used to inform control decisions for real-world items. To ensure the effectiveness of the DT, a well-established model of the object, a method for interacting with the object, and a suitable environment for simulating and visualizing it are essential [57]. The evolutionary digital model of the DT must be continuously updated with actual data from the objects to maintain the integrity of corresponding physical elements throughout their lifecycles. As a result, the DT remains a faithful representation of its physical counterpart. The integration of a wide range of technologies is key to achieving this level of fidelity in a digital twin [58]. It is important that these aspects are addressed for the digital twin to become a product that is responsive, and capable of achieving significant improvements in various industrial domains [14].

A progressive depiction of DT implementation is presented in Fig. 15.4, separated into three stages. In level 1, the DT reflects the fundamental characteristics of a real-world object. At this initial stage, a virtual duplication of the physical system is created to capture its basic characteristics. As the DT advances to level 2, it gains the capability to monitor and control the real-world object. Consequently, the DT establishes a two-way connection to its physical counterpart, enabling real-time data exchange and the execution of control actions based on insights gained. Level 3 represents the highest level of sophistication for a digital twin. With cutting-edge technologies, such as AI and other advanced tools, it can conduct complex simulations and in-depth analyses, providing valuable insights into the behavior of the physical system [14].

Table 15.2 provides a summary of research employing DT in the realm of BSHM. The overarching theme across these studies is the transformative potential of DT in revolutionizing bridge maintenance systems and enhancing workplace safety. This is achieved through the provision of capabilities such as real-time data collection, monitoring, simulation, and forecasting.

The utilization of DT technology for bridge monitoring has shown to be advantageous. The studies reviewed highlight diverse approaches, such as synchronizing physical and virtual domains, integrating 3D models with geospatial data, and utilizing cloud computing and deep learning techniques. These applications enable

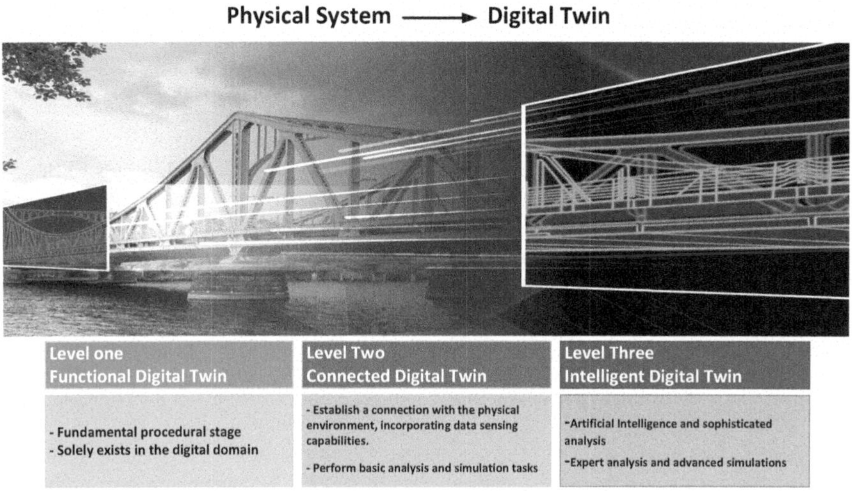

Fig. 15.4 Step-by-step definition of DT [14]

real-time monitoring, simulation, and accurate damage identification. The combination of DT with risk assessment techniques enhances the platform's effectiveness in evaluating and mitigating risks. The studies collectively affirm the versatility of DT in revolutionizing bridge maintenance systems, offering precise data representation and forecasting. The multifaceted nature of DT applications, from structural impact monitoring to AIoT-informed communication frameworks, underscores its efficacy in addressing various challenges associated with bridge monitoring. The integration of DT in managing transportation infrastructure further establishes its effectiveness in dynamic testing and seismic upgrading.

15.4 3D Model of the Bridge

There are common ways to reconstruct a bridge's 3D model to detect damages and also to make a DT including LiDAR [15], photogrammetry [28], TLS [19–27], BIM [37], GPR [23], and UAV/drones [16–18]. These technologies enable engineers and researchers to create a 3D model, assessing the structural integrity of bridges and identifying the possible issues, without causing damage, strategize maintenance or repair plans. LiDAR uses laser pulses to illuminate a target and then analyzes the reflected light to measure distance. Using these systems, the bridge's geometry and surroundings are represented as a 3D point cloud with high accuracy. Due to its ability to cover large areas quickly, LiDAR is suitable for assessing bridges at a regional level. The application of TLS involves using a laser scanner to capture detailed 3D images of bridges and their surroundings using a stationary laser scanner. A point

Table 15.2 Applications of DT in bridge monitoring

References	Description	Results
Kang et al. [14]	• Proposed approach involved the synchronization of the physical and virtual domains by capturing sensor's data	• DT was created, which represents the bridge's behavior and condition precisely • Timely maintenance actions can minimize costs and extend the lifespan of the bridge
Yu et al. [59]	• Focusing on welding residual stresses, fatigue performance, and the joint between the U-ribs and the top deck	• Based on continuum damage mechanics, a nonlinear fatigue damage model was developed
Dan et al. [60]	• Designing a DT of a network of bridges within a regional transportation infrastructure system	• A group of bridges near Shanghai, China, were used to demonstrate the feasibility of the proposed DT system
Dang et al. [61]	• Designing a next-generation bridge maintenance system that uses 3D DT models to model bridge structures	• They were able to create detailed DT by combining alignment-based parametric modeling approaches with 3D scanning
Sofia et al. [62]	• Modeling the Mohammed VI Bridge in Morocco • DT was integrated into geospatial data, with GIS data captured using automated mobile mapping (MMS)	• Demonstrated its potential to revolutionize maintenance systems through real-time data gathering, monitoring, simulation, and forecasting
Dang et al. [63]	• Utilizing cloud computing and deep learning techniques	• Using relevant data, these algorithms established a 92% accuracy rate for identifying damage
Kaewunruen et al. [64]	• Combining DT capabilities with risk assessment techniques, the researchers developed a bridge risk inspection model (Zhongcheng village bridge in Zhejiang province, China)	• Improved the effectiveness the platform by providing stakeholders with a comprehensive framework for evaluating and mitigating risks and uncertainties
Lei et al. [65]	• A study of a bridge swivel construction monitoring and management KNN (K-nearest neighbors) model	• Monitoring the structural impact of shock and vibration allows for the early detection of damage accumulation, preventing potential threats to the safety of the structure
Nicoletti et al. [66]	• Analyzing the advantages of DT in managing transportation infrastructure	• Two case studies of the real bridges were presented in the paper, involving dynamic testing after restoration and seismic upgrading
Gao et al. [67]	• An AIoT (artificial intelligence of things)-informed DT communication framework was proposed to address data synchronization and fault tolerance issues associated with bridge DT services	• Analyses, models, development, and validation were provided in the paper to support the effectiveness of the proposed framework

(continued)

Table 15.2 (continued)

References	Description	Results
Adibfar et al. [68]	• Combining weigh-in-motion (WIM) data with a bridge information model (BrIM), testing the feasibility of developing a DT of a mock-up bridge	• Creating digital twins and testing them in real-time demonstrated the viability of their application in bridge monitoring and maintenance
Broo et al. [69]	• Analyzing the implementation case study • A systems perspective, an information perspective, and an organizational perspective were identified	• Gained a deeper understanding of the multidisciplinary nature of the design and implementation of DT
Rios et al. [70]	• 76 pertinent studies from 2017–2022 were reviewed and analyzed	• A thorough analysis of the literature on DT management, design, and operation was carried out
Jiang et al. [71]	• Establishing a framework for steel bridge fatigue lifetime management using probabilistic optimization	• Inspection/monitoring planning was optimized by incorporating uncertainties and utilizing decision tree analysis
Febrianto et al. [72]	• The statistical finite element method (statFEM) was used to develop a DT of a self-sensing structure, such as a steel railway bridge	• With sensor placement optimization and reliable what-if analyses, DT predicted the system response
Hidayat et al. [73]	• Using "Smart X Platforms", and UAV laser scanner data, the method involved designing an architecture using a six-step monitoring and maintenance cycle	• A DT architecture for monitoring and maintaining road and bridge construction in West Java province
Lin et al. [74]	• Shake table tests were performed on a scaled long-span cable-stayed bridge to demonstrate the efficacy of DT-based evaluation	• FE models were created: one nonlinearly updated, one linearly updated, and one based on the design document
Jiang et al. [75]	• Highlighting the distinctions between DT and BIM and CPS and providing a thorough analysis of 134 works on DT in the field of civil engineering	• The study discusses the potential uses and advancements of DT in civil engineering
Wenner et al. [76]	• Implementation of DT on the Köhlbrandbrücke in Hamburg, Germany	• Optimize the maintenance process and integrate predictive capabilities

cloud is created when TLS devices emit laser in a horizontal and vertical plane. Moreover, UAVs, commonly known as drones, equipped with LiDAR sensors or cameras, can be used to monitor bridges in a flexible and cost-effective manner. By accessing hard-to-reach areas and performing regular inspections without disrupting traffic or requiring elaborate equipment setup, they can conduct inspections seamlessly. With GPR, structures, including bridges, can be imaged, and investigated using electromagnetic waves. Using the photogrammetry method, bridges are photographed from different angles and positions. By analyzing the images, recognizing shared features, and generating a 3D point cloud, the collected data is transformed into a 3D mesh model for monitoring bridges and constructing digital twins. Table 15.3 provides examples of using these techniques in BSHM. The technologies have proven to be efficient, cost-effective, and objective.

The various methods explored for creating 3D models of bridges demonstrate their effectiveness in bridge monitoring. LiDAR and TLS were proven to quantitatively assess concrete damages, detect cracks and deformations, monitor deformation and under clearance, and efficiently obtain as-built data. The integration of TLS with other technologies, such as synthetic aperture radar and ground-based microwave interferometry, highlighted its potential in identifying potential damage areas. TLS complemented traditional approaches in vibration analysis. Studies also explored TLS for monitoring bridge construction, including pose measurement for precast bridge piers and structural assessment during underpinning works.

15.5 Conclusion

This paper explores the use of IoT and DT technologies in bridge monitoring. First, basic IoT concepts and applications are outlined. The paper then discusses the interpretation and applications of DT and its correlation with IoT, highlighting how this combination can improve damage detection and bridge health monitoring with greater accuracy. This synthesis not only enhances monitoring capabilities, but also provides a dynamic framework for capturing and analysing data in real time. The paper concludes with an examination of recent methods that improve bridge monitoring practices. Technologies such as LiDAR, Photogrammetry, TLS, BIM, GPR, and UAVs/drones are evaluated for their ability to create detailed 3D models of bridges. These methods not only increase accuracy but improve the practicality and effectiveness of the monitoring process. The process described illustrates the progression from foundational concepts to state-of-the-art technologies, leading towards a future where bridges are monitored with unprecedented accuracy, practicality, and efficiency. As the bridge monitoring landscape continues to evolve, this review positions itself as a guidepost for researchers, and practitioners, navigating the intricate intersection of IoT, DT, and advanced monitoring techniques in the pursuit of safer and more resilient infrastructure.

The monitoring of bridges is set to undergo a significant transformation through the integration and improvement of AI algorithms. The combination of AI with IoT and DT has the potential to greatly impact data analysis and predictive modeling, enabling more proactive maintenance. In order to successfully navigate this changing landscape, researchers and practitioners must work together to embrace and explore the connections between AI, IoT and DT. This joint effort is essential for advancing bridge monitoring into a new era of innovation and effectiveness. Academic research is now developing and validating innovative approaches to analytical and scientific methodologies. However, it is crucial to guarantee that these advancements are used

Table 15.3 Applications of other technologies to create 3D model of the bridge for structural monitoring

References	Description	Results
Chen et al. [77]	• Quantitatively assessing concrete material damages in bridges using optical-photonic images generated from laser scans (3D LiDAR)	• Provided with critical measurements for damage analysis that photogrammetry and plan photography were not able
Kaartinen et al. [78]	• Exploring LiDAR applications in SHM, especially for detecting cracks, deformations, and defects in civil infrastructure • Mobile laser scanning (MLS) and TLS were reviewed	• The summary highlights the current state of development and acknowledges existing limitations
Cha et al. [19]	• Using TLS to monitor deformation or deflection of bridge structures may be a feasible use of shape information models	• In bridge 1 and bridge 2, laser scanning resulted in 91.7% and 89.3% data reduction, respectively • Compared with LVDT (linear variable differential transducer) for a similar steel box-girder bridge (bridge 3), the shape information model provided reasonable results
Puri et al. [79]	• Obtaining as-built data from a construction site in a timely manner using LiDAR • Identify deviations between performed and planned work, a framework was developed based on mobile LiDAR data and 4D design models	• The percentage of completion (POC) was compared to the POC calculated for the as-planned elements • Most elements were tracked accurately and efficiently with a difference ranging from -7% to 6%
Spadavecchia et al. [15]	• An integrated dataset of point clouds and radiometric information obtained through TLS is used for semi-automatic detection of deterioration • The random forest (RF) machine learning classifier is found to be the most effective	• Demonstrated promising performance in detecting deterioration
Valença et al. [80]	• The MCrack-TLS approach was employed to automate the assessment of cracks in concrete bridges, utilizing TLS technology and image processing techniques	• All processed data was recorded, and 3D models of the bridge's condition were created

(continued)

Table 15.3 (continued)

References	Description	Results
Watson et al. [81]	• Using LiDAR to assess the effect of vehicle crossings and temperature variations on bridge underclearance measurement	• LiDAR technology can provide accurate assessments of underclearance and monitoring of potential changes due to vehicle crossings and temperature variations
Lee et al. [82]	• Using LiDAR to measure long-term displacements of bridge structures with short and medium spans	• A lab validation of the proposed method demonstrates LiDAR's independence from position • The method is subsequently applied in real-world bridge monitoring
Psomas et al. [20]	• Scanning the steel bridge multiple times over eight months and creating 3D point clouds with TLS • CloudCompare3D software was used	• Assessments were made quickly, intuitively, and accurately using this approach
Lõhmus et al. [21]	• Two bridge loads tests in Estonia (Loobu highway bridge and Tartu railway overpass) were executed, using a Leica ScanStation C10 (a time-of-flight type TLS)	• The Loobu investigation found a maximum discrepancy of 3.4 mm between TLS and accurate leveling, whereas the Tartu study reported a difference of 0.8 mm
Zogg et al. [83]	• A 33-year-old Swiss viaduct under load was evaluated for fatigue resistance and an analytical model was refined, using TLS	• Terrestrial laser scanning with precise leveling were compared
Truong-Hong et al. [22]	• Studying vertical bridge deflections during dynamic loading using TLS • In Northern Ireland, the Loughbrickland Bridge was tested using this technique	• With millimeter-level accuracy, TLS generated a large amount of data points, making it a promising alternative to traditional surveying techniques
Cafiso et al. [23]	• In the absence of bridge design records, an assessment was made of the condition of a bridge deck in central Sicily • With GPR system equipped with two antennae, the thickness of the asphalt overlay, concrete cover depth, deck thickness, and rebar reinforcement can be determined	• The 3D output model with high accuracy was produced, allowing the interpretation of GPR data to be improved and verified

(continued)

Table 15.3 (continued)

References	Description	Results
Chu et al. [24]	• A long-span steel truss arch bridge was monitored using laser scanning for high-temperature deformation • Two scans were conducted twice daily under 9 °C temperature difference on the bridge, whose main span is 432 m • Using 3D point cloud data and non-uniform rational basis spline (NURBS) surfaces for overall deformation measurement, a new deformation measurement method was proposed	• Compared to traditional single-point deformation monitoring methods, the TLS method enabled recording linear deformation and presenting overall deformation features
Minehane et al. [25]	• A 3D laser scanning approach was used • Case study 1, restore the Chetwynd Viaduct as a pedestrian/cycle bridge • Case study 2, monitor a masonry arch river bridge in Killorglin, Co. Kerry, during underpinning works	• Successful in case study 1 to create a 3D finite element model for the Chetwynd Viaduct structural assessment • A masonry arch bridge underpinning project in Case Study 2 could be monitored accurately by 3D laser scanning
Zhang et al. [26]	• Use of TLS as a monitoring tool for suspension bridges with long spans • Using field laser scanning data from the Ma'anshan Yangtze Bridge in China, three iterative closest point (ICP) registration methods, namely traditional ICP, kd-tree-based ICP, and feature point-based ICP, were cross-evaluated	• Using the kd-tree algorithm for ICP was more time-efficient than using the traditional algorithm for a single iteration
Neitzel et al. [27]	• Use of a low-cost accelerometer sensor system and a TLS for vibration analysis in the context of structural monitoring, with a focus on bridges	• TLS contributes to the dynamic perspective of structural health assessment, complementing traditional geodetic approaches that typically offer a static viewpoint
Xu et al. [84]	• Employing TLS to provide a precise and effective method for measuring the position of precast bridge piers during installation	• During the construction of an expressway viaduct, the proposed system was evaluated

(continued)

Table 15.3 (continued)

References	Description	Results
Liu et al. [85]	• TLS was employed to create a digital surface model (DSM), ground-based microwave interferometry was used to capture dynamic displacement over time, and permanent scatterer interferometry synthetic aperture radar (PS-InSAR) was used to analyze surface deformations	• The proposed framework identified potential damage areas on the Beishatan bridge, China • When applied to time-series displacement data, the extreme-point symmetric mode decomposition (ESMD) method effectively detected damages
Sánchez-Rodríguez et al. [86]	• Using laser scanner to diagnose structural problems in the piers of masonry arch bridges • The method involved segmenting the global point cloud into different structural elements	• A sample of representative masonry arch bridges was used to validate the methodology
Yan et al. [87]	• Using 3D laser scanners, making an automated approach for creating high-fidelity FEM of aging structures and infrastructures	• This method reduced the need for manual intervention in the FEM creation process by integrating computer vision techniques and prior information on structural components
Tysiac et al. [88]	• An innovative method for integrating laser scanning into acceptability testing to facilitate non-destructive measurements	• The conducted tests confirmed that the structure's reaction aligns with the predictions made by the FEM model, leading to the approval of the object for use
Mader et al. [16]	• Inspecting bridges using UAVs equipped with laser scanners and other sensors	• Developed and constructed UAVs that can navigate, avoid obstacles, and acquire 3D object data
Mohammadi et al. [17]	• An aerial photogrammetry technique employing UAV and TLS was employed to create a digital representation of an Australian heritage bridge • Various methods were used to assess and evaluate the point clouds, taking into account factors such as point distribution, outlier noise, surface deviation, data completeness, and geometric precision	• The TLS-based point clouds had a higher density and were more consistent with as-is measurements

(continued)

Table 15.3 (continued)

References	Description	Results
Moreu et al. [89]	• Bridge displacements during train crossings were to be measured using a contact-free, reference-free method. Laser doppler vibrometer (LDV) sensors were integrated with UAV	• Bridge displacements could be measured more efficiently and effectively without the need for sensors
Zollini et al. [18]	• Assessing the surface deterioration of concrete bridges and viaduct • An overpass in L'Aquila, Italy, was visually inspected using UAV and terrestrial photogrammetry • Deteriorated areas and decay forms were identified and classified using object-based image analysis (OBIA)	• Metric information was quantified, such as crack widths and lengths and weathered areas' extension • Periodic checks enabled easy and fast evaluation of deterioration over time, allowing them to plan preservation or maintenance measures accordingly
Han et al. [90]	• Optimizing UAV flight plans for oblique photography with feature waypoints and utilizing point cloud data for bridge alignment	• The bridge alignment errors, measured over a length of 142 m, were determined by comparing point cloud data collected from TLS and UAV devices to total station data • The errors for TLS and UAV devices were found to be 4.41 mm and 10.97 mm, respectively
Bolourian et al. [91]	• A novel 3D path planning approach for bridge inspection utilizing a UAV equipped with a LiDAR scanner • This approach utilizes a combination of a genetic algorithm (GA) and A* algorithm to solve the traveling salesman problem (TSP), while also considering the possible locations of surface defects such as cracks on the bridge	• Accurate and efficient data collection can be achieved by taking into account overlapping views according to the criticality of zones and utilizing perpendicular views for all viewpoints

in practical settings. Merely existing in theory is inadequate for these developments; they must also be accessible and relevant in practical scenarios. Furthermore, there is a continuous requirement to deliver extensive instruction to professionals about the utilization of these novel methodologies in the field of civil engineering. It is crucial to establish a connection between academic understanding and actual application. This training guarantees that technicians are not only knowledgeable about the latest developments in the industry, but also proficient in integrating them into their professional practices. The industry may fully utilize the possibilities of artificial intelligence, the internet of things, and digital technologies to improve the effectiveness and dependability of bridge monitoring.

Conflicts of Interest The authors assert that they have no conflicts of interests.

References

1. Malekjafarian A, McGetrick PJ, Obrien EJ (2015) A review of indirect bridge monitoring using passing vehicles. Shock Vib 1–16. https://doi.org/10.1155/2015/286139
2. Sarmadi H, Entezami A, Yuen K-V, Behkamal B (2023) Review on smartphone sensing technology for structural health monitoring. Measurement 223:113716
3. Wenzel H (2009) Health monitoring of bridges. Austria, Vienna
4. Chupanit P, Phromsorn C (2012) The importance of bridge health monitoring. Proceedings of world academy of science, vol 6, 135–138
5. Rashidi M, Mohammadi M, Sadeghlou Kivi S, Abdolvand MM, Truong-Hong L, Samali B (2020) A decade of modern bridge monitoring using terrestrial laser scanning: review and future directions. Remote Sens (Basel) 12:3796. https://doi.org/10.3390/rs12223796
6. Soleymani A, Jahangir H, Nehdi ML (2023) Damage detection and monitoring in heritage masonry structures: systematic review. Constr Build Mater 397:132402. https://doi.org/10.1016/j.conbuildmat.2023.132402
7. Hasani H, Freddi F (2023) Operational modal analysis on bridges: a comprehensive review. Infrastructures (Basel) 8:172. https://doi.org/10.3390/infrastructures8120172
8. Abbasi A, Khatibinia M, Jahangir H, Dos Santos JVA, Lopes HMR (2023) Shearography-wavelet-based damage detection methodology for aluminum beams. Rehabil Civil Eng 11:22–43. https://doi.org/10.22075/jrce.2023.29435.1780
9. Daneshvar MH, Saffarian M, Jahangir H, Sarmadi H (2023) Damage identification of structural systems by modal strain energy and an optimization-based iterative regularization method. Eng Comput 39:2067–2087. https://doi.org/10.1007/s00366-021-01567-5
10. Rathore MM, Ahmad A, Paul A, Rho S (2016) Urban planning and building smart cities based on the internet of things using big data analytics. Comput Netw 101:63–80. https://doi.org/10.1016/j.comnet.2015.12.023
11. Jiang D (2020) The construction of smart city information system based on the internet of things and cloud computing. Comput Commun 150:158–166. https://doi.org/10.1016/j.comcom.2019.10.035
12. Mishra M, Lourenço PB, Ramana GV (2022) Structural health monitoring of civil engineering structures by using the internet of things: a review. J Build Eng 48:103954. https://doi.org/10.1016/j.jobe.2021.103954
13. He Z, Li W, Salehi H, Zhang H, Zhou H, Jiao P (2022) Integrated structural health monitoring in bridge engineering. Autom Constr 136:104168. https://doi.org/10.1016/j.autcon.2022.104168
14. Kang JS, Chung K, Hong EJ (2021) Multimedia knowledge-based bridge health monitoring using digital twin. Multimed Tools Appl 80:34609–34624. https://doi.org/10.1007/s11042-021-10649-x
15. Spadavecchia C, Belcore E, Di Pietra V (2023) Preliminary test on structural elements health monitoring with a LiDAR-based approach. Int Arch Photogramm Remote Sens Spat Inf Sci XLVIII-2/W:247–253. https://doi.org/10.5194/isprs-archives-XLVIII-2-W3-2023-247-2023
16. Mader D, Blaskow R, Westfeld P, Weller C (2016) Potential Of Uav-based laser scanner and multispectral camera data in building inspection. Int Arch Photogramm Remote Sens Spat Inf Sci XLI-B1:1135–1142. https://doi.org/10.5194/isprs-archives-XLI-B1-1135-2016

17. Mohammadi M, Rashidi M, Mousavi V, Karami A, Yu Y, Samali B (2021) Quality evaluation of digital twins generated based on UAV photogrammetry and TLS: bridge case study. Remote Sens (Basel) 13:3499. https://doi.org/10.3390/rs13173499
18. Zollini S, Alicandro M, Dominici D, Quaresima R, Giallonardo M (2020) UAV photogrammetry for concrete bridge inspection using object-based image analysis (OBIA). Remote Sens (Basel) 12:3180. https://doi.org/10.3390/rs12193180
19. Cha G, Park S, Oh T (2019) A terrestrial LiDAR-based detection of shape deformation for maintenance of bridge structures. J Constr Eng Manag 145. https://doi.org/10.1061/(ASCE)CO.1943-7862.0001701
20. Psomas L, Trainor R, Alzraiee H Case study: assessing the structural condition of steel bridges using terrestrial laser scanner (TLS). Presented at the
21. Lõhmus H, Ellmann A, Märdla S, Idnurm S (2018) Terrestrial laser scanning for the monitoring of bridge load tests–two case studies. Surv Rev 50:270–284. https://doi.org/10.1080/00396265.2016.1266117
22. Truong-Hong L, Laefer DF (2018) Using terrestrial laser scanning for dynamic bridge deflection measurement. Angewandte Chemie Int Edit 6(11):951–952. 3:10–27
23. Cafiso S, Di Graziano A, Goulias D, Mangiameli M, Mussumeci G (2019) Bridge monitoring combining laser scanning and ground penetrating radar. Presented at the
24. Xi CX, Zhou Z, Xiang X, He S, Hou X (2018) Monitoring of long-span bridge deformation based on 3D laser scanning. Instrum Mesure Métrol 18:113–130. https://doi.org/10.3166/i2m.17.113-130
25. Minehane MJ, O'Donovan R, Ruane KD, O'Keeffe B (2014) The use of 3D laser scanning technology for bridge inspection and assessment. Struct Health Monit 13
26. Zhang H, Zhu Y, Xiong W, Cai CS (2023) Point cloud registration methods for long-span bridge spatial deformation monitoring using terrestrial laser scanning. Struct Control Health Monit 2023:1–16. https://doi.org/10.1155/2023/2629418
27. Neitzel F, Niemeier W, Weisbrich S, Lehmann M (2012) Investigation of low-cost accelerometer, terrestrial laser scanner and ground-based radar interferometer for vibration monitoring of bridges. In: Proceedings of the 6th European workshop-structural health monitoring 2012, EWSHM 2012. 1, pp 542–551
28. Jiang R, Jáuregui DV, White KR (2008) Close-range photogrammetry applications in bridge measurement: literature review. Measurement 41:823–834. https://doi.org/10.1016/j.measurement.2007.12.005
29. Zhang L, Zhou G, Han Y, Lin H, Wu Y (2018) Application of internet of things technology and convolutional neural network model in bridge crack detection. IEEE Access 6:39442–39451. https://doi.org/10.1109/ACCESS.2018.2855144
30. Tong X, Yang H, Wang L, Miao Y (2019) The development and field evaluation of an IoT System of low-power vibration for bridge health monitoring. Sensors 19:1222. https://doi.org/10.3390/s19051222
31. Lee JL, Tyan YY, Wen MH, Wu YW (2017) Development of an ioT-based bridge safety monitoring system. In: Proceedings of the 2017 IEEE international conference on applied system innovation: applied system innovation for modern technology, ICASI 2017, pp 84–86. https://doi.org/10.1109/ICASI.2017.7988352
32. Hou S, Wu G (2019) A low-cost IoT-based wireless sensor system for bridge displacement monitoring. Smart Mater Struct 28:085047. https://doi.org/10.1088/1361-665X/ab2a31
33. Yi L, Deng X, Yang LT, Wu H, Wang M, Situ Y (2021) Reinforcement-learning-enabled partial confident information coverage for IoT-based bridge structural health monitoring. IEEE Internet Things J 8:3108–3119. https://doi.org/10.1109/JIOT.2020.3028325
34. Sun Y (2014) Research on the railroad bridge monitoring platform based on the internet of things. Int J Control Autom 7
35. Naraharisetty V, Talari VS, Neridu S, Kalapatapu P, Pasupuleti VDK (2021) Cloud architecture for IOT based bridge monitoring applications. In: Proceedings of the 2021 international conference on emerging techniques in computational intelligence, ICETCI 2021, pp 39–42. https://doi.org/10.1109/ICETCI51973.2021.9574044

36. Thedy J, Liao K-W, Tseng C-C, Liu C-M (2020) Bridge health monitoring via displacement reconstruction-based NB-IoT technology. Appl Sci 10:8878. https://doi.org/10.3390/app10248878
37. Zhao Z, Gao Y, Hu X, Zhou Y, Zhao L, Qin G, Guo J, Liu Y, Yu C, Han D (2019) Integrating BIM and IoT for smart bridge management. IOP Conf Ser Earth Environ Sci 371:022034. https://doi.org/10.1088/1755-1315/371/2/022034
38. Farhangdoust S, Mederos C, Farkiani B, Mehrabi A, Taheri H, Younesian D (2020) A creative vibration energy harvesting system to support a self-powered internet of thing (IoT) network in smart bridge monitoring. In: Volume 2B: advanced manufacturing. American Society of Mechanical Engineers
39. Lazo C, Gallardo P, Céspedes S (2015) A bridge structural health monitoring system supported by the internet of things. In: 2015 IEEE Colombian conference on communications and computing, COLCOM 2015-Conference Proceedings. https://doi.org/10.1109/ColComCon.2015.7152094
40. Mohapatra AG, Khanna A, Gupta D, Mohanty M, de Albuquerque VHC (2022) An experimental approach to evaluate machine learning models for the estimation of load distribution on suspension bridge using FBG sensors and IoT. Comput Intell 38:747–769. https://doi.org/10.1111/coin.12406
41. Moallemi A, Burrello A, Brunelli D, Benini L (2022) Exploring scalable, distributed real-time anomaly detection for bridge health monitoring. IEEE Internet Things J 9:17660–17674. https://doi.org/10.1109/JIOT.2022.3157532
42. Zinno R, Haghshenas SS, Guido G, Rashvand K, Vitale A, Sarhadi A (2022) The state of the art of artificial intelligence approaches and new technologies in structural health monitoring of bridges. Appl Sci 13:97. https://doi.org/10.3390/app13010097
43. Pawar PG, Suryawanshi SR (2022) Use of IoT for structural health monitoring of bridge prototype using force sensitive resistor 5:137–143
44. Fernando H, Heykoop ., Woods J, Hoult N (2022) Development of an IoT monitoring system for bridge bearing movement using a MEMS accelerometer-based inclination sensing. In: Canadian conference on electrical and computer engineering. 2022-Septe, pp 314–318. https://doi.org/10.1109/CCECE49351.2022.9918468
45. Li T, Hongyan Y (2020) Design of data acquisition system for highway bridge health monitoring based on internet of things. In: 2020 IEEE international conference on industrial application of artificial intelligence (IAAI). IEEE, pp 113–118
46. Aono K, Hasni H, Pochettino O, Lajnef N, Chakrabartty S (2018) Quasi-self-powered infrastructural internet of things. In: Proceedings of the 2018 on great lakes symposium on VLSI. ACM, New York, NY, USA, pp 335–340
47. Raj A, Tanwar R, Gupta A, Goyal A (2020) Bridge health monitoring using IoT. SSRN Electron J. https://doi.org/10.2139/ssrn.3547798
48. Xu Y (2022) Research on fault monitoring device of highway bridge expansion joint based on internet of things technology. Presented at the
49. Tahir M, Enam RN, Mustafa SMN, Javed R (2020) EMAS: environment monitoring and smart alert system for internet of things (IoT). In: 2020 global conference on wireless and optical technologies (GCWOT). IEEE, pp 1–4
50. Lamonaca F, Sciammarella PF, Scuro C, Carni DL, Olivito RS (2018) Internet of things for structural health monitoring. In: 2018 Workshop on metrology for industry 4.0 and IoT. IEEE, pp 95–100
51. Paul P, Dutta N, Biswas BA, Das M, Biswas S, Khalid Z, Saha HN (2018) An internet of things (IoT) based system to analyze real-time collapsing probability of structures. In: 2018 IEEE 9th annual information technology, electronics and mobile communication conference (IEMCON). IEEE, pp 1070–1075
52. Malik H, Khattak KS, Wiqar T, Khan ZH, Altamimi AB (2019) Low cost internet of things platform for structural health monitoring. In: 2019 22nd international multitopic conference (INMIC). IEEE, pp 1–7

53. Ding K, Shi H, Hui J, Liu Y, Zhu B, Zhang F, Cao W (2018) Smart steel bridge construction enabled by BIM and internet of things in industry 4.0: a framework. In: 2018 IEEE 15th international conference on networking, sensing and control (ICNSC), pp 1–5. https://doi.org/10.1109/ICNSC.2018.8361339
54. Aono K, Lajnef N, Faridazar F, Chakrabartty S (2016) Infrastructural health monitoring using self-powered internet-of-things. In: 2016 IEEE international symposium on circuits and systems (ISCAS). IEEE, pp 2058–2061
55. Vachalek J, Bartalsky L, Rovny O, Sismisova D, Morhac M, Loksik M (2017) The digital twin of an industrial production line within the industry 4.0 concept. In: 2017 21st International conference on process control (PC). IEEE, pp 258–262
56. Kritzinger W, Karner M, Traar G, Henjes J, Sihn W (2018) Digital Twin in manufacturing: a categorical literature review and classification. IFAC-PapersOnLine 51:1016–1022. https://doi.org/10.1016/j.ifacol.2018.08.474
57. Söderberg R, Wärmefjord K, Carlson JS, Lindkvist L (2017) Toward a digital twin for real-time geometry assurance in individualized production. CIRP Ann 66:137–140. https://doi.org/10.1016/j.cirp.2017.04.038
58. Maria A (1997) Introduction to modeling and simulation. In: Proceedings of the 29th conference on winter simulation-WSC '97. ACM Press, New York, New York, USA, pp 7–13
59. Yu S, Li D, Ou J (2022) Digital twin-based structure health hybrid monitoring and fatigue evaluation of orthotropic steel deck in cable-stayed bridge. Struct Control Health Monit 29. https://doi.org/10.1002/stc.2976
60. Dan D, Ying Y, Ge L (2022) Digital twin system of bridges group based on machine vision fusion monitoring of bridge traffic load. IEEE Trans Intell Transp Syst 23:22190–22205. https://doi.org/10.1109/TITS.2021.3130025
61. Dang NS, Kang H, Lon S, Shim CS (2018) 3D digital twin models for bridge maintenance. In: Proceedings of 10th international conference on short and medium span bridges, pp 1–9
62. Sofia H, Anas E, Faiz O (2020) Mobile mapping, machine learning and digital twin for road infrastructure monitoring and maintenance: case study of mohammed VI bridge in Morocco. In: Proceedings-2020 IEEE international conference of moroccan geomatics, MORGEO 2020. https://doi.org/10.1109/Morgeo49228.2020.9121882
63. Dang H, Tatipamula M, Nguyen HX (2022) Cloud-based digital twinning for structural health monitoring using deep learning. IEEE Trans Industr Inform 18:3820–3830. https://doi.org/10.1109/TII.2021.3115119
64. Kaewunruen S, Sresakoolchai J, Ma W, Phil-Ebosie O (2021) Digital twin aided vulnerability assessment and risk-based maintenance planning of bridge infrastructures exposed to extreme conditions. Sustainability 13:2051. https://doi.org/10.3390/su13042051
65. Lei Z, Zhu L, Fang Y, Niu C, Zhao Y (2022) Fiber bragg grating smart material and structural health monitoring system based on digital twin drive. J Nanomater 2022:1–13. https://doi.org/10.1155/2022/4356974
66. Nicoletti V, Martini R, Carbonari S, Gara F (2023) Operational modal analysis as a support for the development of digital twin models of bridges. Infrastructures (Basel) 8:24. https://doi.org/10.3390/infrastructures8020024
67. Gao Y, Li H, Xiong G, Song H (2023) AIoT-informed digital twin communication for bridge maintenance. Autom Constr 150:104835. https://doi.org/10.1016/j.autcon.2023.104835
68. Adibfar A, Costin AM (2022) Creation of a mock-up bridge digital twin by fusing intelligent transportation systems (ITS) data into bridge information model (BrIM). J Constr Eng Manag 148. https://doi.org/10.1061/(ASCE)CO.1943-7862.0002332
69. Gürdür Broo D, Bravo-Haro M, Schooling J (2022) Design and implementation of a smart infrastructure digital twin. Autom Constr 136:104171. https://doi.org/10.1016/j.autcon.2022.104171
70. Jiménez Rios A, Plevris V, Nogal M (2023) Bridge management through digital twin-based anomaly detection systems: a systematic review. Front Built Environ 9. https://doi.org/10.3389/fbuil.2023.1176621

71. Jiang F, Ding Y, Song Y, Geng F, Wang Z (2022) Digital twin-driven framework for fatigue lifecycle management of steel bridges. Struct Infrastruct Eng 1–21. https://doi.org/10.1080/15732479.2022.2058563
72. Febrianto E, Butler L, Girolami M, Cirak F (2000) Digital twinning of self-sensing structures using the statistical finite element method. Data Cent Eng 3. https://doi.org/10.1017/dce.2022.28
73. Hidayat F, Supangkat SH, Hanafi K (2022) Digital twin of road and bridge construction monitoring and maintenance. In: ISC2 2022-8th IEEE international smart cities conference, pp 1–7. https://doi.org/10.1109/ISC255366.2022.9922473
74. Lin K, Xu Y-L, Lu X, Guan Z, Li J (2021) Digital twin-based collapse fragility assessment of a long-span cable-stayed bridge under strong earthquakes. Autom Constr 123:103547. https://doi.org/10.1016/j.autcon.2020.103547
75. Jiang F, Ma L, Broyd T, Chen K (2021) Digital twin and its implementations in the civil engineering sector. Autom Constr 130:103838. https://doi.org/10.1016/j.autcon.2021.103838
76. Wenner M, Meyer-Westphal M, Herbrand M, Ullerich C (2021) The concept of the digital twin to revolutionise the infrastructure maintenance: the pilot project smartBRIDGE Hamburg. 27th ITS World Congress, pp 11–15
77. Chen S-E, Liu W, Bian H, Smith B (2012) 3D LiDAR scans for bridge damage evaluations. In: Forensic engineering 2012. American Society of Civil Engineers, Reston, VA, pp 487–495
78. Kaartinen E, Dunphy K, Sadhu A (2022) LiDAR-based structural health monitoring: applications in civil infrastructure systems. Sensors 22:4610. https://doi.org/10.3390/s22124610
79. Puri N, Turkan Y (2020) Bridge construction progress monitoring using lidar and 4D design models. Autom Constr 109:102961. https://doi.org/10.1016/j.autcon.2019.102961
80. Valença J, Puente I, Júlio E, González-Jorge H, Arias-Sánchez P (2017) Assessment of cracks on concrete bridges using image processing supported by laser scanning survey. Constr Build Mater 146:668–678. https://doi.org/10.1016/j.conbuildmat.2017.04.096
81. Watson C, Chen S-E, Bian H, Hauser E (2012) Three-dimensional terrestrial LIDAR for operational bridge clearance measurements. J Perform Constr Facil 26:803–811. https://doi.org/10.1061/(ASCE)CF.1943-5509.0000277
82. Lee J, Lee K, Lee S, Lee Y, Sim S (2019) Long-term displacement measurement of bridges using a LiDAR system. Struct Control Health Monit 26. https://doi.org/10.1002/stc.2428
83. Zogg H-M, Ingensand H (2008) Terrestrial laser scanning for deformation monitoring. In: International archives of the photogrammetry, remote sensing and spatial information sciences XXXVII (B5)
84. Xu Y, Luo Y, Zhang J (2022) Laser-scan based pose monitoring for guiding erection of precast concrete bridge piers. Autom Constr 140:104347. https://doi.org/10.1016/j.autcon.2022.104347
85. Liu X, Wang P, Lu Z, Gao K, Wang H, Jiao C, Zhang X (2019) Damage detection and analysis of urban bridges using terrestrial laser scanning (TLS), ground-based microwave interferometry, and permanent scatterer interferometry synthetic aperture radar (PS-InSAR). Remote Sens (Basel) 11:580. https://doi.org/10.3390/rs11050580
86. Sánchez-Rodríguez A, Riveiro B, Conde B, Soilán M (2018) Detection of structural faults in piers of masonry arch bridges through automated processing of laser scanning data. Struct Control Health Monit 25:e2126. https://doi.org/10.1002/stc.2126
87. Yan Y, Guldur B, Hajjar JF (2017) Automated structural modelling of bridges from laser scanning. In: Structures congress 2017: bridges and transportation structures - selected papers from the structures congress 2017, pp 457–468
88. Tysiac P, Miskiewicz M, Bruski D (2022) Bridge non-destructive measurements using a laser scanning during acceptance testing: case study. Materials 15:8533. https://doi.org/10.3390/ma15238533
89. Moreu F, Taha MR (2018) Railroad bridge inspections for maintenance and replacement prioritization using unmanned aerial vehicles (UAVs) with laser scanning capabilities. Transportation Research Board

90. Han Y, Feng D, Wu W, Yu X, Wu G, Liu J (2023) Geometric shape measurement and its application in bridge construction based on UAV and terrestrial laser scanner. Autom Constr 151:104880. https://doi.org/10.1016/j.autcon.2023.104880
91. Bolourian N, Hammad A (2020) LiDAR-equipped UAV path planning considering potential locations of defects for bridge inspection. Autom Constr 117:103250. https://doi.org/10.1016/j.autcon.2020.103250

The manufacturer's authorised representative in the EU is Springer Nature Customer Service Centre GmbH, Europaplatz 3, 69115 Heidelberg, Germany. If you have any concerns regarding our products, please contact ProductSafety@springernature.com

Printed and bound by CPI Group (UK) Ltd, Croydon, CR0 4YY
01/10/2026
02225963-0006